ADJUSTMENT COMPUTATIONS

WILEY SERIES IN SURVEYING AND BOUNDARY CONTROL

Series Editor: Roy Minnick

Cole/WATER BOUNDARIES

Smith/INTRODUCTION TO GEODESY: THE HISTORY AND CONCEPTS OF MODERN GEODESY

Wolf and Ghilani/ADJUSTMENT COMPUTATIONS: STATISTICS AND LEAST SQUARES IN SURVEYING AND GIS

ADJUSTMENT COMPUTATIONS

Statistics and Least Squares in Surveying and GIS

Paul R. Wolf, Ph.D.

Professor Emeritus
Department of Civil and Environmental Engineering
University of Wisconsin–Madison

Charles D. Ghilani, Ph.D.

Associate Professor
Surveying Program
Pennsylvania State University

A WILEY-INTERSCIENCE PUBLICATION

JOHN WILEY & SONS, INC.

New York Chichester Brisbane Toronto Singapore Weinheim

This text is printed on acid-free paper.

Library of Congress Cataloging in Publication Data:

Wolf, Paul R.

Adjustment computations : statistics and least squares in surveying and GIS / by Paul R. Wolf and Charles D. Ghilani.—3rd ed.

p. cm.—(Wiley series in surveying and boundary control)

Includes bibliographical references and index.

ISBN 0-471-16833-5 (alk. paper)

1. Surveying—Mathematics. 2. Surveying—Statistical methods. 3. Least squares. 4. Error analysis (Mathematics) I. Ghilani, Charles D. II. Title. III. Series.

TA556.M38W65 1996

526.9'072—dc20 96-22146

Printed in the United States of America

10 9 8 7 6 5 4 3

CONTENTS

PREFACE

No measurement is ever exact. As a corollary, every measurement contains error. These statements are fundamental and universally accepted. It follows logically, therefore, that surveyors and GIS managers who are measurement specialists should have a thorough understanding of errors. They must be familiar with the different types of errors, their sources, and their expected magnitudes. Armed with this knowledge they will be able to (1) adopt procedures for reducing error sizes when making their measurements, and (2) account rigorously for the presence of errors as they analyze and adjust their measured data. This book is devoted to creating a better understanding of these topics.

In recent years, least squares has been gaining popularity rapidly as the method used for analyzing and adjusting surveying data. This should not be surprising, because the method is the most rigorous adjustment procedure available. It is soundly based on the mathematical theory of probability, it allows for appropriate weighting of all observations in accordance with their expected precisions, and it enables complete statistical analyses to be made following adjustments so that the expected precisions of adjusted quantities can be determined. Procedures for employing the method of least squares and then statistically analyzing the results are major topics covered in this book.

In years past, least squares was seldom used for adjusting surveying data because the time required to set up and solve the necessary equations was too great for manual methods. Now computers have eliminated that disadvantage. Besides advances in computer technology, some other recent developments have also led to increased use of least squares. Prominent among these are the global positioning system (GPS), and geographic information systems and land information systems (GISs and LISs). These systems rely heavily on rigorous adjustment of data and statistical analysis of the results. But perhaps the most compelling of all reasons for the recent increased interest in least-squares adjustment is that new accuracy standards for surveys are being developed that are based on quantities obtained from least-squares adjustments. Thus surveyors of the future will not be able to test their measurements for compliance with these standards unless they adjust their data using least squares. Clearly, modern surveyors must be able to apply the method of least squares to adjust their measured data, and they must also be able to perform a statistical evaluation of the results after making the adjustments.

This book originated in 1968 as a set of lecture notes for a course taught to a group of practicing surveyors in the San Francisco Bay area by Professor Paul R. Wolf. The notes were subsequently bound and used as the text for formal courses in adjustment computations taught at both the University of California–Berkeley and the University of Wisconsin–Madison. In 1980, a second edition was produced which incorporated many changes and suggestions from students and others who had used the notes. The second edition, published by Landmark Enterprises, has been distributed widely to practicing surveyors and GIS managers, and has also been used as a textbook for adjustment computations courses in several colleges and universities.

For this new third edition, Professor Charles D. Ghilani of the Pennsylvania State University has joined Professor Wolf as coauthor. Professor Ghilani has used the second edition in his classes for several years and has incorporated substantial amounts of new material into the book which he has found beneficial in his teaching.

This third edition has been revised completely and substantially expanded. New chapters have been added on confidence intervals and statistical testing, propagation of errors, constraint equations, blunder detection, and general least squares. A new chapter has also been added on the use of least squares to adjust GPS network surveys.

This current edition now consists of 22 chapters and several appendixes. The chapters are arranged in the order found most convenient in teaching college courses on adjustment computations. It is believed that this order also best facilitates those practicing surveyors who use the book for self-study. Earlier chapters define terms and introduce students to the fundamentals of errors and methods for analyzing them. The next several chapters are devoted to the subject of error propagation in the different types of basic surveying measurements. Then chapters follow that describe observation weighting and introduce the least-squares method for adjusting observations. Application of least squares in adjusting basic types of surveys are then presented in separate chapters. Adjustment of level nets, trilateration, triangulation, traverses and horizontal networks, and GPS networks are included. The subject of error ellipses is covered in a separate chapter. Procedures for applying least squares in curve fitting and in computing coordinate transformations are also presented. The more advanced topics of blunder detection, the method of general least squares, and computer optimization are covered in the later chapters.

As with the previous editions, matrix methods which are so well adapted to adjustment computations continue to be used in this edition. For those students who have never studied matrices, or those who wish to review this topic, an introduction to matrix methods is given in Appendixes A and B.

Those students who have already studied matrices can conveniently skip those sections.

Least-squares adjustments often require the formation and solution of nonlinear equations. Procedures for linearizing nonlinear equations by Taylor's theorem are therefore important in adjustment computations, and this topic is presented in Appendix C. Appendix D contains several new statistical tables. These include the standard normal error distribution, the chi-squared distribution, the Student's *t* Distribution, and a set of *F* distribution tables. These tables are described at appropriate locations in the text, and their use is demonstrated with examples.

Examples are presented throughout the book that illustrate statistical analyses and least-squares adjustment of typical surveying measurements. The examples progress from the analysis of a single observation to adjustment of complex networks of data.

A diskette is included with the book. It contains three major computer program packages. The first package, called STATS, performs basic statistical analyses. For any given set of measured data, it will compute the mean, median, mode, and standard deviation, and develop and plot the histogram and normal distribution curve. The second package, called ADJUST, contains programs for performing specific least-squares adjustments. Level nets, horizontal surveys (trilateration, triangulation, traverses, and horizontal network surveys), and GPS networks can be adjusted using programs in this package. It also contains programs to compute the least-squares solution for a variety of coordinate transformations, and determine the least-squares fit of a line, parabola, or circle to a set of data points. Each of these programs computes residuals and standard deviations following the adjustment. The third package, called MATRIX, performs a collection of basic matrix operations such as addition, subtraction, transpose, multiplication, inverse, and more. Using this program, systems of simultaneous linear equations can be solved quickly and conveniently, and the basic algorithm for doing least-squares adjustments can be performed in a stepwise fashion. All programs on this diskette are menu driven and very easy to use. Examples within the text are solved with the programs to demonstrate their use. Documentation on the use of these programs is given in Appendix F. For those who wish to develop their own software, the book provides several helpful computer algorithms in the languages BASIC, C, FORTRAN, and PASCAL.

Homework problems are given at the end of each chapter so that students can test their understanding of the material in the text. More than 300 problems are included. An Instructor's Manual is available upon request to instructors who adopt the book for use in their classes. It contains answers to all after-chapter problems, and diskettes which contain the figures from this textbook. Instructors can make overheads from this diskette.

A course in calculus is a necessary prerequisite for a complete understanding of some of the theoretical coverage and equation derivations given herein. Nevertheless, those who do not have this background but who wish to learn how to apply least squares in adjusting basic surveying measurements can use this book by ignoring these theories and derivations.

Besides being appropriate for use as a textbook in college classes, this book should also be valuable to practicing surveyors and GIS managers. The authors hope that through the publication of this book, least-squares adjustment and rigorous statistical analyses of surveying data will become more commonplace, as they should.

Paul R. Wolf
Charles D. Ghilani

ACKNOWLEDGMENTS

Through the years many people have contributed to the development of this book. As noted, the book has been used in continuing education classes taught to practicing surveyors as well as in classes taken by students at the University of California–Berkeley, the University of Wisconsin–Madison, and the Pennsylvania State University at Wilkes-Barre. The students in these classes have provided data for many of the examples, and have supplied numerous helpful suggestions for improvements throughout the book. The authors gratefully acknowledge their contributions.

Earlier editions of the book benefited specifically from the contributions of Joseph Dracup of the National Geodetic Survey, Professor Harold Welch of the University of Michigan, Professor Sandor Veress of the University of Washington, and Professor David Mezera of the University of Wisconsin–Madison. The manuscript for this current edition was reviewed in total by Charles Schwarz of the National Geodetic Survey and editor of the ACSM journal *Surveying and Land Information Systems*, and by Earl Burkholder, a consultant in surveying and mapping and editor of the *ASCE Journal of Surveying Engineering*. The suggestions and contributions of these people were extremely valuable and are very much appreciated. Special thanks are also expressed to Paul Hartzheim of the Wisconsin Department of Transportation and to Eduardo Fernandez-Falcon of Pennsylvania State University for providing the field data used in Chapter 16 and for their review of that chapter and also to Jesse Kozlowski and Bryn Fosburgh of Trimble Navigation for their review of Chapter 16. Finally the authors thank Susan Burford for her contributions.

To improve future editions, the authors will gratefully accept any constructive criticisms of this edition and suggestions for its improvement.

1

INTRODUCTION

1.1 INTRODUCTION

We currently live in what is often termed the *information age*. Aided by new and emerging technologies, data are being collected at unprecedented rates in all walks of life. In the field of surveying, for example, total station instruments, global positioning system (GPS) equipment, digital metric cameras, and satellite imaging systems are only some of the new instruments that are now available for rapid generation of vast quantities of measured data.

Geographic information systems (GISs) have evolved concurrently with the development of these new data acquisition instruments. GISs are now used extensively for management, planning, and design. They are being applied worldwide at all levels of government, in business and industry, by public utilities, and in private engineering and surveying offices. Implementation of a GIS depends upon large quantities of data from a variety of sources, many of them consisting of measurements made with new instruments such as those noted above.

Before measured data can be utilized, however, whether for surveying and mapping projects, for engineering design, or for use in a geographic information system, they must be processed. One of the most important aspects of this is to account for the fact that *no measurements are exact. That is, they always contain errors.*

The steps involved in accounting for the existence of errors in measurements consist of (1) performing statistical analyses of the measurements to assess the magnitudes of their errors, and studying their distributions to determine whether or not they are within acceptable tolerances; and (2) if the measurements are acceptable, adjusting them so that they conform to exact geometric conditions or other required constraints. Procedures for performing these two steps in processing measured data are principal subjects of this book.

1.2 DIRECT AND INDIRECT MEASUREMENTS

Measurements are defined as observations made to determine unknown quantities. They may be classified as either direct or indirect. *Direct measurements* are made by applying an instrument directly to the unknown quantity and observing its value, usually by reading it directly from graduated scales on the device. Determining the distance between two points by making a direct measurement using a graduated tape, or measuring an angle by making a direct observation from the graduated circle of a theodolite or total station instrument, are examples of direct measurement.

Indirect measurements are obtained when it is not possible or practical to make direct measurements. In such cases the quantity desired is determined from its mathematical relationship to direct measurements. Surveyors may, for example, measure angles and lengths of lines between points directly and use these measurements to compute station coordinates. From these coordinate values, other distances and angles that were not measured directly may be derived indirectly by computation. During this procedure, the errors that were present in the original direct observations are *propagated* (distributed) by the computational process into the indirect values. Thus the indirect measurements (computed station coordinates, distances, and angles) contain errors that are functions of the original errors. This distribution of errors is known as *error propagation*. The analysis of how errors propagate is also a principal topic of this book.

1.3 MEASUREMENT ERROR SOURCES

It can be stated unconditionally that (1) *no measurement is exact*, (2) *every measurement contains errors*, (3) *the true value of a measurement is never known, and thus* (4) *the exact sizes of the errors present are always unknown*. These facts can be illustrated by the following. If an angle is measured with a scale divided into degrees, its value can be read only to perhaps the nearest tenth of a degree. If a better scale graduated in minutes were available and read under magnification, however, the same angle might be estimated to tenths of a minute. With a scale graduated in seconds, a reading to the nearest tenth of a second might be possible. From the foregoing, it should be clear that no matter how well the measurement is taken, a better one may be possible. Obviously in this example, measurement accuracy depends on the division size of the scale. But accuracy depends on many other factors, including the overall reliability and refinement of the equipment used, environmental conditions that exist when the measurements are taken, and human limitations (e.g., the ability to estimate fractions of a scale division). As better equipment is developed, environmental conditions improve, and observer ability increases, measurements will approach their true values more closely, but they can never be exact.

By definition, an *error* is the difference between a measured value for any quantity and its true value, or

$$\epsilon = y - \mu \tag{1.1}$$

where ϵ is the error in a measurement, y the measured value, and μ its true value.

As discussed above, errors stem from three sources, which are classified as instrumental, natural, and personal. These sources are described as follows:

1. *Instrumental Errors*. These errors are caused by imperfections in instrument construction or adjustment. For example, the divisions on a theodolite or total station instrument may not be spaced uniformly. These error sources are present whether the equipment is read manually or digitally.
2. *Natural Errors*. These errors are caused by changing conditions in the surrounding environment. These include variations in atmospheric pressure, temperature, wind, gravitational fields, and magnetic fields.
3. *Personal Errors*. These errors arise due to limitations in human senses, such as the ability to read a micrometer or to center a level bubble. The sizes of these errors are affected by personal ability to see and by manual dexterity. These factors may be influenced further by temperature, insects, and other physical conditions that cause humans to behave in a less precise manner than they would under ideal conditions.

1.4 DEFINITIONS

From the discussion thus far it can be stated with absolute certainty that all measured values contain errors, whether due to lack of refinement in readings, instabilities in environmental conditions, instrumental imperfections, or human limitations. Some of these errors result from physical conditions that cause them to occur in a systematic way, whereas others occur with apparent randomness. Accordingly, errors are classified as either systematic or random. But before defining systematic and random errors, it is helpful to define mistakes. These three terms are defined as follows:

1. *Mistakes*. These are caused by confusion or by an observer's carelessness. They are not classified as errors and must be removed from any set of observations. Examples of mistakes include (a) forgetting to

set the proper parts-per-million (ppm) correction on an EDM instrument or failure to read the correct air temperature, (b) mistakes in reading graduated scales, and (c) blunders in recording (i.e., writing down 27.55 for 25.75). Mistakes are also known as *blunders* or *gross errors*.

2. *Systematic Errors*. These errors follow some physical law and thus can be predicted. Some systematic errors are removed by following correct measurement procedures (e.g., balancing backsight and foresight distances in differential leveling to compensate for earth curvature and refraction). Others are removed by deriving corrections based on the physical conditions that were responsible for their creation (e.g., applying a computed correction for earth curvature and refraction on a trigonometric leveling observation. Additional examples of systematic errors are (a) temperature not being standard while taping, (b) an index error of the vertical circle of a theodolite or total station instrument, and (c) use of a level rod that is not standard length. Corrections for systematic errors can be computed and applied to measurements to eliminate their effects.
3. *Random Errors*. These are the errors that remain after all mistakes and systematic errors have been removed from the measured values. In general, they are the result of human and instrument imperfections. They are generally small and are as likely to be negative as positive. They usually do not follow any physical law and therefore must be dealt with according to the mathematical laws of probability. Examples of random errors are (a) imperfect centering over a point during distance measurement with an EDM instrument, (b) bubble not centered at the instant a level rod is read, and (c) small errors in reading graduated scales. It is impossible to avoid random errors in measurements entirely. Although they are often called accidental errors, their occurrence should not be considered an accident.

1.5 PRECISION VERSUS ACCURACY

Due to errors, repeated measurement of the same quantity will often yield different values. A *discrepancy* is defined as the algebraic difference between two measurements of the same quantity. When small discrepancies exist between repeated measurements, it is generally believed that only small errors exist. Thus the tendency is to give higher credibility to such data and to call the measurements *precise*. Precise values are not necessarily accurate values, however. To clarify the difference between precision and accuracy, the following definitions are given:

1. *Precision*: degree of consistency between measurements and is based on the sizes of the discrepancies in a data set. The degree of precision

attainable is dependent on the stability of the environment during the time of measurement, the quality of the equipment used to make the measurements, and the observer's skill with the equipment and measurement procedures.

2. *Accuracy*: measure of the absolute nearness of a measured quantity to its true value. Since the true value of a quantity can never be determined, accuracy is always an unknown.

The difference between precision and accuracy can be demonstrated using distance measurements. Assume that the distance between two points is paced, taped, and measured electronically and that each procedure is repeated five times. The resulting measurements are:

Observation	Pacing, p	Taping, t	EDM, e
1	571	567.17	567.133
2	563	567.08	567.124
3	566	567.12	567.129
4	588	567.38	567.165
5	557	567.01	567.114

The arithmetic means for these data sets are 569, 567.15, and 567.133, respectively. A line plot illustrating relative values of the electronically measured distances, denoted by e, and the taped distances, denoted by t, is shown in Figure 1.1. Notice that although the means of the EDM data set and of the taped measurements are relatively close, the EDM set has smaller discrepancies. This indicates that the EDM instrument produced a higher precision. However, this higher precision does not necessarily prove that the mean of the electronically measured data set is implicitly more accurate than the mean of the taped values. In fact, the opposite may be true if, for example, the reflector constant was entered incorrectly, causing a large systematic error to be present in all the electronically measured distances. Because of the larger discrepancies, it is unlikely that the mean of the paced distances is as accurate as either of the other two values. But its mean could be more accurate if large

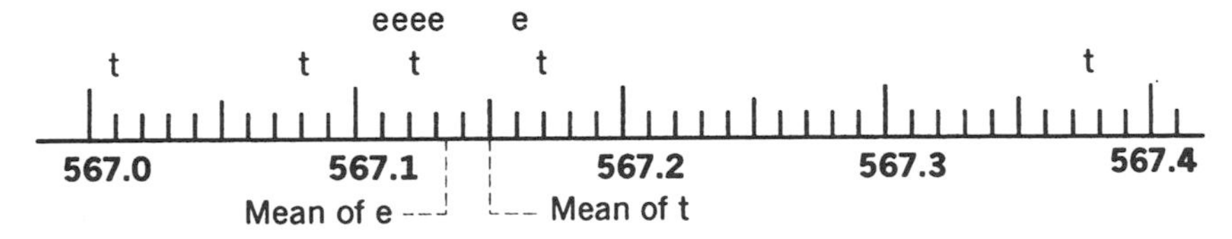

Figure 1.1 Line plot of distance quantities.

systematic errors were present in both the taped and electronically measured distances.

Another illustration explaining differences between precision and accuracy involves target shooting, depicted in Figure 1.2. As shown, four situations can occur. If accuracy is considered as closeness of shots to the center of a target at which a marksman shoots, and precision as the closeness of the shots to each other, then (1) the data may be both precise and accurate, as shown in Figure 1.2(*a*); (2) the data may produce an accurate mean but not be precise, as shown in Figure 1.2(*b*); (3) the data may be precise but not accurate, as shown in Figure 1.2(*c*); or (4) the data may be neither precise nor accurate, as shown in Figure 1.2(*d*).

Obviously, Figure 1.2(*a*) is the desired result when observing quantities. The other cases can be attributed to the following situations. The results shown in Figure 1.2(*b*) occur when there is little refinement in the measurement process. Someone skilled at pacing may achieve these results. Figure 1.2(*c*) generally occurs when systematic errors are present in the measurement process. This can occur for example in taping if corrections are not made for tape length and temperature, or with electronic distance measurements when using the wrong combined instrument–reflector constant. Figure 1.2(*d*) shows results obtained when the observations are not corrected for systematic errors and are taken carelessly by the observer (or the observer is unskilled at the particular measurement procedure).

In general, when making measurements, data such as those shown in Figure 1.2(*b*) and (*d*) are undesirable. Rather, results similar to those shown in Figure 1.2(*a*) are preferred. However, in making measurements the results of

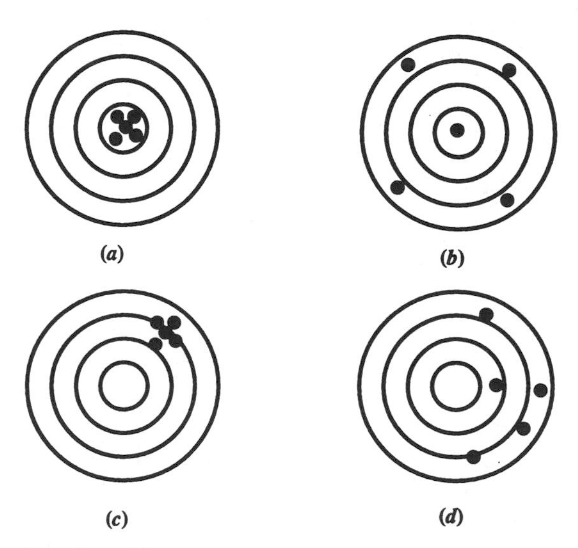

Figure 1.2 Examples of precision versus accuracy.

Figure 1.2(*c*) can be just as acceptable if proper steps are taken to correct for the presence of the systematic errors. (This correction would be equivalent to the marksman realigning the sights after taking the shots.) To make these corrections, (1) the specific types of systematic errors that have occurred in the observations must be known, and (2) the procedures used in correcting them must be understood.

1.6 REDUNDANT MEASUREMENTS IN SURVEYING AND THEIR ADJUSTMENT

As noted earlier, errors exist in all measurements. In surveying, the presence of errors is obvious in many situations where the measurements must meet certain conditions. In level loops that begin and close on the same benchmark, for example, the elevation difference for the loop must equal zero. In practice this is hardly ever the case, however, due to the presence of random errors. (For this discussion it is assumed that all mistakes have been eliminated from the measurements and that corrections have been applied to remove all systematic errors.) Other conditions that disclose errors in surveying measurements are that (1) the three measured angles in a plane triangle must total 180°, (2) the sum of the angles measured around the horizon at any point must equal 360°, and (3) the algebraic sum of the latitudes (and departures) must equal zero for closed polygon traverses that begin and end on the same station. Many other conditions could be cited; however, in any of them, the measurements rarely, if ever, meet the required conditions, due to the presence of random errors.

The examples above not only demonstrate that errors are present in surveying measurements but also illustrate the importance of *redundant measurements*, those made that are in excess of the minimum number needed to determine the unknowns. Two measurements of the length of a line, for example, yield one redundant measurement. The first measurement would be sufficient to determine the unknown length, and the second is redundant. This second measurement is very valuable, however. First, by examining the discrepancy between the two values, an assessment of the size of the error in the measurements can be made. If a large discrepancy exists, a blunder or large error is likely to have occurred. In that case, measurements of the line would be repeated until two values are obtained that have an acceptably small discrepancy. Second, the redundant measurement permits an adjustment to be made to obtain a final value for the unknown line length, and that final adjusted value will be more precise statistically than either of the individual measurements. In this case, if the two measurements were of equal precision, the adjusted value would be the simple mean.

Each of the specific conditions cited in the first paragraph of this section involves one redundant measurement. There is one redundant measurement, for example, when the three angles of a plane triangle are measured. This is true because with two measured angles, say A and B, the third could be computed as $C = 180° - A - B$, and thus measurement of C is unnecessary. Measuring angle C, however, enables an assessment of the errors in the angles to be made, and it also makes possible an adjustment to obtain final angles with statistically improved precision. Assuming that the angles were of equal precision, the adjustment would enforce a 180° sum for the three angles by distributing the total discrepancy in equal parts to each angle.

Although the examples cited here are indeed simple, they help to define redundant measurements and illustrate their importance. In large surveying networks, the number of redundant measurements can become extremely large, and the adjustment process is somewhat more involved than it was for the simple examples given here.

Prudent surveyors always make redundant measurements in their work, for the two important reasons indicated above: (1) to enable assessing errors and making decisions regarding acceptance or rejection of the measurements, and (2) to make possible an adjustment whereby final values with higher precisions are determined for the unknowns.

1.7 ADVANTAGES OF LEAST-SQUARES ADJUSTMENT

As indicated in Section 1.6, in surveying it is recommended that redundant measurements always be made and that adjustments of the measurements always be performed. These adjustments account for the presence of errors in the observations and increase the precisions of the final values computed for the unknowns. When an adjustment is completed, all measurements are corrected so that they are consistent throughout the survey network [i.e., the same values for the unknowns are determined no matter which corrected observation(s) are used to compute them].

Many different methods have been devised for making adjustments in surveying. The method of least squares should be used, however, because it has significant advantages over all other procedures. The advantages of least squares over other methods can be summarized with the following four general statements: (1) it is the most rigorous of adjustments, (2) it can be applied with greater ease than other adjustments, (3) it enables rigorous postadjustment analyses to be made, and (4) it can be used to perform presurvey planning. These advantages are discussed further below.

Least-squares adjustment is rigorously based on the theory of mathematical probability, whereas in general, the other methods do not have this rigorous base. As described later in the book, in a least-squares adjustment, the following condition of mathematical probability is enforced: *The sum of the squares of the errors times their respective weights are minimized.* By enforcing this condition in any adjustment, that set of errors is computed which has the highest probability of occurrence. Another aspect of least-squares adjustment that adds to its rigor is that it permits all observations, regardless of their number or type, to be entered into the adjustment and used simultaneously in the computations. Thus an adjustment can combine distances, horizontal angles, azimuths, zenith or vertical angles, height differences, coordinates, and even GPS observations. One important additional asset of least-squares adjustment is that it enables "relative weights" to be applied to the observations in accordance with their expected relative reliabilities. These reliabilities are based on expected precisions. Thus if distances were measured in the same survey by pacing, taping, and using an EDM instrument, they could all be combined in an adjustment by assigning appropriate relative weights.

Years ago, because of the comparatively heavy computational effort involved in least squares, non-rigorous or so-called "rule-of-thumb" adjustments were most often applied. Now, however, because computers have eliminated the computing problem, the reverse is true and least squares is performed more easily than these rule of thumb techniques. Least squares is less complicated because the same fundamental principles are followed regardless of the type of survey or the kind of measurements. Also, the same basic procedures are used regardless of the geometric figures involved (e.g., triangles, closed polygons, quadrilaterals, or other more complicated networks). Rules of thumb, on the other hand, are not the same for all types of surveys (e.g., level nets use one rule and traverses use another), and they vary for different geometric shapes. Furthermore, the rule of thumb applied for a particular survey by one surveyor may be different from that applied by another surveyor. A favorable characteristic of least-squares adjustments is that there is only one rigorous approach to the procedure, and thus no matter who performs the adjustment for any particular survey, the same results should be obtained.

Least squares has the advantage that after an adjustment has been finished, a complete statistical analysis can be made of the results. Based on the sizes and distribution of the errors, various tests can be conducted to determine if a survey meets acceptable tolerances or whether the measurements must be repeated. If blunders exist in the data, these can be detected and eliminated. Least squares enables precisions for the adjusted quantities to be determined easily, and these precisions can be expressed in terms of error ellipses for clear and lucid depiction. Procedures for accomplishing these tasks are described in subsequent chapters.

Besides its advantages in adjusting survey data, least squares also can be used to plan surveys. In this application, prior to conducting a needed survey, simulated surveys can be run in a trial-and-error procedure. For any project, an initial trial geometric figure for the survey is selected. Based on the figure, trial measurements are either computed or scaled. Relative weights are assigned to the observations in accordance with the precisions that can be expected using different combinations of equipment and field procedures. A least-squares adjustment of this initial network is then performed and the results analyzed. If goals have not been met, the geometry of the figure and the measurement precisions are varied and the adjustment run again. In this process different types of measurements can be used, and observations can be added or deleted. These different combinations of geometric figures and measurements are varied until one is achieved that produces either optimum or satisfactory results. The survey crew can then proceed to the field, confident that if the project is conducted according to the optimum design, satisfactory results will be obtained. This technique of applying least squares in survey planning is discussed further in later chapters.

1.8 OVERVIEW OF THE BOOK

In the remainder of the book the interrelationship between measurement errors and their adjustment is explored. In Chapters 2 through 4, methods used to determine the reliability of measurements are described. In these chapters, the ways that errors of multiple measurements tend to be distributed are illustrated, and techniques used to compare the quality of different sets of measured values are examined. In Chapters 5 through 8 and in Chapter 11, methods used to model error propagation in observed and computed quantities are discussed. In particular, error sources present in traditional surveying techniques are examined, and the ways these errors propagate throughout the measurement and computational processes are explained. In the remainder of the book, the *principles of least squares* are applied to adjust measurements in accordance with random error theory, and techniques used to locate mistakes in measurements are examined.

PROBLEMS

1.1 Describe an example in which directly measured quantities are used to obtain an indirect measurement.

1.2 Explain the difference between systematic and random errors.

1.3 What systematic errors may be present in a taped distance?

1.4 List three examples of mistakes that can be made when measuring distances with EDM instruments.

1.5 In your own words, define the difference between precision and accuracy.

1.6 Identify each of the following errors according to its source (natural, instrumental, personal):

(a) tape length
(b) EDM–reflector constant
(c) tape temperature
(d) reading a tape graduation
(e) earth curvature in EDM measurements
(f) index error of a vertical circle

1.7 Why do surveyors measure angles using both faces of a theodolite (i.e., direct and reverse)?

1.8 When measuring an angle, what possible error sources might affect the precision of the measurement?

1.9 Give an example of compensating systematic errors in a horizontal angle measurement when the angle is doubled.

1.10 What systematic errors could possibly exist in differential leveling? In trigonometric leveling?

1.11 Discuss the importance of making redundant measurements in surveying.

1.12 What are the advantages of making adjustments by the method of least squares?

2

MEASUREMENTS AND THEIR ANALYSIS

2.1 INTRODUCTION

Sets of data can be represented and analyzed using either graphical or numerical methods. Simple graphical analyses to depict trends commonly appear in newspapers or on television. A plot of the daily variation of the closing Dow Jones industrial average over the past year is an example. A bar chart showing daily high temperatures over the past month is another. Data can also be presented in numerical form and be subjected to numerical analysis. As a simple example, instead of using the bar chart, the daily high temperatures could be tabulated and the mean computed. In surveying, measurement data can also be represented and analyzed either graphically or numerically. In this chapter some rudimentary methods for doing so are explored.

2.2 SAMPLE VERSUS POPULATION

Due to time and financial constraints, in statistical analyses, generally only a small data sample is collected from a much larger, possibly infinite population. For example, political parties may wish to know the percentage of voters who support their candidate. It would be prohibitively expensive to query the entire voting population to obtain the desired information. Instead, polling agencies select a sample subset of voters from the voting population. This is an example of *population sampling*.

As another example, suppose that an employer wishes to determine the relative measuring capabilities of two prospective new employees. The candidates could theoretically spend days or even weeks demonstrating their abilities. Obviously, this would not be very practical, so instead, the employer could have each person record a sample of readings, and from the readings predict the person's abilities. The employer could, for instance, have each candidate read a micrometer 30 times. The 30 readings would represent a *sample* of the entire *population* of possible readings. In fact, in surveying, every time that distances, angles, or elevation differences are measured, samples are being collected from a population of measurements.

From the preceding discussion, the following definitions can be made:

1. *Population.* A population consists of all possible measurements that can be made of a particular quantity. Often, a population has an infinite number of data elements.
2. *Sample.* A sample is a subset of data selected from the population.

2.3 RANGE AND MEDIAN

Suppose that a 1-second (1″) instrument is used to read a direction 50 times. The seconds portions of the readings are shown in Table 2.1. These readings constitute what is called a *data set.* How can these data be organized to make them more meaningful? How can one answer the question: Are the data representative of readings that should reasonably be expected with this instrument and a competent operator? What statistical tools can be used to represent and analyze this data set?

One quick numerical method used to analyze data is to compute its *range,* also called *dispersion.* Range is the difference between the highest and lowest values. It provides an indication of the precision of the data. From Table 2.1, the lowest value is 20.1 and the highest is 26.1. Thus the range is 26.1–20.1, or 6.0. The range for this data set can be compared with ranges of other sets, but this comparison has little value when the two sets differ in size. For instance, would a set of 100 data points with a range of 8.5 be better than

Table 2.1 Fifty readings

22.7	25.4	24.0	20.5	22.5
22.3	24.2	24.8	23.5	22.9
25.5	24.7	23.2	22.0	23.8
23.8	24.4	23.7	24.1	22.6
22.9	23.4	25.9	23.1	21.8
22.2	23.3	24.6	24.1	23.2
21.9	24.3	23.8	23.1	25.2
26.1	21.2	23.0	25.9	22.8
22.6	25.3	25.0	22.8	23.6
21.7	23.9	22.3	25.3	20.1

the set in Table 2.1? Clearly, other methods of statistically analyzing data sets would be useful.

To assist in analyzing data, it is often helpful to list the values in order of increasing size. This was done with the data of Table 2.1 to produce the results shown in Table 2.2. By looking at this ordered set, it is possible to determine quickly the data's middle value or *midpoint*. In this example it lies between the values of 23.4 and 23.5. The midpoint value is also known as the *median*. Since there are an even number of values in this example, the median is given by the average of the two values closest to (which straddle) the midpoint. That is, the median is assigned the average of the 25th and 26th entries in the ordered set of 50 values, and thus for the data set of Table 2.2, the median is the average of 23.4 and 23.5, or 23.45.

2.4 GRAPHICAL REPRESENTATION OF DATA

Although an ordered numerical tabulation of data allows for some data distribution analysis, it can be improved with a *frequency histogram*, usually simply called a *histogram*. Histograms are bar graphs that show the frequency distributions in data. To create a histogram, the data are divided into *classes*. These are subregions of data that usually have a uniform range in values, or *class width*. Although there are no universally applicable rules for the selection of class width, generally 5 to 20 classes are used. As a rule of thumb, a data set of 30 values may have only five or six classes, whereas a data set of 100 values may have as many as 15 to 20 classes. In general, the smaller the data set, the lower the number of classes used.

The histogram *class width* (range of data represented by each histogram bar) is determined by dividing the total range by the number of classes to be used. Consider, for example, the data of Table 2.2. If they were divided into seven classes, the *class width* would be $6.0/7 = 0.857 \approx 0.86$. The first *class interval* is found by adding the class width to the lowest data value. For the

Table 2.2 Arranged data

20.1	20.5	21.2	21.7	21.8
21.9	22.0	22.2	22.3	22.3
22.5	22.6	22.6	22.7	22.8
22.8	22.9	22.9	23.0	23.1
23.1	23.2	23.2	23.3	23.4
23.5	23.6	23.7	23.8	23.8
23.8	23.9	24.0	24.1	24.1
24.2	24.3	24.4	24.6	24.7
24.8	25.0	25.2	25.3	25.3
25.4	25.5	25.9	25.9	26.1

data in Table 2.2, the first class interval is from 20.1 to (20.1+ 0.86), or 20.96. This class interval includes all data from 20.1 up to (but not including) 20.96. The next class interval is from 20.96 up to (20.96 + 0.86), or 21.82. Remaining class intervals are found by adding the class width to the upper boundary value of the preceding class. The class intervals for the data of Table 2.2 are listed in column (1) of Table 2.3.

After creating class intervals, the number of data values in each interval are tallied. This is called the *class frequency*. Obviously, having data ordered consecutively as shown in Table 2.2 aids greatly in this counting process. Column (2) of Table 2.3 shows the class frequency for each class interval of the data in Table 2.2.

Often, it is also useful to calculate the *class relative frequency* for each interval. This is found by dividing the class frequency by the total number of measurements. For the data in Table 2.2, the class relative frequency for the first class interval is 2/50 = 0.04. Similarly, the class relative frequency of the fourth interval (from 22.67 to 23.53) is 13/50 = 0.26. The class relative frequencies for the data of Table 2.2 are given in column (3) of Table 2.3. *Notice that the sum of all class relative frequencies is always 1*. Class relative frequency enables easy determination of percentages. For instance, the class interval from 21.82 to 22.67 contains 16% (0.16 × 100%) of the sample measurements.

A *histogram* is a bar graph plotted with either class frequencies or class relative frequencies on the ordinate, versus values of the class interval bounds on the abscissa. Using the data from Table 2.3, the histogram shown in Figure 2.1 was constructed. Notice that in this figure, relative frequencies have been plotted as ordinates.

Histograms drawn with the same ordinate and abscissa scales can be used to compare two different data sets. If one data set is more precise than the other, it will have comparatively tall bars in the center of the histogram, with

Table 2.3 Frequency count

(1) Class Interval	(2) Class Frequency	(3) Class Relative Frequency
20.10–20.96	2	2/50 = 0.04
20.96–21.82	3	3/50 = 0.06
21.82–22.67	8	8/50 = 0.16
22.67–23.53	13	13/50 = 0.26
23.53–24.38	11	11/50 = 0.22
24.38–25.24	6	6/50 = 0.12
25.24–26.1	7	7/50 = 0.14
		Σ = 50/50 = 1

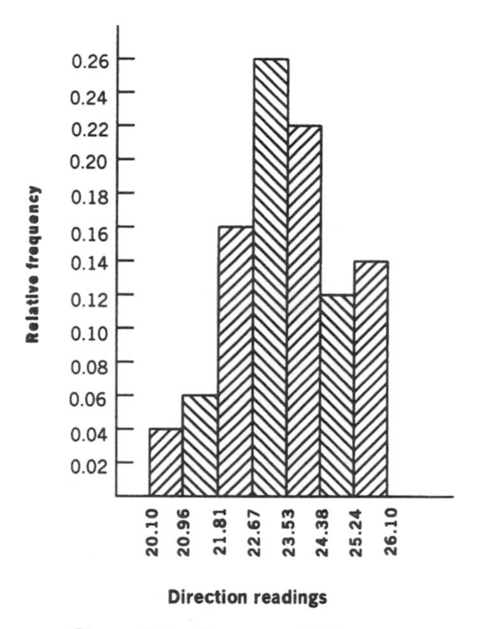

Figure 2.1 Frequency histogram.

relatively short bars near its edges. Conversely, the less precise data set will yield a wider range of abscissa values, with shorter bars at the center.

A summary of items easily seen on a histogram include:

- Whether the data are symmetrical about a central value
- The range or dispersion in the measured values
- The frequency of occurrence of the measured values
- The steepness of the histogram, which is an indication of measurement precision

Figure 2.2 shows several possible histogram shapes. Figure 2.2(*a*) depicts a histogram that is symmetric about its central value with a single peak in

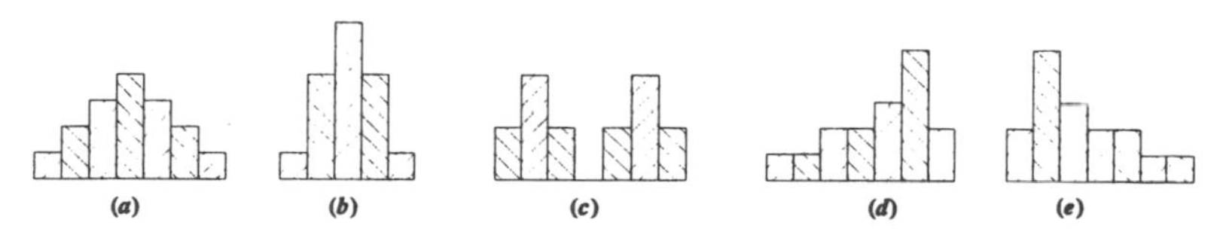

Figure 2.2 Common histogram shapes.

the middle. Figure 2.2(*b*) is also symmetric about the center but has a steeper slope than Figure 2.2(*a*), with a higher peak for its central value. Assuming the ordinate and abscissa scales to be equal, the data used to plot Figure 2.2(*b*) are more precise than those used for Figure 2.2(*a*). Symmetric histogram shapes are common in surveying practice as well as in many other fields. In fact, they are so common that the shapes are said to be examples of a *normal* distribution. In Chapter 3, reasons why these shapes are so common are discussed.

Figure 2.2(*c*) has two peaks and is said to be a *bimodal* histogram. In the histogram of Figure 2.2(*d*), there is a single peak with a long tail to the left. This results from a *skewed* data set, and in particular, these data are said to be *skewed to the right*. The data of histogram Figure 2.2(*e*) are *skewed to the left*.

In surveying, the varying histogram shapes just described result from variations in personnel, physical conditions, and equipment. For example, consider histograms resulting from repeated measurements of a long distance made with an EDM instrument and by taping. The EDM procedure would probably produce data having a very narrow range, and thus the resulting histogram would be narrow and steep with a tall central bar like that in Figure 2.2(*b*). The histogram of the same distance measured by tape and plotted at the same scales would probably be wider, with the sides not as steep nor the central value as great, like that shown in Figure 2.2(*a*). Since measurements in surveying practice tend to be normally distributed, bimodal or skewed histograms from measured data are not expected. The appearance of such a histogram should lead to an investigation for the cause of this shape. For instance, if a data set from an EDM calibration plots as a bimodal histogram, it could raise questions about whether the instrument or reflector were moved during the measuring process, or if atmospheric conditions changed dramatically during the session. Similarly, a skewed histogram in EDM work may indicate the appearance of a weather front that stabilized over time. The existence of multipath errors in GPS observations could also produce these types of histogram plots.

2.5 NUMERICAL METHODS OF DESCRIBING DATA

Numerical descriptors are values computed from a data set that are used to interpret the data's precision or quality. Numerical descriptors fall into three categories: (1) *measures of central tendency*, (2) *measures of data variation*, and (3) *measures of relative standing*. These categories are all called *statistics*. *Simply described, a statistic is a numerical descriptor computed from sample data.*

2.6 MEASURES OF CENTRAL TENDENCY

Measures of central tendency are computed statistical quantities that give an indication of the value within a data set that tends to exist at the center. The arithmetic mean, the median, and the mode are three such measures. They are described as follows:

1. *Arithmetic Mean.* For a set of n measurements, $y_1, y_2, \ldots, y_n$, this is the average of the measurements. Its value, $\bar{y}$, is computed from the following equation:

$$\bar{y} = \frac{\sum_{i=1}^{n} y_i}{n} \tag{2.1}$$

   Typically, the symbol $\bar{y}$ is used to represent the sample's arithmetic mean and the symbol μ is used to represent the population mean. Otherwise, the same equation applies. Using Equation (2.1), the mean of the measurements in Table 2.2 is 23.5.
2. *Median.* As mentioned previously, this is the midpoint of a sample set when arranged in ascending or descending order. One-half of the data are above the median, and one-half are below it. When there is an odd number of quantities, only one such value satisfies this condition. For a data set with an even number of quantities, the average of the two measurements that straddle the midpoint is used to represent the median.
3. *Mode.* Within a sample of data, the mode is the most frequently occurring value. It is seldom used in surveying because of the relatively small number of values observed in a typical set of measurements. In small sample sets, several different values may occur with the same frequency, and hence the mode can be meaningless as a measure of central tendency. The mode for the data in Table 2.2 is 23.8.

2.7 ADDITIONAL DEFINITIONS

Several other terms, which are pertinent to the study of measurements and their analysis, are listed and defined below.

1. *True Value.* A quantity's theoretically correct or exact value. As noted in Section 1.3, the true value can never be determined.
2. *Error,* ϵ. The difference between any individual measured quantity and its true value. The true value is simply the population's arithmetic mean if all repeated measurements have equal precision. Since the true value

of a measured quantity is indeterminate, errors are also indeterminate and are therefore only theoretical quantities. As given in Equation (1.1) and repeated for convenience here, errors are expressed as

$$\epsilon_i = y_i - \mu \tag{2.2}$$

where y_i is the individual measurement associated with error ϵ_i and μ is the true value for that quantity.

3. *Most Probable Value*, $\bar{y}$. That value for a measured quantity which, based upon the observations, has the highest probability. It is derived from a sample set of data rather than the population, and is simply the mean if the repeated measurements have the same precision.
4. *Residual*, v. The difference between any individual measured quantity and the most probable value for that quantity. Residuals are the values that are used in adjustment computations since most probable values can be determined. The term *error* is frequently used when residual is meant, and although they are very similar and behave in the same manner, there is this theoretical distinction. The mathematical expression for a residual is

$$v_i = \bar{y} - y_i \tag{2.3}$$

where v_i is the residual in the ith measurement y_i, and $\bar{y}$ is the most probable value for the unknown.

5. *Degrees of Freedom*. The number of observations that are in excess of the number necessary to solve for the unknowns. In other words, the number of degrees of freedom equals the number of *redundant* observations (see Section 1.6). As an example, if a distance between two points is measured three times, one observation would determine the unknown distance and the other two are redundant. These redundant observations reveal the discrepancies and inconsistencies in observed values. This, in turn, makes possible the practice of adjustment computations for obtaining most probable values based on the measured quantities.
6. *Variance*, σ^2. This is a value by which the precision for a set of data is given. *Population variance* applies to a data set consisting of an entire *population*. It is the mean of the squares of the *errors* and is given by

$$\sigma^2 = \frac{\sum_{i=1}^{n} \epsilon_i^2}{n} \tag{2.4}$$

Sample variance applies to a sample set of data. It is an unbiased estimate for the population variance given in Equation (2.4), and is calculated as

$$S^2 = \frac{\sum_{i=1}^{n} v_i^2}{n - 1} \tag{2.5}$$

Note that Equations (2.4) and (2.5) are identical except that ϵ has been changed to v, and n has been changed to $(n - 1)$ in Equation (2.5). The validity of these modifications is demonstrated in Section 2.10.

It is important to note that the simple algebraic average of all errors in a data set cannot be used as a meaningful precision indicator. This is because random errors are as likely to be positive as negative, and thus the algebraic average will equal zero. This fact is shown for a population of data in the following simple proof. Summing Equation (2.2) for n samples gives

$$\sum_{i=1}^{n} \epsilon_i = \sum_{i=1}^{n} (y_i - \mu) = \sum_{i=1}^{n} y_i - \sum_{i=1}^{n} \mu = \sum_{i=1}^{n} y_i - n\mu \tag{a}$$

Then substituting Equation (2.1) into Equation (a) yields

$$\sum_{i=1}^{n} \epsilon_i = \sum_{i=1}^{n} y_i - n \frac{\sum_{i=1}^{n} y_i}{n} = \sum_{i=1}^{n} y_i - \sum_{i=1}^{n} y_i = 0$$

Similarly, it can be shown that the mean of all residuals of a sample data set equals zero.

7. *Standard Error*, σ. This is the square root of the population variance. From Equation (2.4) and this definition, the following equation is written for the standard error:

$$\sigma = \pm\sqrt{\frac{\sum_{i=1}^{n} \epsilon_i^2}{n}} \tag{2.6}$$

where n is the number of measurements and $\Sigma_{i=1}^{n} \epsilon_i^2$ is the sum of the squares of the errors.

Note that both population variance, σ^2, and standard error, σ, are indeterminate because true values, and hence errors, are indeterminate.

As will be explained in Section 3.5, 68.3% of all measurements in a population data set lie within $\pm\sigma$ of the true value, μ. Thus the larger the standard error, the more dispersed are the values in the data set and the less precise is the measurement.

8. *Standard Deviation, S.* This is the square root of the *sample variance*. It is calculated using the following expression:

$$S = \pm\sqrt{\frac{\sum_{i=1}^{n} v_i^2}{n-1}} \tag{2.7}$$

where S is the standard deviation, $n - 1$ is the degrees of freedom or number of redundancies, and $\Sigma_{i=1}^{n} v_i^2$ the sum of the squares of the residuals. Standard deviation is an estimate for the standard error of the population. Since the standard error cannot be determined, the standard deviation is a practical expression for the precision of a *sample* set of data. Residuals are used rather than errors because they can be calculated from most probable values, whereas errors cannot be determined. Again, as discussed in Section 3.5, for a sample data set 68.3% of the measurements will theoretically lie between the most probable value, plus and minus the standard deviation, S. The meaning of this statement will be clarified in examples which follow.

9. *Standard Deviation of the Mean.* Because all measured values contain errors, the mean that is computed from a sample set of measured values will also contain error. The standard deviation of the mean is computed from the sample standard deviation according to the following equation:

$$S_{\bar{y}} = \pm\frac{S}{\sqrt{n}} \tag{2.8}$$

Notice that as $n \rightarrow \infty$, then $S_{\bar{y}} \rightarrow 0$. This illustrates that as the size of the sample set approaches the total population, the computed mean $\bar{y}$ will approach the true mean μ. This equation is derived in Chapter 4.

2.8 ALTERNATIVE FORMULA FOR DETERMINING VARIANCE

An alternative formula for computing sample variance can be derived which does not require the calculation of residuals. The use of this formula results in a savings of time and effort. From the definition of residuals, Equation (2.5) is rewritten as

$$S^2 = \frac{\sum_{i=1}^{n} (\bar{y} - y_i)^2}{n - 1} \tag{2.9}$$

Expanding Equation (2.9) gives

$$S^2 = \frac{1}{n-1} [(\bar{y} - y_1)^2 + (\bar{y} - y_2)^2 + \cdots + (\bar{y} - y_n)^2] \tag{b}$$

Substituting $\Sigma_{i=1}^{n} y_i/n$ for $\bar{y}$ into Equation (*b*) and dropping the bounds for the summation yields

$$S^2 = \frac{1}{n-1}\left[\left(\frac{\sum y_i}{n} - y_1\right)^2 + \left(\frac{\sum y_i}{n} - y_2\right)^2 + \cdots + \left(\frac{\sum y_i}{n} - y_n\right)^2\right] \tag{c}$$

Expanding Equation (*c*), we obtain

$$\begin{aligned} S^2 = \frac{1}{n-1}\Bigg[&\left(\frac{\sum y_i}{n}\right)^2 - 2y_1\frac{\sum y_i}{n} + y_1^2 + \left(\frac{\sum y_i}{n}\right)^2 - 2y_2\frac{\sum y_i}{n} \\ &+ y_2^2 + \cdots + \left(\frac{\sum y_i}{n}\right)^2 - 2y_1\frac{\sum y_n}{n} + y_n^2\Bigg] \end{aligned} \tag{d}$$

Rearranging Equation (*d*) and recognizing that $(\Sigma\ y_i/n)^2$ occurs n times in Equation (*d*) gives us

$$\begin{aligned} S^2 = \frac{1}{n-1}\Bigg[&n\left(\frac{\sum y_i}{n}\right)^2 - 2\frac{\sum y_i}{n}(y_1 + y_2 + \cdots + y_n) \\ &+ y_1^2 + y_2^2 + \cdots + y_n^2\Bigg] \end{aligned} \tag{e}$$

Adding the summation symbol to Equation (*e*) yields

$$S^2 = \frac{1}{n-1}\left[n\left(\frac{\sum y_i}{n}\right)^2 - \frac{2}{n}\left(\sum y_i\right)^2 + \sum (y_i)^2\right] \tag{f}$$

Factoring and regrouping similar summations in Equation (*f*), we have

$$S^2 = \frac{1}{n-1}\left[\sum (y_i)^2 - \left(\frac{2}{n} - \frac{1}{n}\right)\left(\sum y_i\right)^2\right]$$

$$= \frac{1}{n-1}\left[\sum (y_i)^2 - \left(\frac{1}{n}\right)\left(\sum y_i\right)^2\right] \tag{g}$$

Multiplying the last term in Equation (g) by n/n gives

$$S^2 = \frac{1}{n-1}\left[\sum (y_i)^2 - n\left(\frac{\sum y_i}{n}\right)^2\right] \tag{h}$$

Finally, recognizing that $\bar{y} = \Sigma\, y_i/n$, the following expression for the variance results:

$$S^2 = \frac{\sum (y_i)^2 - n\bar{y}^2}{n-1} \tag{2.10}$$

Using Equation (2.10), the variance of a sample data set can be computed by subtracting n times the square of the data's mean from the summation of the squared individual measurements. With this equation, the variance and the standard deviation can be computed directly from the data. However, it should be stated that with large numerical values, Equation (2.10) may overwhelm a hand-held calculator or a computer working in single precision. If this problem should arise, the data should be *centered* or Equation (2.5) used. Centering a data set involves subtracting a constant value (usually, the arithmetic mean) from all values in a data set. By doing this, the values are modified to a smaller, more manageable size.

2.9 NUMERICAL EXAMPLES

Example 2.1 Using the data from Table 2.2, determine the sample set's mean, median, and mode and the standard deviation using both Equations (2.7) and (2.10). Also plot its histogram. (Recall that the data of Table 2.2 result from the seconds portion of 50 measured directions.)

SOLUTION

Mean: From Equation (2.1) and using the $\Sigma\, y_i$ value from Table 2.4, we have

$$\bar{y} = \frac{\sum_{i=1}^{n} y_i}{n} = \frac{\sum_{i=1}^{50} y_i}{50} = \frac{1175}{50} = 23.5''$$

Table 2.4 Data arranged for the solution of Example 2.1

No.	y	v	v^2	No.	y	v	v^2	No.	y	v	v^2	No.	y	v	v^2
1	20.1	3.4	11.56	14	22.7	0.8	0.64	27	23.6	−0.1	0.01	40	24.7	−1.2	1.44
2	20.5	3.0	9.00	15	22.8	0.7	0.49	28	23.7	−0.2	0.04	41	24.8	−1.3	1.69
3	21.2	2.3	5.29	16	22.8	0.7	0.49	29	23.8	−0.3	0.09	42	25.0	−1.5	2.25
4	21.7	1.8	3.24	17	22.9	0.6	0.36	30	23.8	−0.3	0.09	43	25.2	−1.7	2.89
5	21.8	1.7	2.89	18	22.9	0.6	0.36	31	23.8	−0.3	0.09	44	25.3	−1.8	3.24
6	21.9	1.6	2.56	19	23.0	0.5	0.25	32	23.9	−0.4	0.16	45	25.3	−1.8	3.24
7	22.0	1.5	2.25	20	23.1	0.4	0.16	33	24.0	−0.5	0.25	46	25.4	−1.9	3.61
8	22.2	1.3	1.69	21	23.1	0.4	0.16	34	24.1	−0.6	0.36	47	25.5	−2.0	4.00
9	22.3	1.2	1.44	22	23.2	0.3	0.09	35	24.1	−0.6	0.36	48	25.9	−2.4	5.76
10	22.3	1.2	1.44	23	23.2	0.3	0.09	36	24.2	−0.7	0.49	49	25.9	−2.4	5.76
11	22.5	1.0	1.00	24	23.3	0.2	0.04	37	24.3	−0.8	0.64	50	26.1	−2.6	6.76
12	22.6	0.9	0.81	25	23.4	0.1	0.01	38	24.4	−0.9	0.81				
13	22.6	0.9	0.81	26	23.5	0.0	0.00	39	24.6	−1.1	1.21				
												$\Sigma =$	1175.0	0.0	92.36

Median: Since there are an even number of observations, the data's midpoint lies between the values that are numerically 25th and 26th from the beginning of the ordered set. These values are 23.4 and 23.5, respectively. Averaging them yields 23.45.

Mode: The mode, which is the most frequently occurring value, is 23.8. It appears three times in the sample.

Range, Class Width, Histogram: These data were developed in Section 2.4, with the histogram plotted in Figure 2.1.

Standard Deviation: Table 2.4 lists the residuals [computed using Equation (2.3)], and their squares, for each measurement.

From Equation (2.7) and using the value of 92.36 from Table 2.4 as the sum of the squared residuals, we obtain

$$S = \pm\sqrt{\frac{\sum_{i=1}^{50} v_i^2}{n - 1}} = \sqrt{\frac{92.36}{50 - 1}} = \pm 1.37''$$

If the individual y values of Table 2.4 are squared and summed, they yield

$$\sum (y_i)^2 = 27{,}704.86$$

By Equation (2.10),

$$S^2 = \frac{27{,}704.86 - 50(23.5)^2}{50 - 1} = \frac{92.36}{49} = 1.88''^2$$

Thus $S = \pm\sqrt{1.88} = \pm 1.37''$.

Notice from Table 2.4 that the number of measurements which lie within a standard deviation of the mean, that is, between (23.5 − 1.37) and (23.5 + 1.37), or between 22.13 and 24.87, is 34. This represents 34/50 × 100%, or 68% of all measurements in the sample and matches the theory noted earlier. The reader can also verify that the algebraic sum of residuals is zero.

The histogram shown in Figure 2.1 plots class relative frequencies versus class values. Notice how the values tend to be grouped about the central point. This is an example of a rather precise data set.

Example 2.2 The data set shown below also represents the seconds portion of 50 measurements of a direction. Compute the mean, median, and mode, and use Equation (2.10) to determine the standard deviation. Also construct a histogram. Compare the data of this example with those of Example 2.1.

34.2	33.6	35.2	30.1	38.4	34.0	30.2	34.1	37.7	36.4
37.9	33.0	33.5	35.9	35.9	32.4	39.3	32.2	32.8	36.3
35.3	32.6	34.1	35.6	33.7	39.2	35.1	33.4	34.9	32.6
36.7	34.8	36.4	33.7	36.1	34.8	36.7	30.0	35.3	34.4
33.7	34.1	37.8	38.7	33.6	32.6	34.7	34.7	36.8	31.8

SOLUTION: Table 2.5, which arranges each measurement and its square in ascending order, is first prepared.

Mean:

$$\bar{y} = \frac{\sum_{i=1}^{50} y_i}{50} = \frac{1737}{50} = 34.74$$

Median: The median is between the 25th and 26th values, which are both 34.7. Thus the median is 34.7.

Mode: The data have three different values that occur with a frequency of 3. Thus the modes for the data set are 32.6, 33.7, and 34.1.

Range: The range of the data is 39.3 − 30.0 = 9.3.

Class Width: For comparison purposes, the class width of 0.86 is taken because it was used for the data in Table 2.2. Since it is desired that the histogram be centered about the data's mean value, the central interval is determined by adding and subtracting one-half of the class width (0.43) to the mean. Thus the central interval is from (34.74 − 0.43), or 34.31, to (34.74 + 0.43) or 35.17. To compute the remaining class intervals, the class width is subtracted, or added, to the bounds of the central interval as necessary until all the data are contained within the bounds of the intervals. Thus the interval immediately preceding the central interval will be from (34.31 − 0.86), or 33.45, to 34.31, and the interval immediately following the central interval will be from 35.17 to (35.17 + 0.86), or 36.03. In a similar fashion, the remaining class intervals were determined and a class frequency chart was constructed as shown in Table 2.6. Using this table, the histogram of Figure 2.3 was constructed.

Variance: By Equation (2.10), using the sum of measurements squared in Table 2.5, the sample variance is

$$S^2 = \sum_{i=1}^{50} y_i^2 - 50\,\frac{\bar{y}^2}{49} = \frac{60{,}584.48 - 50(34.74)^2}{50 - 1} = 4.92$$

and the sample *standard deviation* is

$$S = \pm\sqrt{4.92} = \pm 2.22$$

Table 2.5 Data arranged for the solution of Example 2.2

No.	y	y^2	No.	y	y^2	No.	y	y^2	No.	y	y^2
1	30.0	900.00	14	33.6	1128.96	27	34.8	1211.04	39	36.4	1324.96
2	30.1	906.01	15	33.6	1128.96	28	34.8	1211.04	40	36.4	1324.96
3	30.2	912.04	16	33.7	1135.69	29	34.9	1218.01	41	36.7	1346.89
4	31.8	1011.24	17	33.7	1135.69	30	35.1	1232.01	42	36.7	1346.89
5	32.2	1036.84	18	33.7	1135.69	31	35.2	1239.04	43	36.8	1354.24
6	32.4	1049.76	19	34.0	1156.00	32	35.3	1246.09	44	37.7	1421.29
7	32.6	1062.76	20	34.1	1162.81	33	35.3	1246.09	45	37.8	1428.84
8	32.6	1062.76	21	34.1	1162.81	34	35.6	1267.36	46	37.9	1436.41
9	32.6	1062.76	22	34.1	1162.81	35	35.9	1288.81	47	38.4	1474.56
10	32.8	1075.84	23	34.2	1169.64	36	35.9	1288.81	48	38.7	1497.69
11	33.0	1089.00	24	34.4	1183.36	37	36.1	1303.21	49	39.2	1536.64
12	33.4	1115.56	25	34.7	1204.09	38	36.3	1317.69	50	39.3	1544.49
13	33.5	1122.25	26	34.7	1204.09						
										Σ = 1737.0	60,584.48

Table 2.6 Frequency table for Example 2.2

Class	Class Frequency	Class Relative Frequency
29.15–30.01	1	0.02
30.01–30.87	2	0.04
30.87–31.73	0	0.00
31.73–32.59	3	0.06
32.59–33.45	6	0.12
33.45–34.31	11	0.22
34.31–35.17	7	0.14
35.17–36.03	6	0.12
36.03–36.89	7	0.14
36.89–37.75	1	0.02
37.75–38.61	3	0.06
38.61–39.47	3	0.06
	$\Sigma = 50$	$\Sigma = 1.00$

The number of measurements that actually fall within the bounds of the mean $\pm$ S (i.e., between 37.74 $\pm$ 2.22) is 30. This is 60% of all the measurements, and closely approximates the theoretical value of 68.3%. These bounds and the mean value are shown as dashed lines in Figure 2.3.

Comparison: The data set of Example 2.2 has a larger standard deviation ($\pm$2.22) than that of Example 2.1 ($\pm$1.37). The range for the data of Example 2.2 (9.3) is also larger than that of Example 2.1 (6.0). Thus the data set of

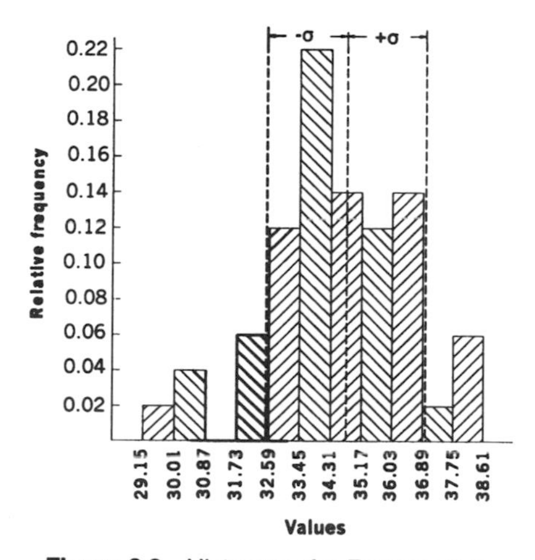

Figure 2.3 Histogram for Example 2.2.

Example 2.2 is less precise than that of Example 2.1. A comparison of the two histograms shows this precision difference graphically. Note, for example, that the histogram in Figure 2.1 is narrower in width and taller at the center than the histogram in Figure 2.3.

2.10 DERIVATION OF THE SAMPLE VARIANCE (BESSEL'S CORRECTION)

Recall from Section 2.7 that the denominator of the equation for sample variance was $n - 1$, while the denominator of the population variance was n. A simple explanation for this difference is that one observation is necessary to compute the mean ($\bar{y}$), and thus only $n - 1$ observations remain for the variance's computation. A derivation of Equation (2.5) will clarify.

Consider a sample size of n drawn from a population with a mean μ and standard error of σ. Let y_i be an observation from the sample. Then

$$\begin{aligned} y_i - \mu &= (y_i - \bar{y}) + (\bar{y} - \mu) \\ &= (y_i - \bar{y}) + \epsilon \end{aligned} \tag{i}$$

where $\epsilon = \bar{y} - \mu$ is the *error* or deviation of the sample mean. Squaring and expanding Equation (*i*) yields

$$(y_i - \mu)^2 = (y_i - \bar{y})^2 + \epsilon^2 + 2\epsilon(y_i - \bar{y})$$

Summing all the observations in the sample from i equaling 1 to n yields

$$\sum_{i=1}^{n} (y_i - \mu)^2 = \sum_{i=1}^{n} (y_i - \bar{y})^2 + n\epsilon^2 + 2\epsilon \sum_{i=1}^{n} (y_i - \bar{y}) \tag{j}$$

Since by definition of $\bar{y}$,

$$\sum_{i=1}^{n} (y_i - \bar{y}) = \sum_{i=1}^{n} y_i - n\bar{y} = \sum_{i=1}^{n} y_i - \sum_{i=1}^{n} y_i = 0 \tag{k}$$

Equation (*j*) becomes

$$\sum_{i=1}^{n} (y_i - \mu)^2 = \sum_{i=1}^{n} (y_i - \bar{y})^2 + n\epsilon^2 \tag{l}$$

Repeating this calculation for many samples, the mean value of the left-hand side of Equation (*l*) (by definition of σ^2) tends to $n\sigma^2$. Similarly, the mean value of $n\epsilon^2 = n(\mu - \bar{y})^2$ will tend to n times the variance of $\bar{y}$ since ϵ represents the deviation of the sample mean from the population mean. Thus $n\epsilon^2 \rightarrow n(\sigma^2/n)$, where σ^2/n is the variance in $\bar{y}$ as $n \rightarrow \infty$. From the discussion above, and using Equation (l), we obtain

$$n\sigma^2 \longrightarrow \sum_{i=1}^{n} (y_i - \bar{y})^2 + \sigma^2 \qquad (m)$$

Rearranging Equation (m) gives

$$\sum_{i=1}^{n} (y_i - \bar{y})^2 \longrightarrow (n - 1)\sigma^2 \qquad (n)$$

Thus from Equation (n), it follows that

$$S^2 = \frac{\sum_{i=1}^{n} (y_i - \bar{y})^2}{n - 1} \longrightarrow \sigma^2 \qquad (o)$$

In other words, for a large number of random samples, the mean value of $\Sigma_{i=1}^{n} (y_i - \bar{y})^2/n - 1$ tends to σ^2. That is, S^2 is an unbiased estimate of the population's variance.

2.11 STATS PROGRAM

A statistical software package called STATS is included on the diskette that is provided with this book. It can be used to quickly perform statistical analysis of data sets as presented in this chapter. Directions on its use are given in Appendix F.

PROBLEMS

2.1 Following are 100 readings of the seconds portion of a direction made using a 1″ instrument. All readings were taken to the tenth of a second.

31.6	35.4	34.3	33.9	36.1	33.9	33.4	41.0	36.4	39.4
35.2	35.1	33.8	36.4	34.5	39.0	36.1	32.5	34.7	33.4
34.8	38.0	40.2	37.6	31.7	33.5	33.2	36.1	33.9	37.2
39.1	39.1	32.6	34.8	38.9	30.3	35.5	30.6	36.1	33.2
35.0	31.5	39.0	33.8	35.2	33.4	32.2	31.5	33.6	35.8

36.4	36.8	37.0	35.0	37.5	32.5	31.5	32.5	34.4	36.0
34.9	35.0	36.6	39.1	37.6	33.5	36.3	33.3	39.3	31.8
32.9	37.1	33.1	32.3	33.7	34.2	36.7	38.1	35.4	35.4
37.7	32.8	33.1	40.3	30.2	36.2	36.2	35.4	37.3	33.3
37.8	32.2	41.6	38.8	37.2	37.9	34.8	33.7	36.8	38.3

(a) What is the mean of the data set?

(b) Construct a frequency histogram of the data using seven uniform-width class intervals.

(c) What are the variance and standard deviation of the data?

(d) What is the standard deviation of the mean?

2.2 An EDM instrument and reflector are set at the ends of a baseline that is 400.781 m long. Its length is measured 25 times, with the following results:

400.806 400.824 400.814 400.793 400.804 400.803 400.816 400.820 400.811
400.807 400.825 400.820 400.809 400.800 400.813 400.813 400.817 400.812
400.815 400.805 400.807 400.810 400.828 400.808 400.799

(a) What are the mean, median, and standard deviation of the data?

(b) Construct a histogram of the data and describe its properties. On the histogram, lay off the standard deviation from both sides of the mean.

(c) How many measurements are between $\bar{y} \pm S$, and what percentage of measurements does this represent?

2.3 A distance was measured in two parts with a 100-ft steel tape and then in its entirety with a 200-ft steel tape. Ten repetitions were made by each method.

Distances measured with 100-ft tape: Distances measured with 200-ft tape:

Part 1:

100.001 100.018 99.974 99.992 99.972 149.326 149.397 149.357 149.294 149.337
99.990 99.950 99.984 99.979 99.988 149.338 149.329 149.331 149.370 149.363

Part 2:

49.329 49.365 49.346 49.300 49.327
49.324 49.349 49.357 49.341 49.333

(a) What are the mean, variance, and standard deviation for the data sets of the two partial distance measurements, and the data set for the full distance measurement?

(b) Create a class frequency table and histogram for each of the three sets of data using a class width of 0.009. Center the data as explained in Example 2.2.

2.4 During a triangulation project, an observer made 50 readings for each direction. The seconds portion of the directions to station Red are listed below. Analyze the data and state whether this set appears to be reasonable. Using a 1″ class interval, plot the histogram using relative frequencies for the ordinates. Note any abnormalities in the data and suggest possible reasons for any detected. As a supervisor, would you suggest that the station be reobserved?

36.7 38.4 34.0 35.8 33.8 35.6 31.6 35.0 34.5 35.4 32.8 38.0 36.2 32.6 35.8 36.2
32.9 35.7 33.4 33.6 36.7 33.8 34.5 37.0 35.9 46.9 50.5 48.6 44.0 46.7 46.4 48.9
49.7 48.1 47.3 50.7 49.7 47.6 46.0 48.4 48.5 49.2 48.2 48.7 46.8 47.2 47.8 51.2
47.5 45.7

2.5 Use the program STATS to compute the mean, median, mode, and standard deviation, and plot a histogram of the data in Table 2.2.

Use the program STATS to do problems 2.6 through 2.9.

2.6 Problem 2.1

2.7 Problem 2.2

2.8 Problem 2.3

2.9 Problem 2.4

2.10 Given the data in Problem 2.2, what is the combined EDM–reflector offset?

Practical Exercises

2.11 Using a 1″ least-count theodolite, lock the horizontal circle, center the micrometer, and read the horizontal circle to the nearest 0.1″. Move the micrometer off center, recenter it, and read the circle again. Repeat this process 50 times. Perform the calculations of Problem 2.1 using the resulting data set.

2.12 Same as Problem 2.11 except that, in addition, make a pointing on some well-defined distant target. Between each reading, back the instrument off the point with the tangent screw, reestablish the pointing, and recenter the micrometer. Construct a histogram of the resulting data set using a class interval equal to one-half of the data set's standard deviation.

2.13 Determine your EDM–reflector constant, K, by observing the distances between three points that are on-line, as shown in the figure. Distance AC should be roughly 1 mile long with B situated at some location between A and C. From measured values AC, AB, and BC, the constant K can be determined as follows:

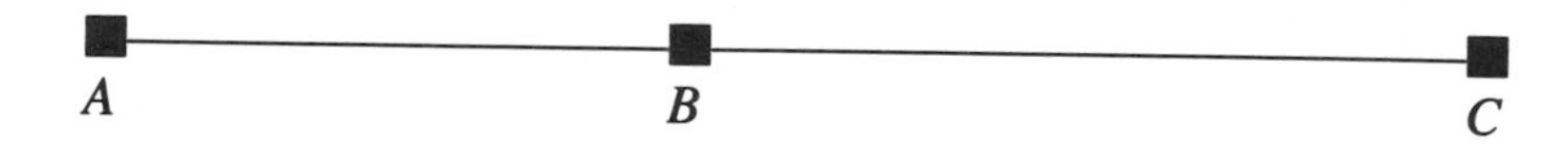

Since

$$AC + K = (AB + K) + (BC + K)$$

thus

$$K = AC - (AB + BC)$$

When establishing the line, be sure that $AB \neq BC$ and that all three points are precisely on a straight line. Use three tripods and tribrachs to minimize setup errors and be sure that all are in adjustment. Measure each line 20 times with the instrument in the metric mode. Be sure to adjust the distances for the appropriate temperature and pressure, and for differences in elevation. Determine the 20 values of K and analyze the sample set. What is the mean value for K, and what is its standard deviation?

3

RANDOM ERROR THEORY

3.1 INTRODUCTION

As noted earlier, the adjustment of measured quantities containing random errors is a major concern to people involved in the measurement sciences. In the remaining chapters it is assumed that all systematic errors have been removed from the measured values and that only random errors, and blunders which have escaped detection, remain. In this chapter the general theory of random errors is developed, and some simple methods that can be used to isolate remaining blunders in sets of data are discussed.

3.2 THEORY OF PROBABILITY

Probability is the ratio of the number of times that an event should occur to the total number of possibilities. For example, the probability of tossing a two with a fair die is 1/6 since there are six total possibilities (faces on a die) and only one of these is a two. When an event can occur in m ways and fail to occur in n ways, the probability of its occurrence is $m/(m + n)$, and the probability of its failure is $n/(m + n)$.

Probability is always a fraction ranging between zero and 1. Zero denotes impossibility, and 1 indicates certainty. Since an event must either occur or fail to occur, the sum of all probabilities for any event is 1, and thus if 1/6 is the probability of throwing a two with one throw of a die, then $1 - 1/6$, or 5/6, is the probability that a two will not appear.

In probability terminology, a *compound event* is the simultaneous occurrence of two or more independent events. This is the situation encountered most frequently in surveying. For example, random errors from angles and distances (compound events) cause traverse misclosures. The probability of the simultaneous occurrence of two independent events is the product of their individual probabilities.

To illustrate, consider the simple example of having two boxes containing combinations of red and white balls. Box *A* contains four balls, one red and three white. Box *B* contains five balls, two red and three white. What is the probability that two red balls would be drawn if one ball is drawn randomly from each box? The total number of possible pairs is 4 × 5, or 20, since by drawing one ball from box *A*, any of the five balls in box *B* would complete the pair. Now, there are only two ways to draw two red balls. That is, box *A*'s red ball which has a $1/4^{th}$ probability of being drawn, can be matched with either red ball from box *B*. The probability of drawing a red ball from Box B is 2/5. Thus the probability of this compound event is the product of the individual probabilities of drawing a red ball from each box, or:

$$P = 1/4 \times 2/5 = 2/20$$

Similarly the probability of simultaneously drawing two white balls is 3/4 × 3/5, or 9/20, and the probability of getting one red and one white is 1 − (2/20 + 9/20), or 9/20.

From the foregoing it is seen that the probability of the simultaneous occurrence of two independent events is the product of the individual probabilities of those two events. This principle is extended to include any number of events:

$$P = P_1 \times P_2 \times \cdots \times P_n \tag{3.1}$$

where P is the probability of the simultaneous occurrence of events having individual probabilities $P_1, P_2, \ldots, P_n$.

To develop the principle of how random errors occur, consider a very simple example where a single tape measurement is taken between points A and B. Assume that this measurement contains a single random error of size 1. Since the error is random, there are two possibilities for the value of the resultant error, +1 or −1. Let t be the number of ways each resultant error can occur, and T be the total number of possibilities, which is two. The probability of obtaining +1, which can occur only one way (i.e., $t = 1$), is thus t/T or 1/2. This is also the probability of obtaining −1. Suppose now that in measuring a distance AB, the tape must be placed end to end so that the result depends on the combination of two of these tape measurements. Then the possible error combinations in the result are −1 *and* −1, −1 *and* +1, +1 *and* −1, and +1 *and* +1, with $T = 4$. The final errors are −2, 0, and +2, and their t values are 1, 2, and 1, respectively. This produces probabilities of 1/4, 1/2, and 1/4, respectively. In general, as n, the number of

Table 3.1 Occurrence of random errors

(1) Number of Combining Measurements	(2) Value of Resultant Error	(3) Frequency, t	(4) Total number of Possibilities, T	(5) Probability
1	+1	1	2	1/2
	−1	1		1/2
	+2	1		1/4
2	0	2	4	1/2
	−2	1		1/4
	+3	1		1/8
3	+1	3	8	3/8
	−1	3		3/8
	−3	1		1/8
	+4	1		1/16
	+2	4		1/4
4	0	6	16	3/8
	−2	4		1/4
	−4	1		1/16
	+5	1		1/32
	+3	5		5/32
5	+1	10	32	5/16
	−1	10		5/16
	−3	5		5/32
	−5	1		1/32

single combined measurements, is increased, T increases according to the function $T = 2^n$, and thus for three combined measurements, $T = 2^3 = 8$, for four measurements, $T = 2^4 = 16$, etc.

The analysis of the preceding paragraph can be continued to obtain the results shown in Table 3.1. Figures 3.1(*a*) through (*e*) are histogram plots of

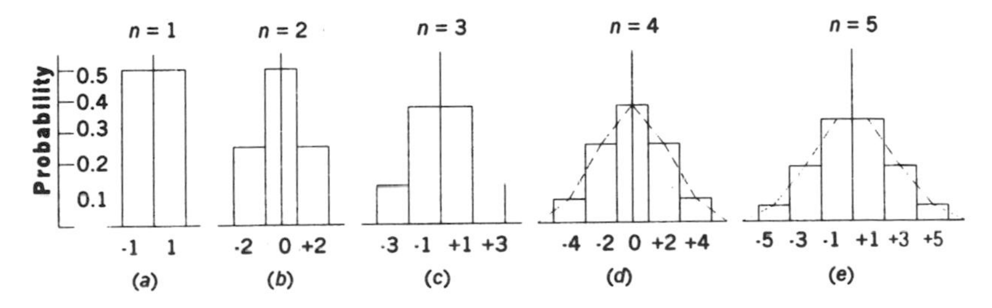

Figure 3.1 Plots of probability versus size of resultant errors.

the results in Table 3.1 where the values of the resultant errors are plotted as the abscissas, and the probabilities are plotted as ordinates of equal-width bars.

If the number of combining measurements, n, is increased progressively to larger values, the plot of error sizes versus probabilities would approach a smooth curve of the characteristic bell-shape shown in Figure 3.2. This curve is known as the *normal error distribution curve*. It is also called the *probability density function of a normal random variable*. Notice that when n is 4 as illustrated in Figure 3.1(*d*), and when $n = 5$ as shown in Figure 3.1(*e*), the dashed lines are already beginning to take on this form.

It is important to notice that the total area of the vertical bars for each plot equals 1. This is true no matter the value of n, and thus *the area under the smooth normal error distribution curve is equal to 1*. If an event has a probability of 1, it is certain to occur, and therefore *the area under the curve represents the sum of all the probabilities of the occurrence of errors*.

As derived in Section D.1 of Appendix D, the equation of the normal distribution curve, also called the *normal probability density function*, is

$$f(x) = \frac{1}{\sigma\sqrt{2\pi}} e^{-x^2/2\sigma^2} \tag{3.2}$$

where $f(x)$ is the probability density function, e the base of natural logarithms, x the error, and σ the standard error as defined in Chapter 2.

3.3 PROPERTIES OF THE NORMAL DISTRIBUTION CURVE

In Equation (3.2), $f(x)$ is the probability of occurrence of an error of a size between x and $x + dx$, where dx is an infinitesimally small value. The error's probability is equivalent to the area under the curve between the limits of x

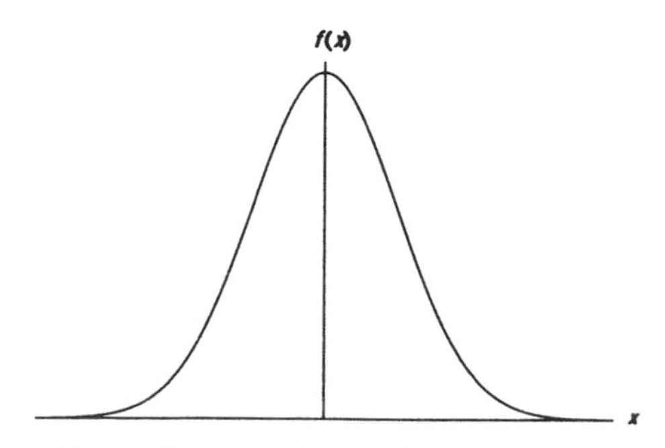

Figure 3.2 Normal error distribution curve.

and $x + dx$, which is shown crosshatched in Figure 3.3. As stated previously, the total area under the probability curve represents the total probability, which is 1. In equation form this is represented as

$$\text{area} = \int_{-\infty}^{+\infty} \frac{1}{\sigma\sqrt{2\pi}} e^{-x^2/2\sigma^2}\, dx = 1 \tag{3.3}$$

Let y represent $f(x)$ in Equation (3.2) and differentiate:

$$\frac{dy}{dx} = -\frac{x}{\sigma^2}\left(\frac{1}{\sigma\sqrt{2\pi}} e^{-x^2/2\sigma^2}\right) \tag{3.4}$$

Recognizing the term in parentheses of Equation (3.4) as y gives

$$\frac{dy}{dx} = -\frac{x}{\sigma^2} y \tag{3.5}$$

Taking the second derivative of Equation (3.2), we obtain

$$\frac{d^2y}{dx^2} = -\frac{x}{\sigma^2}\frac{dy}{dx} - \frac{y}{\sigma^2} \tag{3.6}$$

Substituting Equation (3.5) into Equation (3.6) gives

$$\frac{d^2y}{dx^2} = \frac{x^2}{\sigma^4} y - \frac{y}{\sigma^2} \tag{3.7}$$

Equation (3.7) can be simplified to

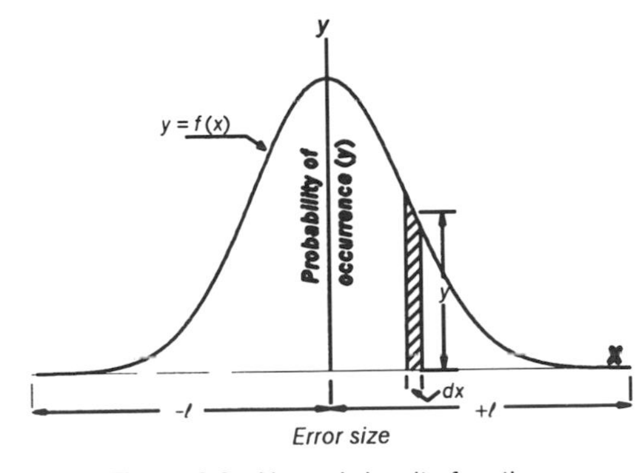

Figure 3.3 Normal density function.

$$\frac{d^2y}{dx^2} = \frac{y}{\sigma^2}\left(\frac{x^2}{\sigma^2} - 1\right) \tag{3.8}$$

From calculus, the first derivative of a function gives the slope of the function when evaluated at a point. In Equation (3.5), $dy/dx = 0$ when the values of x or y equal zero. This implies that the curve is parallel to the x axis at the center of the curve when x is zero and is asymptotic to the x axis as y approaches zero.

Also from calculus, a function's second derivative provides the rate of change in a slope when evaluated at a point. The curve's inflection points (points where the algebraic sign of the slope changes) can be located by finding where the function's second derivative equals zero. In Equation (3.8), $d^2y/dx^2 = 0$ when $x^2/\sigma^2 - 1 = 0$, and thus the curve's inflection point occurs when x equals $\pm\sigma$.

Since $e^0 = 1$, if x is set equal to zero in Equation (3.2), then

$$y = \frac{1}{\sigma\sqrt{2\pi}} \tag{3.9}$$

This is the curve's central ordinate, and as can be seen, it is inversely proportional to σ. According to Equation (3.9), a group of measurements having small σ must have a large central ordinate. Thus the area under the curve will be concentrated near the central ordinate, and the errors will be correspondingly small. This indicates that the set of measurements is precise. Since σ bears this relationship to the precision, it is a numerical measure for the precision of a measurement set. In Section 2.7 we defined σ as the *standard error* and gave equations for computing its value.

3.4 STANDARD NORMAL DISTRIBUTION FUNCTION

In Section 3.2 we defined the probability density function of a normal random variable as $f(x) = (1/\sigma\sqrt{2\pi})\, e^{-x^2/2\sigma^2}$. From this we develop the *normal distribution function*:

$$F_x(t) = \int_{-\infty}^{t} \frac{1}{\sigma\sqrt{2\pi}}\, e^{-x^2/2\sigma^2}\, dx \tag{3.10}$$

where t is the upper bound of the integration, as shown in Figure 3.4.

As stated in Section 3.3, the area under the normal density curve represents the probability of occurrence. Furthermore integration of this function yields the area under the curve. Unfortunately, the integration called for in Equation (3.10) cannot be carried out in closed form, and thus numerical integration techniques must be used to tabulate values for this function. This has been

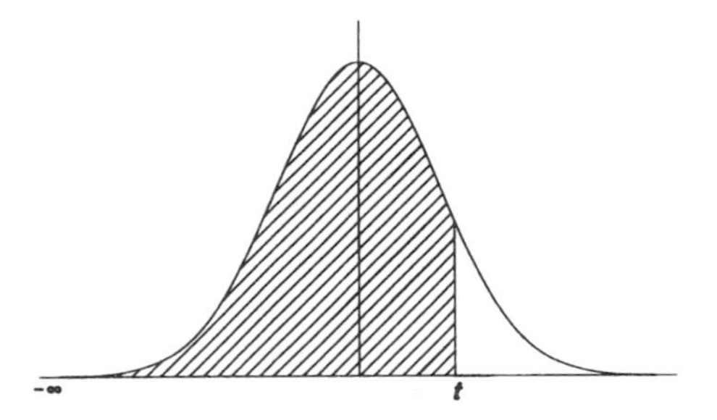

Figure 3.4 Area determined using Equation (3.10).

done for the function when the mean is zero ($\mu = 0$) and the variance is 1 ($\sigma^2 = 1$). The results of this integration are shown in the *standard normal distribution table*, Table D.1 of Appendix D. In this table the leftmost column with a heading of t is the value shown in Figure 3.4 in units of σ. The top row (with headings 0 through 9) represents the hundredths decimal places for the t values. The tabulated values in the body of Table D.1 represent areas under the standard normal distribution curve from $-\infty$ to t. For example, to determine the area under the curve from $-\infty$ to 1.68, first find the row with 1.6 in the t column. Then scan along the row to the column with a heading of 8. At the intersection of row 1.6 and column 8 (1.68), the value 0.95352 occurs. This is the area under the standard normal distribution curve from $-\infty$ to a t value of 1.68. Similarly, other areas under the standard normal distribution curve can be found for various values for t. Since the area under the curve represents probability, and its maximum area is 1, this means that there is a 95.352% (0.95352 $\times$ 100) probability that t is less than or equal to 1.68. Alternatively, it can be stated that there is a 4.648% (1 $-$ 0.95352) $\times$ 100 probability that t is greater than 1.68.

Once available, Table D.1 can be used to evaluate the distribution function for any mean, μ, and variance, σ^2. For example, if y is a normal random variable with a mean of μ and a variance of σ^2, an equivalent normal random variable $z = (y - \mu)/\sigma$ can be defined that has a mean of zero and a variance of 1. Substituting the definition for z with $\mu = 0$ and $\sigma^2 = 1$ into Equation (3.2), its density function is

$$N_z(z) = \frac{1}{\sqrt{2\pi}} e^{-z^2/2} \tag{3.11}$$

and its distribution function, known as the *standard normal distribution function*, becomes

$$N_z(z) = \int_{-\infty}^{t} \frac{1}{\sqrt{2\pi}} e^{-z^2/2}\, dz \tag{3.12}$$

For any group of normally distributed measurements, the probability of the normal random variable can be computed by analyzing the integration of the distribution function. Again, as stated previously, the area under the curve in Figure 3.4 represents probability. Let z be a normal random variable; then the probability that z is less than some value of t is given by

$$P(z < t) = N_z(t) \tag{3.13}$$

To determine the area (probability) between t values of a and b (the cross-hatched area in Figure 3.5), the difference in the areas going to b and a, respectively, can be taken. By Equation (3.13), the area from $-\infty$ to b is $P(z < b) = N_z(b)$. By the same equation, the area from $-\infty$ to a is $P(z < a) = N_z(a)$. Thus the area between a and b is the difference in these values and is expressed as

$$P(a < z < b) = N_z(b) - N_z(a) \tag{3.14}$$

If the bounds are equal in magnitude but opposite in sign (i.e., $-a = b = t$), the probability is

$$P(|z| < t) = N_z(t) - N_z(-t) \tag{3.15}$$

From the symmetry of the normal distribution in Figure 3.6, it is seen that

$$P(z > t) = P(z < -t) \tag{3.16}$$

for any $t > 0$. This symmetry can be also shown with Table D.1. The tabular value (area) for a t value of -1.00 is 0.15866. Furthermore, the tabular value for a t value of $+1.00$ is 0.84134. Since the maximum probability (area) is 1, the area above $+1.00$ is $1 - 0.84134$, or 0.15866, which is the same as the area below -1.00. Thus since the total probability is always 1, we can define the following relationship:

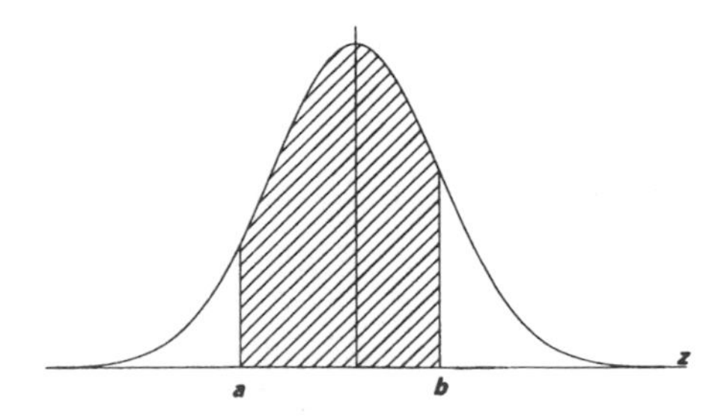

Figure 3.5 Area representing probability in Equation (3.14).

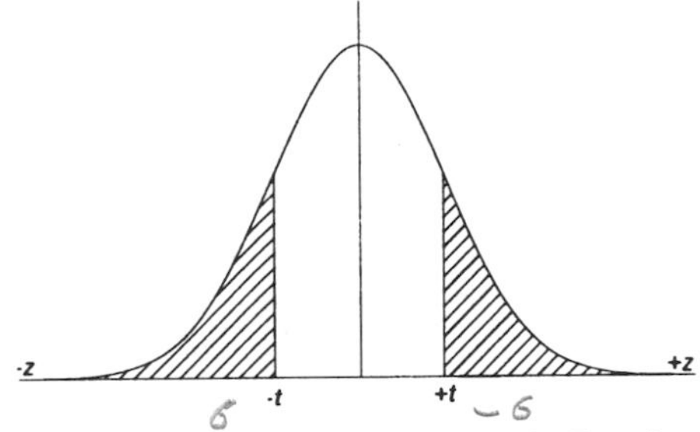

Figure 3.6 Area representing probability in Equation (3.16).

$$1 - N_z(t) = N_z(-t) \tag{3.17}$$

Now substituting Equation (3.17) into Equation (3.15), we have

$$P(|z| < t) = 2N_z(t) - 1 \tag{3.18}$$

3.5 PROBABILITY OF THE STANDARD ERROR

The equations above can be used to determine the probability of the standard error, which from previous discussion is the area under the normal distribution curve between the limits of $\pm\sigma$. For the standard normal distribution when σ^2 is 1, it is necessary to locate the values of $t = -1$ ($\sigma = -1$) and $t = +1$ ($\sigma = 1$) in Table D.1. As seen previously, the appropriate value from the table for $t = -1.00$ is 0.15866. Also, the tabular value for $t = 1.00$ is 0.84134, and thus, according to Equation (3.15), the area between $-\sigma$ and $+\sigma$ is

$$P(-\sigma < z < +\sigma) = N_z(+\sigma) - N_z(-\sigma)$$

$$P(-\sigma < z < +\sigma) = 0.84134 - 0.15866 = 0.68268$$

From this it has been determined that approximately 68.3% of all measurements from any data set are expected to lie between $-\sigma$ and $+\sigma$. It also means that for any group of measurements there is approximately a 68.3% chance that any single observation has an error between $\pm\sigma$. The crosshatched area of Figure 3.7 illustrates that approximately 68.3% of the area exists between $\pm\sigma$. This is true for any set of measurements having normally distributed errors. Note that as discussed in Section 3.3, the inflection points of the normal distribution curve occur at $\pm\sigma$. This is illustrated in Figure 3.7.

3.5.1 The 50% Probable Error

For any group of observations, the 50% probable error establishes the limits within which 50% of the errors should fall. In other words, any measurement

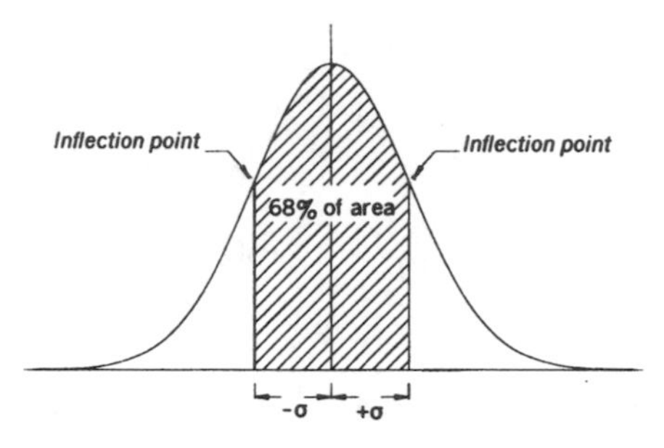

Figure 3.7 Normal distribution curve.

has the same chance of coming within these limits as it has of falling outside them. Its value can be obtained by multiplying the observations' standard deviation by the appropriate *t value*. Since the 50% probable error has a probability of 1/2, Equation (3.18) is set equal to 0.50 and the *t value* corresponding to this area is determined:

$$P(|z| < t) = 0.5 = 2N_z(t) - 1$$

$$1.5 = 2N_z(t)$$

$$0.75 = N_z(t)$$

From Table D.1 it is apparent that 0.75 is between a *t value* of 0.67 and 0.68; that is,

$$N_z(0.67) = 0.7486 \quad \text{and} \quad N_z(0.68) = 0.7517$$

The t value can be found by linear interpolation, as follows:

$$\frac{\Delta t}{0.68 - 0.67} = \frac{0.75 - 0.7486}{0.7517 - 0.7486} = \frac{0.0014}{0.0031} = 0.4516$$

or

$$\Delta t = 0.01 \times 0.4516 = 0.0045$$

and $t = 0.67 + 0.0045 = 0.6745$.

For any set of observations, therefore, the 50% probable error can be obtained by computing the standard error and then multiplying it by 0.6745, or

$$E_{50} = 0.6745\sigma \tag{3.19}$$

3.5.2 The 95% Probable Error

The 95% probable error, or E_{95}, is the bound within which, theoretically, 95% of the observation group's errors should fall. This error category is popular with surveyors for expressing precision. Using the same reasoning as in developing the equation for the 50% probable error, substituting into Equation (3.18) gives

$$
\begin{aligned}
0.95 = P(|z| < t) &= 2N_z(t) - 1 \\
1.95 &= 2N_z(t) \\
0.975 &= N_z(t)
\end{aligned}
$$

Again from Table D.1 it is determined that 0.975 occurs with a *t value* of 1.960. Thus to find the 95% probable error for any group of measurements, the following equation is used:

$$E_{95} = 1.960\sigma \tag{3.20}$$

3.5.3 Other Percent Probable Errors

Using the same computation techniques as in Sections 3.5.1 and 3.5.2, other percent probable errors can be calculated. One other percent error worthy of particular note is $E_{99.7}$. It is obtained by multiplying the standard error by 2.965, or

$$E_{99.7} = 2.965\sigma \tag{3.21}$$

This value is often used for detecting blunders, as discussed in Section 3.6. A summary of probable errors with varying percentages, together with their multipliers, is given in Table 3.2.

Table 3.2 Multipliers for various percent probable errors

Symbol	Multiplier	Percent Probable Errors
E_{50}	0.6745σ	50
E_{90}	1.6449σ	90
E_{95}	1.960σ	95
E_{99}	2.576σ	99
$E_{99.7}$	2.965σ	99.7
$E_{99.9}$	3.29σ	99.9

3.6 USES FOR PERCENT ERRORS

Standard errors and errors of other percent probabilities are commonly used to evaluate measurements for acceptance. Project specifications and contracts often require that acceptable errors be within specified limits such as the 90% or 95% errors. The 95% error, sometimes called the *two-sigma* (2σ) *error* because it is computed from 1.960σ, is most often specified. Standard error is also frequently used. The probable error, E_{50}, is seldom employed.

Higher percent errors are used to help isolate outliers (very large errors) and blunders in data sets. Since outliers seldom occur in a data set, measurements outside a selected high percentage range can be rejected as possible blunders. Generally, any data that differ from the mean by more than 3σ can be considered as blunders and removed from a data set. As seen in Table 3.2, rejecting observations greater that 3σ means that about 99.7% of all measurements should be retained. In other words, only about 0.3% of the measurements in a set (or 3 observations in 1000) should lie outside the range of $\pm 3\sigma$.

Note that as explained in Chapter 2, standard error and standard deviation are often used interchangeably, when in practice, it is actually the standard deviation that is computed, not the standard error. Thus for practical applications, σ in the equations of the preceding sections is replaced by S.

3.7 PRACTICAL EXAMPLES

Example 3.1 Suppose that the following values (in feet) were obtained in 15 independent distance measurements, D_i: 212.22, 212.25, 212.23, 212.15, 212.23, 212.11, 212.29, 212.34, 212.22, 212.24, 212.19, 212.25, 212.27, 212.20, and 212.25. Calculate the mean, S, E_{50}, E_{95}, and check for any measurements outside the 99.7% certainty level.

SOLUTION: From Equation (2.1), the mean is

$$\bar{y} = \frac{\sum D_i}{n} = \frac{3183.34}{15} = 212.22 \text{ ft}$$

From Equation (2.10), S is

$$S = \sqrt{\frac{675{,}576.955 - 15 \times 212.223^2}{15 - 1}} = \sqrt{\frac{0.051298}{14}} = \pm 0.06 \text{ ft}$$

where $\Sigma\, D_i^2 = 675{,}576.955$. By scanning the data, it is seen that 10 measurements are between 212.22 ± 0.06 or within the range of (212.16, 212.28).

This corresponds to 10/15 × 100, or 66.7% of the measurements. For the set, this is what is expected if it conforms to normal error distribution theory.

From Equation (3.19), E_{50} is

$$E_{50} = 0.6745S = \pm 0.6745\ (0.060) = \pm 0.04 \text{ ft}$$

Again by scanning the data, nine measurements lie between 212.22 ± 0.04, which is within the range of (212.18, 212.26). This corresponds to 9/15 × 100, or 60% of the measurements. Although this should be 50% and thus is a little high for a normal distribution, it must be remembered that this is only a sample of the population and should not be considered a reason to reject the entire data set. (In Chapter 4, statistical intervals involving sample sets are discussed.)

From Equation (3.20), E_{95} is

$$E_{95} = 1.960S = \pm 1.960(0.060) = \pm 0.12 \text{ ft}$$

Note that all measurements actually lie between 212.22 ± 0.12 (212.10, 212.34).

At the 99.7% level of confidence, the range of $\pm 2.965S$ corresponds to an interval of ±0.18. From the data, all values lie in this range, and thus there is no reason to believe that any measurement may be a blunder or outlier.

Example 3.2 The seconds portion of 50 direction readings from a 1″ instrument are listed below. Find the mean, standard deviation, and E_{95}. Check the observations at a 99% level of certainty for blunders.

41.9	46.3	44.6	46.1	42.5	45.9	45.0	42.0	47.5	43.2	43.0	45.7	47.6
49.5	45.5	43.3	42.6	44.3	46.1	45.6	52.0	45.5	43.4	42.2	44.3	44.1
42.6	47.2	47.4	44.7	44.2	46.3	49.5	46.0	44.3	42.8	47.1	44.7	45.6
45.5	43.4	45.5	43.1	46.1	43.6	41.8	44.7	46.2	43.2	46.8		

SOLUTION: The sum of the 50 measurements is 2252, and thus the mean is 2252/50 = 45.04″.

Using Equation (2.10), the standard deviation is

$$S = \sqrt{\frac{101{,}649.94 - 50(45.04)^2}{50 - 1}} = \pm 2.12''$$

where $\Sigma y^2 = 101{,}649.94$. There are 35 measurements between 45.04 ± 2.12, or within the range from 42.92 to 47.16. This corresponds to 35/50 × 100 = 70% of the measurements, and correlates well with the anticipated level of 68.3%.

From Equation (3.20), $E_{95} = \pm 1.960(2.12) = \pm 4.16$. The data actually contain three values that deviate from the mean by more than 4.16 (i.e., that are outside the range 40.88 to 49.20). They are 49.5(2) and 52.0. No values are less than 40.88, and therefore 47/50 × 100, or 94% of the measurements lie in the E_{95} range.

From Table 3.2, $E_{99} = \pm 2.576(2.12) = \pm 5.46$, and thus 99% of the data should fall between 45.04 ± 5.46, or (39.58, 50.50). Actually, one value is greater than 50.50, and thus 98% of all the measurements fall in this range.

By the analysis above it is seen that the data set is skewed to the left. That is, values higher than the range always came from the right side of the data. The histogram shown in Figure 3.8 depicts this skewness. This suggests that it may be wise to reject the value of 52.0 as a mistake. The recomputed values for the data set (minus the observation 52.0) are

$$\text{mean} = \frac{2252 - 52}{49} = 44.90''$$

$$\sum y^2 = 101{,}649.94 - 52.0^2 = 98{,}945.94$$

$$S = \sqrt{\frac{98{,}945.94 - 49(44.90)^2}{49 - 1}} = \pm 1.83''$$

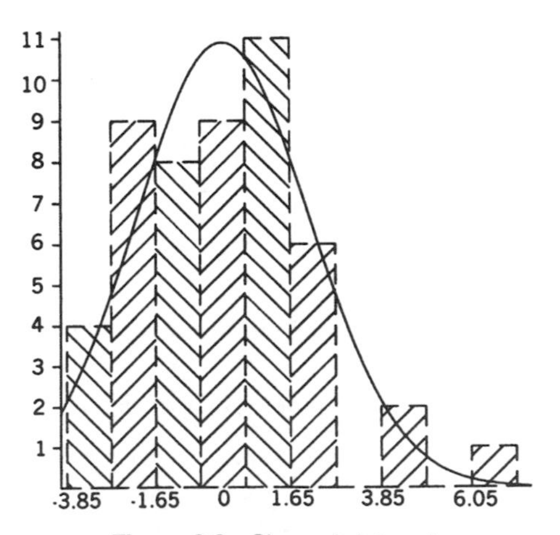

Figure 3.8 Skewed data set.

Now after recomputing errors, 32 measurements lie between $\pm S$, which represents 65.3% of the measurements, 47 measurements lie in the E_{95} range, which represents 95.9% of the data, and no values are outside the E_{99} range. Thus there is no reason to reject any additional data at a 99% certainty level.

PROBLEMS

3.1 Determine the *t value* for E_{80}.

3.2 Determine the *t value* for E_{75}.

3.3 Use Program STATS to do Problem 3.1.

3.4 Use Program STATS to do Problem 3.2.

3.5 Assuming a normal distribution, explain the statement: As the standard deviation of the group of measurements decreases, the precision of the group increases.

3.6 If the mean of a population is 2.456 and its variance of 2.042, what is the peak value for the normal distribution curve, and at what abscissas do the points of inflection occur?

3.7 If the mean of a population is 13.4 and its variance is 5.8, what is the peak value for the normal distribution curve, and at what abscissas do the points of inflection occur?

3.8 Plot the curve in Problem 3.6 using Equation (3.2) to determine ordinate and abscissa values.

3.9 Plot the curve in Problem 3.7 using Equation (3.2) to determine ordinate and abscissa additional values.

3.10 The following data represent 60 planimeter measurements of the area within a plotted traverse.

1.677	1.676	1.657	1.667	1.673	1.671	1.673	1.670	1.675	1.664	1.664	1.668
1.664	1.651	1.663	1.665	1.670	1.671	1.651	1.665	1.667	1.662	1.660	1.667
1.660	1.667	1.667	1.652	1.664	1.690	1.649	1.671	1.675	1.653	1.654	1.665
1.668	1.658	1.657	1.690	1.666	1.671	1.664	1.685	1.667	1.655	1.679	1.682
1.662	1.672	1.667	1.667	1.663	1.670	1.667	1.669	1.671	1.660	1.683	1.663

(a) Calculate: (1) the mean and (2) the standard deviation.

(b) Plot the relative frequency histogram (of residuals) for the data above using a class interval of one-half of the standard deviation.

(c) Calculate the E_{50} and E_{90} intervals.

(d) Can any measurements be rejected at a 99% level of certainty?

(e) What is the peak value for the normal distribution curve, and where are the points of inflection on the curve?

3.11 Discuss the normality of each set of data below and whether any measurements may be removed at the 99% level of certainty as blunders or outliers. Determine which set is more precise after apparent blunders and outliers are removed. Plot the relative frequency histograms to defend your decisions.

Set 1:

468.09 468.13 468.11 468.13 468.10 468.13 468.12 468.09 468.14 468.10
468.10 468.12 468.14 468.16 468.12 468.10 468.10 468.11 468.13 468.12
468.18

Set 2:

750.82 750.86 750.83 750.88 750.88 750.86 750.86 750.86 750.85 750.86
750.86 750.88 750.84 750.84 750.88 750.86 750.87 750.86 750.83 750.90
750.84

3.12 Using the following data set, answer the questions below.

17.5	15.0	13.4	23.9	25.2	19.5	25.8	30.0	22.5	35.3
39.5	23.5	26.5	21.3	22.3	21.6	27.2	21.1	24.0	23.5
32.5	32.2	24.2	35.7	28.0	24.0	16.8	21.1	19.0	30.7
30.2	33.7	19.7	19.7	25.1	27.9	28.5	22.7	31.0	28.4
31.2	24.6	30.2	16.8	26.9	23.3	21.5	18.8	21.4	20.7

(a) What are the mean and standard deviation of the data set?

(b) Construct a relative frequency histogram of the data, and discuss whether it appears to be a normal data set.

(c) What is the E_{95} interval for this data set?

(d) Would there be any reason to question the validity of any measurements at the 95% level?

3.13 Repeat Problem 3.12 using the following data:

2.898 2.918 2.907 2.889 2.901 2.901 2.899 2.899 2.911 2.909 2.895 2.904
2.905 2.900 2.920 2.899 2.896 2.897 2.907 2.897

Use Program STATS to do each problem.

3.14 Problem 3.10

3.15 Problem 3.11

3.16 Problem 3.12

3.17 Problem 3.13

4

CONFIDENCE INTERVALS AND STATISTICAL TESTING

4.1 INTRODUCTION

Table 4.1 contains a finite population of 100 values. The mean (μ) and variance (σ^2) of that population are 26.1 and 17.5, respectively. By randomly selecting 10 values from Table 4.1, an estimate for the mean and variance of the population can be computed. However, it would not be expected that these estimates ($\bar{y}$ and S^2) would exactly match the mean and variance of the population. Sample sets of 10 values each could continue to be selected from the population to determine additional estimates for the mean and variance of the population. However, it is just as unlikely that these additional values would match those obtained from either the population or the first sample set.

Now if the sample size were increased, it would be expected that the mean and variance of the sample would more nearly match the values of the population. In fact, as the sample size becomes very large, the mean and variance of the samples tend to approach those of the population. This procedure was done for various sample sizes starting at 10 and increasing them by 10, with the results shown in Table 4.2. Note that the value computed for the mean of the sample approaches the value of the population as the sample size is increased. Similarly, the value computed for the variance of the sample also tends to approach the value of the population as the sample size is increased.

Since the mean of a sample set, $\bar{y}$, and its variance, S^2, are computed from random variables, they are also random variables. This means that even if the size of the sample is kept constant, varying values for the mean and variance can be expected from the samples. Thus it is concluded that the values computed from a sample also contain errors. To illustrate this, an experiment was run for four randomly selected sets of 10 values from Table 4.1. Table 4.3

Table 4.1 Population of 100 values

18.2	26.4	20.1	29.9	29.8	26.6	26.2
25.7	25.2	26.3	26.7	30.6	22.6	22.3
30.0	26.5	28.1	25.6	20.3	35.5	22.9
30.7	32.2	22.2	29.2	26.1	26.8	25.3
24.3	24.4	29.0	25.0	29.9	25.2	20.8
29.0	21.9	25.4	27.3	23.4	38.2	22.6
28.0	24.0	19.4	27.0	32.0	27.3	15.3
26.5	31.5	28.0	22.4	23.4	21.2	27.7
27.1	27.0	25.2	24.0	24.5	23.8	28.2
26.8	27.7	39.8	19.8	29.3	28.5	24.7
22.0	18.4	26.4	24.2	29.9	21.8	36.0
21.3	28.8	22.8	28.5	30.9	19.1	28.1
30.3	26.5	26.9	26.6	28.2	24.2	25.5
30.2	18.9	28.9	27.6	19.6	27.9	24.9
21.3	26.7					

lists the samples, their means, and variances. Notice the variation in the computed values for the four sets. As discussed above, this variation is expected.

Fluctuations in means and variances computed for sample sets raises questions about the reliability of these estimates. For example, a higher confidence value is likely to be placed on a sample set with a small variance than on one with a large variance. Thus in Table 4.3, because of its small variance, one is more likely to believe that the mean of the second set is the most reliable estimate for the mean of the population from among the four sample sets. In reality this is not the case, however, as the means of the other three sets are actually closer to the population mean of 26.1.

As noted earlier, sample size also factors into considerations about reliability in a computed mean and variance. If the mean were computed from a

Table 4.2 Increasing sample sizes

No.	$\bar{y}$	S^2
10	26.9	28.1
20	25.9	21.9
30	25.9	20.0
40	26.5	18.6
50	26.6	20.0
60	26.4	17.6
70	26.3	17.1
80	26.3	18.4
90	26.3	17.8
100	26.1	17.5

Table 4.3 Random sample sets from population

Set 1: 29.9, 18.2, 30.7, 24.4, 36.0, 25.6, 26.5 29.9, 19.6, 27.9	$\bar{y} = 26.9$, $S^2 = 28.1$
Set 2: 26.9, 28.1, 29.2, 26.2, 30.0, 27.1, 26.5, 30.6, 28.5, 25.5	$\bar{y} = 27.9$, $S^2 = 2.9$
Set 3: 32.2, 22.2, 23.4, 27.9, 27.0, 28.9, 22.6, 27.7, 30.6, 26.9	$\bar{y} = 26.9$, $S^2 = 10.9$
Set 4: 24.2, 36.0, 18.2, 24.3, 24.0, 28.9, 28.8, 30.2, 28.1, 29.0	$\bar{y} = 27.2$, $S^2 = 23.0$

sample of five values, and then another calculated from a sample of 30 values, more confidence is likely to be placed in the larger sample rather than in the smaller one, even if both sample sets had similar variances.

In statistics, this relationship between the sample sets, their sizes, and the values computed for their means and variances is part of *sampling distribution theory*. This theory recognizes that estimates for the mean and variance do vary from sample to sample. *Estimators* are the functions used to compute these estimates. Examples of estimator functions are Equations (2.1) and (2.5), which are used to compute estimates of the mean and variance for a population, respectively. As demonstrated and discussed, these estimates vary from sample to sample and thus have their own population distributions. In Section 4.2 three distributions are defined that are used for describing or qualifying the reliability of mean and variance estimates. By applying these distributions, statements can be made about how good we can expect these computed estimates to be at any given level of confidence. In other words, a range or interval, called the "confidence interval," can be determined within which the mean and variance can be expected to fall for selected levels of probability.

4.2 DISTRIBUTIONS USED IN SAMPLING THEORY

4.2.1 χ^2 Distribution

The *chi-square distribution*, symbolized as χ^2, compares the relationship between the population variance and the variance of a sample set, based on the number of redundancies in the sample. If a random sample of n observations $y_1, y_2, \ldots, y_n$, is selected from a population that has a normal distribution with mean μ and variance σ^2, then by definition the χ^2 sampling distribution is

$$\chi^2 = \frac{vS^2}{\sigma^2} \tag{4.1}$$

where v is the number of degrees of freedom in the sample and the other terms are as defined previously.

A plot of the distribution is shown in Figure 4.1. The number of redundancies (degrees of freedom) in sample set statistics such as those for the mean or variance are $v = n - 1$. Table D.2 is a tabulation of χ^2 distribution curves having various degrees of freedom from 1 to 120. To find the area

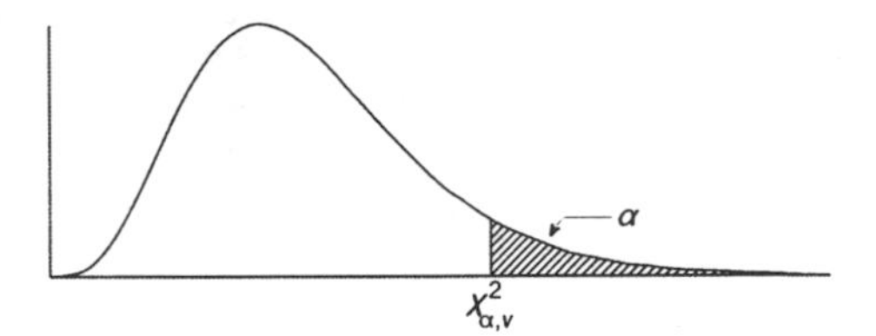

Figure 4.1 χ^2 distribution.

under the upper tail of the curve (right side shown crosshatched in Figure 4.1) starting at some specific χ^2 value and going to infinity (∞), simply intersect the row corresponding to the appropriate degrees of freedom, v, with the column corresponding to the desired area under the curve. For example, to find the specific χ^2 value relating to 1% ($\alpha = 0.010$) of the area under a curve having 10 degrees of freedom, intersect the row headed by 10 with the column headed by 0.010 and find a χ^2 value of 23.21. This means that 1% of the area under this curve is between the values of 23.21 and ∞.

Due to the nonsymmetric nature of the distribution, the percentage points[1] (α's) of the lower tail (left side of the curve) must be computed from those tabulated for the upper tail. A specific area under the left side of the curve starting at zero and going to some specific χ^2 value is found by subtracting the tabulated α (right-side area) from 1. This is due to the fact that the table lists α (areas) starting at the χ^2 value and going to ∞, and the total area under the curve is 1. For example, if there are 10 degrees of freedom, and the χ^2 value relating to 1% of the area under the left side of the curve is needed, the row corresponding to v equal to 10 is intersected with the column headed by $\alpha = 0.990$ $(1 - 0.010)$, and a value of 2.56 is obtained. This means that 1% of the area under the curve occurs from 0 to 2.56.

The χ^2 distribution is used in sampling statistics to determine the range in which the variance of the population can be expected to occur based on (1) some specified percentage probability, (2) the variance of a sample set, and (3) the number of degrees of freedom in the sample. In example 4.3 of Section 4.6, this distribution is used to construct probability statements about the variance S^2 of the population being in a range centered about the variance of a sample having v degrees of freedom. In Example 4.7 of Section 4.10 a test is presented using the χ^2 distribution to check if the variance of a sample is a valid estimate for the variance of a population.

4.2.2 *t* (Student) Distribution

This distribution is used to compare a population mean with the mean of a sample set based on the number of redundancies (v) in the sample set. It is

[1] Percentage points are decimal equivalents of percent probability; that is, a percent probability of 95% is equivalent to 0.95 percentage points.

similar to the normal distribution (discussed in Chapter 3), except that the normal distribution applies to an entire population, whereas the t distribution applies to a sample. This distribution is preferred over the normal distribution when samples contain fewer than 30 values. Thus it is an important distribution in analyzing surveying data.

If z is a standard normal random variable as defined in Section 3.4, χ^2 is a chi-square random variable with v degrees of freedom, and z and χ^2 are both independent variables, then by definition,

$$t = \frac{z}{\sqrt{\chi^2/v}} \tag{4.2}$$

The t values for selected upper-tail percentage points (crosshatched area in Figure 4.2) versus the *t distributions* with various degrees of freedom (v) are listed in Table D.3. For specific degrees of freedom (v) and percentage points (α), the table lists specific t values that correspond to the areas α under the curve between the tabulated t values and ∞. Similar to the normal distribution, the t distribution is symmetric. Generally in statistics, only percentage points in the range 0.0005 to 0.4 are necessary, and thus these t values are tabulated in Table D.3. To find the t value relating to $\alpha = 0.01$ for a curve developed with 10 degrees of freedom ($v = 10$), intersect the row corresponding to $v = 10$ with the column corresponding to $\alpha = 0.01$. At this intersection the t value of 2.764 is obtained. This means that 1% ($\alpha = 0.01$) of the area exists under the t distribution curve having 10 degrees of freedom in the interval between 2.764 and ∞. Due to the symmetry of this curve, it can also be stated that 1% ($\alpha = 0.01$) of the area under the curve developed for 10 degrees of freedom also lies between $-\infty$ and -2.764.

As described in Section 4.3, this distribution is used to construct confidence intervals for the population mean (μ) based upon the mean ($\bar{y}$) and variance (S^2) of a sample set, and the degrees of freedom (v). Example 4.1 in that section illustrates the procedure. Furthermore, in Section 4.9 it is shown that this distribution can be used to determine whether the mean of a sample set is a reliable estimate for its population mean. Example 4.6 in that section demonstrates that procedure.

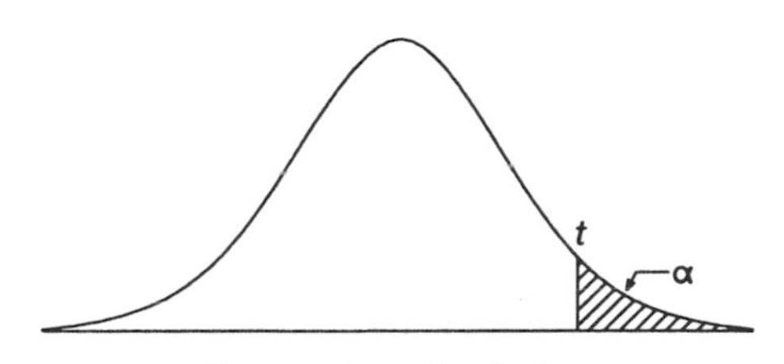

Figure 4.2 *t* distribution.

4.2.3 *F* Distribution

This distribution is used when comparing the computed variances from two sample sets. If χ_1^2 and χ_2^2 are two chi-square random variables with v_1 and v_2 degrees of freedom, respectively, and both variables are independent, then by definition

$$F = \frac{\chi_1^2/v_1}{\chi_2^2/v_2} \tag{4.3}$$

Various percentage points (areas under the upper tail of the curve, shown crosshatched in Figure 4.3) of the F distribution are tabulated in Table D.4. Notice that this distribution has v_1 numerator degrees of freedom, and v_2 denominator degrees of freedom which correspond to the two sample sets. Thus unlike the χ^2 and t distributions, each desired α percentage point must be represented in a separate table. In Appendix D, tables for the more commonly used values of α (0.20, 0.10, 0.05, 0.025, 0.01, 0.005, and 0.001) are listed.

To illustrate the use of the tables, suppose that the F value that marks a 1% area under the upper tail of the curve is needed. Also assume that 5 is

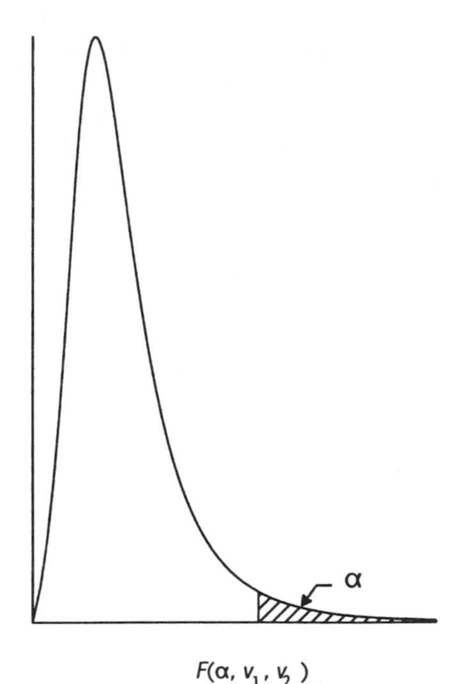

Figure 4.3 *F* distribution.

the numerator degrees of freedom relating to S_1, and 10 is the denominator degrees of freedom relating to S_2. In this example, α equals 0.01, and thus the F table in Table D.4 that is written for $\alpha = 0.01$ must be used. In that table, intersect the row headed by v_2 equal to 10 with the column headed by v_1 equal to 5, and find the F value of 5.64. This means that 1% of the area under the curve constructed using these degrees of freedom lies in the region going from 5.64 to ∞.

The F distribution is used to answer the question of whether two sample sets come from the same population. For example, suppose that two samples have variances of S_1^2 and S_2^2. If these two sample variances represent the same population variance, the ratio of their population variances (σ_1^2/σ_2^2) should equal 1, (i.e., $\sigma_1^2 = \sigma_2^2$). As we discuss in Section 4.7, this distribution enables confidence intervals to be established for this ratio. Also, as discussed in Section 4.11, the distribution can be used to test whether the ratio of the two variances is statistically equal to 1.

4.3 CONFIDENCE INTERVAL FOR THE MEAN: *t* STATISTIC

In Chapter 3 the standard normal distribution was used to predict the range in which the mean of a population mean may exist. This was based on the mean and standard deviation for a sample set. However, as noted previously, the normal distribution is based on an entire population, and as was demonstrated, variations from the normal distribution are expected for sample sets having a small number of values. From this expectation, the *t distribution* was developed. As we demonstrate later in this section with an example, the t distribution (in Table D.3) for samples having an infinite number of values uses the same t values as those listed in Table 3.2 for the normal distribution. It is generally accepted that when the number of observations is greater than about 30, the values in Table 3.2 are valid for constructing intervals about the population mean. However, when the sample set is less than 30, a t value from the t distribution should be used to construct the confidence interval for the population mean.

To derive an expression for a confidence interval, a sample mean $(\bar{y})$ is computed from a sample set of normally distributed data having a population mean of μ and variance in the mean of σ^2/n. Then let $z = (\bar{y} - \mu)/(\sigma/\sqrt{n})$ be a normal random variable, and substitute it and Equation (4.1) into Equation (4.2). This yields

$$t = \frac{z}{\sqrt{\chi^2/v}} = \frac{(\bar{y} - \mu)/(\sigma/\sqrt{n})}{\sqrt{(vS^2/\sigma^2)/v}} = \frac{(\bar{y} - \mu)/(\sigma/\sqrt{n})}{S/\sigma} = \frac{\bar{y} - \mu}{S/\sqrt{n}} \tag{4.4}$$

Thus to compute a confidence interval for the population mean (μ) given a sample set mean and variance, it is necessary to determine the area of a

$(1 - \alpha)$ region. For example, in a 95% confidence interval (noncrosshatched area in Figure 4.4), center the percentage point of 0.95 on the t distribution. This leaves 0.025 in each of the upper and lower tail areas (crosshatched areas in Figure 4.4). The t value that locates an $\alpha/2$ area in both the upper and lower tails of the distribution is given in Table D.3 as $t_{\alpha/2,v}$. For sample sets having a mean of $\bar{y}$ and variance of S^2, the correct probability statement to locate this area is

$$P(|z| < t) = 1 - \alpha \tag{a}$$

Substituting Equation (4.4) into Equation (a) gives

$$P\left(\left|\frac{\bar{y} - \mu}{S/\sqrt{n}}\right| < t\right) = 1 - \alpha$$

which after rearranging yields

$$P\left(\bar{y} - t\frac{S}{\sqrt{n}} < \mu < \bar{y} + t\frac{S}{\sqrt{n}}\right) = 1 - \alpha \tag{4.5}$$

Thus, given $\bar{y}$, $t_{\alpha/2,v}$, n, and S, it is seen from Equation (4.5) that a $(1 - \alpha)$ probable error interval for the population mean μ is computed as

$$\bar{y} - t_{\alpha/2}\frac{S}{\sqrt{n}} < \mu < \bar{y} + t_{\alpha/2}\frac{S}{\sqrt{n}} \tag{4.6}$$

where $t_{\alpha/2}$ is the t value, which is from the t distribution based on v degrees of freedom and $\alpha/2$ percentage points.

The following example illustrates the use of Equation (4.6) and Table D.3 for determining the 95% confidence interval for the population mean, based on a sample set having a small number of values (n) with a mean of $\bar{y}$ and a variance of S.

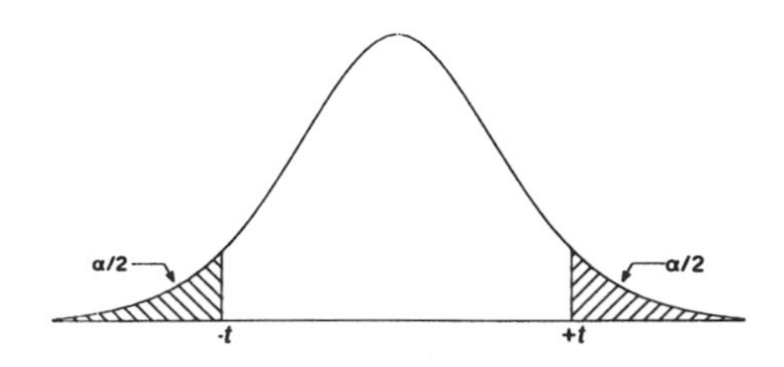

Figure 4.4 $t_{\alpha/2}$ plot.

Example 4.1 In carrying out a control survey, 16 pointings were measured for a single direction. The mean (seconds portion only) of the readings was 25.4″ with a standard deviation of ±1.3″. Determine the 95% confidence interval for the population mean. Compare this with the interval determined by using a t value determined from the standard normal distribution tables (Table 3.2).

SOLUTION: In this example the confidence level $(1 - \alpha)$ is 0.95, and thus α is 0.05. Since the interval is to be centered about the population mean μ, a value of $\alpha/2$ in Table D.3 is used. This yields equal areas in both the lower and upper tails of the distribution, as shown in Figure 4.4. Thus for this example, $\alpha/2$ is 0.025. The appropriate t value for this percentage point with v degrees of freedom is found in Table D.2 as follows:

Step 1: In the leftmost column of Table D.3, find the row with the correct number of degrees of freedom (v) for the sample, which in this case is 16 − 1 or 15.

Step 2: Find the column headed by 0.025 for $\alpha/2$.

Step 3: Locate the intersection of this row and column, which is 2.131.

Step 4: Then by Equation (4.6), the appropriate 95% confidence interval is

$$24.7 = 25.4 - 2.131\left(\frac{1.3}{\sqrt{16}}\right) = \bar{y} - t_{0.025}\frac{S}{\sqrt{n}} < \mu$$

$$< \bar{y} + t_{0.025}\frac{S}{\sqrt{n}} = 25.4 + 2.131\left(\frac{1.3}{\sqrt{16}}\right) = 26.1$$

This computation can be written more compactly by stating that μ has a 95% probability of being in the range

$$\bar{y} \pm t_{0.025}\frac{S}{\sqrt{n}} \quad \text{or} \quad 25.4 \pm 2.131\left(\frac{1.3}{\sqrt{16}}\right) = 25.4 \pm 0.7$$

After making the above calculation, it can be stated that for this sample, with 95% confidence, the population mean (μ) lies in the range (24.7, 26.1). If this were a large sample, the t value from Table 3.2 could be used for 95%. That t value for 95% is 1.960, and the standard error in the mean then would be $\pm 1.3/\sqrt{16} = \pm 0.325$. Thus the population's mean would be in the range 25.4 ± 1.960 × 0.325″, or (24.8, 26.0). Notice that due to the small sample size, the t distribution gives a larger range for the true mean than does the standard normal distribution. Notice also that in the t distribution of Table D.3, for a sample of infinite size (i.e., $v = \infty$), the tabulated t value for α equal to 0.025 is 1.960, which matches Table 3.2.

4.4 TESTING THE VALIDITY OF THE CONFIDENCE INTERVAL

A test that demonstrates the validity of the theory of the confidence interval is illustrated as follows. Using a computer and normal random number generating software, 1000 sample data sets of 16 values each were collected randomly from a population with mean $\mu = 25.4$ and standard error $\sigma = \pm 1.3$. Using a 95% confidence interval ($\alpha = 0.05$) and Equation (4.6), the interval for the population mean derived for each sample set was compared with the actual population mean. If the theory is valid, the interval constructed would be expected to contain the population's mean 95% of the time based on the confidence level of 0.05. Table E.1 shows the 95% intervals computed for the 1000 samples. The intervals not containing the population mean of 25.4 are marked with an asterisk. From the data tabulated it is seen that 50 of 1000 sample sets failed to contain the population mean. This corresponds to exactly 5% of the samples. In other words, the proportion of samples that do enclose the mean is exactly 95%. This demonstrates that the bounds calculated by Equation (4.6) do, in fact, enclose the true value of μ at the confidence level selected.

4.5 SELECTING A SAMPLE SIZE

A common problem encountered in surveying practice is determination of the necessary number of repeated measurements needed to meet a specific measurement precision. In practice, the size of S cannot be controlled absolutely. Rather, as seen in Equation (4.6), the confidence interval can be controlled only by varying the number of repeated measurements. In general, the larger the sample size, the smaller the confidence interval. From Equation (4.6), the range in which the population mean (μ) resides at a selected level of confidence (α) is

$$\bar{y} \pm t_{\alpha/2} \frac{S}{\sqrt{n}} \tag{b}$$

Now let I represent one-half of the interval in which the population mean lies. Then from Equation (*b*), I is

$$I = t_{\alpha/2} \frac{S}{\sqrt{n}} \tag{4.7}$$

Rearranging Equation (4.7), we have

$$n = \left(\frac{t_{\alpha/2} S}{I} \right)^2 \tag{4.8}$$

where n is the number of repeated measurements, I the desired confidence interval, $t_{\alpha/2}$ the t value based on the number of degrees of freedom (v), and S the sample set standard deviation. In the practical application of Equation (4.8), $t_{\alpha/2}$ and S are unknown since the data set has yet to be collected. Also, the number of measurements, and thus the number of redundancies, are unknown, since they are the computational objectives in this problem. Therefore, Equation (4.8) must be modified to use the standard normal random variable, z, and its value for t, which is not dependent on v or n; that is,

$$n = \left(\frac{t_{\alpha/2}\sigma}{I}\right)^2 \tag{4.9}$$

where n is the number of repetitions, $t_{\alpha/2}$ the t value determined from the standard normal distribution table (Table D.1), σ an estimated value for the standard error of the measurement, and I the desired confidence interval.

Example 4.2 From the preanalysis of a horizontal control network, it is known that all angles must be measured to within $\pm 2''$ at the 95% confidence level. How many repetitions will be needed if the standard deviation for a single angle measurement has been determined to be $\pm 2.6''$?

SOLUTION: In this problem, a final 95% confidence interval of $\pm 2''$ is desired. From previous experience, or analysis[2], the standard error for a single angle observation is estimated as $\pm 2.6''$. From Table 3.2, the multiplier, (or t value) for a 95% confidence level is found to be 1.960. Substituting this into Equation (4.9) gives

$$n = \left(\frac{1.960 \times 2.6}{2}\right)^2 = 6.49$$

Thus, eight repetitions are selected, since this is the closest even number above 6.49. [Note that it is necessary to select an even number of repetitions because an equal number of face-left (direct) and face-right (reverse) readings must be taken to eliminate instrumental systematic errors.]

4.6 CONFIDENCE INTERVAL FOR A POPULATION VARIANCE

From Equation (4.1), $\chi^2 = vS^2/\sigma^2$, and thus confidence intervals for the variance of the population, σ^2, are based on the χ^2 statistic. Percentage points (areas) for the upper and lower tails of the χ^2 distribution are tabulated in Table D.2. This table lists values (denoted by χ^2_α) that determine the upper boundary for areas from χ^2_α to $+\infty$ of the distribution, such that

[2] See Chapter 6 for a methodology to estimate the variance in an angle observation.

$$P(\chi^2 > \chi^2_\alpha) = \alpha$$

for a given number of redundancies, v. Unlike the normal distribution and the t distribution, the χ^2 distribution is not symmetric about zero. To locate an area in the lower tail of the distribution, the appropriate value of $\chi^2_{1-\alpha}$ must be found, where $P(\chi^2 > \chi^2_{1-\alpha}) = 1 - \alpha$. These facts are used to construct a probability statement for χ^2 as

$$P(\chi^2_{1-\alpha/2} < \chi^2 < \chi^2_{\alpha/2}) = 1 - \alpha \tag{4.10}$$

where $\chi^2_{1-\alpha/2}$ and $\chi^2_{\alpha/2}$ are tabulated in Table D.2. Substituting Equation (4.1) into Equation (4.10) gives

$$P\left(\chi^2_{1-\alpha/2} < \frac{vS^2}{\sigma^2} < \chi^2_{\alpha/2}\right) = P\left(\frac{\chi^2_{1-\alpha/2}}{vS^2} < \frac{1}{\sigma^2} < \frac{\chi^2_{\alpha/2}}{vS^2}\right) \tag{4.11}$$

Recalling a property of mathematical inequalities, that in taking the reciprocal of a function the inequality is reversed, it follows that

$$P\left(\frac{vS^2}{\chi^2_{\alpha/2}} < \sigma^2 < \frac{vS^2}{\chi^2_{1-\alpha/2}}\right) = 1 - \alpha \tag{4.12}$$

Thus a $(1 - \alpha)100\%$ confidence interval for the population variance (σ^2) is

$$\frac{vS^2}{\chi^2_{\alpha/2}} < \sigma^2 < \frac{vS^2}{\chi^2_{1-\alpha/2}} \tag{4.13}$$

Example 4.3 An observer's pointing and reading error with a 1″ instrument is estimated by collecting 20 readings while pointing at a distant, well-defined target. The sample standard deviation is determined to be $\pm 1.8''$. What is the 95% confidence interval for σ^2?

SOLUTION: For this example the desired area enclosed by the confidence interval $(1 - \alpha)$ is 0.95. Thus α is 0.05, and $\alpha/2$ is 0.025. The tabulated values of $\chi^2_{0.025}$ and $\chi^2_{0.975}$ with v equal to $(20 - 1) = 19$ degrees of freedom are needed. They are found in the χ^2 table (Table D.2) as follows:

Step 1: Find the row with 19 degrees of freedom and intersect it with the column headed by 0.975. The value at the intersection is 8.91.

Step 2: Follow this procedure for 19 degrees of freedom and 0.025. The value is 32.85.

Using Equation (4.13), the 95% confidence interval for σ^2 is

$$\frac{(20 - 1)1.8^2}{32.85} < \sigma^2 < \frac{(20 - 1)1.8^2}{8.91}$$

$$1.87 < \sigma^2 < 6.91$$

Thus 95% of the time the population's variance should lie between 1.87 and 6.91.

4.7 CONFIDENCE INTERVAL FOR THE RATIO OF TWO POPULATION VARIANCES

Another common statistical procedure is used to compare the ratio of two population variances. The sampling distribution of the ratio σ_1^2/σ_2^2 ratio is well known when samples are collected randomly from a normal population. The confidence interval for σ_1^2/σ_2^2 is based on the F distribution using Equation (4.3) as follows:

$$F = \frac{\chi_1^2/\upsilon_1}{\chi_2^2/\upsilon_2}$$

Substituting Equation (4.1) and reducing gives

$$F = \frac{\upsilon_1 S_1^2/(\sigma_1^2/\upsilon_1)}{\upsilon_2 S_2^2/(\sigma_2^2/\upsilon_2)} = \frac{S_1^2/\sigma_1^2}{S_2^2/\sigma_2^2} = \frac{S_1^2}{S_2^2} \times \frac{\sigma_2^2}{\sigma_1^2} \tag{4.14}$$

To establish a confidence interval for the ratio, the lower and upper values corresponding to the distribution's tails must be found. Notice that the lower tail values are not listed in the F distribution table (Table D.4). Listing them is not necessary because

$$F_l = F_{1-\alpha,\upsilon_1,\upsilon_2} = \frac{1}{F_{\alpha,\upsilon_2,\upsilon_1}} \tag{4.15}$$

where F_l is the lower tail.

A probability statement to find the confidence interval for the ratio is constructed as follows:

$$P(F_{1-\alpha/2,\upsilon_1,\upsilon_2} < F < F_{\alpha/2,\upsilon_1,\upsilon_2}) = 1 - \alpha$$

Rearranging yields

$$P(F_l < F < F_u) = P\left(F_l < \frac{S_1^2}{S_2^2} \times \frac{\sigma_2^2}{\sigma_1^2} < F_u\right)$$

$$P(F_l < F < F_u) = P\left(F_l \frac{S_2^2}{S_1^2} < \frac{\sigma_2^2}{\sigma_1^2} < \frac{S_2^2}{S_1^2} F_u\right)$$

$$= P\left(\frac{1}{F_u} \times \frac{S_1^2}{S_2^2} < \frac{\sigma_1^2}{\sigma_2^2} < \frac{S_1^2}{S_2^2} \times \frac{1}{F_l}\right) = 1 - \alpha \qquad (4.16)$$

By substituting Equation (4.15) into (4.16), we obtain

$$P\left(\frac{S_1^2}{S_2^2} \times \frac{1}{F_{\alpha/2,v_1,v_2}} < \frac{\sigma_1^2}{\sigma_2^2} < \frac{S_1^2}{S_2^2} \frac{1}{F_{1-\alpha/2,v_1,v_2}}\right)$$

$$= P\left(\frac{S_1^2}{S_2^2} \times \frac{1}{F_{\alpha/2,v_1,v_2}} < \frac{\sigma_1^2}{\sigma_2^2} < \frac{S_1^2}{S_2^2} \times F_{\alpha/2,v_2,v_1}\right) = 1 - \alpha \qquad (4.17)$$

Thus, from Equation (4.17), the $(1 - \alpha)$ confidence interval for the ratio σ_1^2/σ_2^2 is

$$\frac{S_1^2}{S_2^2} \times \frac{1}{F_{\alpha/2,v_1,v_2}} < \frac{\sigma_1^2}{\sigma_2^2} < \frac{S_1^2}{S_2^2} \times F_{\alpha/2,v_2,v_1} \qquad (4.18)$$

Notice that the degrees of freedom for the upper and lower limits in Equation (4.18) are opposite each other, and thus v_2 is the numerator degrees of freedom and v_1 is the denominator degrees of freedom in the upper limit.

An important situation where Equation (4.18) can be applied occurs in the analysis and adjustment of horizontal control surveys. During least-squares adjustments of these types of surveys, control stations fix the data both positionally and rotationally in space. When observations tie into more than a minimal amount of control, the control station coordinates must be mutually consistent. If they are not, adjusting the measurements to the control will warp the data to fit the discrepancies in the control. A method for isolating control stations that are not consistent is to first do a least-squares adjustment using only enough control to fix the data both positionally and rotationally in space. This is known as a *minimally constrained* adjustment. In traverse surveys this means that one station must have fixed coordinates, and one line must be fixed in direction. This adjustment is then followed with an adjustment using all available control. If the control information is consistent, the reference variance (S_1^2) from the minimally constrained adjustment should be statistically equivalent to the reference variance (S_2^2) obtained when using all control information. That is, the S_1^2/S_2^2 ratio should be equal to 1.

Example 4.4 Assume that a minimally constrained trilateration network adjustment with 24 degrees of freedom has a reference variance of 0.49 and that the fully constrained network adjustment with 30 degrees of freedom has a reference variance of 2.25. What is the 95% $(1 - \alpha)$ confidence interval for the ratio of the variances, and does this surround the numerical value of 1 or is there reason to be concerned about the control having values that are not consistent?

SOLUTION: In this example the objective is to determine whether the two reference variances are statistically equal. Equation 4.18 is used. Let the variance in the numerator be 2.25 and the variance in the denominator be 0.49. Thus the numerator has 30 degrees of freedom ($v_1 = 30$) and corresponds to the adjustment using all the control. The denominator has 24 degrees of freedom ($v_2 = 24$) and corresponds to the minimally constrained adjustment[3]. Because α is equal to 0.05 the values $F_{\alpha/2,v_1,v_2}$ and $F_{\alpha/2,v_2,v_1}$ are taken from Table D.4 with $\alpha/2 = 0.025$. Using Equation (4.18), the 95% confidence interval for this ratio is

$$2.08 = \frac{2.25}{0.49} \times \frac{1}{2.21} < \frac{\sigma_1^2}{\sigma_2^2} < \frac{2.25}{0.49} \times 2.14 = 9.83$$

Note from the calculations above that 95% of the time, the ratio of the population variances is in the range 2.08 to 9.83. Since this interval does not contain 1, it can be said that $\sigma_1^2/\sigma_2^2 \neq 1$ and that $\sigma_1^2 \neq \sigma_2^2$ at a 95% level of certainty. Recalling from Equation (2.4) that the size of the variance is dependent of the sizes of the errors, it can be stated that the fully constrained adjustment revealed discrepancies between the observations and the control. This could be caused by inconsistencies in the coordinates of the control stations, but it could also stem from the presence of uncorrected systematic errors in the observations.

4.8 HYPOTHESIS TESTING

In Example 4.4 the concern was not so much about the actual bounds of the constructed interval but rather, about whether the constructed interval contained the expected ratio of the variances. This is often the case in statistics. That is, the actual values of the interval are not as important as is answering the question: Is the statistic consistent with what is expected from the pop-

[3] For confidence intervals, it is not important which variance is selected as the numerator. In this case, the larger variance was selected arbitrarily as the numerator.

ulation? The procedures used to test the validity of a statistic are known as *hypothesis testing*. The basic elements of an hypothesis test are as follows:

1. The *null hypothesis*, H_0, is a statement that compares a population statistic with a sample statistic. This implies that the sample statistic is part of the "expected" population. In Example 4.4 this would be that the ratio of the variances is statistically equivalent to 1.
2. The *alternative hypothesis*, H_a, is what is accepted when a decision is made to reject the null hypothesis, and thus represents an alternative population from which the sample statistic was determined. In Example 4.4 the alternative hypothesis would be that the ratio of the variances is not equal to 1.
3. The *test statistic* is computed from the sample data and is the value used to determine whether the null hypothesis should be rejected. When the null hypothesis is rejected, it can be said that the sample statistic computed is not consistent with what is expected from its population. In Example 4.4 a rejection of the null hypothesis would occur when the ratio of the variances is not statistically equivalent to 1.
4. The *rejection region* is the value for the test statistic where the null hypothesis is rejected. In reference to confidence intervals, this number takes the place of the confidence interval bounds. That is, when the computed test statistic is greater than the value defining the rejection region, it is equivalent to the sample statistic of the null hypothesis being outside the bounds of the confidence interval.

Whenever a decision is made concerning the null hypothesis, there is the possibility of making a wrong decision since there can never be 100% certainty about a statistic or a test. Returning to Example 4.4, a confidence interval of 95% was constructed. With this interval, there is a 5% chance that the decision was wrong. That is, it is possible that the larger than expected ratio of the variances is consistent with the population of observations. This reasoning suggests that further analysis of statistical testing is needed.

Two basic errors can occur when a decision is made about a statistic. A valid statistic could be rejected, or an invalid statistic could be accepted. These two errors can be stated in terms of statistical testing elements as *type I* and *type II* errors. If the null hypothesis is rejected when in fact it is true, a type I error is committed. If the null hypothesis is not rejected when in fact it is false, a type II error occurs. Since these errors are not from the same population, the probability of committing each error is not directly related. A decision must be made as to the type of error that is more serious for the situation, and that decision should be based on the consequences of committing each error. For instance, if a contract calls for positional accuracies on 95% of the stations to be within ± 0.3 ft, the surveyor is more inclined to commit a type I error to ensure that the contract specifications are met. How-

Table 4.4 Relationships in statistical testing

	Decision	
Situation	Accept H_0	Reject H_0
H_0 true	Correct decision: $P = 1 - \alpha$ (confidence level)	Type I error: $P = \alpha$ (significance level)
H_0 false (H_a true)	Type II error: $P = \beta$	Correct decision: $P = 1 - \beta$ (power of test)

ever, the same surveyor needing only 1-ft accuracy on control to support a small-scale mapping project may be more inclined to commit a type II error. In either case it is important to compute the probabilities of committing both type I and type II errors to assess the reliability of the inferences derived from an hypothesis test. For emphasis the two basic hypothesis-testing errors are repeated.

- *Type I Error:* rejecting the null hypothesis when it is in fact true (symbolized by α).
- *Type II Error:* not rejecting the null hypothesis when it is in fact false (symbolized by β).

Table 4.4 shows the relationship between the decision, the probabilities of α and β, and the acceptance or rejection of the null hypothesis, H_0. In Figure 4.5, the left distribution represents the data from which the null hypothesis is derived. That is, this distribution represents a true null hypothesis. Similarly, the distribution on the right represents the data for the true alternative. These two distributions could be attributed to measurements that contain only random errors (left distribution) versus measurements containing blunders (right distribution). In the figure it is seen that valid measurements in the α region

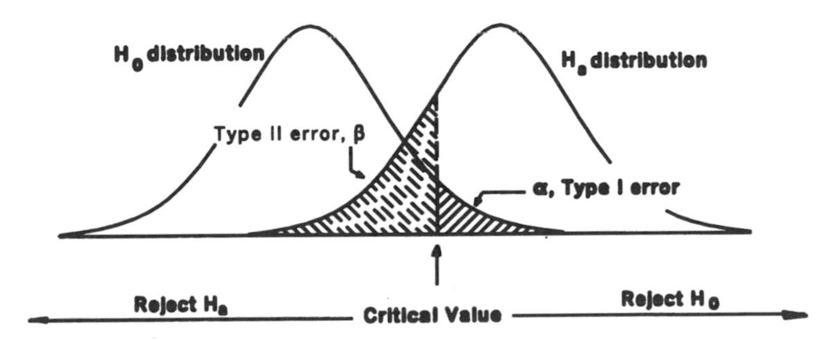

Figure 4.5 Graphical interpretation of type I and II errors.

of the left distribution are being rejected at a significance level of α. Thus, α represents the probability of committing a type I error. This is known as the *significance level of the test.* Furthermore, data from the right distribution are being accepted at a β level of significance. The *power of the test* is $1 - \beta$, and corresponds to a true alternative hypothesis. Methods of computing β or $1 - \beta$ are not clear, or are often difficult, since nothing is generally known about the distribution of the alternative. Consequently, in statistical testing, the objective is to prove the alternative hypothesis true by showing that the data does not support the statistic coming from the null hypothesis distribution. In doing this, only a type I error can be made, for which a known probability of making a wrong decision is α.

Example 4.5 As an example, assume that for a population of 10,000 people, a flu virus test has a 95% confidence level and thus a significance level, α, of 0.05. Suppose that 9200 people test negative for the flu virus and 800 test positive. Of the 800 people who tested positive, 5%, or 40 people, will test incorrectly (false positive). That is, they will test positive for the flu but do not have it. This is an example of committing a type I error at an α significance level. Similarly, 0.05×9200, or 460 people, will test negative for the flu when, in fact, they do have it (a false-negative case). This is an example of a type II error at a probability of β, which is 0.046, (460/10,000). Furthermore, the power of the test is $1 - \beta$, or 0.954, in this example.

From the foregoing it is seen that it is possible to set the probability of committing a type I error for a given H_0. However, for a fixed level of α and sample size n, the probability of a type II error, β, may be unknown. If the null hypothesis, H_0, and α are fixed, the power of the test can only be increased by increasing the sample size, n. Since the power of the test may be low or unknown, statisticians always say that the test *failed to reject* the null hypothesis rather than making any statements about its acceptance.

A similar situation exists with surveying measurements. If a distance measurement contains a large systematic error, it is possible to detect this with a fully constrained adjustment and thus reject the null hypothesis. However, if a distance contains a very small systematic error, the ability to detect this situation is low. Thus, while some confidence can be placed in the rejection of the null hypothesis, it can never be said that the null hypothesis should be accepted since the probability of undetected small systematic errors cannot be determined.

4.9 TEST OF HYPOTHESIS FOR THE POPULATION MEAN

At times it may be desirable to test a sample mean against a known value. The null hypothesis for this test can take two forms: one-, or two-tailed tests. In the *one-tailed test*, the concern is whether the sample mean is either statistically greater or less than the population mean. In the *two-tailed test*, the concern is whether the sample mean is statistically different from the population mean. These tests are:

	One-Tailed Test	*Two-Tailed Test*
The null hypothesis is	H_0: $\mu = \bar{y}$	H_0: $\mu = \bar{y}$
The alternative hypothesis is	H_a: $\mu > \bar{y}$ ($\mu < \bar{y}$)	H_a: $\mu \neq \bar{y}$

The test statistic is

$$t = \frac{\bar{y} - \mu}{S/\sqrt{n}} \tag{4.19}$$

The region where the null hypothesis is rejected is

One-Tailed Test	Two-Tailed Test
$t > t_\alpha$ (or $t < t_\alpha$)	$\lvert t \rvert > t_{\alpha/2}$

It should be stated that for large samples ($n > 30$), the *t value* can be replaced by the standard normal variate, z.

Example 4.6 A baseline of calibrated length 400.008 m is measured repeatedly with an EDM instrument. After 20 measurements, the average of the measured distances is 400.012 m with a standard deviation of ± 0.002 m. Is the measured distance significantly different from the calibrated distance at a 0.05 level of significance?

SOLUTION: Assuming that proper field procedures were followed, the fundamental question is whether the EDM instrument is calibrated within specifications and thus is providing distance measurements in a population of calibrated values. To answer this question, a two-tailed test is conducted to determine whether the distance is either the same or different than the calibrated distance at a 0.05 level of significance. That is, the measured distance will be rejected if it is either statistically too short or too long to be considered the same as the calibration value. The rationale behind using a two-tailed test is similar to that used when constructing a confidence interval, as in Example 4.1. That is, 2.5% of the area from the lower and upper tails of the t distribution is to be excluded from the interval constructed, or in this case, the test.

The null hypothesis is

$$H_0: \mu = 400.012$$

and the alternative hypothesis is

$$H_a: \mu \neq 400.012$$

By Equation (4.19), the test statistic is

$$t = \frac{\bar{y} - \mu}{S/\sqrt{n}} = \frac{400.012 - 400.008}{0.002/\sqrt{20}} = 8.944$$

and the rejection region is

$$t = 8.944 > t_{\alpha/2}$$

Since a two-tailed test is being done, the $\alpha/2$ (0.025) column in the t distribution table is intersected with the ($v = n - 1$) or 19 degrees of freedom row. From the t distribution (Table D.3) $t_{0.025,\ 19}$ is found to be 2.093, and thus the rejection region is satisfied. In other words, the value computed for t is greater than the tabulated value, and thus the null hypothesis can be rejected. That is,

$$t = 8.944 > t_{\alpha/2} = 2.093$$

Based upon the foregoing, there is reason to believe that the average measured distance is significantly different from the calibrated distance at a 5% significance level. This implies that at least 5% of the time, the decision will be wrong. As stated earlier, a 95% confidence interval for the population mean could also have been constructed to derive the same results. Using Equation (4.6), that interval would be

$$400.011 \approx 400.012 - 2.093 \times \frac{0.002}{\sqrt{20}} \leq \mu$$

$$\leq 400.012 + 2.093 \times \frac{0.002}{\sqrt{20}} \approx 400.013$$

Note that this 95% confidence interval fails to contain the baseline value of 400.008, and similarly, there is reason to be concerned about the calibration status of the instrument. That is, it may not be measuring distances from the population of calibrated measurements.

4.10 TEST OF HYPOTHESIS FOR THE POPULATION VARIANCE, σ^2

In Example 4.6, the procedure for checking whether a measured distance compares favorably with a calibrated distance was discussed. The surveyor may also want to check if the instrument is measuring at its published precision. The χ^2 distribution is used when comparing the variance of a sample set against that of a population. This test involves checking the variance computed from a sample set of measurements against the published value (the expected variance of the population).

As stated in Section 4.2.1, the confidence interval for the variance of a population can be determined from the computed variance of a sample set using a χ^2 distribution. By using Equation (4.1), the following statistical test can be written:

	One-Tailed Test	*Two-Tailed Test*
The null hypothesis is	$H_0: S^2 = \sigma^2$	$H_0: S^2 = \sigma^2$
and the alternative hypothesis is	$H_a: S^2 > \sigma^2$ (or $H_a: S^2 < \sigma^2$)	$H_a: S^2 \neq \sigma^2$

The test statistic is

$$\chi^2 = \frac{vS^2}{\sigma^2} \tag{4.20}$$

from which the null hypothesis is rejected when the following rejection region is satisfied:

$$\chi^2 > \chi^2_\alpha \text{ (or } \chi^2 < \chi^2_{1-\alpha}) \qquad \chi^2 < \chi^2_{1-\alpha/2} \text{ or } \chi^2 > \chi^2_{\alpha/2}$$

The rejection region is determined from Equation (4.11). Graphically, the null hypothesis is rejected in the one-tailed test when the computed χ^2 value is greater value than that tabulated. This rejection region is the shaded region shown in Figure 4.6(*a*). In the two-tailed test, the null hypothesis is rejected when the computed value is either less than $\chi^2_{1-\alpha/2}$ or greater than $\chi^2_{\alpha/2}$. This is similar to the computed variance being outside the constructed confidence interval for the population variance. Again in the two-tailed test, the probability selected is evenly divided between the upper and lower tails of the distribution such that the acceptance region is centered on the distribution. These rejection regions are shown graphically in Figure 4.6(*b*).

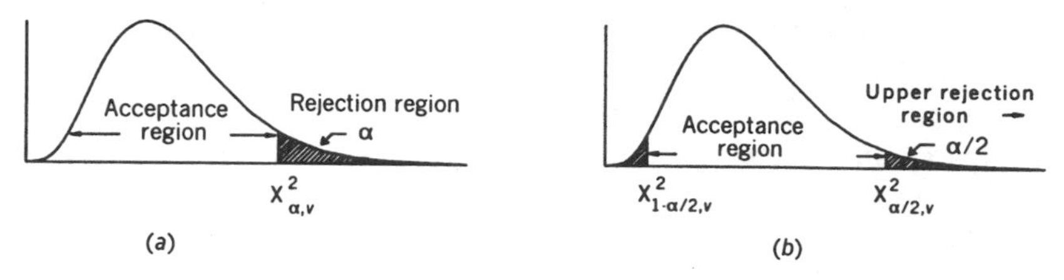

Figure 4.6 Graphical interpretation of (*a*) one- and (*b*) two-tailed tests.

Example 4.7 The manager of a surveying firm wants all surveying technicians to be able to read a particular instrument to within ±1.5″. To test this value the senior field crew chief is asked to perform a reading test with the instrument. The crew chief reads the circle 30 times and obtains a $S_r = \pm 0.9''$. Does this support the 1.5″ limit at a 5% significance level?

SOLUTION: In this case, the manager wishes to test the hypothesis that the computed sample variance is the same as the population variance, versus its being greater than the population variance. That is, all standard deviations that are equal to or less than 1.5″ will be accepted. Thus a one-tailed test is constructed as follows. (note that $v = 30 - 1$, or 29.)

The null hypothesis is

$$H_0\colon S^2 = \sigma^2$$

The alternative hypothesis is

$$H_a\colon S^2 > \sigma^2$$

The test statistic is

$$\chi^2 = \frac{(30 - 1)(0.9)^2}{1.5^2} = 10.44$$

and the null hypothesis is rejected when the computed test statistic exceeds the tabulated value, or when

$$\chi^2 = 10.44 > \chi^2_{\alpha,v} = \chi^2_{0.05,29} = 42.56$$

where 42.56 is from Table D.2 for $\chi^2_{0.05,29}$. Since the computed χ^2 value (10.44) is less than the tabulated value (42.56), the null hypothesis is not rejected.

However, simply failing to reject the null hypothesis does not mean that the value of ±1.5″ is valid. This example demonstrates a common problem in statistical testing when results are interpreted incorrectly. By selecting only one employee, a valid sample set from the population of all surveying employees has not been obtained. Furthermore, every instrument reads differently, and thus a new employee may initially have problems reading an instrument due to lack of experience with it. To account properly for this lack of experience, the employer could test a random sample of prospective employees during the interview process, and again after several months of employment. A check could then be made for correlation between the company's satisfaction with the employee, and the employee's initial ability to read the instrument.

This example illustrates an important point to be made when using statistics. *The interpretation of statistical testing requires judgment by the person performing the test.* It should always be remembered that with a test, the objective is to reject and not accept an hypothesis.

4.11 TEST OF HYPOTHESIS FOR THE RATIO OF TWO POPULATION VARIANCES

When adjusting data, surveyors have generally considered control to be absolute and without error. However, it is a known fact that control, like any other quantities derived from measurements, contains errors. As discussed in Example 4.4, one method of detecting either errors in control, or possible systematic errors in horizontal network measurements, is to do both a minimally constrained and fully constrained least-squares adjustment with the data. After doing both adjustments, the postadjustment reference variances can be compared. If the control is without error and no systematic errors are present in the data, the ratio of the two reference variances should be close to 1. Using Equation (4.16), a hypothesis test can be constructed to compare the ratio of variances for two sample sets as follows:

One-Tailed Test | *Two-Tailed Test*

The null hypothesis is

$$H_0: \frac{S_1^2}{S_2^2} = 1 \text{ (i.e., } S_1^2 = S_2^2) \qquad H_0: \frac{S_1^2}{S_2^2} = 1 \text{ (i.e., } S_1^2 = S_2^2)$$

The alternative hypothesis is

$$H_a: \frac{S_1^2}{S_2^2} > 1 \text{ (i.e., } S_1^2 > S_2^2) \qquad H_a: \frac{S_1^2}{S_2^2} \neq 1 \text{ (i.e., } S_1^2 \neq S_2^2)$$

or

$$\left(\mathrm{H_a}: \frac{S_1^2}{S_2^2} < 1 \text{ (i.e., } S_1^2 < S_2^2)\right)$$

The test statistic that will be used to determine rejection of the null hypothesis is

One-Tailed Test	*Two-Tailed Test*
$F = \frac{S_1^2}{S_2^2}$ or $F = \frac{S_2^2}{S_1^2}$	$F = \frac{\text{larger sample variance}}{\text{smaller sample variance}}$

and the null hypothesis should be rejected when the following statement is satisfied.

$F > F_\alpha$	$F > F_{\alpha/2}$

F_α and $F_{\alpha/2}$ are values that locate the α and $\alpha/2$ areas, respectively, in the upper tail of the F distribution with v_1 numerator degrees of freedom and v_2 denominator degrees of freedom. Notice that in the two-tailed test, the degrees of freedom of the numerator are taken from the numerically larger sample variance and the degrees of freedom of the denominator are from the smaller variance.

Example 4.8 Using the same data as in Example 4.4, would the null hypothesis be rejected?

SOLUTION: In this example, a two-tailed test is appropriate since the only concern is whether the two reference variances are statistically equal. In this problem, the interval is centered on the F distribution with an $\alpha/2$ area in the lower and upper tails. In the analysis, the 30 degrees of freedom in the numerator corresponds to the larger sample variance, and the degrees of freedom in the denominator are 24, so that the following test is constructed.

The null hypothesis is

$$\mathrm{H_0}: \frac{S_1^2}{S_2^2} = 1 \ (\sigma_1^2 = \sigma_2^2)$$

The alternative hypothesis is

$$\mathrm{H_a}: \frac{S_1^2}{S_2^2} \neq 1 \ (\sigma_1^2 \neq \sigma_2^2)$$

The test statistic for checking rejection of the null hypothesis is

$$F = \frac{2.25}{0.49} = 4.59$$

Rejection of the null hypothesis occurs when the following statement is satisfied:

$$F = 4.59 > F_{\alpha/2,v1,v2} = F_{0.025,30,24} = 2.21$$

Here it is seen that the computed F value (4.59) is greater than its tabulated value (2.21) from Table D.4. Thus the (null) hypothesis can be rejected. In other words, the fully constrained adjustment does not have the same variance as its minimally constrained counterpart at the selected 0.05 level of significance. Notice that the same result was obtained here as was obtained in Example 4.4 with the 95% confidence interval. Again, the network should be inspected for the presence of systematic errors, followed by an analysis of possible mistakes in the control stations. This postadjustment analysis is revisited in greater detail in Chapter 20.

Example 4.9 Ron and Kathi continually debate who measures angles more precisely with a particular instrument. Their supervisor, after hearing enough, describes a test where each is to measure a particular direction by pointing and reading the instrument 51 times. They must then compute the variance for their data. At the end of the 51 readings, Kathi determines her variance to be 0.81, and Ron finds his to be 1.21. Is Kathi a better instrument operator at a 0.01 level of significance?

SOLUTION: In this situation, even though Kathi's variance implies that her measurements are more precise than Ron's, a determination must be made to see if the reference variances are statistically equal, versus Kathi's being better than the Ron's. This test requires a one-tailed F test with a significance level of $\alpha = 0.01$.

The null hypothesis is

$$H_0: \frac{S_R^2}{S_K^2} = 1 \; (S_R^2 = S_K^2)$$

and the alternative hypothesis is

$$H_a: \frac{S_R^2}{S_K^2} > 1 \; (S_R^2 > S_K^2)$$

The test statistic is

$$F = \frac{1.21}{0.81} = 1.49$$

The null hypothesis is rejected when the computed value for F (1.49) is greater than the tabulated value of $F_{0.01,50,50}$ (1.95). Here it is seen that the computed value for F is less than its tabulated value, and thus the test statistic does not satisfy the rejection region. That is,

$$F = 1.49 > F_{\alpha,50,50} = 1.95 \text{ is false.}$$

Therefore, there is no statistical reason to believe that Kathi is better than Ron at a 0.01 level of significance.

Example 4.10 A baseline is repeatedly measured using an EDM instrument over a five-day period. Each day, 10 measurements are taken and averaged. The variances for the measurements are listed below. At a significance level of 0.05, are the results of day 2 significantly different from those of day 5?

Day	1	2	3	4	5
Variance, S^2 (mm^2)	50.0	61.0	51.0	53.0	54.0

SOLUTION: This problem involves checking whether the variances of day 2 and day 5 are statistically equal versus their being different. This is the same as constructing a confidence interval involving the ratio of the variances. Because the concern is about equality or inequality, this will require a two-tailed test. Since there were 10 values collected each day, both variances are based on 9 degrees of freedom (v_1 and v_2). Assume that the variance for day 2 is S_2^2 and the variance for day 5 is S_5^2. The test is constructed as follows:

The null hypothesis is

$$H_0: \frac{S_2^2}{S_5^2} = 1$$

and the alternative hypothesis is

$$H_a: \frac{S_2^2}{S_5^2} \neq 1$$

The computed test statistic is

$$F = \frac{61}{54} = 1.13$$

The null hypothesis is rejected when the computed F value (1.13) is greater than the tabulated value (4.03) in Table D.4. In this case the rejection region is, $F > F_{0.025,9,9} = 4.03$, and is not satisfied because $F = 1.13$. Consequently, the test fails to reject the null hypothesis, and there is no statistical reason to believe that the data of day 2 are statistically different from those of day 5.

PROBLEMS

4.1 Use the χ^2 distribution table (Table D.2) to determine the values of $\chi^2_{\alpha/2}$ that would be used to construct confidence intervals for a population variance for the following combinations:

(a) $\alpha = 0.10$, $v = 25$
(b) $\alpha = 0.05$, $v = 15$
(c) $\alpha = 0.05$, $v = 10$
(d) $\alpha = 0.01$, $v = 30$

4.2 Use the t distribution table (Table D.3) to determine the values of $t_{\alpha/2}$ that would be used to construct confidence intervals for a population mean for each of the following combinations:

(a) $\alpha = 0.10$, $v = 25$
(b) $\alpha = 0.05$, $v = 15$
(c) $\alpha = 0.01$, $v = 10$
(d) $\alpha = 0.01$, $v = 40$

4.3 Use the F distribution table (Table D.4) to determine the values of F_{α,v_1,v_2} that would be used to construct confidence intervals for the ratio of two sample variances for each of the following combinations:

(a) $\alpha = 0.20$, $v_1 = 24$, $v_2 = 2$
(b) $\alpha = 0.01$, $v_1 = 15$, $v_2 = 8$
(c) $\alpha = 0.05$, $v_1 = 60$, $v_2 = 20$
(d) $\alpha = 0.80$, $v_1 = 2$, $v_2 = 24$

Use Program STATS to do each problem.

4.4 Problem 4.1

4.5 Problem 4.2

4.6 Problem 4.3

4.7 Compare the variances of day 2 and day 5 in Example 4.10 at a level of significance of 0.20 ($\alpha = 0.20$). Would testing the variances of day 1 and day 2 result in a different finding?

4.8 Using the data given in Example 4.9, determine if Kathi is statistically better with the equipment than Ron at a significance level of:

(a) 0.05.
(b) 0.10.

4.9 A least-squares adjustment is computed twice on a data set. When the data are minimally constrained with 24 degrees of freedom, a variance of 0.67 is obtained. In the second run, the fully constrained network has 30 degrees of freedom with a standard deviation of 1.25. The a priori estimate for the variance in both adjustments is 1; that is, $\sigma_1^2 = \sigma_2^2 = 1$.

(a) What is the 95% confidence interval for the ratio of the two variances?

(b) Are these reference variances statistically equal at a 5% level of signficance?

4.10 A surveying company decides to base a portion of their employee's salary raises on improvement in use of equipment. To determine their improvement, the employees measure their ability to point on a target and read the circles of a theodolite every six months. One employee is tested six weeks after starting employment and obtains a standard deviation of ±1.5″ with 25 measurements. Six months later this same employee obtains a standard deviation of ±1.2″ with 30 measurements. Did the employee statistically improve over six months at a 5% level of significance? Is this test an acceptable method of determining improvements in quality? What suggestions, if any, would you give to modify the test?

4.11 The calibrated length of a baseline is 402.167 m. An average distance of 402.251 m with a standard deviation of ±0.025 is computed after the line is measured 10 times with an EDM.

(a) Is the measured distance statistically different from the calibrated distance at a 5% significance level?

(b) What is the 95% confidence interval for the measurement? Does this range contain the calibrated value?

4.12 An observer's pointing and reading standard deviation is determined to be ±1.8″ after pointing and reading the circles of a particular instrument 20 times ($n = 20$). What is the 99% confidence interval for the population variance?

4.13 If the data sets in Problem 3.11 were distances measured with an EDM instrument that had a manufacturer's specified accuracy of ±(5 mm + 5 ppm), are they statistically justifiable measurements? (Assume that the distances in Problem 3.11 are in units of feet.)

4.14 Use Program STATS to determine at what level of significance the variance of day 2 in Example 4.10 would have been rejected.

5

PROPAGATION OF RANDOM ERRORS IN INDIRECTLY MEASURED QUANTITIES

5.1 BASIC ERROR PROPAGATION EQUATION

As discussed in Section 1.2, unknown values are often determined indirectly by making direct measurements of other quantities which are functionally related to the desired unknowns. Then the unknowns are computed. Examples in surveying include computing station coordinates from distance and angle measurements, obtaining station elevations from rod readings in differential leveling, and determining the azimuth of a line from astronomical observations. But as noted in Section 1.2, because all directly measured quantities contain errors, any values computed from them will also contain errors. This intrusion, or *propagation*, of errors that occurs in quantities computed from direct measurements is called *error propagation*. This topic is one of the most important discussed in this book.

In this chapter it is assumed that all systematic errors have been eliminated, so that only random errors remain in the direct observations. To derive the basic error propagation equation, consider the simple function, $z = a_1x_1 + a_2x_2$, where x_1 and x_2 are two independently observed quantities with standard errors σ_1 and σ_2, and a_1 and a_2 are constants. By analyzing how errors propagate in this function, a general expression can be developed for the propagation of random errors through any function.

Since x_1 and x_2 are two independently observed quantities, they each have different probability density functions. Let the errors in n determinations of x_1 be ϵ_1^i, ϵ_1^{ii}, ϵ_1^{iii}, ..., ϵ_1^n, and the errors in n determinations of x_2 be ϵ_2^i, ϵ_2^{ii}, ϵ_2^{iii}, ..., ϵ_2^n. Then z_T, the true value of z for each independent measurement, is

$$
\begin{aligned}
z_T &= a_1(x_1^i - \epsilon_1^i) + a_2(x_2^i - \epsilon_2^i) = a_1x_1^i + a_2x_2^i - (a_1\epsilon_1^i + a_2\epsilon_2^i) \\
z_T &= a_1(x_1^{ii} - \epsilon_1^{ii}) + a_2(x_2^{ii} - \epsilon_2^{ii}) = a_1x_1^{ii} + a_2x_2^{ii} - (a_1\epsilon_1^{ii} + a_2\epsilon_2^{ii}) \\
z_T &= a_1(x_1^{iii} - \epsilon_1^{iii}) + a_2(x_2^{iii} - \epsilon_2^{iii}) = a_1x_1^{iii} + a_2x_2^{iii} - (a_1\epsilon_1^{iii} + a_2\epsilon_2^{iii}) \\
&\vdots
\end{aligned}
\tag{5.1}
$$

The values for z computed from the observations are

$$
\begin{aligned}
z^i &= a_1x_1^i + a_2x_2^i \\
z^{ii} &= a_1x_1^{ii} + a_2x_2^{ii} \\
z^{iii} &= a_1x_1^{iii} + a_2x_2^{iii} \\
&\vdots
\end{aligned}
\tag{5.2}
$$

Substituting Equations (5.2) into Equations (5.1) and regrouping Equations (5.1) to isolate the errors for each computed value, we have

$$
\begin{aligned}
z^i - z_T &= a_1\epsilon_1^i + a_2\epsilon_2^i \\
z^{ii} - z_T &= a_1\epsilon_1^{ii} + a_2\epsilon_2^{ii} \\
z^{iii} - z_T &= a_1\epsilon_1^{iii} + a_2\epsilon_2^{iii} \\
&\vdots
\end{aligned}
\tag{5.3}
$$

From Equation (2.4), for the variance in a population, $n\sigma^2 = \Sigma_{i=1}^n \epsilon_i^2$, and thus for the case under consideration, the sum of the squared errors for the value computed is

$$
\sum_{i=1}^{n} \epsilon^2 = (a_1\epsilon_1^i + a_2\epsilon_2^i)^2 + (a_1\epsilon_1^{ii} + a_2\epsilon_2^{ii})^2 + (a_2\epsilon_1^{iii} + a_2\epsilon_2^{iii})^2 + \cdots = n\sigma_z^2
\tag{5.4}
$$

Expanding the terms in Equation (5.4) yields

$$
n\sigma_z^2 = (a_1\epsilon_1^i)^2 + 2a_1a_2\epsilon_1^i\epsilon_2^i + (a_2\epsilon_2^i)^2 + (a_1\epsilon_1^{ii})^2 + 2a_1a_2\epsilon_1^{ii}\epsilon_2^{ii} + (a_2\epsilon_2^{ii})^2 + \cdots
\tag{5.5}
$$

Factoring terms in Equation (5.5) gives us

$$n\sigma_z^2 = a_1^2(\epsilon_1^{i2} + \epsilon_1^{ii2} + \epsilon_1^{iii2} + \cdots) + a_2^2(\epsilon_2^{i2} + \epsilon_2^{ii2} + \epsilon_2^{iii2} + \cdots)$$
$$+ 2a_1a_2(\epsilon_1^i\epsilon_2^i + \epsilon_1^{ii}\epsilon_2^{ii} + \epsilon_1^{iii}\epsilon_2^{iii} + \cdots) \tag{5.6}$$

Inserting summation symbols for the error terms in Equation (5.6) gives

$$\sigma_z^2 = a_1^2\left(\frac{\sum_{i=1}^{n}\epsilon_1^2}{n}\right) + 2a_1a_2\left(\frac{\sum_{i=1}^{n}\epsilon_1\epsilon_2}{n}\right) + a_2^2\left(\frac{\sum_{i=1}^{n}\epsilon_2^2}{n}\right) \tag{5.7}$$

Recognizing that the terms in parentheses in Equation (5.7) are by definition σ_{x1}^2, $\sigma_{x_1x_2}$, and σ_{x2}^2, respectively, Equation (5.7) can be rewritten as

$$\sigma_z^2 = a_1^2\sigma_{x_1}^2 + 2a_1a_2\sigma_{x_1x_2} + a_2^2\sigma_{x2}^2 \tag{5.8}$$

In Equation (5.8), the middle term, $\sigma_{x_1x_2}$ is known as the *covariance*. This term shows the interdependence between the two unknown variables x_1 and x_2. Its importance is discussed in more detail in later chapters.

Equations (5.7) and (5.8) can be written in matrix form as

$$\Sigma_{zz} = \begin{bmatrix} a_1 & a_2 \end{bmatrix} \begin{bmatrix} \sigma_{x1}^2 & \sigma_{x1x2} \\ \sigma_{x_1x_2} & \sigma_{x2}^2 \end{bmatrix} \begin{bmatrix} a_1 \\ a_2 \end{bmatrix} \tag{5.9}$$

where Σ_{zz} is the *variance–covariance matrix* (also simply called the *covariance matrix*) for the function z. It follows logically from this simple derivation that, in general, if z is a function of n independently measured quantities, x_1, x_2, ..., x_n, then Σ_{zz} is

$$\Sigma_{zz} = \begin{bmatrix} a_1 & a_2 & \cdots & a_n \end{bmatrix} \begin{bmatrix} \sigma_{x1}^2 & \sigma_{x1x2} & \cdots & \sigma_{x1xn} \\ \sigma_{x1x2} & \sigma_{x2}^2 & & \sigma_{x2xn} \\ \vdots & & \ddots & \vdots \\ \sigma_{x1xn} & \sigma_{x2xn} & \cdots & \sigma_{xn}^2 \end{bmatrix} \begin{bmatrix} a_1 \\ a_2 \\ \vdots \\ a_n \end{bmatrix} \tag{5.10}$$

Further, for a set of m functions with n independently measured quantities, x_1, x_2, ..., x_n, Equation (5.10) expands to

$$\Sigma_{zz} = \begin{bmatrix} a_{11} & a_{12} & \cdots & a_{1n} \\ a_{21} & a_{22} & \cdots & a_{2n} \\ \vdots & \vdots & \cdots & \vdots \\ a_{m1} & a_{m2} & \cdots & a_{mn} \end{bmatrix} \begin{bmatrix} \sigma_{x1}^2 & \sigma_{x1x2} & \cdots & \sigma_{x1xn} \\ \sigma_{x1x2} & \sigma_{x2}^2 & & \sigma_{x2xn} \\ \vdots & & \ddots & \vdots \\ \sigma_{x1xn} & \sigma_{x2xn} & \cdots & \sigma_{xn}^2 \end{bmatrix} \begin{bmatrix} a_{11} & a_{21} & \cdots & a_{m1} \\ a_{12} & a_{22} & \cdots & a_{m2} \\ \vdots & \vdots & \cdots & \vdots \\ a_{1n} & a_{2n} & \cdots & a_{mn} \end{bmatrix} \tag{5.11}$$

Similarly, if the functions are nonlinear, a first-order Taylor series expansion can be used to linearize them.[1] Thus a_{11}, a_{12}, ... are replaced by the partial derivatives of the z_1, z_2, ..., with respect to the measurements, x_1, x_2, ..., etc. Therefore, after linearizing a set of nonlinear equations, the covariance matrix for the function z can be written in linear form as

$$\Sigma_{zz} = \begin{bmatrix} \frac{\partial Z_1}{\partial x_1} & \frac{\partial Z_1}{\partial x_2} & \cdots & \frac{\partial Z_1}{\partial x_n} \\ \frac{\partial Z_2}{\partial x_1} & \frac{\partial Z_2}{\partial x_2} & \cdots & \frac{\partial Z_2}{\partial x_n} \\ \vdots & \vdots & \cdots & \cdots \\ \frac{\partial Z_m}{\partial x_1} & \frac{\partial Z_m}{\partial x_2} & \cdots & \frac{\partial Z_m}{\partial x_n} \end{bmatrix} \begin{bmatrix} \sigma_{x_1}^2 & \sigma_{x_1x_2} & \cdots & \sigma_{x_1x_n} \\ \sigma_{x_1x_2} & \sigma_{x_2}^2 & & \sigma_{x_2x_n} \\ \vdots & & \ddots & \vdots \\ \sigma_{x_1x_n} & \sigma_{x_2x_n} & \cdots & \sigma_{x_n}^2 \end{bmatrix}$$

$$\times \begin{bmatrix} \frac{\partial Z_1}{\partial x_1} & \frac{\partial Z_2}{\partial x_1} & \cdots & \frac{\partial Z_m}{\partial x_1} \\ \frac{\partial Z_1}{\partial x_2} & \frac{\partial Z_2}{\partial x_2} & \cdots & \frac{\partial Z_m}{\partial x_2} \\ \vdots & \vdots & \cdots & \vdots \\ \frac{\partial Z_1}{\partial x_n} & \frac{\partial Z_2}{\partial x_n} & \cdots & \frac{\partial Z_m}{\partial x_n} \end{bmatrix} \tag{5.12}$$

Equations (5.11) and (5.12) are known as the *general law of propagation of variances* (GLOPOV) for linear and nonlinear equations, respectively. Both Equations (5.11) and (5.12) can be written symbolically in matrix notation as

$$\Sigma_{zz} = A\Sigma A^T \tag{5.13}$$

where Σ_{zz} is the covariance matrix for the function Z, and Σ is the covariance matrix for the measurements. For a nonlinear set of equations that has been linearized using Taylor's Theorem, the coefficient matrix (A) is called a *Jacobian matrix*. That is, it is a matrix of partial derivatives with respect to each unknown, as shown in Equation (5.12).

If the measurements, x_1, x_2, ..., x_n, are unrelated, that is, if they are statistically independent, then the covariance terms (off-diagonal elements), $\sigma_{x_1x_2}$, $\sigma_{x_1x_3}$, ... are equal to zero, and thus Equations (5.11) and (5.12) can be rewritten, respectively, as

[1] Readers who are unfamiliar with solving nonlinear equations should refer to Appendix C.

$$\begin{bmatrix} a_{11} & a_{12} & \cdots & a_{1n} \\ a_{21} & a_{22} & \cdots & a_{2n} \\ \vdots & \vdots & \cdots & \vdots \\ a_{m1} & a_{m2} & \cdots & a_{mn} \end{bmatrix} \begin{bmatrix} \sigma_{x1}^2 & 0 & \cdots & 0 \\ 0 & \sigma_{x2}^2 & & 0 \\ \vdots & & \ddots & \vdots \\ 0 & 0 & \cdots & \sigma_{xn}^2 \end{bmatrix} \begin{bmatrix} a_{11} & a_{21} & \cdots & a_{m1} \\ a_{12} & a_{22} & \cdots & a_{m2} \\ \vdots & \vdots & \cdots & \vdots \\ a_{1n} & a_{2n} & \cdots & a_{mn} \end{bmatrix} \tag{5.14}$$

$$\begin{bmatrix} \frac{\partial Z_1}{\partial x_1} & \frac{\partial Z_1}{\partial x_2} & \cdots & \frac{\partial Z_1}{\partial x_n} \\ \frac{\partial Z_2}{\partial x_1} & \frac{\partial Z_2}{\partial x_2} & \cdots & \frac{\partial Z_2}{\partial x_n} \\ \vdots & \vdots & \cdots & \vdots \\ \frac{\partial Z_m}{\partial x_1} & \frac{\partial Z_m}{\partial x_2} & \cdots & \frac{\partial Z_m}{\partial x_n} \end{bmatrix} \begin{bmatrix} \sigma_{x1}^2 & 0 & \cdots & 0 \\ 0 & \sigma_{x2}^2 & & 0 \\ \vdots & & \ddots & \vdots \\ 0 & 0 & \cdots & \sigma_{xn}^2 \end{bmatrix} \begin{bmatrix} \frac{\partial Z_1}{\partial x_1} & \frac{\partial Z_2}{\partial x_1} & \cdots & \frac{\partial Z_m}{\partial x_1} \\ \frac{\partial Z_1}{\partial x_2} & \frac{\partial Z_2}{\partial x_2} & \cdots & \frac{\partial Z_m}{\partial x_2} \\ \vdots & \vdots & \cdots & \vdots \\ \frac{\partial Z_1}{\partial x_n} & \frac{\partial Z_2}{\partial x_n} & \cdots & \frac{\partial Z_m}{\partial x_n} \end{bmatrix} \tag{5.15}$$

If there is only one function Z, involving n unrelated quantities, x_1, x_2, ..., x_n, then Equation (5.15) can be rewritten in algebraic form as

$$\sigma_z = \sqrt{\left(\frac{\partial Z}{\partial x_1}\sigma_{x_1}\right)^2 + \left(\frac{\partial Z}{\partial x_2}\sigma_{x_2}\right)^2 + \cdots + \left(\frac{\partial Z}{\partial x_n}\sigma_{x_n}\right)^2} \tag{5.16}$$

Equations (5.14), (5.15), and (5.16) express the *special law of propagation of variances* (SLOPOV). These equations govern the manner in which errors from statistically independent measurements (i.e., $\sigma_{x_ix_j} = 0$) propagate in a function. In these equations, individual terms $(\partial Z/\partial x_i)\sigma_{x_i}$ represent the individual contributions to the total error that occur as the result of measurement errors in each independent variable. When the size of a function's estimated error is too large, inspection of these individual terms will indicate the largest contributors to the error. Then the most efficient method to reduce the overall error in the function is to examine closely ways to reduce those largest terms in Equation (5.16).

Generic Example. Let $A = B + C$, and assume that B and C are independently measured quantities. Note that $\partial A/\partial B = 1$ and $\partial A/\partial C = 1$. Substituting these into Equation (5.16) gives

$$\sigma_A = \sqrt{(1\sigma_B)^2 + (1\sigma_C)^2} \tag{5.17}$$

Using Equation (5.15) yields

$$\Sigma_{AA} = \begin{bmatrix} 1 & 1 \end{bmatrix} \begin{bmatrix} \sigma_B^2 & 0 \\ 0 & \sigma_C^2 \end{bmatrix} \begin{bmatrix} 1 \\ 1 \end{bmatrix} = [\sigma_B^2 + \sigma_C^2]$$

which produces the same results as Equation (5.16). In the equations above, standard error (σ) and standard deviation (S) can be interchanged.

5.2 FREQUENTLY ENCOUNTERED SPECIFIC FUNCTIONS

5.2.1 Standard Deviation of a Sum

Let $A = B_1 + B_2 + \cdots + B_n$, where the B's are n independently observed quantities having standard errors of $S_{B_1}, S_{B_2}, \ldots, S_{B_n}$; then by Equation (5.16),

$$S_A = \sqrt{S_{B_1}^2 + S_{B_2}^2 + \cdots + S_{B_n}^2} \tag{5.18}$$

5.2.2 Standard Deviation in a Series

Assume that the error for each measured value in Equation (5.18) is equal; that is, $S_{B_1}, S_{B_2}, \ldots, S_{B_n} = S_B$. Then Equation (5.18) simplifies to

$$S_A = S_B \sqrt{n} \tag{5.19}$$

5.2.3 Standard Deviation of the Mean

Let $\bar{y}$ be the mean obtained from n independently observed quantities $y_1, y_2, \ldots, y_n$, each of which has the same standard deviation S. The mean is expressed as

$$\bar{y} = \frac{y_1 + y_2 + \cdots + y_n}{n}$$

An equation for $S_{\bar{y}}$, the standard deviation of $\bar{y}$, is obtained by substituting the expression above into Equation (5.16). Note that $\partial\bar{y}/\partial y_1 = 1/n = \partial\bar{y}/\partial y_2 = \cdots = \partial\bar{y}\partial y_n$.

$$S_{\bar{y}} = \sqrt{\left(\frac{1}{n}S_{y_1}\right)^2 + \left(\frac{1}{n}S_{y2}\right)^2 + \cdots + \left(\frac{1}{n}S_{y_n}\right)^2} = \sqrt{\frac{nS^2}{n^2}} = \frac{S}{\sqrt{n}} \tag{5.20}$$

Notice that Equation (5.20) is the same as Equation (2.8).

5.3 NUMERICAL EXAMPLES

Example 5.1 The dimensions of a rectangular tank (Figure 5.1) are measured as

$$L = 40.00 \text{ ft} \qquad S_L = \pm 0.05 \text{ ft}$$
$$W = 20.00 \text{ ft} \qquad S_W = \pm 0.03 \text{ ft}$$
$$H = 10.00 \text{ ft} \qquad S_H = \pm 0.02 \text{ ft}$$

Find the tank's volume and its estimated standard deviation using the measurements above.

SOLUTION: The volume of the tank is found by the formula

$$V = LWH = 40.00 \times 20.00 \times 10.00 = 8{,}000 \text{ ft}^3$$

Given that $\partial V/\partial L = WH$, $\partial V/\partial W = LH$, and $\partial V/\partial H = LW$, the standard deviation in the computed volume is determined by using Equation (5.16):

$$\begin{aligned} S_V &= \sqrt{\left(\frac{\partial V}{\partial L} S_L\right)^2 + \left(\frac{\partial V}{\partial W} S_W\right)^2 + \left(\frac{\partial V}{\partial H} S_H\right)^2} \\ &= \sqrt{(WH)^2(0.05)^2 + (LH)^2(0.03)^2 + (LW)^2(0.02)^2} \\ &= \sqrt{(200 \times 0.05)^2 + (400 \times 0.03)^2 + (800 \times 0.02)^2} \\ &= \sqrt{100 + 144 + 256} = \sqrt{500} = \pm 22 \text{ ft}^3 \qquad (a) \end{aligned}$$

In Equation (*a*), the third term is the largest contributor to the total error, and thus to reduce the overall estimated error in the computed volume, it would be prudent first to try to make S_H smaller, to achieve the greatest effect on the estimated error of the function.

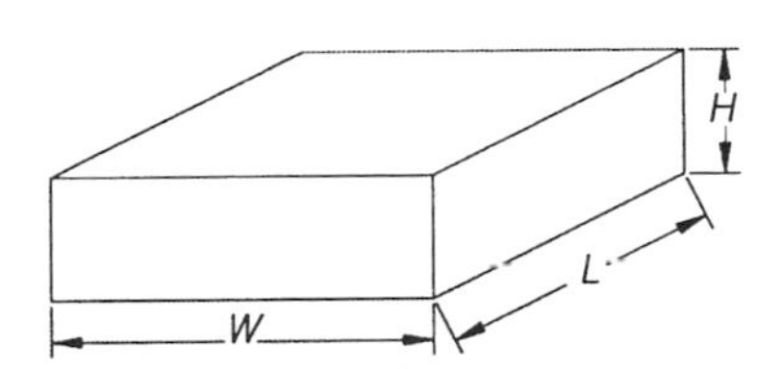

Figure 5.1 Rectangular tank.

Example 5.2 A vertical angle α to point B is measured at point A as 3°00, with S_α being $\pm 1'$, and the slope distance, D, from A to B is measured as 1000.00 ft, with S_D being ± 0.05 ft (Figure 5.2). Find the horizontal distance H and its standard deviation.

SOLUTION: The horizontal distance can be determined using the equation

$$H = D\cos(\alpha) = 1000.00\cos(3°00') = 998.63 \text{ ft}$$

Given that $\partial H/\partial D = \cos(\alpha)$ and $\partial H/\partial\alpha = -D\sin(\alpha)$, the estimated error in the function is determined by using Equation (5.16) as

$$S_H = \sqrt{\left(\frac{\partial H}{\partial D}S_D\right)^2 + \left(\frac{\partial H}{\partial \alpha}S_\alpha\right)^2} \tag{5.21}$$

In Equation (5.21), S_α must be converted to its equivalent radian value to achieve agreement in the units, thus:

$$S_H = \sqrt{[\cos(\alpha) \times (0.05)]^2 + \frac{[-\sin(\alpha) \times D]^2 (1')^2}{60' \times 180°/\pi}}$$

$$= \sqrt{(0.9986 \times 0.05)^2 + \left(-0.0523 \times 1000 \times \frac{1'}{10{,}800'/\pi}\right)^2}$$

$$= \sqrt{0.04993^2 + 0.0152^2} = \pm 0.052 \text{ ft}$$

Notice in this example that the major contributing error source (largest number under the radical) is 0.04993^2. This is the error associated with the distance measurement, and thus if the resulting error of ± 0.052 ft is too large, the logical way to improve the results (reduce the overall error) is to adopt a more precise method of measuring the distance.

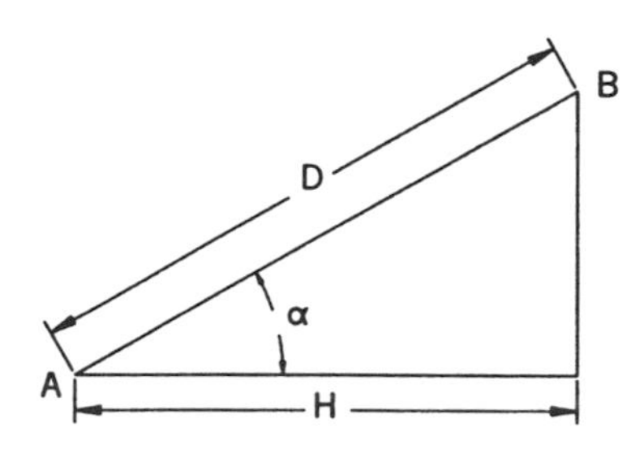

Figure 5.2 Horizontal distance from slope measurements.

Example 5.3 The elevation of a point C on a chimney stack is desired. Field angles and distances are measured as shown in Figure 5.3. Station A has an elevation of 1298.65 ± 0.006, and station B has an elevation of 1301.53 ± 0.004. The instrument height, hi, at station A is 5.25 ± 0.005, and the instrument height, hi, at station B is 5.18 ± 0.005. The other field measurements and their estimated errors are:

$$AB = 136.45 \pm 0.018$$

$$\measuredangle A = 44°12'34'' \pm 8.6'' \qquad \measuredangle B = 39°26'56'' \pm 11.3''$$

$$\measuredangle v_1 = 8°12'47'' \pm 4.1'' \qquad \measuredangle v_2 = 5°50'10'' \pm 5.1''$$

What are the elevation of the chimney stack and the estimated error in its computed value?

SOLUTION: Normally, this problem is worked in several steps. The steps include computing distances AI and BI, and then solving for the average elevation of C using values obtained from both v_1 and v_2 of Figure 5.1. However, caution must be exercised when doing error analysis in a stepwise fashion since the computed values could be correlated, and thus the stepwise method may lead to an incorrect analysis of the errors. To avoid this, either GLOPOV can be used or a single function can be derived that includes all quantities measured in the calculation of the elevation. This procedure is demonstrated here.

From the sine law we can derive the solution of AI and BI as

$$AI = \frac{AB \sin(\measuredangle B)}{\sin[180° - (\measuredangle A + \measuredangle B)]} = \frac{AB \sin(\measuredangle B)}{\sin(\measuredangle A + \measuredangle B)} \tag{5.22}$$

$$\text{also } BI = \frac{AB \sin(\measuredangle A)}{\sin(\measuredangle A + \measuredangle B)} \tag{5.23}$$

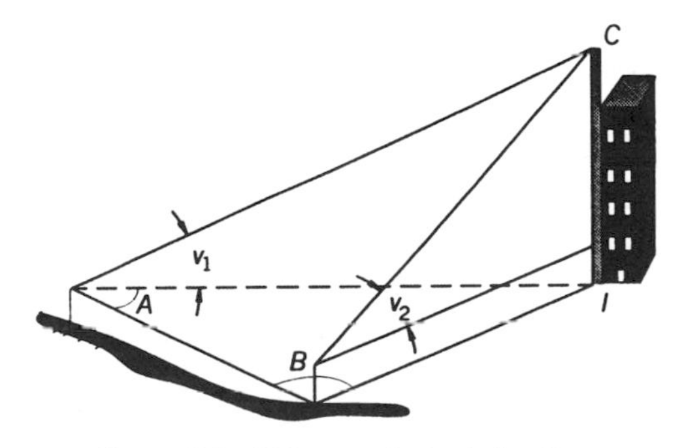

Figure 5.3 Chimney stacked elevation.

Using Equations (5.22) and (5.23), the elevations for C from both v_1 and v_2 are

$$\text{Elev}_{C_{\text{Left}}} = AI \tan(\measuredangle v_1) + \text{Elev}_A + \text{hi}_A \tag{5.24}$$

$$\text{Elev}_{C_{\text{Right}}} = BI \tan(\measuredangle v_2) + \text{Elev}_B + \text{hi}_B \tag{5.25}$$

Thus the stack's elevation, given as the average Equations (5.24) and (5.25), is

$$\text{Elev}_C = \frac{1}{2}(\text{Elev}_{C_{\text{Left}}} + \text{Elev}_{C_{\text{Right}}}) \tag{5.26}$$

Substituting Equations (5.22) through (5.25) into (5.26), a single expression for the chimney elevation is

$$\text{Elev}_C = \frac{1}{2}\left[\text{Elev}_A + \text{hi}_A + \frac{AB \sin(\measuredangle B) \tan(\measuredangle v_1)}{\sin(\measuredangle A + \measuredangle B)} + \text{Elev}_B + \text{hi}_B + \frac{AB \sin(\measuredangle A) \tan(\measuredangle v_2)}{\sin(\measuredangle A + \measuredangle B)}\right] \tag{5.27}$$

From Equation (5.27), the elevation of C is 1,316.49. To perform the error analysis, Equation (5.16) is used. In this complex problem it is often easier to break the problem into smaller parts. This can be done by solving numerically each partial derivative necessary for Equation (5.16) before squaring and summing the results. From Equation (5.27),

(a) $$\frac{\partial \text{Elev}_C}{\partial \text{Elev}_A} = \frac{\partial \text{Elev}_C}{\partial \text{Elev}_B} = \frac{1}{2}$$

(b) $$\frac{\partial \text{Elev}_C}{\partial \text{hi}_A} = \frac{\partial \text{Elev}_C}{\partial \text{hi}_B} = \frac{1}{2}$$

(c) $$\frac{\partial \text{Elev}_C}{\partial AB} = \frac{1}{2}\left[\frac{\sin(\measuredangle B) \tan(\measuredangle v_1) + \sin(\measuredangle A) \tan(\measuredangle v_2)}{\sin(\measuredangle A + \measuredangle B)}\right] = 0.08199$$

(d) $$\frac{\partial \text{Elev}_C}{\partial \measuredangle A} = \frac{AB}{2} \frac{-\cos(\measuredangle A + \measuredangle B)[\sin(\measuredangle B) \tan(\measuredangle v_1) + \sin(\measuredangle A) \tan(\measuredangle v_2)]}{\sin^2(\measuredangle A + \measuredangle B)} + \frac{AB}{2} \frac{\cos(\measuredangle A) \tan(\measuredangle v_2)}{\sin(\measuredangle A + \measuredangle B)} = 3.78596$$

(e) $\dfrac{\partial \text{Elev}_C}{\partial \measuredangle B}$

$$= \frac{AB}{2}\,\frac{-\cos(\measuredangle A + \measuredangle B)[\sin(\measuredangle B)\tan(\measuredangle v_1) + \sin(\measuredangle A)\tan(\measuredangle v_2)]}{\sin^2(\measuredangle A + \measuredangle B)}$$

$$+ \frac{AB}{2}\,\frac{\cos(\measuredangle B)\tan(\measuredangle v_1)}{\sin(\measuredangle A + \measuredangle B)} = 6.40739$$

(f) $$\frac{\partial \text{Elev}_C}{\partial \measuredangle v_1} = \frac{AB\sin(\measuredangle B)}{2\sin(\measuredangle A + \measuredangle B)\cos^2(\measuredangle v_1)} = 44.52499$$

(g) $$\frac{\partial \text{Elev}_C}{\partial \measuredangle v_2} = \frac{AB\sin(\measuredangle A)}{2\sin(\measuredangle A + \measuredangle B)\cos^2(\measuredangle v_2)} = 48.36511$$

Again for compatibility of the units in this problem, all angular errors are converted to their radian equivalents by dividing by 206,264.8″/rad. Finally, using Equation (5.16), the estimated error in the computed elevation is

$$S^2_{\text{Elev}C} = \left(\frac{\partial \text{Elev}_C}{\partial \text{Elev}_A}S_{\text{Elev}A}\right)^2 + \left(\frac{\partial \text{Elev}_C}{\partial \text{Elev}_B}S_{\text{Elev}B}\right)^2 + \left(\frac{\partial \text{Elev}_C}{\partial \text{hi}_A}S_{\text{hi}A}\right)^2$$
$$+ \left(\frac{\partial \text{Elev}_C}{\partial \text{hi}_B}S_{\text{hi}B}\right)^2 + \left(\frac{\partial \text{Elev}_C}{\partial AB}S_{AB}\right)^2 + \left(\frac{\partial \text{Elev}_C}{\partial \measuredangle A}S_{\measuredangle A}\right)^2$$
$$+ \left(\frac{\partial \text{Elev}_C}{\partial \measuredangle B}S_{\measuredangle B}\right)^2 + \left(\frac{\partial \text{Elev}_C}{\partial \measuredangle v_1}S_{\measuredangle v1}\right)^2 + \left(\frac{\partial \text{Elev}_C}{\partial \measuredangle v_2}S_{\measuredangle v2}\right)^2$$

$$S_{\text{Elev}C} = \left[\left(\frac{1}{2}\,0.004\right)^2 + \left(\frac{1}{2}\,0.006\right)^2 + 2\left(\frac{1}{2}\,0.005\right)^2\right.$$
$$+ (0.08199 \times 0.018)^2$$
$$+ (3.78596 \times 4.1693 \times 10^{-5})^2$$
$$+ (6.40739 \times 5.4783 \times 10^{-5})^2$$
$$+ (44.52499 \times 1.9877 \times 10^{-5})^2$$
$$\left. + (48.36511 \times 2.4725 \times 10^{-5})^2\right]^{1/2}$$
$$= \pm 0.0055 \text{ ft} \simeq \pm 0.01 \text{ ft}$$

Thus the elevation of point C is 1316.49 ± 0.01 ft.

5.4 CONCLUSIONS

Errors associated with any indirect measurement problem can be analyzed as described above. Besides being able to compute the estimated error in a function, the sizes of the individual errors contributing to that functional error can also be analyzed. This identifies those measurements whose errors are most critical to reduce if the functional error is too large. An alternative use of the error propagation equation involves computing the expected error in a function of measured values prior to going to the field. The calculations can be based on the geometry of the problem, and estimates for the measurements can be used in the function. Errors can be assigned to the measured values and they can be varied to correspond with those expected using different combinations of available equipment and procedures. That particular combination which produces the desired accuracy in the final computed function can then be adopted in the field. This analysis falls under the heading of *survey planning and design*. This topic is discussed further in Chapters 7 and 19.

PROBLEMS

5.1 In running a line of levels, 10 instrument setups are required, with a backsight and foresight taken from each. For each rod reading, the estimated error is ± 0.005 ft, due to various causes. What is the estimated error of the measured elevation difference between the origin and terminus?

5.2 In Problem 2.3 compute the estimated error in the overall distance as measured by both the 100-ft and 200-ft tapes. Which tape produced the smallest estimated error?

5.3 A line was measured in sections as follows:

Section	Measured Length (ft)	Standard Deviation (ft)
AB	416.24	± 0.06
BC	1044.16	± 0.08
CD	590.03	± 0.06
DE	714.28	± 0.08

Find the estimated error in length *AE*.

5.4 The volume of a cone is given by $V = (1/12)\pi D^2 h$. The cone's measured height is 10.0 in. with $S_h = \pm 0.20$ in. Its measured diameter is 6.0 in., with $S_D = \pm 0.20$ in. What are the cone's volume and standard deviation?

5.5 An EDM instrument manufacturer publishes the instrument's accuracy as $\pm$(5 mm + 5 ppm). [*Note*: 5 ppm means 5 parts per million. This is a scaling error and is computed as (distance × 5/1,000,000).]

(a) What formula should be used to determine the error in a distance measured with this instrument?

(b) What is the expected error in a 2750.34-ft distance measured with this EDM instrument?

5.6 A racetrack is measured in three simple components; a rectangle and two semicircles (Figure 5.4). Using an EDM instrument with a manufacturer's specified accuracy of ±(5 mm + 10 ppm), the rectangle's dimensions measured at the inside of the track are 5280.02 ft by 840.34 ft. The semicircles are on the shorter sides of the rectangle. Assuming only errors in the distance measurements, what are the:

(a) area enclosed by the track?

(b) standard deviation in the track's dimensions?

(c) standard deviation in the area enclosed by the track?

5.7 Using an EDM instrument with a manufacturer's specified accuracy of ±(3 mm + 5 ppm), the rectangular dimensions of a large building 1435.67 ft by 453.67 ft were laid out. Assuming only errors in distance measurements, what are the:

(a) standard deviations in the building's dimensions?

(b) area enclosed by the building, and its standard deviation?

5.8 A surveyor's error due only to reading is determined to be ±1.5″ when making observations with a particular instrument. After repeatedly pointing on a distant target with the same instrument, the surveyor determines the combined error due to both pointing and reading to be ±2.6″. What is the surveyor's pointing error?

5.9 For each tape correction formula noted below, express the error propagation formula in the form of Equation (5.16) using the variables listed.

(a) $H = L\cos(\alpha)$, where L is the slope length and α the slope angle. Determine the error with respect to L and α.

(b) $C_T = k(T_f - T)L$, where k is the coefficient of thermal expansion, T_f the tape's field temperature, T the calibrated temperature of the

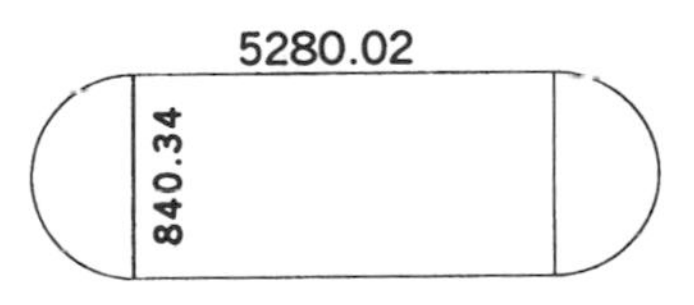

Figure 5.4

tape, and L the measured length. Determine the error with respect to T_f.

(c) $C_p = (P_f - P)L/AE$, where P_f is the field tension, P the calibrated tension for the tape, A its cross-sectional area, E the modulus of elasticity, and L the measured length. Determine the error with respect to P_f.

(d) $C_S = -w^2 l_s^3/24P_f^2$, where w is the weight per unit length of the tape, l_s the length between supports, and P_f the field tension. Determine the error with respect to P_f.

5.10 Given the following tape calibration data:

$A = 0.006\ \text{in}^2$	$l = 99.994$ ft
$l' = 100$ ft	$w = 0.02$ lb
$k = 0.00000645/°F$	$E = 29{,}000{,}000\ \text{lb/in}^2$
$P = 10$ lb	$T = 68°F$

Compute the corrected distance and its expected error if the measured distance is 145.67 ft. Assume that $T_f = 50°F \pm 5°F$, $P_f = 20\ \text{lb} \pm 4$ lb, a reading error of ± 0.005 ft, and that the distance was measured as two end-support distances of 80.00 ft and 65.67 ft. [*Reminder*: Don't forget the correction for length: $C_L = [(l - l')/l']L$, where l is the actual tape length, l' its nominal length and L the measured line length.]

5.11 Show that Equation (5.12) is equivalent to Equation (5.11) for linear equations.

5.12 Derive an expression similar to Equation (5.9) for the function $z = a_1x_1 + a_2x_2 + a_3x_3$.

Practical Exercises

5.13 Use the data gathered in Problems 2.11 and 2.12 to determine your error in pointing.

5.14 With an engineer's scale, measure the radius of the circle in Figure 5.5 (on page 95) ten times using different starting locations on the scale. Use a magnifying glass and interpolate the readings on the scale to the tenth of a 1/60 inch graduation.

(a) What are the mean radius of the circle and its standard deviation?

(b) Compute the area of the circle and its standard deviation.

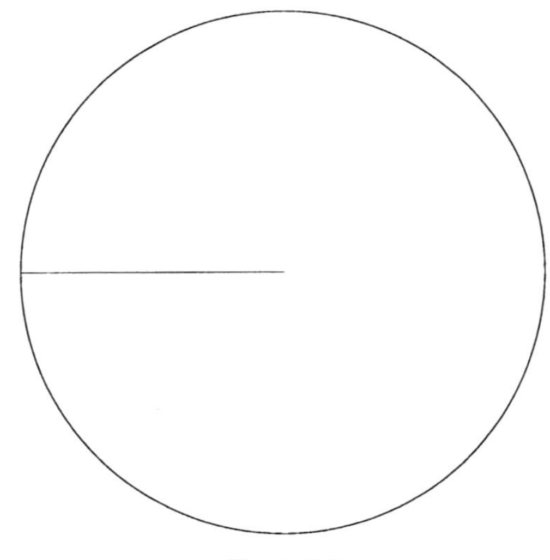

Figure 5.5

(c) Calibrate a planimeter by measuring a 2-in. square. Calculate the mean constant for the planimeter (k = units/4 in^2), and based on 10 measurements, determine the standard deviation in the constant.

(d) Using the same planimeter, measure the area of the circle and determine its standard deviation.

6

ERROR PROPAGATION IN ANGLE AND DISTANCE MEASUREMENTS

6.1 INTRODUCTION

All surveying measurements are subject to errors from varying sources. For example, when measuring an angle, the major error sources include instrument placement and leveling, target placement, circle reading, and target pointing. Although great care may be taken in measuring an angle, these error sources will nonetheless render inexact results. To appreciate fully the need for adjustments, surveyors must be able to identify the major measurement error sources, know their effects on the measurements, and understand how they can be modeled. In this chapter, emphasis is placed on analyzing the errors in measured horizontal angles and horizontal distances. In Chapter 7 the manner in which these errors propagate to produce traverse misclosures is studied. In Chapter 8 the propagation of errors in elevation determination is covered.

6.2 ERROR SOURCES IN HORIZONTAL ANGLES

Whether a transit, theodolite, or total station instrument is used, errors are present in every horizontal angle measurement. Whenever an instrument's circles are read, a small error is introduced into the final measured angle. Also, in pointing to a target, a small amount of error always occurs. Other major error sources in angle measurement include instrument and target setup errors, and instrument leveling. Each of these sources produces random errors. They may be small or large, depending on the instrument, the operator, and the conditions at the time of the angle measurement. The effects of reading, pointing, and leveling errors can be reduced by increasing the number of angle repetitions. However, the effects of instrument and target setup errors can only be reduced by increasing sight distances.

6.3 READING ERRORS

Errors in reading conventional transits and theodolites are dependent on the quality of the instrument's optics, the size of the smallest division of the scale, and the observer's abilities, for example the ability to set and read the micrometer of a theodolite. Typical reading errors for a 1″ micrometer theodolite can range from tenths of a second to several seconds. Reading errors also occur with digital instruments, their size being dependent on the sensitivity of the particular electronic angular resolution system. Manufacturers of digital reading instruments quote the expected combined pointing and reading precision for an individual direction measured face left (direct) and face right (reverse) with their instruments in terms of standard deviations. Typical values range from $\pm 1''$ for the more precise instruments, to $\pm 10''$ for the less expensive ones. These errors are random, and their effects on an angle are dependent on the measurement method and the number of angle repetitions.

6.3.1 Angles Measured by the Repetition Method

When measuring a horizontal angle by the *repetition method* using a repeating instrument, the circle is first zeroed so that angles can be accumulated. Then the angle is turned a number of times, and finally, the cumulative angle is read and divided by the number of repetitions to determine the average measured value. In this method, a reading error exists in just two of the circle readings regardless of the number of repetitions. The first occurs when the circle is zeroed, and the second when reading the final cumulative angle. For this procedure, the average angle is calculated as

$$\alpha = \frac{\alpha_1 + \alpha_2 + \cdots + \alpha_n}{n} \tag{a}$$

where α is the average angle and $\alpha_1, \alpha_2, \cdots, \alpha_n$ are the n repetitions of the angle. Recognizing that readings occur only in α_1 and α_n, and applying Equation (5.16) to the function of Equation (*a*), the standard error σ_{α_r} in the average angle due only to reading errors is

$$\sigma_{\alpha_r} = \frac{\sqrt{\sigma_0^2 + \sigma_r^2}}{n} \tag{6.1}$$

where σ_{α_r} is the estimated error in the average angle due to reading, σ_0 the estimated error in setting zero on the circle, σ_r the estimated error in the final reading, and n the number of repetitions of the angle. Note that the number of repetitions should always be an even number, with half being turned face left (direct) and half face right (reverse). This procedure compensates for systematic instrumental errors.

If an operator's ability to set zero and to read the circle are assumed to be equal (a reasonable assumption), Equation (6.1) can be simplified to

$$\sigma_{\alpha_r} = \frac{\sigma_r \sqrt{2}}{n} \tag{6.2}$$

Example 6.1 Suppose that an angle is read six times using the repetition method. For an operator having a personal reading error of $\pm 1.5''$, what is the estimated error in the angle due to circle reading?

SOLUTION: From Equation (6.2),

$$\sigma_{\alpha_r} = \pm \frac{1.5'' \sqrt{2}}{6} = \pm 0.4''$$

6.3.2 Angles Measured by the Direction Method

When a horizontal angle is measured by the *directional method*, the horizontal circle is read in both the backsight and foresight directions. The angle is then the difference between the two readings. Multiple measurements of the angle are made, with the circle being advanced prior to each reading to compensate for the instrument's systematic errors. The final angle is taken as the mean of all measured values. Again, an even number of repetitions are made, with half taken in face left and half in face right. Since each repetition of the angle requires two readings, the error in the average angle due only to reading, by Equation (5.16), is

$$\sigma_{\alpha_r} = \frac{\sqrt{(\sigma^2_{r_{1b}} + \sigma^2_{r_{1f}}) + (\sigma^2_{r_{2b}} + \sigma^2_{2_{2f}}) + \cdots + (\sigma^2_{r_{nb}} + \sigma^2_{r_{nf}})}}{n} \tag{6.3}$$

where $\sigma_{r_{ib}}$ and $\sigma_{r_{if}}$ are the estimated reading errors in the backsight and foresight directions, respectively, and n is the number of repetitions. Assuming that one's ability to read the circle is independent of the particular direction, so that, $\sigma_{r_{1b}} = \sigma_{r_{1f}} = \sigma_r$, Equation (6.3) simplifies to

$$\sigma_{\alpha_r} = \frac{\sigma_r \sqrt{2}}{\sqrt{n}} \tag{6.4}$$

Example 6.2 Using the same parameters of six repetitions and an observer reading error of $\pm 1.5''$ as given in Example 6.1, find the estimated angular error due to reading when the directional method is used.

SOLUTION

$$\sigma_{\alpha_r} = \pm \frac{1.5'' \sqrt{2}}{\sqrt{6}} = \pm 0.9''$$

Note that the additional readings needed in the directional method produced a larger error in the average angle than that obtained with the repetition method.

6.4 POINTING ERRORS

Accuracy in pointing to a target is dependent on several factors. These include the optical qualities of the telescope, target size, the observer's personal ability to place the crosswires on a target, and weather conditions at the time of observation. Pointing errors are random, and they will occur in every angle measurement no matter what instrument is used. Since each repetition of an angle consists of two pointings, the pointing error for an angle that is the mean of n repetitions can be estimated, using Equation (5.16) as

$$\sigma_{\alpha_p} = \frac{\sqrt{2\sigma_{p1}^2 + 2\sigma_{p2}^2 + \cdots + 2\sigma_{pn}^2}}{n} \tag{6.5}$$

where σ_{α_p} is the estimated contribution to the overall angular error due to pointing, and $\sigma_{p1}, \sigma_{p2}, \cdots, \sigma_{pn}$ are the estimed errors in pointings for the first repetition, second repetition, and so on. Again for a given instrument and observer the pointing error can be assumed the same for each repetition (i.e., $\sigma_{p1} = \sigma_{p2} = \cdots = \sigma_{pn} = \sigma_p$) and Equation (6.5) simplifies to

$$\sigma_{\alpha_p} = \frac{\sigma_p \sqrt{2}}{\sqrt{n}} \tag{6.6}$$

Example 6.3 An angle is measured six times by an observer whose ability to point on a well-defined target is estimated to be $\pm 1.8''$. What is the estimated error in the angle due to pointing?

SOLUTION: From Equation (6.6),

$$\sigma_{\alpha_p} = \pm \frac{1.8'' \sqrt{2}}{\sqrt{6}} = \pm 1.0''$$

6.5 ESTIMATED POINTING AND READING ERRORS WITH TOTAL STATIONS

With the introduction of digital theodolites and subsequently, total station instruments, new standards were developed for estimating errors in angle measurements. The new standards, called DIN18723, provide values for estimated errors in the mean of two direction measurements, one each in face left and face right. Thus, in terms of a single pointing and reading error, σ_{pr}, the DIN value, σ_{DIN}, can be expressed as

$$\sigma_{\text{DIN}} = \frac{\sigma_{pr}\sqrt{2}}{2} = \frac{\sigma_{pr}}{\sqrt{2}}$$

Using this equation, the expression for the estimated error in the measurement of a single direction due to pointing and reading with a digital theodolite is

$$\sigma_{pr} = \sigma_{\text{DIN}}\sqrt{2} \tag{b}$$

Using a procedure similar to that given in Equation (6.6), the estimated error in an angle measured n times and averaged due to pointing and reading is

$$\sigma_{\alpha_{pr}} = \frac{\sigma_{pr}\sqrt{2}}{\sqrt{n}} \tag{c}$$

Substituting Equation (*b*) into Equation (*c*) yields

$$\sigma_{\alpha_{pr}} = \frac{2\,\sigma_{\text{DIN}}}{\sqrt{n}} \tag{6.7}$$

Example 6.4 An angle is measured six times (3 direct and 3 reverse) by an observer with a total station instrument having a published DIN18723 value for pointing and reading of $\pm 5''$. What is the estimated error in the average angle due to pointing and reading?

SOLUTION: From Equation (6.7),

$$\sigma_{\alpha_{pr}} = \frac{2 \times 5''}{\sqrt{6}} = \pm 4.1''$$

6.6 TARGET-CENTERING ERRORS

Whenever a target is set over a station, there will be some error due to faulty centering. It can be attributed to environmental conditions, optical plummet errors, quality of the optical plummet optics, plumb bob centering error, personal abilities, and so on. When care is taken, the instrument is usually within 0.001 to 0.01 ft of the true station location. Although these sources produce a constant centering error for any particular angle, target centering errors will appear as random in the adjustment of a network involving many stations.

An estimate of the effect of this error in an angle measurement can be made by analyzing its contribution to a single direction. As shown in Figure 6.1, the angular error due to centering is dependent on the position of the target. If the target is on line but off center, as shown in Figure 6.1(*a*), target decentering contributes no angular error. However, as the target moves to either side of the sight line, the error size increases. As shown in Figure 6.1(*d*), the largest error occurs when the target is offset perpendicular to the line of sight. Letting σ_d represent the distance the target is from the *true* station location, from Figure 6.1(*d*) the maximum error in an individual direction due to target decentering is

$$e = \frac{\pm\sigma_d}{D} \quad \text{rad} \tag{6.8}$$

where e is the uncertainty in the direction due to target decentering, σ_d the amount of a centering error at the time of pointing, and D the distance from the instrument center to the target (Figure 6.2).

Since two directions are required for each angle measurement, an estimate of the angular error is,

$$\sigma_{\alpha_t} = \sqrt{\left(\frac{\sigma_{d_1}}{D_1}\right)^2 + \left(\frac{\sigma_{d_2}}{D_2}\right)^2} \tag{6.9}$$

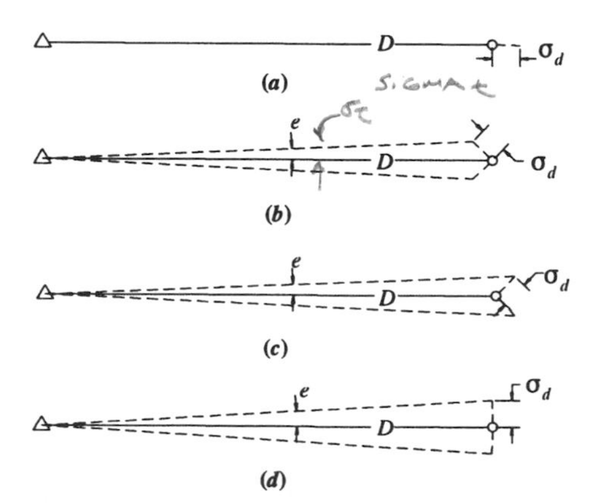

Figure 6.1 Possible target positions.

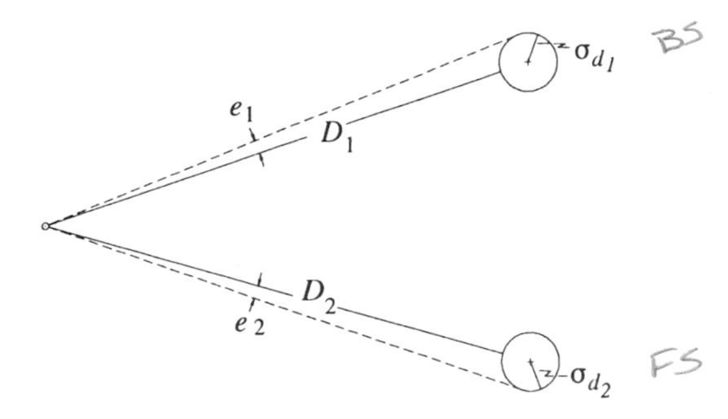

Figure 6.2 Error in angle due to target centering.

where σ_{α_t} is the angular error due to target centering, σ_{d_1} and σ_{d_2} are the target-centering errors at stations 1 and 2, respectively, and D_1 and D_2 are the distances from the target to the instrument at stations 1 and 2, respectively. Again, assuming the ability to center the target over a point is independent of the particular direction, and thus, $\sigma_{d_1} = \sigma_{d_2} = \sigma_t$, and converting the result of Equation (6.9) to arc seconds by multiplying by ρ, it can be rewritten as

$$\sigma_{\alpha_t} = \pm \frac{\sqrt{D_1^2 + D_2^2}}{D_1 D_2} \sigma_t \rho \tag{6.10}$$

Notice that since the same target-centering error occurs on each pointing, it cannot be reduced in size by making multiple pointings. Thus Equation (6.10) is not divided by the number of angle repetitions. This makes the target-centering error one of the more significant ones in angle measurement. Note also that Equations (6.9) and (6.10) are unitless, and thus the answers obtained using them must be converted from their radian values to some appropriate units. This conversion is generally to seconds, and is made using the factor $\rho = 206{,}264.8''/\text{rad}$.

Example 6.5 An observer's estimated ability at centering targets over a station is ±0.003 ft. For a particular angle measurement, the backsight and foresight distances from the instrument station to the targets are approximately 250 ft and 450 ft, respectively. What is the expected contribution to the overall angular error due to target centering?

SOLUTION: From Equation (6.10) the estimated error is

$$\sigma_{\alpha_t} = \pm \frac{\sqrt{250^2 + 450^2}}{250 \times 450} 0.003 \times 206{,}264.8''/\text{rad} = \pm 2.8''$$

If hand-held range poles were used in this example with an estimated centering error of ±0.01 ft, the estimated angular error due to the target centering would be

$$\sigma_{\alpha_t} = \pm\frac{\sqrt{250^2 + 450^2}}{250 \times 450}\,0.01 \times 206{,}264.8''/\text{rad} = \pm 9.4''$$

Obviously, this error source can be significant if care is not taken in target centering.

6.7 INSTRUMENT CENTERING ERRORS

Every time an instrument is centered over a point, there is some error in its position with respect to the *true* station location. This error is dependent on the quality of the instrument and the state of adjustment of its optical plummet, the quality of the tripod, and the skill of observer. The error can be compensating as shown in Figure 6.3(*a*), or it can be maximum when the instrument is on the angle bisector as shown in Figure 6.3(*b*) and (*c*). For any individual setup this error is a constant, however, since the instrument's location is random with respect to the *true* station location, it will appear to be random in the adjustment of a network involving many stations. From Figure 6.3, the true angle α is

$$\alpha = (P_2 + \epsilon_2) - (P_1 + \epsilon_1) = (P_2 - P_1) + (\epsilon_2 - \epsilon_1)$$

where P_1 and P_2 are the true directions and ϵ_1 and ϵ_2 are errors in those directions due to faulty instrument centering. Thus the error size for any setup is

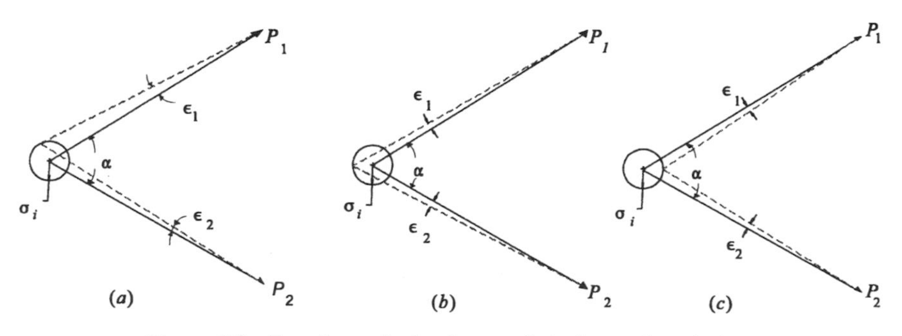

Figure 6.3 Error in angle due to error in instrument centering.

$$\epsilon = \epsilon_2 - \epsilon_1 \tag{6.11}$$

The analysis of the error in a measured angle due to instrument-centering errors is accomplished by propagating errors in a formula based on (x, y) coordinates. In Figure 6.4 a coordinate system has been constructed with the x axis going from the *true* station to the foresight station. The y axis passes through the instrument's vertical axis and is perpendicular to the x axis. From the figure the following equations can be derived:

$$ih = ip - qr = iq \cos(\alpha) - sq \sin(\alpha) \tag{6.12}$$

Letting $sq = x$ and $iq = y$, Equation (6.12) can be rewritten as:

$$ih = y \cos(\alpha) - x \sin(\alpha) \tag{6.13}$$

Furthermore, in Figure 6.4,

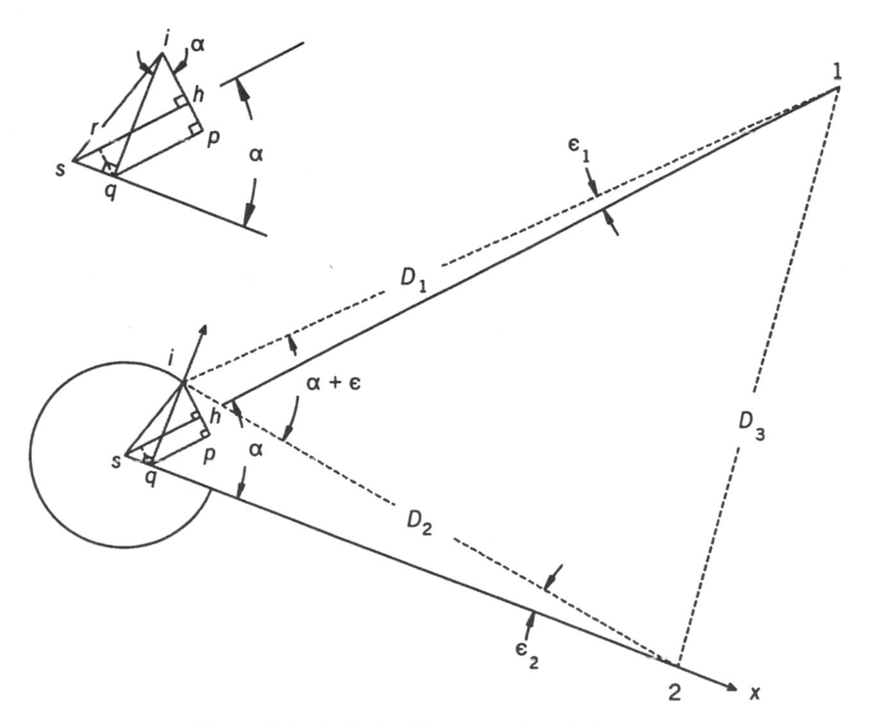

Figure 6.4 Analysis of instrument-centering error.

$$\epsilon_1 = \frac{ih}{D_1} = \frac{y\cos(\alpha) - x\sin(\alpha)}{D_1} \tag{6.14}$$

and

$$\epsilon_2 = \frac{y}{D_2} \tag{6.15}$$

By substituting Equations (6.14) and (6.15) into Equation (6.11), the estimated error in a measured angle due to instrument centering is

$$\epsilon = \frac{y}{D_2} - \frac{y\cos(\alpha) - x\sin(\alpha)}{D_1} \tag{6.16}$$

Reorganizing Equation (6.16) gives

$$\epsilon = \frac{D_1 y + D_2 x\sin(\alpha) - D_2 y\cos(\alpha)}{D_1 D_2} \tag{6.17}$$

Now because the instrument's position is truly random, Equation (5.16) can be used to find the angular uncertainty due to instrument centering. Taking the partial derivative of Equation (6.17) with respect to both x and y gives

$$\frac{\partial \epsilon}{\partial x} = \frac{D_2\sin(\alpha)}{D_1 D_2}$$

and

$$\frac{\partial \epsilon}{\partial y} = \frac{D_1 - D_2\cos(\alpha)}{D_1 D_2} \tag{6.18}$$

Substituting the partial derivatives of Equation (6.18) into Equation (5.16) yields

$$\sigma_\epsilon^2 = \left(\frac{D_2\sin(\alpha)}{D_1 D_2}\right)^2 \sigma_x^2 + \left[\frac{D_1 - D_2\cos(\alpha)}{D_1 D_2}\right]^2 \sigma_y^2 \tag{6.19}$$

Because this error is a constant for any setup, the mean angle has the same error as a single angle, and thus it is not reduced by making repetitions. The estimated error in the position of a station is derived from a *bivariate* distribution,[1] where the coordinate components are independent and have equal magnitudes. Assuming that estimated errors in the x and y axes are σ_x and σ_y, from Figure 6.5 it is seen that

[1] The bivariate distribution is discussed in Chapter 19.

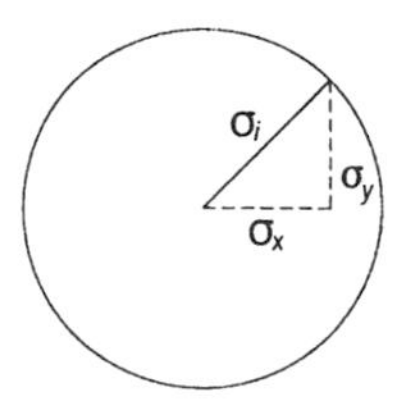

Figure 6.5 Centering errors at a station.

$$\sigma_x = \sigma_y = \frac{\sigma_i}{\sqrt{2}}$$

Letting $\sigma_\epsilon = \sigma_{\alpha_i}$, expanding the squares of Equation (6.19), and rearranging yields

$$\sigma_{\alpha_i}^2 = \frac{D_1^2 + D_2^2[\cos^2(\alpha) + \sin^2(\alpha)] - 2D_1D_2\cos(\alpha)}{D_1^2D_2^2}\frac{\sigma_i^2}{2} \tag{6.20}$$

In Equation (6.20), recognize that $\cos^2(\alpha) + \sin^2(\alpha) = 1$, and that $D_1^2 + D_2^2 - 2D_1D_2\cos(\alpha) = D_3^2$ (refer to Figure 6.4 for D_3). Making those substitutions, and taking the square root of both sides and multiplying by ρ to convert the results to arc seconds, Equation (6.20) is rewritten as

$$\sigma_{\alpha_i} = \pm\frac{D_3}{D_1D_2\sqrt{2}}\sigma_i\,\rho \tag{6.21}$$

Example 6.6 An observer centers the instrument to within ±0.005 ft of a station for an angle with backsight and foresight distances of 250 ft and 450 ft, respectively. The measured angle is 50°. What is the estimated contribution to the overall angular error due to the instrument-centering error?

SOLUTION: Using the cosine law, $D_3^2 = D_1^2 + D_2^2 - 2D_1D_2\cos(\measuredangle)$, and substituting in the appropriate values, we find D_3 to be

$$D_3 = \sqrt{250^2 + 450^2 - 2 \times 250 \times 450 \times \cos(50°)} = 346.95 \text{ ft}$$

Substituting this value into Equation (6.21), the estimated contribution of the instrument centering error to the overall angular error is

$$\sigma_{\alpha_i} = \pm\frac{346.95}{250 \times 450 \times \sqrt{2}}\,0.005 \times 206{,}264.8''/\text{rad} = \pm 2.2''$$

6.8 EFFECTS OF LEVELING ERROR IN ANGLE MEASUREMENT

If an instrument is imperfectly leveled, its vertical axis is not truly vertical, and its horizontal circle and horizontal axis are both inclined. If while an instrument is imperfectly leveled it is used to measure horizontal angles, the angles will be measured in some plane other than horizontal. Errors that result from this source are most severe when the backsights and foresights are steeply inclined, for example in making astronomical observations. If the bubble of a theodolite were to remain off center by the same amount during the entire angle-measuring process at a station, the resulting error would be systematic. However, because an operator normally carefully monitors the bubble and attempts to keep it centered while turning angles, the amount and direction by which the instrument is out of level become random, and hence the resulting errors tend to be random.

In Figure 6.6, ϵ represents the angular error that occurs in either the backsight or foresight of a horizontal angle measurement made with an imperfectly leveled instrument located at station *I*. The line of sight *IS* is elevated by vertical angle ν. In the figure, *IS* is shown perpendicular to the instrument's horizontal axis. The amount by which the instrument is out of level is $f_d\mu$, where f_d is the number of divisions the bubble is off center and μ is the bubble sensitivity. From the figure,

$$SP = D\tan(\nu) \tag{d}$$

and

$$PP' = D\epsilon \tag{e}$$

where *D* is the horizontal component of the sighting distance and the angular error ϵ is in radians. Because the amount of leveling error is small, *PP'* can be approximated as a circular arc, and thus

$$PP' = f_d\mu(SP) \tag{f}$$

Substituting Equation (*d*) into Equation (*f*) gives

$$PP' = f_d\mu D\tan(\nu) \tag{6.22}$$

Now substituting Equation(6.22) into Equation (*e*) and reducing, the error in an individual pointing due to an imperfectly leveled instrument is

$$\epsilon = f_d\mu\tan(\nu) \tag{6.23}$$

As noted above, Figure 6.6 shows the line of sight oriented perpendicular to the instrument's horizontal axis. Also, the direction a bubble runs is random. Thus Equation (5.16) can be used to estimate the combined angular error that results from *n* repetitions of an angle made with an imperfectly

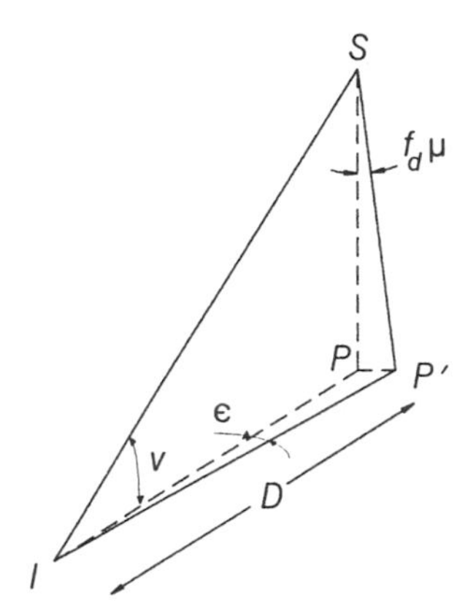

Figure 6.6 Effects of instrument leveling error.

leveled instrument (note that each angle measurement involves both a backsight and a foresight pointing):

$$\sigma_{\alpha_l} = \pm \frac{\sqrt{[f_d\mu \tan(v_b)]^2 + [f_d\mu \tan(v_f)]^2}}{\sqrt{n}} \tag{6.24}$$

where f_d is the fractional division the instrument is off level; v_b and v_f are the vertical angles to the backsight and foresight targets, respectively; and n is the number of repetitions of the angle.

Example 6.7 A horizontal angle is measured on a mountainside where the backsight is to the peak and the foresight is in the valley. The average zenith angle to the backsight and foresight are 80° and 95°, respectively. The instrument has a level bubble with a sensitivity of 30″/div and is leveled to within 0.3 div. For the average angle obtained from six repetitions, what is the estimated contribution of this leveling error to the overall angular error?

SOLUTION: The zenith angles converted to vertical angles are +10° and −5°, respectively.
Substituting the appropriate values into Equation (6.24) gives

$$\sigma_{\alpha_l} = \pm \frac{\sqrt{[0.3 \text{ div} \times 30''/\text{div} \times \tan(+10°)]^2 + [0.3 \text{ div} \times 30''/\text{div} \times \tan(-5)]^2}}{\sqrt{6}}$$

$$= \pm 0.7''$$

This error is generally small for traditional surveying work when normal care in leveling the instrument is taken. Thus it can generally be ignored for all but the most precise work. However, as noted earlier, for astronomical observations, this error can become quite large, due to the steeply inclined sights to celestial objects. Thus, for astronomical observations, it is extremely important to keep the instrument precisely leveled for each observation.

6.9 NUMERICAL EXAMPLE OF COMBINED ERROR PROPAGATION IN A SINGLE HORIZONTAL ANGLE

Example 6.8 Assume that an angle is measured four (2D and 2R) circle with a directional-type instrument. The observer has an estimated reading error of $\pm 1''$ and a pointing error of $\pm 1.5''$. The targets are well defined and placed on an optical plummet tribrach with an expected centering error of ± 0.003 ft. The instrument is in adjustment and centered over the station to within ± 0.003 ft. The horizontal distances from the instrument to the backsight and foresight targets are approximately 251 ft and 347 ft, respectively. The measured angle is $65°37'12''$. What is the standard deviation in the angle?

SOLUTION: The best way to solve this type of problem is to develop each item in Sections 6.3 to 6.8 individually and then apply Equation (5.18).

Error Due to Reading: Substituting the appropriate values into Equation (6.4) gives

$$\sigma_{\alpha_r} = \pm \frac{1''\sqrt{2}}{\sqrt{4}} = \pm 0.71''$$

Error Due to Pointing: Substituting the appropriate values into Equation (6.6) yields

$$\Sigma_{\alpha_p} = \pm \frac{1.5''\sqrt{2}}{\sqrt{4}} = \pm 1.06''$$

Error Due to Target Centering: Substituting the appropriate values into Equation (6.10) gives

$$\sigma_{\alpha_t} = \frac{\sqrt{(251)^2 + (347)^2}}{251 \times 347} (\pm 0.003) \times 206{,}264.8''/\text{rad} = \pm 3.04''$$

Error Due to Instrument Centering: From the cosine law we have

$$D_3^2 = (251)^2 + (347)^2 - 2(251)(347)\cos(65°37'12'')$$

$$D_3 = 334 \text{ ft}$$

Substituting the appropriate values into Equation (6.21) yields

$$\sigma_{\alpha_i} = \pm\frac{334}{251 \times 347 \times \sqrt{2}}\, 0.003 \times 206{,}264.8''/\text{rad} = \pm 1.68''$$

Combined Error: From Equation (5.18), the estimated angular error is

$$\sigma_\alpha = \sqrt{0.71^2 + 1.06^2 + 3.04^2 + 1.68^2} = \pm 3.7''$$

In Example 6.8, notice that the largest error sources are target and instrument centering, respectively. This is true even when the estimated centering error is only ±0.003 ft. Since these two error sources do not decrease with increased repetitions, there is a limit to what can be expected from the *typical* survey. And often surveyors place more confidence in their measurements than is warranted. For instance, assume that the targets were hand-held reflector poles with an estimated positional error of ±0.01 ft. Then the error due to target centering becomes ±10.1″. This results in a total estimated angular error of ±10.3″. If the 99% probable error were computed, a value as large as ±30″ could be expected!

6.10 USE OF ESTIMATED STANDARD ERRORS TO CHECK ANGULAR MISCLOSURE IN A TRAVERSE

When a traverse is geometrically closed, the angles are usually checked for misclosure. By computing the estimated standard errors for each angle in the traverse as described in Section 6.9 and summing these errors with Equation (5.18), an estimate for the size of the angular misclosure is obtained. The procedure is best demonstrated with an example.

Example 6.9 Assume that each of the angles in Figure 6.7 were measured using two repetitions (1D and 1R) and that their *estimated* standard errors were computed as shown in Table 6.1. Does this traverse meet acceptable angular closure at a 95% level?

SOLUTION: The *actual* angular misclosure of the traverse is 30″. The *estimated* angular misclosure of the traverse, at a 95% level of confidence, is found by applying Equation (5.18) with the estimated standard errors. That is, the estimated angular misclosure is

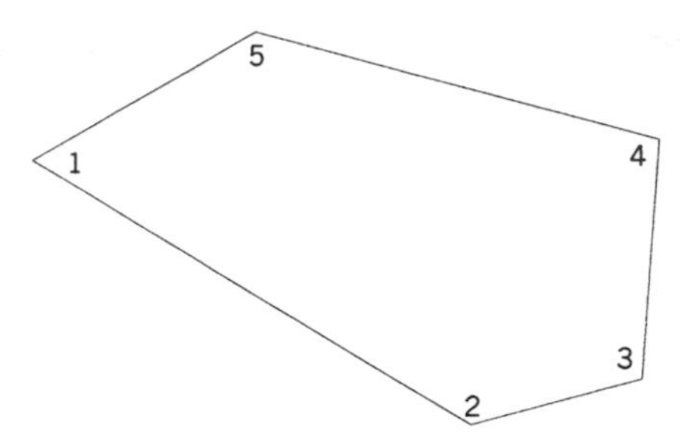

Figure 6.7 Closed polygon traverse.

$$\sigma_{\Sigma\measuredangle} = \sqrt{8.9^2 + 12.1^2 + 13.7^2 + 10.0^2 + 9.9^2} = \pm 24.7''$$

Thus the actual angular misclosure of 30″ is greater than the value estimated (24.7″) at a 68.3% probable error level. However, since each angle was turned only twice, a 95% probable error must be computed by using the appropriate t value (Table D.3). Since $t_{0.025,1} = 12.705$, the 95% probable error for the angular sum is found to be

$$\sigma_{95\%} = 12.705 \times 24.7 = \pm 314''$$

Thus the traverse angles meet the minimum level of angular closure at a 95%o confidence level.

6.11 ERRORS IN ASTRONOMICAL OBSERVATIONS FOR AZIMUTH

The total error in an azimuth determined from astronomical observations is dependent on errors from several sources, including those in timing, the ob-

Table 6.1 Data for Example 6.9

Angle	Observed Value	Estimated Standard Error
1	60°50′48″	± 8.9″
2	134°09′24″	±12.1″
3	109°00′12″	±13.7″
4	100°59′54″	±10.0″
5	135°00′12″	±9.9″

server's latitude and longitude, the celestial object's position at observation time, watch accuracy, observer response time, instrument optics, atmospheric conditions, and others as identified in Section 6.2. The resultant error in an astronomical azimuth measurement can be estimated by analyzing the hour-angle formula, which is

$$z = \tan^{-1}\left(\frac{\sin(t)}{\cos(\phi)\tan(\delta) - \sin(\phi)\cos(t)}\right) \tag{6.25}$$

where z is the azimuth of the celestial object at the time of the observation, t the angle at the pole in the *PZS* triangle at the time of observation, ϕ the observer's latitude, and δ the celestial object's declination at the time of the observation.

The t angle is a function of the *local hour angle* (LHA) of the sun or star at the time of observation. That is, when LHA <180°, t = LHA; otherwise, $t = 360° -$ LHA. Furthermore, LHA is a function of the *Greenwich hour angle* (GHA) of the celestial body, and the observer's longitude; that is,

$$\text{LHA} = \text{GHA} - \lambda \tag{6.26}$$

where λ is the observer's longitude, considered positive if west and negative if east. The GHA increases approximately 15° per hour of time, and thus an estimate of the error in the GHA is approximately

$$\sigma_t = 15 \times \sigma_T \text{ degrees}$$

where σ_T is the estimated error in time (in hours). Similarly, by using the declination at 0^h and 24^h, the amount of change in declination per second can be derived, and thus the estimated error in declination determined.

Using Equation (5.16), the error in a star's azimuth is estimated by taking the partial derivative of Equation (6.25) with respect to t, δ, ϕ, and λ. To do this, simplify Equation (6.25) by letting

$$F = \cos(\phi)\tan(\delta) - \sin(\phi)\cos(t), \text{ and} \tag{6.27a}$$

$$u = \sin(t) \times F^{-1} \tag{6.27b}$$

Substituting Equations (6.27), Equation (6.25) is rewritten as

$$z = \tan^{-1}\left(\frac{\sin(t)}{F}\right) = \tan^{-1}(u) \tag{6.28}$$

From calculus it is known that

$$\frac{d\tan^{-1}(u)}{dx} = \frac{1}{1+u^2}\frac{du}{dx}$$

Applying this fundamental relation to Equation (6.28) and letting G represent GHA gives

$$\frac{\partial z}{\partial G} = \frac{1}{1 + [\sin(G-\lambda)/F]^2}\frac{du}{dG} = \frac{F^2}{F^2 + \sin^2(G-\lambda)}\frac{du}{dG} \tag{6.29}$$

Now du/dG is

$$\frac{du}{dG} = \frac{\cos(G-\lambda)}{F} - \frac{\sin(G-\lambda)}{F^2}\sin(\phi)\sin(G-\lambda)$$

$$= \frac{\cos(G-\lambda)}{F} - \frac{\sin^2(G-\lambda)\sin(\phi)}{F^2}$$

and thus

$$\frac{du}{dG} = \frac{F\cos(G-\lambda) - \sin^2(G-\lambda)\sin(\phi)}{F^2} \tag{6.30}$$

Substituting Equation (6.30) into Equation (6.29) and substituting in t for $G-\lambda$ yields

$$\frac{\partial z}{\partial G} = \frac{F\cos(t) - \sin^2(t)\sin(\phi)}{F^2 + \sin^2(t)} \tag{6.31}$$

In a similar fashion, the following partial derivatives are developed from Equation (6.25):

$$\frac{\partial z}{\partial \delta} = -\frac{\sin(t)\cos(\phi)}{\cos^2(\delta)[F^2 + \sin^2(t)]} \tag{6.32}$$

$$\frac{\partial z}{\partial \phi} = \frac{\sin(t)\cos(t)\cos(\phi) + \sin(t)\sin(\phi)\tan(\delta)}{F^2 + \sin^2(t)} \tag{6.33}$$

$$\frac{\partial z}{\partial \lambda} = \frac{\sin^2(t)\sin(\phi) - F\cos(t)}{F^2 + \sin^2(t)} \tag{6.34}$$

where z is the celestial object's azimuth, t the angle at the pole in the *PZS* triangle, δ the celestial object's declination, ϕ the observer's latitude, λ the observer's longitude, and $F = \cos(\phi)\tan(\delta) - \sin(\phi)\cos(t)$.

If the horizontal angle (H∡) is the measured angle to the right from the line to the celestial body, the equation for a line's azimuth is

$$\text{Az} = z + 360° - \text{H}\measuredangle$$

Therefore, the error contributions from the horizonal angle measurement must be included in computing the overall azimuth error. Since the distance to the star is considered infinite, the estimated contribution to the angular error due to instrument centering can be determined with a formula similar to that for target centering with one pointing. That is,

$$\sigma_{\alpha_i} = \frac{\sigma_i}{D} \tag{6.35}$$

where σ_i is the amount of decentering of the instrument and D is the length of the azimuth line in the same units.

Example 6.10 Using Equation (6.25) for a single face left observation, the azimuth to Polaris was found to be 0°01′31.9″. The observation time was 1:00:00 UTC with an estimated standard error of $\sigma_T = \pm 0.5^s$. The Greenwich hour angles to the star at 0^h and 24^h UTC were 243°27′05.0″ and 244°25′50.0″, respectively. The LHA at the time of the observation was 181°27′40.4″. The declinations at 0^h and 24^h were 89°13′38.18″ and 89°13′38.16″, respectively. At the time of observation, the declination was 89°13′38.18″. The clockwise horizontal angle measured from the backsight to a target 450.00 ft away was 221°25′55.9″. The observer's latitude and longitude were scaled from a map as 40°13′54″ and 77°01′51.5″, respectively, with estimated standard errors of ±1″. The vertical angle to the star was 39°27′33.1″. The observer's estimated standard errors in reading and pointing are ±1″ and ±1.5″, respectively, and the instrument was leveled to within 0.3 of a division with a bubble sensitivity of 25″/div. The estimated error in instrument and target centering is ±0.003 ft. What is the line's azimuth and its estimated error? What is the error at the 95% confidence level?

SOLUTION: The line's azimuth is Az = 0°01′31.9″ + 360 − 221°25′55.9″ = 138°35′36″. Using the Greenwich hour angles at 0^h and 24^h, an error of 0.5^s time will result in an estimated error in the GHA of

$$\frac{360° + (244°25'50.0'' - 243°27'05.0'')}{24^h \times 60^m \times 60^s} \times 0.5^s = \pm 7.52''$$

F according to Equation (6.27*a*) is [note: $t = 360° - \text{LHA} = 178°32'19.6''$]

$$\begin{aligned} F &= \cos(40°13'54'')\tan(89°13'38.18'') \\ &\quad - \sin(40°13'54'')\cos(178°32'19.6'') \\ &= 57.249 \end{aligned}$$

Now the error in the measured azimuth can be estimated by computing the individual error terms as follows:

(a) From Equation (6.31) with respect to the GHA, G:

$$\frac{\partial z}{\partial G}\sigma_G = \frac{57.249\cos(178°32'19.6'') - \sin^2(178°32'19.6'')\sin(40°13'54'')}{(57.249)^2 + \sin^2(178°32'19.6'')}7.52''$$

$$= \pm 0.13''$$

(b) By observing the change in declination, it is obvious that for this observation, the error in a time of 0.5 second is insignificant. In fact, for the entire day, the declination changes only 0.02″. This situation is common for stars. However, the sun's declination may change from a few seconds daily to more than 23 minutes per day, and thus for solar observations, this error term should not be ignored.

(c) From Equation (6.33) with respect to latitude, ϕ:

$$\frac{\partial z}{\partial \phi}\sigma_\phi = \frac{\sin(t)\cos(t)\cos(\phi) + \sin(t)\sin(\phi)\tan(\delta)}{F^2 + \sin^2(t)}\sigma_\phi = \pm 1.32''$$

(d) From Equation (6.34) with respect to longitude, λ:

$$\frac{\partial z}{\partial \lambda}\sigma_\lambda = \frac{\sin^2(t)\sin(\phi) - 57.249\cos(t)}{(57.249)^2 + \sin^2(t)} \times 1''$$

$$= \pm 0.02''$$

(e) The circles are read both when pointing on the star and on the azimuth mark. Thus from Equation (6.2), the reading contribution to the estimated error in the azimuth is

$$\sigma_{\alpha_r} = \pm\sigma_r\sqrt{2} = \pm 1''\sqrt{2} = \pm 1.41''$$

(f) Using Equation (6.6), the estimated error in the azimuth due to pointing is

$$\sigma_{\alpha_p} = \pm\sigma_p\sqrt{2} = \pm 1.5''\sqrt{2} = \pm 2.12''$$

(g) From Equation (6.8), the estimated error in the azimuth due to target centering is

$$\sigma_{\alpha_t} = \pm\frac{d}{D} = \pm\frac{0.003}{450} \times 206{,}264.8'' = \pm 1.37''$$

(h) Using Equation (6.35), the estimated error in the azimuth due to instrument centering is

$$\sigma_{\alpha_i} = \pm\frac{d}{D} = \pm\frac{0.003}{450} \times 206{,}264.8'' = \pm 1.37''$$

(i) From Equation (6.23), the estimated error in the azimuth due to the leveling error is

$$\sigma_{\alpha_b} = \pm f_d \mu \tan(\nu) = \pm 0.3 \times 25'' \tan(39°27'33.1'') = \pm 6.17''$$

Parts (a) through (i) are the errors from each individual source. Now applying Equation (5.18), the total estimated error in the azimuth is

$$\begin{aligned}\sigma_{\text{Az}} &= [(0.13'')^2 + (1.32'')^2 + (0.02'')^2 + (1.41'')^2 \\ &\quad + (2.12'')^2 + 2 \times (1.37'')^2 + (6.17'')^2]^{1/2} \\ &= \pm 7.1''\end{aligned}$$

Using the t value of $t_{0.025,1}$ from Table D.3, the 95% probable error is

$$\sigma_{\text{Az}} = \pm 12.705 \times 7.1'' = \pm 90.2''$$

Notice that in this problem, the largest error source in the azimuth error is caused by instrument leveling.

6.12 ERRORS IN ELECTRONIC DISTANCE MEASUREMENTS

All EDM instrument measurements are subject to instrumental errors that manufacturers list as a constant and scalar error. A typical specified accuracy is $\pm(a + b \text{ ppm})$. In this expression, a is generally in the range 3 to 10 mm, and b is a scalar error which typically has the range 3 to 10 parts per million (ppm). Other errors involved in electronic distance measurements stem from target and instrument decentering. Since in any survey involving several stations these errors tend to be random, they should be combined using Equation (5.18). Thus, the estimated error in an EDM measured distance is

$$\sigma_D = \sqrt{\sigma_i^2 + \sigma_t^2 + a^2 + (D \times b \text{ ppm})^2} \tag{6.36}$$

where σ_D is the error in the measured distance D, σ_i the error in instrument setup centering, σ_t the error in reflector setup centering, and a and b are the instrument's specified accuracy parameters.

Example 6.11 A distance of 453.87 ft is measured using an EDM with a manufacturer's specified standard error of ±(5 mm + 10 ppm). The instrument is centered over the station with an estimated standard error of ±0.003 ft, and the reflector, which is mounted on a hand-held prism pole, is centered with an estimated standard error of ±0.01 ft. What is the total error in the measured distance? What is E_{90}?

SOLUTION: Converting millimeters to feet using the survey foot[2] definition gives us

$$0.005 \text{ m} \times 39.37 \text{ in.}/12 \text{ in.} = 0.0164 \text{ ft}$$

The scalar portion of the manufacturer's estimated standard error is computed as

$$\text{distance} \times k/1{,}000{,}000$$

In this example, the error is 453.87 × 10/1,000,000 = 0.0045 ft. Thus according to Equation (6.36), the total estimated distance error is

$$\sigma = \sqrt{(0.003)^2 + (0.01)^2 + (0.0164)^2 + (0.0045)^2} = \pm 0.02 \text{ ft}$$

Using the appropriate t value from Table 3.2, the 90% error is

$$E_{90} = 1.6449 \times \sigma = \pm 0.03 \text{ ft}$$

Notice in this example that the instrument's constant error is the largest single contributor to the total error, and it is followed closely by the target-centering error. Furthermore, since both errors are constants, their contribution to the total error is unchanged regardless of the distance. Thus, for this particular EDM instrument, distances under 200 ft would probably be measured more accurately with a calibrated steel tape.

6.13 USE OF SPREADSHEETS

The computations demonstrated in the chapter are rather tedious and time consuming when done by hand, and this often leads to mistakes. This type of problem, and others in surveying that involve repeated computations of a few equations with different values, can be done conveniently with a spreadsheet.

[2] The survey foot definition is 1 meter = 39.37 inches, exactly.

PROBLEMS

6.1 Plot a graph of vertical angles from 0° to 80°, versus the error in horizontal angle measurement due to a dislevelment in an instrument of 10″.

6.2 For an angle of size 135° with equal sight distances to the targets of 50, 100, 150, 200, 250, 300, 400, 600, 800, 1000, and 1500 ft, construct:

(a) a table of estimated standard deviations due to target centering when $\sigma_d = \pm 0.005$ ft.

(b) a plot of distance versus the standard deviations computed in part (a).

6.3 For an angle of size 135° with equal sight distances to the targets of 50, 100, 150, 200, 250, 300, 400, 600, 800, 1000, and 1500 ft, construct:

(a) a table of standard deviations due to instrument centering when $\sigma_i = \pm 0.005$ ft.

(b) a plot of distance versus the standard deviations computed in part (a).

6.4 Derive Equation (6.32).

6.5 Derive Equation (6.33).

6.6 Derive Equation (6.34).

6.7 For the following traverse data, compute the estimated standard deviation in each angle if $\sigma_r = \pm 1''$, $\sigma_p = \pm 1.8''$, $\sigma_i = \sigma_t = \pm 0.005$ ft, and the angles were each measured four times (twice direct and twice reverse) using the direction method. Does the traverse meet acceptable angular closures at a 95% confidence level?

Station	Angle	Distance (ft)
A	62°33′11″	
		221.85
B	124°56′19″	
		346.55
C	60°44′08″	
		260.66
D	111°46′07″	
		349.17
A		

6.8 A total station with a DIN18723 value of ±5″ was used to turn the angles in Problem 6.7. Repeat Problem 6.7 for this instrument.

6.9 The problem is the same as Problem 6.8 except a total station with a DIN18723 value of ±10″ was used to turn the angles in Problem 6.7.

6.10 For the following traverse data, compute the estimated error in each angle if $\sigma_r = \pm 18''$, $\sigma_p = \pm 5''$, $\sigma_i = \sigma_t = \pm 0.005$ ft, and the angles were measured four times (2 direct and 2 reverse) using the repetition method. Does the traverse meet acceptable angular closures at a 95% confidence level?

Station	Angle	Distance (ft)
A	38°58′24″	
		321.31
B	148°53′30″	
		276.57
C	84°28′06″	
		100.30
D	114°40′24″	
		306.83
E	152°59′18″	
		255.48
A		

6.11 A total station with a DIN18723 value of ±2″ was used to turn the angles in Problem 6.10. Repeat Problem 6.10 for this instrument.

6.12 The problem is the same as Problem 6.11 except a total station with a DIN18723 value of ±5″ was used to turn the angles in Problem 6.10.

6.13 An EDM instrument was used to measure the distances in Problem 6.7. The manufacturer's specified standard error for the instrument was ±(5 mm + 10 ppm). Using σ_i and σ_t from Problem 6.7, calculate the standard errors in each distance.

6.14 Similar to Problem 6.13, but for the distances in Problem 6.10. The manufacturer's specified standard error for the instrument was ±(3 mm + 5 ppm). Use σ_i and σ_t from Problem 6.10.

6.15 A surveyor notices that a crew has obtained perfect angular closure in measuring the interior angles of a 12-sided closed polygon traverse by the repetition method. Upon closer inspection, the first angle and average angle are found to be exactly the same for every angle. Should the surveyor accept or reject these results? What is the basis for your decision?

6.16 The following observations and data apply to a solar observation to determine the azimuth of a line.

Observation	UTC	Horizontal Angle	Vertical Angle	δ	LHA	z
1	16:30:00	41°02′33″	39°53′08″	−3°28′00.58″	339°54′05.5″	153°26′51.8″
2	16:35:00	42°35′28″	40°16′49″	−3°28′05.43″	341°09′06.5″	154°59′42.4″
3	16:40:00	44°09′23″	40°39′11″	−3°28′10.27″	342°24′07.5″	156°33′39.0″
4	16:45:00	45°44′25′	41°00′05″	−3°28′15.11″	343°39′08.5″	158°08′39.2″
5	16:50:00	47°20′21″	41°19′42″	−3°28′19.96″	344°54′09.5″	159°44′40.2″
6	16:55:00	48°57′24″	41°37′47″	−3°28′24.80″	346°09′10.5″	161°21′38.9″

The Greenwich hour angles for the day were 182°34′06.00″ at 0^h UT, and 182°38′53.30″ at 24^h UT. The declinations were −3°12′00.80″ at 0^h and −3°35′16.30″ at 24^h. The observer's latitude and longitude were scaled from a map as 43°15′22″ and 90°13′18″, respectively, with an estimated standard error of ±1″ for both values. Stopwatch times were assumed to be correct to within a standard error of $\pm 0.5^s$. A Roelof's prism was used to take pointings on the center of the sun. The target was 535 ft from the observer's station. The observer's estimated reading and pointing standard errors were ±1.2″ and ±1.8″, respectively. The instrument was leveled to within 0.3 div on a level bubble with a sensitivity of 20″/div. The target was centered to within an estimated standard error of ±0.003 ft of the station. What is the:

(a) average azimuth of the line and its standard deviation?

(b) estimated error of the azimuth at the 95% level of confidence?

(c) largest error contributor in the observation?

6.17 The following observations were made on the sun:

Pointing	Universal Time	Horizontal Angle	Zenith Angle
1	13:01:27	179°16′35″	56°00′01″
2	13:03:45	179°40′25″	55°34′11″
3	13:08:58	180°35′19″	54°35′36″
4	13:11:03	180°57′28″	54°12′12″
5	13:16:53	182°00′03″	53°06′47″
6	13:18:23	182°16′23″	52°50′05″

The Greenwich hour angles for the day were 178°22′55.20″ at 0^h and 178°22′58.70″ at 24^h. The declinations were 19°25′44.40″ at 0^h and 19°12′18.80″ at 24^h. The observer's latitude and longitude were scaled from a map as 41°18′06″ and 75°00′01″, respectively, with an estimated standard error of ±1″ for both. Stopwatch times were assumed to be correct to within an estimated standard error of $\pm 0.5^s$. A Roelof's prism was used to take pointings on the center of the sun. The target was 335 ft from the observer's station. The observer's estimated reading standard error was $\pm 1.1''$ and the estimated pointing standard error was $\pm 1.6''$. The instrument was leveled to within 0.3 div on a level bubble with a sensitivity of 30″/div. The target was centered to within an estimated standard error of ± 0.003 ft of the station. What is the:

(a) average azimuth of the line and its standard deviation?

(b) estimated error of the azimuth at the 95% confidence level ?

(c) largest error contributor in the observation?

Programming Problems

6.18 Create a program or spreadsheet that will compute the errors in angle observations. Use the spreadsheet to compute the standard deviations for the angles in Problem 6.7.

6.19 Create a program or spreadsheet that will compute the errors in distances measured with an EDM instrument. Use the spreadsheet to solve Problem 6.13.

6.20 Create a program or spreadsheet that will compute the standard deviations in astronomical observations. Use the spreadsheet to solve Problem 6.16.

7

ERROR PROPAGATION IN TRAVERSE SURVEYS

7.1 INTRODUCTION

Specifications for projects may allow varying levels of accuracies, but the presence of blunders in measurements is never acceptable. Thus an important question for every surveyor is: How can I tell when blunders are present in the data? In this chapter we begin to address that question, and in particular, we stress traverse analysis. The topic is discussed further in Chapter 19.

In Chapter 5 it was shown that the estimated error in a function of measurements is dependent on the individual errors in the measurements. Generally, measurements in horizontal surveys (e.g., traverses) are independent. That is, the distance measurement of a line is independent of its azimuth measurement. But the latitude and departure of each line, which are computed from the distance and azimuth, are not independent. Figure 7.1 shows the effects of distance and azimuth errors on the computed latitude and departure. In the figure it can be seen that there is correlation between the latitude and departure (i.e., if either distance or azimuth changes, it causes changes in both latitude and departure).

Because the measurements from which latitudes and departures are computed are assumed to be independent with no correlation, the SLOPOV approach, Equation (5.15), can be used to determine the estimated error in their computed values. However, for proper computation of the errors in functions that use these computed values i.e., traverse linear misclosure, the effects of correlation must be considered, and thus the GLOPOV approach, Equation (5.12), must be used.

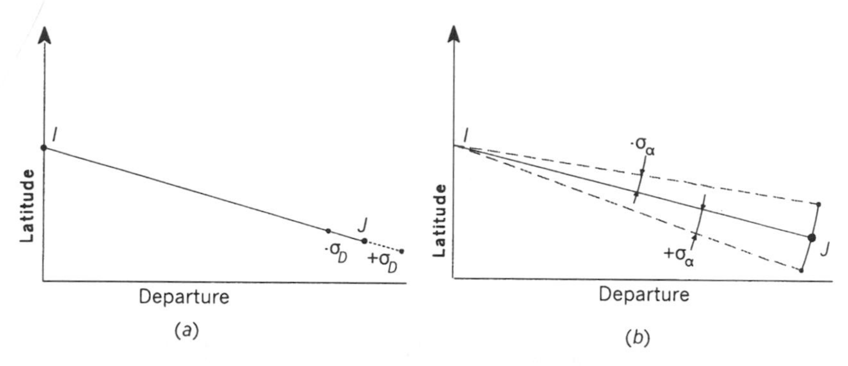

Figure 7.1 Latitude and departure uncertainties due to (*a*) the distance standard error (σ_D), and (*b*) the azimuth standard error (σ_α). Note that if either the distance or azimuth changes, both the latitude and departure are affected.

7.2 DERIVATION OF ESTIMATED ERROR IN LATITUDE AND DEPARTURE

When computing the latitude and departure of a line, the following well-known equations are used.

$$\begin{aligned} \text{Lat} &= D\cos(\text{Az}) \\ \text{Dep} &= D\sin(\text{Az}) \end{aligned} \tag{7.1}$$

where Lat is the latitude, Dep is the departure, Az is the azimuth, and D is the horizontal length of the line. To derive the estimated error in the line's latitude or departure, the following partial derivatives from Equation (7.1) are required in using Equation (5.15):

$$\begin{aligned} \frac{\partial \text{Lat}}{\partial D} &= \cos(\text{Az}) \qquad & \frac{\partial \text{Lat}}{\partial \text{Az}} &= -D\sin(\text{Az}) \\ \frac{\partial \text{Dep}}{\partial D} &= \sin(\text{Az}) \qquad & \frac{\partial \text{Dep}}{\partial \text{Az}} &= D\cos(\text{Az}) \end{aligned} \tag{7.2}$$

Example 7.1 A traverse course has a length of 456.87 ± 0.02 ft and an azimuth of 23°35′26″ ± 9″. What are the latitude and departure, and their estimated errors?

SOLUTION: This problem is solved using matrix equation (5.15) as

$$\Sigma_{\text{Lat,Dep}} = \begin{bmatrix} \dfrac{\partial \text{Lat}}{\partial D} & \dfrac{\partial \text{Lat}}{\partial \text{Az}} \\ \dfrac{\partial \text{Dep}}{\partial D} & \dfrac{\partial \text{Dep}}{\partial \text{Az}} \end{bmatrix} \begin{bmatrix} \sigma_D^2 & 0 \\ 0 & \sigma_{\text{Az}}^2 \end{bmatrix} \begin{bmatrix} \dfrac{\partial \text{Lat}}{\partial D} & \dfrac{\partial \text{Dep}}{\partial D} \\ \dfrac{\partial \text{Lat}}{\partial \text{Az}} & \dfrac{\partial \text{Dep}}{\partial \text{Az}} \end{bmatrix} = \begin{bmatrix} \sigma_{\text{Lat}}^2 & \sigma_{\text{Lat,Dep}} \\ \sigma_{\text{Lat,Dep}} & \sigma_{\text{Dep}}^2 \end{bmatrix}$$

Substituting partial derivatives into the above gives

$$\Sigma_{\text{Lat,Dep}} = \begin{bmatrix} \cos(\text{Az}) & -D\sin(\text{Az}) \\ \sin(\text{Az}) & D\cos(\text{Az}) \end{bmatrix} \begin{bmatrix} 0.02^2 & 0 \\ 0 & (9''/\rho)^2 \end{bmatrix} \times \begin{bmatrix} \cos(\text{Az}) & \sin(\text{Az}) \\ -D\sin(\text{Az}) & D\cos(\text{Az}) \end{bmatrix} \tag{7.3}$$

Entering the appropriate numerical values into Equation (7.3), the covariance matrix is computed to be

$$\Sigma_{\text{Lat,Dep}} = \begin{bmatrix} 0.9164 & -456.87\ (0.4002) \\ 0.4002 & 456.87\ (0.9164) \end{bmatrix} \begin{bmatrix} 0.0004 & 0 \\ 0 & \left(\dfrac{9}{206{,}264.8}\right)^2 \end{bmatrix} \times \begin{bmatrix} 0.9164 & 0.4002 \\ -456.87\ (0.4002) & 456.87\ (0.9164) \end{bmatrix}$$

from which

$$\Sigma_{\text{Lat,Dep}} = \begin{bmatrix} 0.00039958 & 0.00000096 \\ 0.00000096 & 0.00039781 \end{bmatrix} \tag{7.4}$$

In Equation (7.4), σ_{11}^2 is the variance of the latitude, σ_{22}^2 is the variance of the departure, and σ_{12} and σ_{21} are their covariances. Thus the standard deviations are $\sigma_{\text{Lat}} = \sqrt{\sigma_{11}^2} = \sqrt{0.00039958} = \pm 0.020$ ft and $\sigma_{\text{Dep}} = \sqrt{\sigma_{22}^2} = \sqrt{0.00039781} = \pm 0.020$ ft. Note that the off-diagonal of $\Sigma_{\text{Lat,Dep}}$ is not equal to zero, and thus the computed values are correlated as illustrated in Figure 7.1.

7.3 DERIVATION OF ESTIMATED STANDARD ERRORS IN COURSE AZIMUTHS

Equations (7.1) are based on the azimuth of a course. In practice, however, traverse azimuths are normally computed from measured angles rather than observed directly. Thus another level of error propagation exists in calculating the azimuths from angular values. In the following analysis, consider that interior angles are measured and that azimuths are computed in a *counter-clockwise* direction successively around the traverse using the formula

$$\mathrm{Az}_c = \mathrm{Az}_p + 180° + \theta_i \tag{7.5}$$

where Az_c is the azimuth for the current course, Az_p the previous course azimuth, and θ_i the appropriate interior angle to use in computing the current course azimuth. By applying Equation (5.17), the error in the current azimuth, σ_{Az_c}, can be estimated as

$$\sigma_{\mathrm{Az}_c} = \sqrt{\sigma^2_{\mathrm{Az}_p} + \sigma^2_{\theta_i}} \tag{7.6}$$

In Equation (7.6) σ_{θ_i} is the error in appropriate interior angle used to compute the current azimuth, and the other terms are as defined previously.

7.4 COMPUTING AND ANALYZING POLYGON TRAVERSE MISCLOSURE ERRORS

From elementary surveying it is known that the following geometric constraints exist for any closed polygon-type traverse:

$$\Sigma \text{ interior } \measuredangle\text{'s} = (n - 2) \times 180° \tag{7.7}$$

and

$$\Sigma \text{ Lats} = 0 \tag{7.8}$$

$$\Sigma \text{ Deps} = 0 \tag{7.9}$$

Deviations from these conditions, normally called *misclosures*, can be calculated for the measurements of any traverse. Statistical analyses can then be made of these misclosures to determine whether or not they are reasonable and acceptable, or whether they indicate the presence of blunders in the measurements. If blunders appear to be present, the measurements must be rejected and the observations repeated. The following example illustrates methods of making these computations for a closed polygon traverse.

Example 7.2 Compute the angular and linear misclosures for the traverse illustrated in Figure 7.2. The data for the traverse are given in Table 7.1. Distances are in feet. Determine the expected misclosures at the 95% confidence level, and comment on whether or not there is any indication of possible blunders.

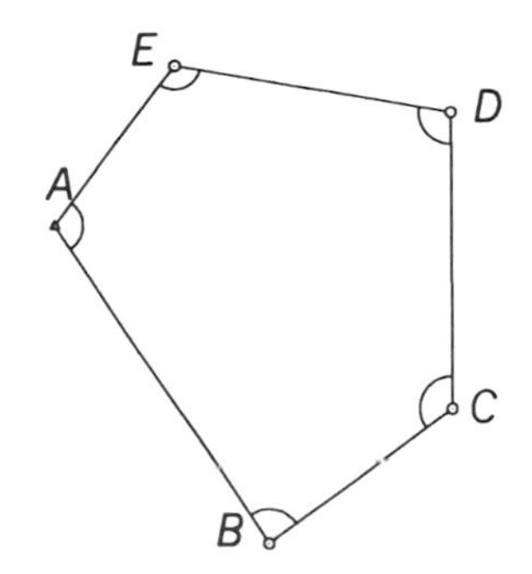

Figure 7.2 Closed polygon traverse.

SOLUTION

Angular Check: First the angular misclosure is checked to see if it is within the tolerances specified. From Equation (5.18), and using the standard deviations given in Table 7.1, the angular sum should have an error within $\pm\sqrt{\sigma^2_{\angle 1} + \sigma^2_{\angle 2} + \cdots + \sigma^2_{\angle n}}$ 68.3% of the time. Since the angles were measured four times, each computed mean has three degrees of freedom, and the appropriate t-value from Table D.3 (the t distribution) is $t_{0.025,3}$ which equals 3.183. Thus the angular misclosure at a 95% confidence level is estimated as

$$\sigma_{\Sigma\angle} = 3.183\sqrt{3.5^2 + 3.1^2 + 3.6^2 + 3.1^2 + 3.9^2} = \pm 24.6''$$

Using the summation of the angles in Table 7.1, the actual angular misclosure in this problem is

$$540°00'19'' - (5 - 2)180° = 19''$$

Thus the actual angular misclosure for the traverse (19″) is within its estimated range (±24.6″), and there is no reason to believe that any blunders exist in the angles at the 95% confidence level.

Azimuth Computation: In this problem, no azimuth is given for the first course. To solve the problem, however, the azimuth of the first course can be assumed as 0°00′00″ and to be free of error. This can be done even when the initial course azimuth is measured, since only geometric closure on the traverse is being checked, not the orientation of the traverse. For the data of Table 7.1, and using Equations (7.5) and (7.6), the values for the course azimuths and their estimated standard errors are computed and listed in Table 7.2.

Computation of Estimated Linear Misclosure: Equation (5.12) properly accounts for correlation in the latitude and departure when computing the linear misclosure of the traverse. Applying the partial derivatives of Equations (7.2) for the latitudes and departures, the Jacobian matrix, A, for this traverse has the form

Table 7.1 Distance and angle observations for Figure 7.2

Occupied	Sighted	Distance (ft)	S (ft)	Backsight	Occupied	Foresight	Angle[a]	S
A	B	1435.67	0.020	E	A	B	110°24′40″	3.5″
B	C	856.94	0.020	A	B	C	87°36′14″	3.1″
C	D	1125.66	0.020	B	C	D	125°47′27″	3.6″
D	E	1054.54	0.020	C	D	E	99°57′02″	3.1″
E	A	756.35	0.020	D	E	A	116°14′56″	3.9″
							Σ = 540°00′19″	

[a]Each angle was measured with four repetitions.

Table 7.2 Estimated errors in the computed azimuths of Figure 7.2

From:	To:	Azimuth	Estimated Error
A	B	0°00′00″	0″
B	C	267°36′14″	±3.1″
C	D	213°23′41″	$\sqrt{3.1^2 + 3.6^2} = \pm 4.8''$
D	E	133°20′43″	$\sqrt{4.8^2 + 3.1^2} = \pm 5.7''$
E	A	69°35′39″	$\sqrt{5.7^2 + 3.9^2} = \pm 6.9''$

$$A = \begin{bmatrix} \cos(\text{Az}_{AB}) & -D_{AB}\sin(\text{Az}_{AB}) & 0 & 0 & 0 & 0 \\ \sin(\text{Az}_{AB}) & D_{AB}\cos(\text{Az}_{AB}) & 0 & 0 & 0 & 0 \\ 0 & 0 & \cos(\text{Az}_{BC}) & -D_{BC}\sin(\text{Az}_{BC}) & 0 & 0 \\ 0 & 0 & \sin(\text{Az}_{BC}) & D_{BC}\cos(\text{Az}_{BC}) & 0 & 0 \\ \vdots & \vdots & \vdots & \vdots & \vdots & \vdots \\ 0 & 0 & 0 & 0 & \cos(\text{Az}_{EA}) & -D_{EA}\sin(\text{Az}_{EA}) \\ 0 & 0 & 0 & 0 & \sin(\text{Az}_{EA}) & D_{EA}\cos(\text{Az}_{EA}) \end{bmatrix} \quad (7.10)$$

Because the lengths and angles were measured independently, they are uncorrelated. Thus the appropriate covariance matrix, Σ, for solving this problem using Equation (5.15) is

$$\Sigma = \begin{bmatrix} \sigma^2_{D_{AB}} & 0 & 0 & 0 & 0 & 0 & 0 & 0 & 0 & 0 \\ 0 & \left(\frac{\sigma_{\text{Az}_{AB}}}{\rho}\right)^2 & 0 & 0 & 0 & 0 & 0 & 0 & 0 & 0 \\ 0 & 0 & \sigma^2_{D_{BC}} & 0 & 0 & 0 & 0 & 0 & 0 & 0 \\ 0 & 0 & 0 & \left(\frac{\sigma_{\text{Az}_{BC}}}{\rho}\right)^2 & 0 & 0 & 0 & 0 & 0 & 0 \\ 0 & 0 & 0 & 0 & \sigma^2_{D_{CD}} & 0 & 0 & 0 & 0 & 0 \\ 0 & 0 & 0 & 0 & 0 & \left(\frac{\sigma_{\text{Az}_{CD}}}{\rho}\right)^2 & 0 & 0 & 0 & 0 \\ 0 & 0 & 0 & 0 & 0 & 0 & \sigma^2_{D_{DE}} & 0 & 0 & 0 \\ 0 & 0 & 0 & 0 & 0 & 0 & 0 & \left(\frac{\sigma_{\text{Az}_{DE}}}{\rho}\right)^2 & 0 & 0 \\ 0 & 0 & 0 & 0 & 0 & 0 & 0 & 0 & \sigma^2_{D_{EA}} & 0 \\ 0 & 0 & 0 & 0 & 0 & 0 & 0 & 0 & 0 & \left(\frac{\sigma_{\text{Az}_{EA}}}{\rho}\right)^2 \end{bmatrix} \quad (7.11)$$

Substituting numerical values for this problem into Equations (7.10) and (7.11), the covariance matrix $\Sigma_{\text{lat,dep}}$ is computed for the latitudes and departures as $A\Sigma A^T$, or:

$$\times \Sigma_{\text{Lat,Dep}} =$$

$$\begin{bmatrix} 0.00040 & 0 & 0 & 0 & 0 & 0 & 0 & 0 & 0 & 0 \\ 0 & 0 & 0 & 0 & 0 & 0 & 0 & 0 & 0 & 0 \\ 0 & 0 & 0.00017 & 0.00002 & 0 & 0 & 0 & 0 & 0 & 0 \\ 0 & 0 & 0.00002 & 0.00040 & 0 & 0 & 0 & 0 & 0 & 0 \\ 0 & 0 & 0 & 0 & 0.00049 & 0.00050 & 0 & 0 & 0 & 0 \\ 0 & 0 & 0 & 0 & 0.00050 & 0.00060 & 0 & 0 & 0 & 0 \\ 0 & 0 & 0 & 0 & 0 & 0 & 0.00064 & -0.00062 & 0 & 0 \\ 0 & 0 & 0 & 0 & 0 & 0 & -0.00062 & 0.00061 & 0 & 0 \\ 0 & 0 & 0 & 0 & 0 & 0 & 0 & 0 & 0.00061 & 0.00034 \\ 0 & 0 & 0 & 0 & 0 & 0 & 0 & 0 & 0.00034 & 0.00043 \end{bmatrix} \tag{7.12}$$

By taking the square roots of the diagonal elements of the $\Sigma_{\text{Lat,Dep}}$ matrix [Equation (7.12)], the estimated errors for the latitude and departure of each course can be found. That is, the estimated error in the latitude for course BC is the square root of the "3,3" element in Equation (7.12), and the estimated error in the departure of BC is the square root of the "4,4" element. In a similar fashion, the estimated errors in latitude and departure can be computed for any other course.

The formula for determining the linear misclosure of a closed polygon traverse is:

$$LC = \sqrt{(\text{Lat}_{AB} + \text{Lat}_{BC} + \cdots + \text{Lat}_{EA})^2 + (\text{Dep}_{AB} + \text{Dep}_{Bc} + \cdots + \text{Dep}_{EA})^2} \tag{7.13}$$

where LC is the linear misclosure. To determine the estimated error in the linear misclosure, Equation (5.15) is applied to the linear misclosure formula, Equation (7.13). The necessary partial derivatives from Equation (7.13) needed for substitution into Equation (5.15) must first be determined. The partial derivatives with respect to the latitude and departure of course AB are

$$\frac{\partial LC}{\partial \text{Lat}_{AB}} = \frac{\sum \text{Lats}}{LC} \qquad \frac{\partial LC}{\partial \text{Dep}_{AB}} = \frac{\sum \text{Deps}}{LC} \tag{7.14}$$

Notice that these partial derivatives are independent of the course. Also, the other courses have the same partial derivatives as given by Equation (7.14), and thus the Jacobian matrix for Equation (5.15) has the form

$$A = \left[\frac{\sum \text{Lats}}{LC} \quad \frac{\sum \text{Deps}}{LC} \quad \frac{\sum \text{Lats}}{LC} \quad \frac{\sum \text{Deps}}{LC} \quad \cdots \quad \frac{\sum \text{Lats}}{LC} \quad \frac{\sum \text{Deps}}{LC} \right] \tag{7.15}$$

Table 7.3 Latitudes and departures for Example 7.2

Course	Latitude (ft)	Departure (ft)
AB	1435.67	0
BC	−35.827	−856.191
CD	−939.811	−619.567
DE	−723.829	766.894
EA	263.715	708.886
	Σ = −0.082	Σ = 0.022

$LC = \sqrt{(-0.082)^2 + (0.022)^2} = 0.085$ ft

As shown in Table 7.3, the sum of the latitudes is −0.082, the sum of the departures is 0.022, and $LC = 0.085$ ft. Substituting these values into Equation (7.15), which in turn is substituted into Equation (5.15), gives

$$\Sigma_{LC} = [-0.9647 \quad 0.2588 \quad -0.9647 \quad 0.2588 \quad \cdots \quad -0.9647 \quad 0.2588] \times \Sigma_{\text{Lat,Dep}} = \begin{bmatrix} -0.9647 \\ 0.2588 \\ -0.9647 \\ 0.2588 \\ -0.9657 \\ 0.2588 \\ -0.9647 \\ 0.2588 \\ -0.9647 \\ 0.2588 \end{bmatrix} = [0.00229] \tag{7.16}$$

In Equation (7.16) Σ_{LC} is a single-element covariance matrix which is the variance of the linear closure and can be called σ_{LC}^2. Also $\Sigma_{\text{Lat,Dep}}$ is the matrix given by Equation (7.12). To compute the E_{95} confidence level, a t value from Table D.3 (the t distribution) must be used, with $\alpha = 0.05$ and 3 degrees of freedom.[1] Thus the estimated error in the traverse closure at a 95% confidence level is

[1] A closed polygon traverse has $2(n - 1)$ unknown coordinates with $2n + 1$ measurements, where n is the number of traverse sides. Thus degrees of freedom is always $2n + 1 - 2(n - 1) = 3$. For a five-sided traverse there are five angle and five distance observations plus one azimuth. Also, there are four stations each having two unknown coordinates, so $11 - 8 = 3$ degrees of freedom.

$$\sigma_{LC} = t_{0.025,3}\sqrt{\sigma^2_{LC}} = 3.183\sqrt{0.00229} = \pm 0.15 \text{ ft}$$

Because this value is well above the actual traverse linear misclosure of 0.085 ft, there is no reason to believe that the traverse contains any blunders at the 95% confidence level.

7.5 COMPUTING AND ANALYZING LINK TRAVERSE MISCLOSURE ERRORS

As illustrated in Figure 7.3, a link traverse begins on one station and ends on a different one. Normally, they are used to establish the positions of intermediate stations, as in *A* through *D* of the figure. The coordinates of the endpoints, stations 1 and 2 of the figure, are known. Angular and linear misclosures are also computed for these types of traverse, and the resulting values used as the basis for accepting or rejecting the measurements. Example 7.3 illustrates the computational methods.

Example 7.3 Compute the angular and linear misclosures for the traverse illustrated in Figure 7.3. The measured data for the traverse are given in Table 7.4. Distances are in feet. Determine the expected misclosures at the 95% confidence level, and comment on whether or not the measurements are expected to contain any blunders.

SOLUTION

Angular Misclosure: In a link traverse, angular misclosure is found by computing initial azimuths for each course, and then subtracting the final computed azimuth from its given counterpart. The initial azimuths and their estimated errors are computed using Equations (7.5) and (7.6) and are shown in Table 7.5.

The difference between the azimuth computed for course *D*2 (84°19′22″) and its actual value (264°19′13″ − 180°) is +9″. Using Equation (5.18), the

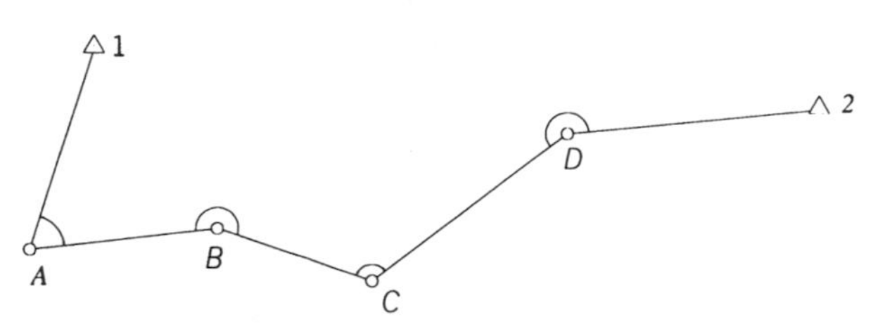

Figure 7.3 Link traverse.

Table 7.4 Link traverse for Example 7.3

Distance observations

From:	To:	Distance (ft)	*S* (ft)
1	*A*	1069.16	±0.021
A	*B*	933.26	±0.020
B	*C*	819.98	±0.020
C	*D*	1223.33	±0.021
D	2	1273.22	±0.021

Control stations

Station	*X* (ft)	*Y* (ft)
1	1248.00	3979.00
2	4873.00	3677.00

Angle observations

Backsight	Occupied	Foresight	Angle	*S*
1	*A*	*B*	66°16′35″	±4.9″
A	*B*	*C*	205°16′46″	±5.5″
B	*C*	*D*	123°40′19″	±5.1″
C	*D*	2	212°00′55″	±4.6″

Azimuth observations

From:	To:	Azimuth	*S*
1	*A*	197°04′47″	±4.3″
2	*D*	264°19′13″	±4.1″

estimated standard error in the difference is $\sqrt{11.0^2 + 4.1^2} = \pm 11.7''$, and thus there is no reason to assume that the angles contain blunders.

Linear Misclosure: First the actual traverse misclosure is computed. From Table 7.6, the total change in latitude for the traverse is −302.128 ft and the total change in departure is 3624.968 ft. From the control coordinates, the cumulative change in *X* and *Y* coordinate values is

$$\Delta X = X_2 - X_1 = 4873.00 - 1248.00 = 3625.00 \text{ ft}$$

$$\Delta Y = Y_2 - Y_1 = 3677.00 - 3979.00 = -302.00 \text{ ft}$$

The actual misclosures in departure and latitude are computed as

Table 7.5 Computed azimuths and their uncertainties

Course	Azimuth	σ
1*A*	197°04′47″	±4.3″
AB	83°21′22″	±6.5″
BC	108°38′08″	±8.5″
CD	52°18′27″	±9.9″
*D*2	84°19′22″	±11.0″

Table 7.6 Computed latitudes and departures

Course	Latitude (ft)	Departure (ft)
1*A*	−1022.007	−314.014
AB	107.976	926.993
BC	−262.022	776.989
CD	747.973	968.025
*D*2	125.952	1266.975
	Σ = −302.128	Σ = 3624.968

$$\begin{aligned} \Delta\text{Dep} &= \sum \text{Dep} - \Delta X = 3624.968 - 3625.00 = -0.032 \text{ ft} \\ \Delta\text{Lat} &= \sum \text{Lat} - \Delta Y = -302.128 - (-302.00) = -0.128 \text{ ft} \end{aligned} \tag{7.17}$$

where ΔDep represents the misclosure in departure and ΔLat represents the misclosure in latitude. Thus the linear misclosure for the traverse is

$$LC = \sqrt{\Delta\text{Dep}^2 + \Delta\text{Lat}^2} = \sqrt{(-0.032)^2 + (-0.128)^2} = 0.132 \text{ ft} \tag{7.18}$$

Expected Misclosure for the Traverse: Following procedures similar to those described earlier for polygon traverses, the expected misclosure in this link traverse is estimated. The Jacobian matrix of the partial derivatives for the latitude and departure with respect to the distance and azimuth measurements is

$$A = \begin{bmatrix} \cos(\text{Az}_{1A}) & -D_{1A}\sin(\text{Az}_{1A}) & 0 & 0 & 0 & 0 \\ \sin(\text{Az}_{1A}) & D_{1A}\cos(\text{Az}_{1A}) & 0 & 0 & 0 & 0 \\ 0 & 0 & \cos(\text{Az}_{AB}) & -D_{AB}\sin(\text{Az}_{AB}) & 0 & 0 \\ 0 & 0 & \sin(\text{Az}_{AB}) & D_{AB}\cos(\text{Az}_{AB}) & 0 & 0 \\ \vdots & \vdots & \vdots & \vdots & \vdots & \vdots \\ 0 & 0 & 0 & 0 & \cos(\text{Az}_{D2}) & -D_{D2}\sin(\text{Az}_{D2}) \\ 0 & 0 & 0 & 0 & \sin(\text{Az}_{D2}) & D_{D2}\cos(\text{Az}_{D2}) \end{bmatrix} \tag{7.19}$$

Similarly, the corresponding covariance matrix Σ in Equation (5.15) has the form

$$\Sigma = \begin{bmatrix} \sigma^2_{D_{1A}} & 0 & 0 & 0 & 0 & 0 & 0 & 0 & 0 & 0 \\ 0 & \left(\frac{\sigma_{Az_{1A}}}{\rho}\right)^2 & 0 & 0 & 0 & 0 & 0 & 0 & 0 & 0 \\ 0 & 0 & \sigma^2_{D_{AB}} & 0 & 0 & 0 & 0 & 0 & 0 & 0 \\ 0 & 0 & 0 & \left(\frac{\sigma_{Az_{AB}}}{\rho}\right)^2 & 0 & 0 & 0 & 0 & 0 & 0 \\ 0 & 0 & 0 & 0 & \sigma^2_{D_{BC}} & 0 & 0 & 0 & 0 & 0 \\ 0 & 0 & 0 & 0 & 0 & \left(\frac{\sigma_{Az_{BC}}}{\rho}\right)^2 & 0 & 0 & 0 & 0 \\ 0 & 0 & 0 & 0 & 0 & 0 & \sigma^2_{D_{CD}} & 0 & 0 & 0 \\ 0 & 0 & 0 & 0 & 0 & 0 & 0 & \left(\frac{\sigma_{Az_{CD}}}{\rho}\right)^2 & 0 & 0 \\ 0 & 0 & 0 & 0 & 0 & 0 & 0 & 0 & \sigma^2_{D_{D2}} & 0 \\ 0 & 0 & 0 & 0 & 0 & 0 & 0 & 0 & 0 & \left(\frac{\sigma_{Az_{D2}}}{\rho}\right)^2 \end{bmatrix} \tag{7.20}$$

Substituting the appropriate numerical values into Equations (7.19) and (7.20) and applying Equation (5.15), the covariance matrix, $\Sigma_{\text{Lat,Dep}}$, is

$$\Sigma_{\text{Lat,Dep}} = \begin{bmatrix} 0.000446 & 0.000263 & 0 & 0 & 0 & 0 & 0 & 0 & 0 & 0 \\ 0.000263 & 0.000492 & 0 & 0 & 0 & 0 & 0 & 0 & 0 & 0 \\ 0 & 0 & 0.000859 & 0.000145 & 0 & 0 & 0 & 0 & 0 & 0 \\ 0 & 0 & 0.000145 & 0.000406 & 0 & 0 & 0 & 0 & 0 & 0 \\ 0 & 0 & 0 & 0 & 0.00107 & -0.000467 & 0 & 0 & 0 & 0 \\ 0 & 0 & 0 & 0 & -0.000467 & 0.000476 & 0 & 0 & 0 & 0 \\ 0 & 0 & 0 & 0 & 0 & 0 & 0.002324 & 0.001881 & 0 & 0 \\ 0 & 0 & 0 & 0 & 0 & 0 & 0.001881 & 0.001565 & 0 & 0 \\ 0 & 0 & 0 & 0 & 0 & 0 & 0 & 0 & 0.004570 & 0.000497 \\ 0 & 0 & 0 & 0 & 0 & 0 & 0 & 0 & 0.000497 & 0.000482 \end{bmatrix}$$

To estimate the expected error in the traverse misclosure, Equation (5.15) must be applied to Equation (7.18). As was the case for closed polygon traverses, the terms of the Jacobian matrix are independent of the course for which they are determined, and thus the Jacobian matrix has the form

$$A = \begin{bmatrix} \frac{\Delta\text{Lat}}{LC} & \frac{\Delta\text{Dep}}{LC} & \frac{\Delta\text{Lat}}{LC} & \frac{\Delta\text{Dep}}{LC} & \cdots & \frac{\Delta\text{Lat}}{LC} & \frac{\Delta\text{Dep}}{LC} \end{bmatrix} \tag{7.21}$$

Following procedures similar to those used in Example 7.2, the expected standard error in the misclosure of this link traverse is

$$\Sigma_{LC} = A\,\Sigma_{\text{Lat,Dep}}\,A^T = [0.01000]$$

From this result and using a t value from Table D.3 for 3 degrees of freedom, the estimated linear misclosure error for a 95% confidence level is

$$\sigma_{95\%} = 3.183\sqrt{0.01000} = \pm 0.32 \text{ ft}$$

Because the actual misclosure of 0.13 ft is less than the misclosure expected at the 95% level (±0.24 ft), there is no reason to believe that the traverse measurements contain any blunders.

When using traditional methods of adjusting link traverse data, such as the compass rule, the control is assumed to be perfect. However, since control coordinates are themselves derived from measurements, they contain errors that are not accounted for in these computations. This fact is apparent in Equation (7.21) where the coordinate values are assumed to have no error and thus are not represented. These equations can easily be modified to consider the control errors, but this is left as an exercise for the student. One of the principal advantages of the least-squares adjustment method is that it enables the control to be included in the adjustment. A full discussion of this subject is presented in Chapter 18.

7.6 CONCLUSIONS

In this chapter propagation of measurement errors through traverse computations has been discussed. Error propagation is a powerful tool for the surveyor, enabling an answer to be obtained for the question: What is an acceptable traverse misclosure? This is an example of *surveying engineering*. Surveyors are constantly designing measurement systems and checking their results against personal or legal standards. The subjects of error propagation and detection of measurement blunders are discussed further in later chapters of the book.

PROBLEMS

7.1 Show that Equation (7.6) is valid for clockwise computations about a traverse.

7.2 Explain the significance of a standard deviation in the azimuth of the first course of a polygon traverse.

7.3 Given a course with an azimuth of 123°15′03″ ± 10″ and a distance of 986.59 ± 0.05, what are the latitude and departure, and their estimated errors?

7.4 A polygon traverse has the following angle measurements and related standard deviations. Each angle was measured two times (1D and 1R). Do the angles meet acceptable closure limits at a 95% confidence level?

Backsight	Occupied	Foresight	Angle	S
A	*B*	*C*	85°37′31″	±6″
B	*C*	*D*	52°46′40″	±4″
C	*D*	*A*	137°02′57″	±2″
D	*A*	*B*	84°32′37″	±9″

7.5 Given an initial azimuth for course *AB* of 116°27′50″, what are the azimuths and their estimated standard errors for the remaining three courses of Problem 7.4?

7.6 Using the distances listed in the following table and the data from Problems 7.4 and 7.5, compute the:

(a) actual linear misclosure of the traverse.

(b) estimated linear traverse misclosure error.

(c) 95% probable traverse misclosure error.

From:	To:	Distance (ft)	S (ft)
A	*B*	227.40	±0.022
B	*C*	377.87	±0.022
C	*D*	237.57	±0.022
D	*A*	220.04	±0.022

7.7 Given the traverse misclosures in Problem 7.6, does the traverse meet acceptable closure limits at a 95% confidence level? Justify your answer numerically.

7.8 Using the data for the link traverse listed below, compute the:

(a) actual angular misclosure and its estimated standard deviation.

(b) actual linear misclosure of the traverse.

(c) estimated linear traverse misclosure error.

(d) 95% probable error in the traverse misclosure.

Distance observations

From:	To:	Distance (ft)	σ (ft)
W	*X*	223.60	±0.02
X	*Y*	424.26	±0.02
Y	*Z*	403.11	±0.02

Angle observations

Backsight	Occupied	Foresight	Angle	σ
W	*X*	*Y*	161°33′58″	±13″
X	*Y*	*Z*	195°15′16″	±9″

Control azimuths

From:	To:	Azimuth	σ
W	*X*	63°26′13″	±11
Y	*Z*	60°15′18″	±6

Control stations

Station	Easting (ft)	Northing (ft)
W	1000.00	1000.00
Z	1850.00	1600.00

7.9 Does the link traverse of Problem 7.8 have acceptable linear traverse misclosure at a 95% level of confidence? Justify your answer numerically.

7.10 Following are the length and azimuth data for a city lot survey.

Course	Distance (ft)	σ_D (ft)	Azimuth	σ_{Az}
AB	134.58	0.02	83°59′54″	±0″
BC	156.14	0.02	353°59′44″	±20″
CD	134.54	0.02	263°59′54″	±28″
DA	156.10	0.02	174°00′04″	±35″

Compute the:

(a) actual linear misclosure of the traverse.

(b) estimated linear traverse misclosure error.

(c) 95% probable linear traverse misclosure error.

7.11 A survey produces the following set of data. The angles were obtained from the average of four measurements (two face left and two face right) made with a directional instrument having a level bubble sensitivity of $\mu = 30''/\text{div}$. The estimated standard errors in the observations are

$$\sigma_r = \pm 1.4'' \qquad \sigma_p = \pm 2.1'' \qquad \sigma_t = \pm 0.010 \text{ ft}$$

$$\sigma_i = \pm 0.005 \text{ ft} \qquad f_d = \pm 0.3 \text{ div}$$

The EDM instrument has a specified accuracy of ±(5 mm + 5 ppm).

Distance observations

From:	To:	Distance (ft)
1	2	999.99
2	3	801.55
3	4	1680.03
4	5	1264.92
5	1	1878.82

Angle observations

Backsight	Occupied	Foresight	Angle
5	1	2	191°40′08″
1	2	3	56°42′22″
2	3	4	122°57′10″
3	4	5	125°02′11″
4	5	1	43°38′10″

Control azimuths

From:	To:	Azimuth	σ
1	2	216°52′11″	±3

Control stations

Station	Easting (ft)	Northing (ft)
1	1000.00	1000.00

Compute the:

(a) estimated standard errors in angles and distances.

(b) actual angular misclosure and its 95% probable error.

(c) actual linear misclosure of the traverse.

(d) estimated linear traverse misclosure error.

(e) 95% probable linear traverse misclosure error.

7.12 Develop new matrices for the link traverse of Example 7.3 that considers the errors in the control. Assume control coordinate standard errors of $\sigma_x = \sigma_y = \pm 0.01$ ft for both stations 1 and 2, and use these new matrices to compute the:

(a) estimated linear traverse misclosure error, and

(b) 95% probable linear traverse misclosure error.

(c) Compare these results with those in Example 7.3.

Programming Problems

7.13 Develop a program or spreadsheet that will compute the course azimuths and their estimated errors given an initial azimuth and measured angles. Use the program or spreadsheet to solve Problem 7.4.

7.14 Develop a program or spreadsheet that will compute estimated linear traverse misclosure error given course azimuths and distances, and their estimated errors. Use the program or spreadsheet to solve Problem 7.10.

7.15 Develop a program or spreadsheet that will solve Problem 7.12, and use the program to solve that problem.

8

ERROR PROPAGATION IN ELEVATION DETERMINATION

8.1 INTRODUCTION

Differential and trigonometric leveling are the two most commonly employed methods for finding elevation differences between stations. Both of these methods are subject to systematic and random errors. The primary systematic errors include earth curvature, atmospheric refraction, and instrument maladjustment. The effects of these systematic errors can be minimized by following proper field procedures. They can also be modeled and corrected for computationally. The random errors in differential and trigonometric leveling occur in instrument leveling, in distance measurements, and in reading graduated scales. These must be treated according to the theory of random errors.

8.2 SYSTEMATIC ERRORS IN DIFFERENTIAL LEVELING

During the differential leveling process, sight distances are limited in length, and balanced, to minimize the effects of systematic errors. Still, it should always be assumed that these errors are present in leveling, and thus corrective measurement procedures should be followed to minimize their effects. These procedures are the subjects of discussions that follow.

8.2.1 Collimation Error

Collimation error occurs when the line of sight of an instrument is not truly horizontal, and is minimized by keeping sight distances balanced. Figure 8.1 shows the effects of a collimation error. For an individual setup, the resulting error in an elevation difference due to collimation error is

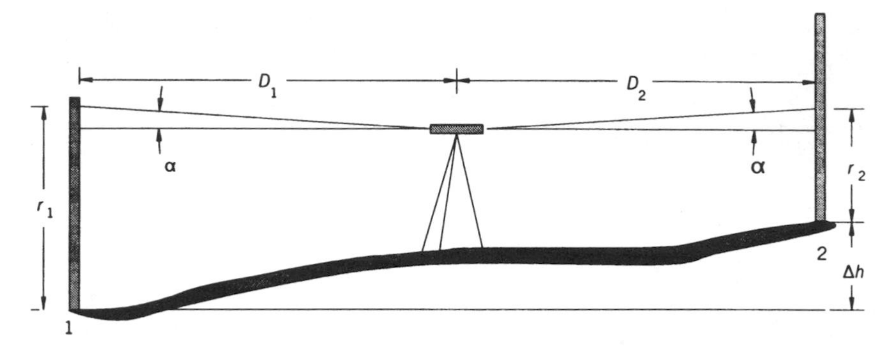

Figure 8.1 Collimation error in differential leveling.

$$e_C = D_1\alpha - D_2\alpha \tag{8.1}$$

where e_C is the error in elevation due to the presence of a collimation error, D_1 and D_2 the distances to the backsight and foresight rods, respectively, and α the amount of collimation error present at the time of the measurement, expressed in radian units. Applying Equation (8.1), the collimation error for a line of levels can be expressed as

$$e_C = \alpha[(D_1 - D_2) + (D_3 - D_4) + \cdots + (D_{n-1} - D_n)] \tag{8.2}$$

where $D_1, D_3, \ldots, D_{n-1}$ are the backsight distances and $D_2, D_4, \ldots, D_n$ are the foresight distances. If the backsight and foresight distances are grouped in Equation (8.2), then

$$e_C = \alpha(\Sigma\, D_{\text{BS}} - \Sigma\, D_{\text{FS}}) \tag{8.3}$$

Thus when the sums of backsights and foresights in a level line are not equal, the collimation error can be determined from Equation (8.3) and subtracted from the measured elevation difference in the level line to obtain the corrected value.

Example 8.1 A level having a collimation error of 0.04 mm/m is used on a level line where the backsight distances sum to 863 m and the foresight distances sum to 932 m. If the measured elevation difference for the line is 22.8654 m, what is the corrected elevation difference?

SOLUTION: Using Equation (8.3), the error due to collimation is

$$e_C = 0.00004(863 - 932)\text{ m} = -0.0028\text{ m},$$

and thus the corrected elevation difference is

$$22.8654 - (-0.0028) = 22.8682 \text{ m}$$

8.2.2 Earth Curvature and Refraction

As the line of sight extends from an instrument, the level surface curves down and away from it. This condition always causes rod readings to be too high. Similarly, as the line of sight extends from the instrument, refraction bends it toward the earth surface, causing readings to be too low. The combined effect of earth curvature and refraction on an individual sight always causes a rod reading to be too high by an amount that may be approximated as

$$h_{CR} = CR\left(\frac{D}{1000}\right)^2 \tag{8.4}$$

where h_{CR} is the error in the rod reading (in feet or meters), CR is 0.0675 when D is in units of meters or 0.0206 when D is in units of feet, and D is the individual sight distance.

The effect of this error on a single elevation difference is minimized by keeping backsight and foresight distances short and equal. For unequal sight distances, the resulting error is expressed by the formula

$$e_{CR} = CR\left(\frac{D_1}{1000}\right)^2 - CR\left(\frac{D_2}{1000}\right)^2 \tag{8.5}$$

where e_{CR} is the error due to earth curvature and refraction on a single elevation difference. Factoring common terms in Equation (8.5), gives

$$e_{CR} = \frac{CR}{1000^2}(D_1^2 - D_2^2) \tag{8.6}$$

Thus if the backsight and foresight at any setup are unequal, the curvature and refraction error can be computed from Equation (8.6) and subtracted from the measured elevation difference to get the corrected value.

Example 8.2 An elevation difference between two stations on a hillside is determined to be 1.256 m. What would be the error in the elevation difference and the corrected elevation difference if the backsight distance were 100 m and the foresight distance only 20 m?

SOLUTION: Substituting the distances into Equation (8.6) and using CR = 0.0675 gives us

$$e_{CR} = \frac{0.0675}{1000^2}(100^2 - 20^2) = 0.0006 \text{ m}$$

From this, the corrected elevation difference is

$$\Delta h = 1.256 - 0.0006 = 1.255 \text{ m}$$

For a line of differential leveling, the combined effect of this error is

$$e_{CR} = \frac{CR}{1000^2}(D_1^2 - D_2^2 + D_3^2 - D_4^2 + \cdots) \tag{8.7}$$

Regrouping backsight and foresight distances, Equation (8.7) becomes

$$e_{CR} = \frac{CR}{1000^2}\left(\sum D_{\text{BS}}^2 - \sum D_{\text{FS}}^2\right) \tag{8.8}$$

The error due to refraction caused by the vertical gradient of temperature can be large when sight lines are allowed to pass through the lower layers of the atmosphere. Since measuring the temperature gradient along the sight line would be cost prohibitive, a field procedure is generally adopted that requires all sight lines to be at least 0.5 m above the ground, and thus the lowest layers of the atmosphere which are difficult to model, are avoided.

8.2.3 Combined Effects of Systematic Errors on Elevation Differences

With reference to Figure 8.1, and by combining Equations (8.1) and (8.5), a corrected elevation difference, Δh, for one instrument setup is

$$\Delta h = (r_1 - r_2) - (D_1\alpha - D_2\alpha) - \frac{CR}{(1000)^2}(D_1^2 - D_2^2) \tag{8.9}$$

where r_1 is the backsight rod reading, r_2 is the foresight rod reading, and the other terms are as defined previously.

8.3 RANDOM ERRORS IN DIFFERENTIAL LEVELING

Differential leveling is subject to several sources of random errors. Included are errors in leveling the instrument and in reading the rod. The sizes of these errors are affected by atmospheric conditions, the quality of the optics of the telescope, the sensitivity of the level bubble or compensator, and the graduation scale on the rods. These errors are discussed below.

8.3.1 Reading Errors

The estimated error in rod readings can be expressed as a ratio of the estimated standard error in the rod reading per unit sight distance length. For example, if an observer's ability to read a rod is within ±0.005 ft per 100 ft, then $\sigma_{r/D}$ is $\pm 0.005/100 = \pm 0.00005$ ft/ft. Using this, rod reading errors for any individual sight distance D can be estimated as

$$\sigma_r = D\sigma_{r/D} \tag{8.10}$$

where $\sigma_{r/D}$ is the estimated error in the rod reading per unit length of sight distance and D is the length of the sight distance.

8.3.2 Instrument Leveling Errors

The estimated error in leveling for an automatic compensator or level vial is generally given in the technical data for each instrument. For precise levels, this information is usually listed in arc seconds or as an estimated elevation error for a given distance. As an example, the estimated error may be listed as ±1.5 mm/km, which corresponds to $\pm 1.5/1{,}000{,}000 \times \rho = \pm 0.3''$. A precise level will usually have a compensator accuracy or setting accuracy between ±0.1″ and ±0.2″, while for a less precise level the value may be as high as ±10″.

8.3.3 Rod Leveling Errors

While a level rod that is held nonvertical always causes the reading to be too high, this error will appear random in a leveling network due to its presence in all backsights and foresights of all lines in the network. This type of error can be modeled. With reference to Figure 8.2, the rod reading error due to imperfect rod leveling is approximately

$$e_{LS}r - r' = \frac{d^2}{2r} \tag{8.11}$$

where d is the linear amount that the rod is out of plumb at the location of the rod reading, r. The size of d is dependent on the rod level bubble centering error and the reading location. If the rod bubble is out of level by β, then d is

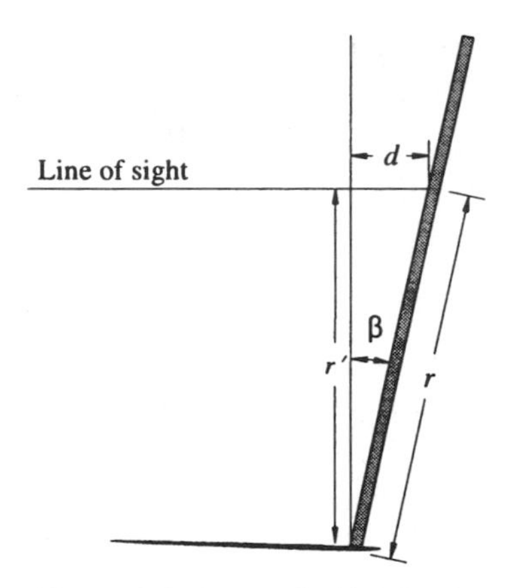

Figure 8.2 Nonvertical level rod.

$$d = r \sin(\beta) \tag{8.12}$$

Substituting Equation (8.12) into (8.11) gives

$$e_{LS} = \frac{r}{2} \sin^2(\beta) \tag{8.13}$$

Example 8.3 Assume that a rod level bubble is within $\pm 1'$ of level and the rod reading is at 3 m. The estimated error in the rod reading is

$$e_{LS} = \frac{3}{2} \sin^2(\pm 1') \times 1000 \text{ mm/m} = 0.0001 \text{ mm}$$

Since this type of error occurs on every sighting, backsight errors will tend to cancel foresight errors. Thus with precise leveling techniques, the combined effect of this error can be written as

$$e = \left[\frac{r_1 \sin^2(\beta)}{2} - \frac{r_2 \sin^2(\beta)}{2}\right] + \left[\frac{r_3 \sin^2(\beta)}{2} - \frac{r_4 \sin^2(\beta)}{2}\right] + \cdots \tag{8.14}$$

Grouping like terms in Equation (8.14) gives

$$e = \frac{1}{2} \sin^2(\beta)\, (r_1 - r_2 + r_3 - r_4 + \cdots) \tag{8.15}$$

Recognizing that the quantity in parentheses in Equation (8.15) is the elevation difference for the leveling line yields

$$e_{LS} = \frac{\Delta\text{Elev}}{2} \sin^2(\beta) \tag{8.16}$$

Example 8.4 If a level rod is maintained to within $\pm 1'$ of level and the elevation difference is 22.8654 m, the estimated error in the final elevation will be

$$e_{LS} = \frac{22.8654}{2} \sin^2(\pm 1') \times 1000 \text{ mm/m} = 0.001 \text{ mm}$$

This error can be practically eliminated by waving the rod or by carefully centering the bubble of a well-adjusted rod level. When these precautions are followed, the error is generally small, and thus will be ignored in subsequent computations.

8.3.4 Estimated Errors in Differential Leveling

The major random error sources in differential leveling are errors in rod readings and errors in instrument leveling. Note that the collimation error, and curvature and refraction errors are considered to be systematic, and are effectively negated by balancing the backsight and foresight distances. However, no matter what method is used to measure the lengths of the sight distances, some random error in these lengths will be present. This will cause random errors in the elevation differences due to the effects of earth curvature, refraction, and instrumental collimation errors. Thus Equation (5.16) can be applied to Equation (8.9) to model the effects of the random errors in rod readings, leveling, and sighting lengths. The following partial derivatives are needed:

$$\begin{aligned}
&\frac{\partial \Delta h}{\partial r_1} = \frac{\partial \Delta h}{\partial r_2} = 1 \qquad \frac{\partial \Delta h}{\partial \alpha_1} = -D_1 \qquad \frac{\partial \Delta h}{\partial \alpha_2} = D_2 \\
&\frac{\partial \Delta h}{\partial D_1} = -\left(\alpha + \frac{CR(D_1)}{500{,}000}\right) \qquad \frac{\partial \Delta h}{\partial D_2} = \alpha + \frac{CR(D_2)}{500{,}000}
\end{aligned} \tag{8.17}$$

By substituting Equations (8.17) into (5.16) with their corresponding estimated standard errors, the standard error in a single elevation difference can be estimated as

$$\begin{aligned}
\sigma_{\Delta h} = \Bigg\{ &(D_1\sigma_{r/D})^2 + (-D_2\sigma_{r/D})^2 + (-D_1\sigma_{\alpha 1})^2 + (D_2\sigma_{\alpha 2})^2 \\
&+ \left[-\left(\alpha + \frac{CR(D_1)}{500{,}000}\right)\sigma_{D_1}\right]^2 + \left[\left(\alpha + \frac{CR(D_2)}{500{,}000}\right)\sigma_{D_2}\right]^2 \Bigg\}^{1/2}
\end{aligned} \tag{8.18}$$

where $\sigma_{r/D}$ is the estimated error in a rod reading, $\sigma_{\alpha 1}$ and $\sigma_{\alpha 2}$ the estimated collimation errors in the backsight and foresight, respectively, and σ_{D_1} and σ_{D_2} the estimated errors in the sight lengths D_1, and D_2, respectively.

In normal differential leveling procedures, $D_1 = D_2 = D$. Also, if the estimated standard errors in the distances are equal, $\sigma_{D_1} = \sigma_{D_2} = \sigma_D$. Furthermore, if the estimated collimation error for the backsight and foresight are assumed equal, $\sigma_{\alpha 1} = \sigma_{\alpha 2} = \sigma_\alpha$. Thus Equation (8.18) simplifies to

$$\sigma_{\Delta h} = \sqrt{2D^2(\sigma_{r/D}^2 + 2\sigma_\alpha^2) + 2\sigma_D^2\left(\alpha + \frac{CR(D)}{500{,}000}\right)^2} \tag{8.19}$$

Equation (8.19) is appropriate for a single elevation difference when the sight distances are approximately equal. In general, if sight distances are kept equal for N instrument setups, the total estimated error in an elevation difference is

$$\sigma_{\Delta h} = \sqrt{2ND^2(\sigma_{r/D}^2 + \sigma_\alpha^2) + 2N\sigma_D^2\left(\alpha + \frac{CR(D)}{500{,}000}\right)^2} \tag{8.20}$$

In Equation (8.20) the estimated error in the elevation difference due to earth curvature and refraction, and collimation are small, and thus the last term can be ignored. Therefore the final equation for the estimated standard error in differential leveling is

$$\sigma_{\Delta h} = D\sqrt{2N(\sigma_{r/D}^2 + \sigma_\alpha^2)} \tag{8.21}$$

Example 8.5 A level line is run from benchmark A to benchmark B. The estimated standard error in rod readings is ± 0.01 mm/m. The instrument is maintained to within $\pm 2.0''$ of level. A collimation test shows that the instrument is within 4 mm per 100 m. Fifty-meter sight distances are maintained within an uncertainty of ± 2 m. The total line length from A to B is 1000 m. What is the estimated error in the elevation difference between A and B? If A had an elevation of 212.345 ± 0.005 m, what is the estimated error in the computed elevation of B?

SOLUTION: The total number of setups in this problem is $1000/(2 \times 50) = 10$ setups. Substituting the appropriate values into Equation (8.20) gives

$$\sigma_{\Delta h} = \left\{2 \times 10 \times 50^2\left[\left(\frac{0.01}{1000}\right)^2 + \left(\frac{2.0''}{\rho}\right)^2\right] + 2 \times 10 \times 2^2\left(\frac{0.004}{100} + \frac{0.0675 \times 50}{500{,}000}\right)^2\right\}^{1/2}$$

$$= \sqrt{0.0031^2 + 0.0004^2} = \pm 0.0031 \text{ m} = \pm 3.1 \text{ mm}$$

From an analysis of the individual error components in the equation above, it is seen that the error caused by the errors in the sight distances is negligible for all but the most precise leveling. Thus this error can be ignored in all but the most precise work. Therefore the simpler Equation (8.21) can be used to solve the problem:

$$\sigma_{\Delta h} = 50\sqrt{2 \times 10\left[\left(\frac{0.01}{1000}\right)^2 + \left(\frac{2.0''}{\rho}\right)^2\right]}$$
$$= \pm 0.0031 \text{ m} = \pm 3.1 \text{ mm}$$

Note that the same answer (± 3.1 mm) is obtained using either Equation (8.20) or (8.21). The estimated error in the elevation of B is found by applying Equation (5.18) as

$$\sigma_{\text{Elev}_\text{B}} = \pm\sqrt{\sigma^2_{\text{Elev}_\text{A}} + \sigma^2_{\Delta\text{Elev}}}$$
$$= \sqrt{5^2 + 3.1^2} = \pm 5.9 \text{ mm}$$

8.4 ERROR PROPAGATION IN TRIGONOMETRIC LEVELING

With the introduction of total station instruments, it is becoming increasingly convenient to measure elevation differences using trigonometric methods. However in this procedure, because sight distances cannot be balanced, it is important that the systematic effects of earth curvature and refraction, and inclination in the instrument's line of sight (collimation error) be removed. From Figure 8.3, the corrected elevation difference, Δh, between two points is

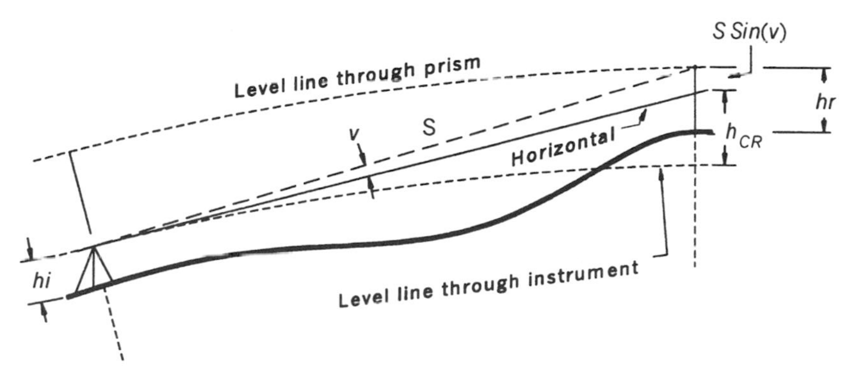

Figure 8.3 Determination of elevation difference by trigonometric leveling.

$$\Delta h = hi + S \sin(v) + h_{CR} - hr \tag{8.22}$$

or for zenith angle reading instruments is

$$\Delta h = hi + S \cos(z) + h_{CR} - hr \tag{8.23}$$

where hi is the instrument height above ground, S the slope distance between the two points, v the vertical angle between the instrument and the prism, z the zenith angle between the instrument and the prism, h_{CR} the earth curvature and refraction correction given in Equation (8.4), and hr the rod reading. Substituting the curvature and refraction formula into Equation (8.23) gives

$$\Delta h = hi + S \cos(z) + CR\left(\frac{S \sin(z)}{1000}\right)^2 - hr \tag{8.24}$$

In developing an error propagation formula for Equation (8.24), not only must errors relating to the height of instrument and prism be considered, but errors in leveling, pointing, reading, and slope distances as discussed in Chapter 6 must also be included. Applying Equation (5.16) to (8.24), the following partial derivatives apply:

$$\frac{\partial \Delta h}{\partial hi} = 1$$

$$\frac{\partial \Delta h}{\partial h_r} = -1$$

$$\frac{\partial \Delta h}{\partial S} = \left(\cos(z) + \frac{CR(S) \sin^2(z)}{500{,}000}\right)$$

$$\frac{\partial \Delta h}{\partial z} = \left(\frac{CR(S^2) \sin(z) \cos(z)}{500{,}000} - S \sin(z)\right)$$

Entering these partial derivatives, and the standard errors of the observations into Equation (5.16), the total error in trigonometric leveling is

$$\sigma_{\Delta h} = \left\{\sigma_{hi}^2 + \sigma_{hr}^2 + \left[\left(\cos(z) + \frac{CR(S) \sin^2(z)}{500{,}000}\right) \sigma_S\right]^2 + \left[\left(\frac{CR(S^2) \sin(z) \cos(z)}{500{,}000} - S \sin(z)\right) \frac{\sigma_z}{\rho}\right]^2\right\}^{1/2} \tag{8.25}$$

where z is the zenith angle, CR is 0.0675 if units of meters are used or 0.0206 if units of feet are used, S is the slope distance, and ρ is the seconds to radian conversion factor of 206,264.8″/rad.

In Equation (8.25), errors from several sources make up the estimated error in the zenith angle. These include the operator's ability to point and read the instrument, the accuracy of the vertical compensator or the operator's ability to center the vertical circle bubble, and the sensitivity of the compensator or vertical circle bubble. For best results, zenith angles should be measured both face left and face right and the average taken. Using Equations (6.4) and (6.6), the estimated error in a zenith angle that is measured in both positions (face left and face right) with a theodolite is

$$\sigma_z = \frac{\sqrt{2\sigma_r^2 + 2\sigma_p^2 + 2\sigma_B^2}}{\sqrt{N}} \tag{8.26a}$$

where σ_r is the error in reading the circle, σ_p is the error in pointing, σ_B is the error in the vertical compensator or in leveling the vertical circle control bubble, and N the number of face-left and face-right measurements of the zenith angle. For digital theodolites or total stations the appropriate formula is

$$\sigma_z = \frac{\sqrt{4\sigma_{\text{DIN}}^2 + 2\sigma_B^2}}{\sqrt{N}} \tag{8.26b}$$

where σ_{DIN} is the DIN18723 value for the instrument and all other values are as above.

Notice that if only a single zenith angle observation is made (i.e., it is measured only in face left), its estimated error is simply

$$\sigma_z = \sqrt{\sigma_r^2 + \sigma_p^2 + \sigma_B^2} \tag{8.27a}$$

For the digital theodolite and total station the estimated error for a single observation is

$$\sigma_z = \sqrt{2\sigma_{\text{DIN}}^2 + \sigma_B^2} \tag{8.27b}$$

Similarly, the estimated error in the slope distance, S, is computed using Equation (6.36).

Example 8.6 A total station instrument has a vertical compensator accurate to within ±0.3″, a digital reading accuracy of ±5″, and a distance measuring accuracy of ±(5 mm + 5 ppm). The measured slope distance was 1256.78 ft. The instrument was set to within ±0.005 ft of the station, and the target was set to within ±0.01 ft. The instrument's height was 5.12 ± 0.01 ft, and the prism height was 6.78 ± 0.01 ft. The zenith angle was measured in only one position and was recorded as 88°13′15″. What are the corrected elevation difference and its estimated error?

SOLUTION: Using Equation (8.24), the corrected elevation difference is

$$\Delta h = 5.12 + 1256.78 \cos(88°13'15'') + 0.0206 \left[\frac{1256.78 \sin(88°13'15'')}{1000}\right]^2 - 6.78 = 37.39 \text{ ft}$$

With Equation (8.27*b*), the zenith angle error is estimated as

$$\sigma_z = \sqrt{2 \times 5^2 + 0.3^2} = \pm 7.1''$$

From Equation (6.36) and converting 5 mm to 0.0164 ft, the estimated error in the distance is

$$\sigma_S = \sqrt{0.005^2 + 0.01^2 + 0.0164^2 + \left(\frac{5}{1{,}000{,}000}\,1256.78\right)^2} = \pm 0.021 \text{ ft}$$

Now substituting numerical values into Equation (8.25), the estimated error in the elevation difference is

$$\sigma_{\Delta h} = \sqrt{0.01^2 + 0.01^2 + (0.031 \times 0.021)^2 + \left(\frac{-1256.172 \times 7.1''}{\rho}\right)^2}$$

$$= \sqrt{0.01^2 + 0.01^2 + 0.00065^2 + .043^2} = \pm 0.045 \text{ ft}$$

Note in this example, that the estimated error in the elevation difference caused by the distance error is negligible (±0.00065 ft), whereas the error in the zenith angle is the largest (±0.043 ft). Furthermore, since the vertical angle was not measured in both face-left and face-right positions, it is possible that an uncompensated systematic error could be present in the final computed value. For example, assume that a 10″ indexing error existed on the vertical circle. If the measurements had been taken in both the face-left and face-right positions, the effects of this error would be negated. However, by making only a face-left measurement, the systematic error due to the vertical indexing error would be

$$1256.78 \sin(10'') = 0.061 \text{ ft}$$

Thus this uncompensated systematic error in the final value would be considerably larger than the estimated error in the measurement. It is thus extremely important to be aware of the causes of systematic errors, and know how to correct for them. Failure to do so can only lead to poor results. In trigonometric leveling, a minimum of one face-left and one face-right reading should always be taken.

PROBLEMS

8.1 A line of three wire differential levels goes from benchmark Gloria to benchmark Carey. The length of the line was determined to be 1487 m. The instrument had a stated compensator accuracy of ± 1.5 mm/km. The instrument–rod combination had an estimated reading error of ± 0.2 mm per 40 m. Sight distances were kept to approximately 50 $\pm$ 5 m. If the observed difference in elevation is 23.4516 m, what is the estimated error in the final elevation difference?

8.2 If in Problem 8.1, benchmark Gloria had a fixed elevation of 431.071 m and Carey had a fixed elevation of 454.520 m, did the job meet acceptable closure limits at a 95% confidence level? Justify your answer statistically.

8.3 An elevation must be established on a bench mark on an island that is 2345 ft from the nearest benchmark on the lake's shore. The surveyor decided to use a total station instrument that has a stated distance measuring accuracy of $\pm$(5mm + 5ppm) and a vertical compensator accurate to within $\pm 5''$. The height of instrument was 5.1 ft with an estimated error of ± 0.05 ft. The prism height was 6.9 ft with an estimated error of ± 0.05 ft. The single zenith angle is read as $88°56'13''$. The estimated errors in instrument and target centering are ± 0.005 ft. If the elevation of the occupied benchmark is 853.09 ft, what is the corrected benchmark elevation on the island? (Assume that the instrument does not correct for earth curvature and refraction.)

8.4 In Problem 8.3, what is the estimated error in the computed benchmark elevation if the observer has a pointing error of $\pm 9''$ and reading error of $\pm 7.5''$?

8.5 After completing the job in Problem 8.3, the surveyor discovered that the instrument had a vertical indexing error that caused the sight line to be inclined by 5 minutes.

(a) How much error would be created in the elevation of the island benchmark if the indexing error were ignored?

(b) What is the corrected benchmark elevation?

8.6 A tilting level was used to run precise levels to a construction project benchmark from benchmark DAM, which had an elevation of

345.901 m. The line was run along a road that goes up a steep incline. To expedite the job, backsight distances were kept to 50 m while foresight distances were 25 m. The total length of differential levels was 5248 m. The measured elevation difference was 21.051 m. What is the corrected project benchmark's elevation if the instrument has a sight line that declines at the rate of 0.5 mm per 50 m?

8.7 In Problem 8.6, the instrument's level bubble was centered to within $\pm 3''$ for each sight and the rod could be read to ± 1 mm per 50 m.

(a) What is the estimated error in the elevation difference?

(b) If a project engineer requests the estimated error in the final established benchmark, what would be your answer?

8.8 Which method of leveling presented in this chapter offers the most precision? Defend your answer statistically.

PROGRAMMING PROBLEMS

8.9 Create a program or spreadsheet that will compute a corrected elevation difference and its estimated error using the method of differential leveling.

8.10 Do Problem 8.9 for the method of trigonometric leveling.

9

WEIGHTS OF OBSERVATIONS

9.1 INTRODUCTION

When surveying data are collected, they usually must conform to a given set of geometric closure conditions, and when they do not, the measurements are adjusted to force those closures. For a set of uncorrelated observations, a measurement with a high precision, as indicated by a small variance, implies a good observation, and in the adjustment it should receive a relatively small portion of the overall correction. Conversely, a measurement with a lower precision as indicated by a larger variance implies an observation with a larger error, and thus it should receive a larger portion of the correction.

The weight of an observation is a measure of its relative worth compared to other measurements. *Weights* are used to control the sizes of corrections applied to measurements in an adjustment. The more precise an observation, the higher its weight, or in other words, the smaller the variance, the higher the weight. From this analysis it can be stated intuitively that *weights are inversely proportional to variances*. Thus it also follows that *correction sizes should be inversely proportional to weights.*

In situations where measurements are correlated, weights are related to the inverse of the covariance matrix, Σ. As discussed in Chapter 5, the elements of this matrix are variances and covariances. Since weights are relative, variances and covariances are often replaced by cofactors. A *cofactor* is related to its covariance by the equation

$$q_{ij} = \frac{\sigma_{ij}}{\sigma_0^2} \tag{9.1}$$

where q_{ij} is the cofactor of the ijth measurement, σ_{ij} the covariance of the ijth measurement, and σ_0^2 the *reference variance*, a value that can be used for scaling. Equation (9.1) can be expressed in matrix notation as

$$Q = \frac{1}{\sigma_0^2}\Sigma \tag{9.2}$$

where Q is defined as the *cofactor matrix*. The structure and individual elements of the Σ matrix are:

$$\Sigma = \begin{bmatrix} \sigma_{x_1}^2 & \sigma_{x_1x_2} & \cdots & \sigma_{x_1x_n} \\ \sigma_{x_2x_1} & \sigma_{x_2}^2 & \cdots & \sigma_{x_2x_n} \\ \vdots & \vdots & \ddots & \vdots \\ \sigma_{x_nx_1} & \sigma_{x_nx_2} & \cdots & \sigma_{x_n}^2 \end{bmatrix}$$

From the discussion above, the weight matrix W is

$$W = Q^{-1} = \sigma_0^2\Sigma^{-1} \tag{9.3}$$

For uncorrelated measurements, the covariances are equal to zero (i.e., all $\sigma_{x_i x_j} = 0$) and the matrix Σ is diagonal. Thus Q is also a diagonal matrix with elements equal to $\sigma_{x_i}^2/\sigma_0^2$. The inverse of a diagonal matrix is also a diagonal matrix, with its elements being the reciprocals of the original diagonals, and therefore Equation (9.3) becomes

$$W = \begin{bmatrix} \frac{\sigma_0^2}{\sigma_{x_1}^2} & 0 & 0 & \cdots & 0 \\ 0 & \frac{\sigma_0^2}{\sigma_{x_2}^2} & 0 & \cdots & 0 \\ 0 & 0 & \frac{\sigma_0^2}{\sigma_{x_3}^2} & \cdots & 0 \\ \vdots & \vdots & & \ddots & \vdots \\ 0 & 0 & 0 & \cdots & \frac{\sigma_0^2}{\sigma_{x_n}^2} \end{bmatrix} = \sigma_0^2\Sigma^{-1} \tag{9.4}$$

From Equation (9.4), any independent measurement with variance equal to σ_i^2 has a weight of

$$w_i = \frac{\sigma_0^2}{\sigma_i^2} \tag{9.5}$$

If the ith observation has a weight, $w_i = 1$, then $\sigma_0^2 = \sigma_i^2$, or $\sigma_0^2 = 1$. Thus, σ_0^2 is often called the *variance of an observation of unit weight* or shortened to the *variance of unit weight*, or simply *unit variance*. Its square root is called the *standard deviation of unit weight*. If σ_0^2 is set equal to 1 in Equation (9.5), then

$$w_i = \frac{1}{\sigma_i^2} \tag{9.6}$$

Note in Equation (9.6) that as earlier stated, the weight of an observation is inversely proportional to its variance.

With correlated observations, it is possible to have a covariance matrix, Σ, and a cofactor matrix, Q, but not a weight matrix. This occurs when the cofactor matrix is singular, and thus an inverse for Q does not exist. Most situations in surveying involve uncorrelated measurements. For the remainder of this chapter, only the uncorrelated case with variance of unit weight will be considered.

9.2 WEIGHTED MEAN

If two measurements are taken of a quantity and the first is twice as good as the second, their relative worth can be expressed by giving the first measurement a weight of 2 and the second a weight of 1. A simple adjustment involving these two measurements would be to compute the mean value. In this calculation, the observation having the weight of 2 could be added twice, and the observation of weight 1 could be added once. As an illustration, suppose that a distance is measured with a tape to be 151.9 ft, and the same distance is measured with an EDM instrument as 152.5 ft. Assume that previous experience has indicated that the electronically measured distance is twice as good as the taped distance, and accordingly, the taped distance is given a weight of 1 and the electronically measured distance is given a weight of 2. Then one method of computing the mean from these measurements is

$$\overline{M} = \frac{151.9 + 152.5 + 152.5}{3} = 152.3$$

As an alternative, the calculation above can be rewritten as:

$$\overline{M} = \frac{1(151.9) + 2(152.5)}{1 + 2} = 152.3$$

Note that the weights of 1 and 2 were entered directly into the second computation and that the result of this calculation is the same as the first. Note also that the computed mean tends to be closer to the measured value having the higher weight (i.e., 152.3 is closer to 152.5 than it is to 151.9). A mean value computed from weighted observations is called the *weighted mean.*

To develop a general expression for computing the weighted mean, suppose that we have m independent, uncorrelated observations (z_1, z_2, ..., z_m) for a quantity z, and each of these observations has standard deviation σ. Then the mean of the observations is

$$\bar{z} = \frac{\sum_{i=1}^{m} z_i}{m} \tag{9.7}$$

If these m observations were now separated into two sets, one of size m_a and the other m_b such that $m_a + m_b = m$, the means for these two sets would be

$$\overline{z_a} = \frac{\sum_{i=1}^{m_a} z_i}{m_a} \tag{9.8}$$

$$\overline{z_b} = \frac{\sum_{i=m_a+1}^{m} z_i}{m_b} \tag{9.9}$$

The mean $\bar{z}$ is found by combining the means of these two sets as

$$\bar{z} = \frac{\sum_{i=1}^{m_a} z_i + \sum_{i=m_a+1}^{m} z_i}{m} = \frac{\sum_{i=1}^{m_a} z_i + \sum_{i=m_a+1}^{m} z_i}{m_a + m_b} \tag{9.10}$$

But from Equations (9.8) and (9.9),

$$\overline{z_a} m_a = \sum_{i=1}^{m_a} z_i \quad \text{and} \quad \overline{z_b} m_b = \sum_{i=m_a+1}^{m} z_i \tag{9.11}$$

Thus

$$\bar{z} = \frac{\overline{z_a} m_a + \overline{z_b} m_b}{m_a + m_b} \tag{9.12}$$

Note the correspondence between Equation (9.12) and the second equation used to compute the weighted mean in our simple illustration given earlier. By intuitive comparison it should be clear that m_a and m_b correspond to weights that could be symbolized as w_a and w_b, respectively. Thus Equation (9.12) can be written as

$$\bar{z} = \frac{w_a\bar{z}_a + w_b\bar{z}_b}{w_a + w_b} = \frac{\sum wz}{\sum w} \tag{9.13}$$

Equation (9.13) is used in calculating the weighted mean for a group of uncorrelated observations having unequal weights. In Chapter 10 it will be shown that the weighted mean is the most probable value for a set of weighted observations.

Example 9.1 As a simple example of computing a weighted mean using Equation (9.13), suppose that a distance d is measured three times with the following results: 92.61 with weight 3, 92.60 with weight 2, and 92.62 with weight 1. Calculate the weighted mean.

SOLUTION: By Equation (9.13):

$$\bar{d} = \frac{3(92.61) + 2(92.60) + 1(92.62)}{3 + 2 + 1} = 92.608$$

Note that if weights had been neglected, the simple mean would have been 92.61.

9.3 RELATION BETWEEN WEIGHTS AND STANDARD ERRORS

By applying the special law of propagation of variances [Equation (5.16)] to Equation (9.8), the variance $\bar{z}_a$ in Equation (9.8) is

$$\sigma_{\bar{z}_a}^2 = \left(\frac{\partial \bar{z}_a}{\partial z_1}\right)^2 \sigma^2 + \left(\frac{\partial \bar{z}_a}{\partial z_2}\right)^2 \sigma^2 + \cdots + \left(\frac{\partial \bar{z}_a}{\partial z_{m_a}}\right)^2 \sigma^2 \tag{9.14}$$

Substituting partial derivatives with respect to the measurements into Equation (9.14) gives

$$\sigma_{\bar{z}_a}^2 = \left(\frac{1}{m_a}\right)^2 \sigma^2 + \left(\frac{1}{m_a}\right)^2 \sigma^2 + \cdots + \left(\frac{1}{m_a}\right)^2 \sigma^2 = m_a \left(\frac{1}{m_a}\right)^2 \sigma^2$$

Thus

$$\sigma_{\bar{z}_a}^2 = \frac{1}{m_a}\sigma^2 \tag{9.15}$$

Using a similar procedure, the variance of $\bar{z}_b$ is

$$\sigma_{\bar{z}_b}^2 = \frac{1}{m_b}\sigma^2 \tag{9.16}$$

In Equations (9.15) and (9.16), σ is a constant, and the weights of $\bar{z}_a$ and $\bar{z}_b$ were established as m_a and m_b, respectively, from Equation (9.13). Since the weights are relative, from Equations (9.15) and (9.16),

$$w_a = \frac{1}{\sigma_{\bar{z}_a}^2}$$

and (9.17)

$$w_b = \frac{1}{\sigma_{\bar{z}_b}^2}$$

Conclusion: With uncorrelated observations, weights of measurements are inversely proportional to their variances.

9.4 STATISTICS OF WEIGHTED OBSERVATIONS

9.4.1 Standard Deviation

By definition, an observation is said to have a weight w when its precision is equal to that of the mean of w observations of unit weight. Let σ_0 be the standard error of an observation of weight 1, or *unit weight*. If $y_1, y_2, \ldots, y_n$ are observations having standard errors $\sigma_1, \sigma_2, \ldots, \sigma_n$ and weights $w_1, w_2, \ldots, w_n$, then by Equation (9.5),

$$\sigma_1 = \frac{\sigma_0}{\sqrt{w_1}}, \quad \sigma_2 = \frac{\sigma_0}{\sqrt{w_2}}, \quad \ldots, \quad \sigma_n = \frac{\sigma_0}{\sqrt{w_n}} \tag{9.18}$$

In Section 2.7, the standard error for a group of observations of equal weight was defined as

$$\sigma = \sqrt{\frac{\sum_{i=1}^{n} \epsilon_i^2}{n}}$$

Now, in the case where the observations are not equal in weight, the equation above becomes

$$\sigma = \sqrt{\frac{w_1\epsilon_1^2 + w_2\epsilon_2^2 + \cdots + w_n\epsilon_n^2}{n}} = \sqrt{\frac{\sum_{i=1}^{n} w_i\epsilon_i^2}{n}} \tag{9.19}$$

When modified for the standard deviation in Equation (2.7), it is

$$S = \sqrt{\frac{w_1v_1^2 + w_2v_2^2 + \cdots + w_nv_n^2}{n-1}} = \sqrt{\frac{\sum_{i=1}^{n} w_iv_i^2}{n-1}} \tag{9.20}$$

9.4.2 Standard Error of Weight, *w*, and Standard Error of the Weighted Mean

The relationship between standard error and standard error of weight w was given in Equation (9.18). Combining this with Equation (9.19) and dropping the summation limits, equations for standard errors of weight w are obtained in terms of σ_0 as follows:

$$\begin{aligned}
\sigma_1 &= \frac{\sigma_0}{\sqrt{w_1}} = \sqrt{\frac{\sum w\epsilon^2}{n}}\frac{1}{\sqrt{w_1}} = \sqrt{\frac{\sum w\epsilon^2}{nw_1}} \\
\sigma_2 &= \frac{\sigma_0}{\sqrt{w_2}} = \sqrt{\frac{\sum w\epsilon^2}{n}}\frac{1}{\sqrt{w_2}} = \sqrt{\frac{\sum w\epsilon^2}{nw_2}} \\
&\vdots \\
\sigma_n &= \frac{\sigma_0}{\sqrt{w_n}} = \sqrt{\frac{\sum w\epsilon^2}{n}}\frac{1}{\sqrt{w_n}} = \sqrt{\frac{\sum w\epsilon^2}{nw_n}}
\end{aligned} \tag{9.21}$$

Similarly, standard deviations of weight w can be expressed as

$$S_1 = \sqrt{\frac{\sum wv^2}{w_1(n-1)}}, \quad S_2 = \sqrt{\frac{\sum wv^2}{w_2(n-1)}}, \quad \ldots, \quad S_n = \sqrt{\frac{\sum wv^2}{w_n(n-1)}} \tag{9.22}$$

Note that if w in the denominator of Equation (9.22) is assigned a value of 1, the *standard deviation of unit* weight, S_0, for a weighted set of observations is obtained. Finally, the reference standard error of the weighted mean is calculated as

$$\sigma_{\overline{M}} = \sqrt{\frac{\sum w\epsilon^2}{\sum wn}} \tag{9.23}$$

and the standard deviation of the weighted mean is

$$S_{\overline{M}} = \sqrt{\frac{\sum wv^2}{\sum w(n - 1)}} \tag{9.24}$$

9.5 WEIGHTS IN ANGLE MEASUREMENTS

Suppose that the three angles α_1, α_2, and α_3 in a plane triangle are measured n_1, n_2, and n_3 times, respectively, with the same instrument under the same conditions. What are the relative weights of the angles?

To analyze the relationship between weights and the number of times an angle is turned, let S be the standard deviation of a single angle measurement. The means of the three angles are

$$\alpha_1 = \frac{\sum \alpha_1}{n_1} \qquad \alpha_2 = \frac{\sum \alpha_2}{n_2} \qquad \alpha_3 = \frac{\sum \alpha_3}{n_3}$$

The variances of the means, as obtained by applying Equation (5.16), are

$$S_{\alpha_1}^2 = \frac{1}{n_1} S^2 \qquad S_{\alpha_2}^2 = \frac{1}{n_2} S^2 \qquad S_{\alpha_3}^2 = \frac{1}{n_3} S^2$$

Again, since the weights of the observations are inversely proportional to the variances and relative, the weights of the three angles are:

$$w_1 = \frac{1}{S_{\alpha_1}^2} = \frac{n_1}{S^2} \qquad w^2 = \frac{1}{S_{\alpha_2}^2} = \frac{n_2}{S^2} \qquad w_3 = \frac{1}{S_{\alpha_3}^2} = \frac{n_3}{S^2}$$

In the above expressions, S is a constant term in each of the weights, and because weights are relative, it can be dropped. Thus the weights of the angles are $w_1 = n_1$, $w_2 = n_2$ and $w_3 = n_3$.

In summary, it has been proven that when all conditions in angle measurement are equal except for the number of turnings, *angle weights are proportional to the number of times the angles are turned.*

9.6 WEIGHTS IN DIFFERENTIAL LEVELING

Suppose that for the leveling network shown in Figure 9.1, the lengths of the lines 1, 2, and 3 are 2 miles, 3 miles, and 4 miles, respectively. For these varying lengths of lines, it can be expected that the errors in their elevation differences will vary, and thus the weights assigned to the elevation differences should also be varied. What relative weights should be used for these lines?

To analyze the relationship of weights and level line lengths, recall from Equation (8.21) that the variance in Δh is estimated as

$$\sigma_{\Delta h}^2 = D^2[2N(\sigma_{r/D}^2 + \sigma_\alpha^2)] \tag{a}$$

where D is the length of the individual sights, N the number of setups, $\sigma_{r/D}$ the estimated error in a rod reading, and σ_α the estimated collimation error for each sight. Let l_i be the length of the ith course between benchmarks; then

$$N = \frac{l_i}{2D} \tag{b}$$

Substituting Equation (b) into Equation (a) yields

$$\sigma_{\Delta h}^2 = l_i D(\sigma_{r/D}^2 + \sigma_\alpha^2) \tag{c}$$

However, D, $\sigma_{r/D}$, and σ_α are all constants, and thus by letting $k = D(\sigma_{r/D}^2 + \sigma_\alpha^2)$, Equation ($c$) becomes

$$\sigma_{\Delta h}^2 = l_i k \tag{d}$$

Thus for this example, it can be said that the three level line weights are:

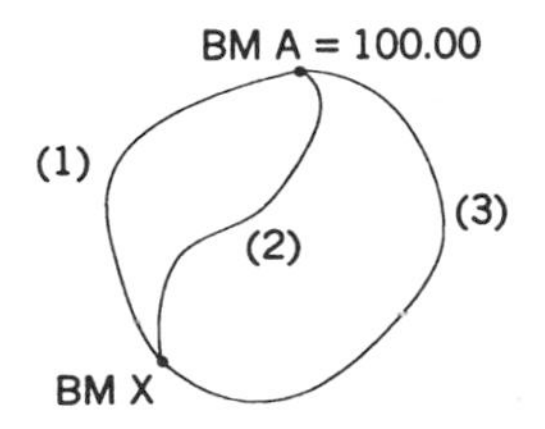

Figure 9.1 Leveling network.

$$w_1 = \frac{1}{l_1 k} \qquad w_2 = \frac{1}{l_2 k} \qquad w_3 = \frac{1}{l_3 k} \tag{e}$$

Now since k is a constant and weights are relative, Equation (e) can be simplified to

$$w_1 = \frac{1}{l_1} = 1/2 \qquad w_2 = \frac{1}{l_2} = 1/3 \qquad w_3 = \frac{1}{l_3} = 1/4$$

In summary, it has been shown that *weights of differential leveling lines are inversely proportional to their lengths*, and since any course length is proportional to its number of setups, *weights are also inversely proportional to the number of setups.*

9.7 PRACTICAL EXAMPLES

Example 9.2. Suppose the angles in an equilateral triangle ABC were each measured by the same observer using the same instrument, but the number of repetitions for each angle varied. The results were: $A = 45°15'25''$, $n = 4$; $B = 83°37'22''$, $n = 8$; and $C = 51°07'39''$, $n = 6$. Adjust the angles.

SOLUTION: Weights proportional to number of repetitions are assigned, and corrections made in inverse proportion to those weights. The sum of the three angles is 180°00′26″, and thus the misclosure that must be adjusted is 26″. The correction process is demonstrated in Table 9.1.

Note that a multiplier of 24 was used in the third column for convenience to avoid fractions in computing correction factors. Because weights are relative, this did not alter the adjustment.

Note also that two computational checks are secured in the above solution; the sum of the individual corrections totaled 26″, and the sum of the corrected angles totaled 180°00′00″.

Table 9.1 Adjustment of Example 9.2

Angle	n (Weight)	Correction Factor	Correction	Corrected Angle
A	4	$(1/4) \times 24 = 6$	$(6/13) \times 26'' = 12''$	45°15′13″
B	8	$(1/8) \times 24 = 3$	$(3/13) \times 26'' = 06''$	83°37′16″
C	6	$(1/6) \times 24 = 4$	$(4/13) \times 26'' = 08''$	51°07′31″
		$\Sigma = 13$	$\Sigma = 26''$	$\Sigma =$ 180°00′00″

Example 9.3 In the leveling network of Figure 9.1, recall that the lengths of lines 1, 2 and 3 were 2, 3 and 4 miles, respectively. If the observed elevation differences in lines 1, 2 and 3 were ±21.20 ft, +21.23 ft and +21.29 ft, respectively, find the weighted mean for the elevation difference, and the adjusted elevation of BMX. (Note: All level lines were run from BMA to BMX).

SOLUTION: The weights of lines 1, 2, and 3 are 1/2, 1/3, and 1/4, respectively. Again since weights are relative, these weights can arbitrarily be multiplied by 12 to obtain weights of 6, 4, and 3, respectively. Applying Equation (9.13), the weighted mean of the elevation difference is

$$\text{Mean } \Delta\text{ELEV} = \frac{6(21.20) + 4(21.23) + 3(21.29)}{6 + 4 + 3} = +21.23$$

Thus the elevation of BMX = 100.00 + 21.23 = 123.23 ft. Note that if the weights had been neglected, the simple average would have given a mean of +21.24.

Example 9.4 A distance is measured as 625.79 ft using a cloth tape and given a weight of 1; it is measured again as 625.71 ft using a steel tape and assigned a weight of 2; and finally, it is measured a third time as 625.69 ft with an EDM instrument and given a weight of 4. Calculate the most probable value of the length (*weighted mean*), and find the standard deviation of the weighted mean.

SOLUTION: By Equation (9.13), the weighted mean is:

$$\overline{M} = \frac{625.79 + 2(625.71) + 4(625.69)}{1 + 2 + 4} = 625.71$$

By Equation (9.24), the standard deviation of the weighted mean is

$$S_{\overline{M}} = \sqrt{\frac{\sum wv^2}{\sum w(n-1)}} = \sqrt{\frac{0.0080}{7(2)}} = \pm 0.024 \text{ ft}$$

where

$$v_1 = 625.71 - 625.79 = -0.08 \qquad w_1v_1^2 = 1(-0.08)^2 = 0.0064$$

$$v_2 = 625.71 - 625.71 = 0.00 \qquad w_2v_2^2 = 2(\ 0.00)^2 = 0.0000$$

$$v_3 = 625.71 - 625.69 = +0.02 \qquad w_3v_3^2 = 4(+0.02)^2 = \underline{0.0016}$$

$$\Sigma\, wv^2 = 0.0080$$

Table 9.2 Route data for Example 9.5

Route	Length (mi)	ΔElev (ft)	w
1	1	+25.35	18
2	2	+25.41	9
3	3	+25.38	6
4	6	+25.30	3

Example 9.5 In leveling from benchmark A to B, four different routes of varying length are taken. The data of Table 9.2 are obtained. (Note that the weights were computed as $18/l$ for computational convenience only.) Calculate the most probable elevation difference (weighted mean), the standard deviation of unit weight, the standard deviation of the weighted mean, and the standard deviations of the weighted observations.

By Equation (9.13), the weighted mean for elevation difference is

$$\overline{M} = \frac{18(25.35) + 9(25.41) + 6(25.38) + 3(25.30)}{36} = +25.366 \text{ ft}$$

Note that the arithmetic mean for this set of observations is 25.335, but the weighted mean is 25.366. To find the standard deviations for the weighted observations, the data in Table 9.3 are first created.

By Equation (9.20), the standard deviation is

$$S = \sqrt{\frac{0.0363}{3}} = \pm 0.11 \text{ ft}$$

By Equation (9.24), the standard deviation of the weighted mean is

Table 9.3 Data for standard deviations in Example 9.5

Route	w	v	v^2	wv^2
1	18	+0.016	0.0002	0.0045
2	9	−0.044	0.0020	0.0176
3	6	−0.014	0.0002	0.0012
4	3	+0.066	0.0043	0.0130
				$\Sigma wv^2 = 0.0363$

$$S_{\overline{M}} = \sqrt{\frac{0.0363}{36(3)}} = \pm 0.018 \text{ ft}$$

By Equation (9.22), the standard deviations for the weighted observations are

$$S_1 = \sqrt{\frac{0.0363}{18(3)}} = \pm 0.026 \text{ ft} \qquad S_2 = \sqrt{\frac{0.0363}{9(3)}} = \pm 0.037 \text{ ft}$$

$$S_3 = \sqrt{\frac{0.0363}{6(3)}} = \pm 0.045 \text{ ft} \qquad S_4 = \sqrt{\frac{0.0363}{3(3)}} = \pm 0.063 \text{ ft}$$

PROBLEMS

9.1 An angle was measured as 49°27′20″ using an engineer's transit, and had a standard deviation of ±15″. It was measured again using a repeating optical theodolite as 49°27′24″ with a standard deviation of ±6″. This angle was measured a third time with a directional theodolite as 49°27′27″ with a standard deviation of ±2″. Calculate the weighted mean of the angle and its standard deviation.

9.2 An angle was measured at four different times with the following results:

Day	Angle	S
1	136°14′45″	±12.2″
2	136°14′30″	±6.7″
3	136°14′30″	±8.9″
4	136°14′16″	±9.5″

What is the most probable value for the angle, and its standard deviation?

9.3 A distance was measured by pacing as 156 ft with a standard deviation of ±2.5 ft. It was then measured as 153.48 ft using a steel tape with a standard deviation of ±0.03 ft. Finally, it was measured as 153.44 ft with an EDM instrument with a standard deviation of ±0.02 ft. What are the most probable value for the distance and its standard deviation?

9.4 A distance was measured by pacing as 267 ft with a standard deviation of ±3 ft. It was then measured as 269.08 ft with a steel tape and had a standard deviation of ±0.05 ft. Finally, it was measured as 268.99 ft with an EDM instrument. The EDM instrument and reflector setup standard deviations were ±0.005 ft and ±0.01 ft, respectively, and the manufacturer's estimated standard deviation for the EDM instrument is ±(5 mm + 5 ppm). What are the most probable value for the distance and its standard deviation?

9.5 What is the computed standard deviation for each weighted observation of Problem 9.4?

9.6 Compute the standard deviation for the taped measurement in Problem 9.4 assuming standard deviations of ±5° in temperature, ±3 lb. in pull, ±0.005 ft in tape length, and ±0.01 ft in reading and marking the tape. The temperature was 78°F, the calibrated tape length is 100.009 ft. and the field tension was recorded as 15 lb. The cross-sectional area of the tape is 0.004 in^2, its modulus of elasticity is 29,000,000 lb/in^2, its coefficient of thermal expansion is 6.45×10^{-5} /°F, and its weight is 2.5 lb. Assume horizontal taping with full tape lengths for all but the last partial distance with ends support only.

9.7 Do Problem 9.3 using the standard deviation information for the EDM instrument in Problem 9.4, and the tape calibration data in Problem 9.6.

9.8 A zenith angle was measured six times direct and 6 times reverse. The average direct reading is 86°34′31″ with a standard deviation of ±15.6″. In the reverse face, it was measured as 273°25′09″ with a standard deviation of ±10.9″. What is the most probable value for the zenith angle in the direct face?

9.9 Three crews level to a benchmark following three different routes. The lengths of the routes and the measured differences in elevation are:

Route	ΔElev (ft)	Length (ft)
1	14.80	2500
2	14.87	5000
3	14.83	3600

What are the:

(a) best value for the difference in elevation?

(b) standard deviation for the elevation difference?

(c) standard deviation for the weighted observations?

9.10 Find the computed standard deviations for the weighted observations in Problem 9.9.

10

PRINCIPLES OF LEAST SQUARES

10.1 INTRODUCTION

In surveying, the measurements must often satisfy established numerical relationships known as *geometric constraints*. As examples, in a closed polygon traverse, horizontal angle and distance measurements should conform to the geometric constraints given in Section 7.4, and in a differential leveling loop, the elevation differences should sum to given a quantity. However, because rarely, if ever, will the geometric constraints be perfectly met, an adjustment of the data must be made.

As discussed in the earlier chapters, errors in measurements conform to the laws of probability, and they follow the normal distribution theory. Thus they should be adjusted in a manner that follows these mathematical laws. While the mean has been used extensively throughout history, the earliest works on least squares started in the late eighteenth century. Its earliest application was primarily for adjusting celestial observations. Laplace first investigated the subject and laid its foundation in 1774. The first published article on the subject, entitled, "Méthode des Moindres Quarrés" ("Method of Least Squares") was written in 1805 by Legendre. However, it is well known that although Gauss did not publish on the subject until 1809, he developed and used the method extensively as a student at the University of Göttingen beginning in 1794 and thus is given credit for its development. In this chapter, equations for performing least-squares adjustments are developed, and their use is illustrated with several simple examples.

10.2 FUNDAMENTAL PRINCIPLE OF LEAST SQUARES

To develop the principle of least squares, a specific case will be considered. Suppose that there are n independent equally weighted measurements, z_1, z_2, ..., z_n, of the same quantity z, which has a *most probable value* denoted by M. By definition,

$$
\begin{aligned}
M - z_1 &= v_1 \\
M - z_2 &= v_2 \\
&\vdots \\
M - z_n &= v_n
\end{aligned}
\tag{10.1}
$$

where the v's are the residual errors. Note that residuals behave in a manner similar to errors, and thus they may be used interchangeably in the normal distribution function given by Equation (3.2). When substituting v for x, there results,

$$
f_x(v) = y = \frac{1}{\sigma\sqrt{2\pi}} e^{-v^2/2\sigma^2} = Ke^{-h^2v^2} \tag{10.2}
$$

where $h = 1/\sigma\sqrt{2}$ and $K = h/\sqrt{\pi}$.

As discussed in Chapter 3, probabilities are represented by areas under the normal distribution curve. Thus the individual probabilities for the occurrence of residuals v_1, v_2, ..., v_n are obtained by multiplying their respective ordinates y_1, y_2, ..., y_n by some infinitesimal increment Δv as follows:

$$
\begin{aligned}
P_1 &= y_1\,\Delta v = Ke^{-h^2v_1^2}\,\Delta v \\
P_2 &= y_2\,\Delta v = Ke^{-h^2v_2^2}\,\Delta v \\
&\vdots \\
P_n &= y_n\,\Delta v = Ke^{-h^2v_n^2}\,\Delta v
\end{aligned}
\tag{10.3}
$$

From Equation (3.1), the probability of the simultaneous occurrence of all the residuals v_1 through v_n is the product of the individual probabilities, and thus:

$$
P = (Ke^{-h^2v_1^2}\,\Delta v)(Ke^{h^2v_2^2}\,\Delta v)\cdots(Ke^{-h^2v_n^2}\,\Delta v) \tag{10.4}
$$

Simplifying Equation (10.4) gives

$$
P = K^n(\Delta v)^n e^{-h^2(v_1^2+v_2^2+\cdots+v_n^2)} \tag{10.5}
$$

M is the quantity that is to be selected in such a way that it gives the greatest probability of occurrence, or stated differently, M is chosen so that the value of P is maximized. Figure 10.1 shows a plot of the e^{-x} versus x. From this plot it is readily seen that e^{-x} is maximized by minimizing x, and thus in relation to Equation (10.5), the probability P is maximized when the quantity $(v_1^2 + v_2^2 + \cdots + v_n^2)$ is minimized. In other words, *to maximize P, the sum of the squares of the residuals must be minimized.* Equation (10.6) expresses this *fundamental principle of least squares*:

$$\sum v^2 = (v_1^2 + v_2^2 + \cdots + v_n^2) = \text{minimum} \tag{10.6}$$

This condition is: *The most probable value (MPV) for a quantity obtained from repeated measurements of equal weight is the value that renders the sum of the residuals squared a minimum.* From calculus, the minimum value of a function can be found by taking the first derivative of the function with respect to the variable, and equating the function to zero. The condition stated in Equation (10.6) is enforced by taking the first derivative of the function with respect to the unknown variable M and setting the result equal to zero. Substituting Equations (10.1) into Equation (10.6) yields

$$\sum v^2 = (M - z_1)^2 + (M - z_2)^2 + \cdots + (M - z_n)^2 \tag{10.7}$$

Taking the derivative of Equation (10.7) with respect to M and setting the resulting equation equal to zero gives

$$\frac{d\left(\sum v^2\right)}{dM} = 2(M - z_1)(1) + 2(M - z_2)(1) + \cdots + 2(M - z_n)(1) = 0 \tag{10.8}$$

Now dividing Equation (10.8) by 2 and simplifying yields

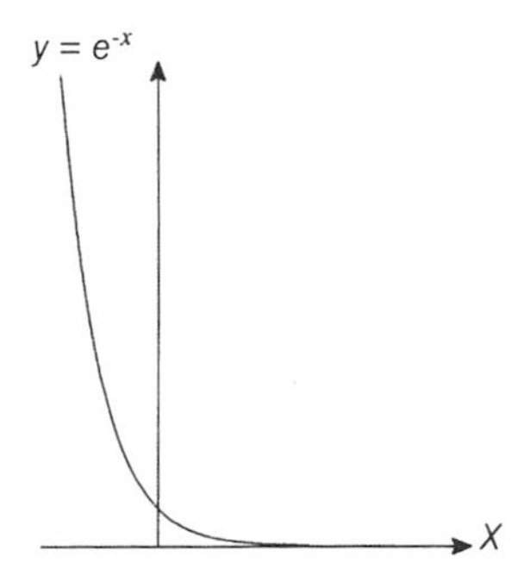

Figure 10.1 Plot of e^{-x}.

$$M - z_1 + M - z_2 + \cdots + M - z_n = 0$$

$$nM = z_1 + z_2 + \cdots + z_n$$

$$M = \frac{z_1 + z_2 + \cdots + z_n}{n} \tag{10.9}$$

In Equation (10.9) the quantity $(z_1 + z_2 + \cdots + z_n)/n$ is the mean of the observed values. *This is proof that when a quantity has been observed independently several times, the MPV is the arithmetic mean.*

10.3 FUNDAMENTAL PRINCIPLE OF WEIGHTED LEAST SQUARES

In Section 10.2 the fundamental principle of a least-squares adjustment was developed for observations having equal (or unit) weights. The more general case of least-squares adjustment assumes that the observations have varying degrees of precision, and thus different weights.

Consider a set of measurements $z_1, z_2, \ldots, z_n$ having relative weights $w_1, w_2, \ldots, w_n$ and residuals $v_1, v_2, \ldots, v_n$. Denote the weighted MPV as M. As in Section 10.2, the residuals are related to the observations through Equations (10.1), and the total probability of their simultaneous occurrence is given by Equation (10.5). However, notice in Equation (10.2) that $h^2 = 1/2\sigma^2$, and since weights are inversely proportional to variances, they are directly proportional to h^2. Thus Equation (10.5) can be rewritten as

$$P = K^n \, (\Delta v)^n e^{-(w_1 v_1^2 + w_2 v_2^2 + \cdots + w_n v_n^2)} \tag{10.10}$$

Again to maximize P in Equation (10.10), the negative exponent must be minimized. To achieve this, the sum of the products of the weights times their respective residuals squared must be minimized, which is the condition imposed in weighted least-squares adjustment. The condition of weighted least-squares adjustment in equation form is

$$w_1 v_1^2 + w_2 v_2^2 + \cdots + w_n v_n^2 = \sum wv^2 = \text{minimum} \tag{10.11}$$

Substituting the values for the residuals given in Equation (10.1) into Equation (10.11) gives

$$w_1(M - z_1)^2 + w_2(M - z_2)^2 + \cdots + w_n(M - z_n)^2 = \text{minimum} \tag{10.12}$$

This condition is: *The most probable value for a quantity obtained from repeated measurements having varying weights is that value which renders the sum of the weights times their respective squared residuals a minimum.*

The minimum condition is imposed by differentiating Equation (10.12) with respect to M and setting the resultant equation equal to zero, and thus

$$2w_1(M - z_1)(1) + 2w_2(M - z_2)(1) + \cdots + 2w_n(M - z_n)(1) = 0 \quad (10.13)$$

Dividing Equation (10.13) by 2, and rearranging yields

$$w_1(M - z_1) + w_2(M - z_2) + \cdots + w_n(M - z_n) = 0 \quad (10.14)$$

$$w_1z_1 + w_2z_2 + \cdots + w_nz_n = w_1M + w_2M + \cdots + w_nM$$

from which $\Sigma\ wz = \Sigma\ wM$. Thus

$$M = \frac{\sum wz}{\sum w} \quad (10.15)$$

Notice that Equation (10.15) is the same as Equation (9.13), which is the formula for computing the weighted mean.

10.4 STOCHASTIC MODEL

The determination of variances, and subsequently the weights of the observations, is known as the *stochastic model* in a least-squares adjustment. In Section 10.3 the inclusion of weights in the adjustment was discussed. It is crucial to the adjustment to select a proper stochastic (weighting) model since, as was discussed in Section 9.1, the weight of an observation controls the amount of correction it receives during the adjustment. However, development of the stochastic model is important not only to weighted adjustments. When doing an unweighted adjustment, all observations are assumed to be of equal weight, and thus the stochastic model is created implicitly. In Chapters 6 to 9 we established the foundations for selecting a proper stochastic model in surveying. It will be shown in Chapter 20 that failure to select the stochastic model correctly will also affect the ability to isolate blunders in a measurement set.

10.5 MATHEMATICAL MODEL

A mathematical model in adjustment computations is an equation or set of equations that represents or defines an adjustment condition. It must either be known, or assumed. If the mathematical model represents the physical situation adequately, the observational errors can be expected to conform to the normal distribution curve. For example, a well-known mathematical model states that the sum of angles in a plane triangle is 180°. This model is adequate if the survey is limited to a small region. However, when the triangles are very large, this model does not account for systematic errors caused by the earth's curvature. In that case the mathematical model is inadequate and needs to be modified to include corrections for spherical excess. In traversing, the mathematical model of plane computations is suitable for smaller surveys, but if the extent of the survey becomes too large, again the model must be changed to account for the systematic errors caused by earth curvature. This can be accomplished by transforming the observations into a plane mapping system such as the state plane coordinate system, or by working with geodetic observation equations. Needless to say, if the model does not fit the physical situation, an incorrect adjustment will result.

There are two basic forms for mathematical models: the *conditional* and *parametric* adjustments. In the conditional adjustment, geometric conditions are enforced upon the observations and their residuals. Examples of conditional adjustment are: (1) the sum of the angles in a closed polygon is $(n - 2) \times 180°$, where n is the number of sides in the polygon; (2) the latitudes, and departures, of a polygon traverse sum to zero; and (3) the sum of the angles in the horizon at any station equal 360°. A least-squares adjustment example using condition equations is given in Section 10.13.

When performing a parametric adjustment, observations are expressed in terms of unknown parameters that were never measured directly. For example, the well-known coordinate equations are used to model the measured angles, directions, and distances in a traverse. The adjustment yields most probable values for the coordinates (parameters), which in turn enable the most probable values for the adjusted observations to be computed.

A primary objective in an adjustment is to ensure that all observations are used to find the most probable values for the unknowns in the model. In least-squares adjustments, no matter if conditional or parametric, the geometric checks at the end of the adjustment are satisfied and the same adjusted observations are obtained. In complicated networks, it is often difficult and time consuming to write the equations to express all conditions that must be met for a conditional adjustment. Therefore this book will focus on the parametric adjustment, which generally leads to larger systems of equations but is straightforward in its development and solution and, as a result, is well suited to computers.

10.6 OBSERVATION EQUATIONS

In the parametric adjustment method, *observation equations* are written that relate measured quantities to both observational residuals and independent, unknown parameters. One equation is written for each observation, and for a unique solution of the unknowns, the number of equations must equal the number of unknowns. Usually, there are more observations (and hence equations) than unknowns, and this permits the determination of the most probable values for the unknowns based on the principle of least squares.

10.6.1 Elementary Example of Observation Equation Adjustment

As an elementary example of a least-squares adjustment by the observation equation method, consider the following three equations:

$$\begin{aligned} &(1)\quad x + y = 3.0 \\ &(2)\quad 2x - y = 1.5 \\ &(3)\quad x - y = 0.2 \end{aligned} \tag{10.16}$$

Equations (10.16) relate the two unknowns, x and y, to the observed quantities (the values on the right side of the equations). One equation is redundant since the values for x and y can be obtained from any two of the three equations. For example, if Equations (1) and (2) are solved, x would equal 1.5 and y would equal 1.5, but if Equations (2) and (3) are solved, x would equal 1.3 and y would equal 1.1, and if Equations (1) and (3) are solved, x would equal 1.6 and y would equal 1.4. Based on the inconsistency of these equations, the measurements apparently contain errors. Therefore new expressions called *observation equations* can be rewritten which include residuals as

$$\begin{aligned} &(4)\quad x + y - 3.0 = v_1 \\ &(5)\quad 2x - y - 1.5 = v_2 \\ &(6)\quad x - y - 0.2 = v_3 \end{aligned} \tag{10.17}$$

Equations (10.17) relate the unknown parameters (x and y) to the observations and their errors. Obviously, it is possible to select values of v_1, v_2, and v_3 that will yield the same values for x and y no matter which pair of equations are used. For example, to obtain consistencies through all of the equations, arbitrarily let $v_1 = 0$, $v_2 = 0$, and $v_3 = -0.2$. In this arbitrary solution, x would equal 1.5 and y would equal 1.5, no matter which pair of equations is solved. This is a consistent solution, however, there are other values for the v's that will produce a smaller sum of squares.

To find the least-squares solution for x and y, the residual equations arc squared, and these squared expressions are added to give a function, $f(x,y)$ which equals the $\Sigma\, v^2$. Doing this for Equations (10.17) yields

$$f(x,y) = \Sigma\, v^2 = (x + y - 3.0)^2 + (2x - y - 1.5)^2 + (x - y - 0.2)^2 \tag{10.18}$$

As discussed previously, to minimize a function, its derivatives must be set equal to zero. Thus in Equation (10.18), the partial derivatives with respect to each unknown must be taken and set equal to zero. This leads to the two equations

$$\frac{\partial \sum v^2}{\partial x} = 2(x + y - 3.0) + 2(2x - y - 1.5)(2) + 2(x - y - 0.2) = 0$$
$$\frac{\partial \sum v^2}{\partial y} = 2(x + y - 3.0) + 2(2x - y - 1.5)(-1) + 2(x - y - 0.2)(-1) = 0 \quad (10.19)$$

Equations (10.19) are called *normal equations*. Simplifying them gives reduced normal equations of

$$\begin{aligned} 6x - 2y - 6.2 &= 0 \\ -2x + 3y - 1.3 &= 0 \end{aligned} \quad (10.20)$$

The simultaneous solution of Equations (10.20) yields x equal to 1.514 and y equal to 1.442. By substituting these adjusted values into Equations (10.17), numerical values for the three residuals can be computed. Table 10.1 provides a comparison of the arbitrary solution to the least squares solution. The tabulated summations of residuals squared shows that the least squares solution yields the smaller total, and thus the better solution. In fact, it is the most probable solution for the unknowns based on the measurements.

10.7 SYSTEMATIC FORMULATION OF THE NORMAL EQUATIONS

10.7.1 Equal-Weight Case

In large systems of observation equations, it would be helpful to use systematic procedures to formulate the normal equations. In developing these pro-

Table 10.1 Comparison of an arbitrary and least-squares solution

Arbitrary Solution		Least-Squares Solution	
$v_1 = 0$	$v_1^2 = 0$	$v_1 = -0.044$	$v_1^2 = 0.002$
$v_2 = 0$	$v_2^2 = 0$	$v_2 = 0.086$	$v_2^2 = 0.007$
$v_3 = -0.2$	$v_3^2 = 0.04$	$v_3 = -0.128$	$v_3^2 = 0.016$
	$\Sigma v^2 = 0.04$		$\Sigma v^2 = 0.025$

cedures, consider the following generalized system of linear observation equations having variables of $(A, B, C, ..., N)$:

$$\begin{aligned} a_1A + b_1B + c_1C + \cdots + n_1N &= l_1 + v_1 \\ a_2A + b_2B + c_2C + \cdots + n_2N &= l_2 + v_2 \\ &\vdots \\ a_mA + b_mB + c_mC + \cdots + n_mN &= l_m + v_m \end{aligned} \tag{10.21}$$

The squares of the residuals for Equations (10.21) are

$$\begin{aligned} v_1^2 &= (a_1A + b_1B + c_1C + \cdots + n_1N - l_1)^2 \\ v_2^2 &= (a_2A + b_2B + c_2C + \cdots + n_2N - l_2)^2 \\ &\vdots \\ v_m^2 &= (a_mA + b_mB + c_mC + \cdots + n_mN - l_m)^2 \end{aligned} \tag{10.22}$$

Summing Equations (10.22), the function, $f(A,B,C, ..., N) = \Sigma\, v^2$ is obtained, which expresses the equal-weight least-squares condition as

$$\begin{aligned} \Sigma\, v^2 = {} & (a_1A + b_1B + c_1C + \cdots + n_1N - l_1)^2 \\ & + (a_2A + b_2B + c_2C + \cdots + n_2N - l_2)^2 \\ & + \cdots + (a_mA + b_mB + c_mC + \cdots + n_mN - l_m)^2 \end{aligned} \tag{10.23}$$

According to least-squares theory, the minimum for Equation (10.23) is found by setting the partial derivatives of the function, with respect to each unknown, equal to zero. This results in the following normal equations:

$$\begin{aligned} \frac{\partial \sum v^2}{\partial A} = {} & 2(a_1A + b_1B + c_1C + \cdots + n_1N - l_1)a_1 \\ & + 2(a_2A + b_2B + c_2C + \cdots + n_2N - l_2)a_2 + \cdots \\ & + 2(a_mA + b_mB + c_mC + \cdots + n_mN - l_m)a_m = 0 \\ \frac{\partial \sum v^2}{\partial B} = {} & 2(a_1A + b_1B + c_1C + \cdots + n_1N - l_1)b_1 \\ & + 2(a_2A + b_2B + c_2C + \cdots + n_2N - l_2)b_2 + \cdots \\ & + 2(a_mA + b_mB + c_mC + \cdots + n_mN - l_m)b_m = 0 \end{aligned} \tag{10.24}$$

$$\frac{\partial \sum v^2}{\partial C} = 2(a_1A + b_1B + c_1C + \cdots + n_1N - l_1)c_1$$
$$+ 2(a_2A + b_2B + c_2C + \cdots + n_2N - l_2)c_2 + \cdots$$
$$+ 2(a_mA + b_mB + c_mC + \cdots + n_mN - l_m)c_m = 0$$
$$\vdots$$
$$\frac{\partial \sum v^2}{\partial N} = 2(a_1A + b_1B + c_1C + \cdots + n_1N - l_1)n_1$$
$$+ 2(a_2A + b_2B + c_2C + \cdots + n_2N - l_2)n_2 + \cdots$$
$$+ 2(a_mA + b_mB + c_mC + \cdots + n_mN - l_m)n_m = 0$$

Dividing each expression of Equations (10.24) by 2 removes the constant, and, after grouping terms, the following are obtained:

$$[(a_1)^2 + (a_2)^2 + \cdots + (a_m)^2]A + [a_1b_1 + a_2b_2 + \cdots + a_mb_m]B$$
$$+ [a_1c_1 + a_2c_2 + \cdots + a_mc_m)C + \cdots + [a_1n_1 + a_2n_2 + \cdots + a_mn_m]N$$
$$- [a_1l_1 + a_2l_2 + \cdots + a_ml_m] = 0$$
$$[b_1a_1 + b_2a_2 + \cdots + b_ma_m]A + [(b_1)^2 + (b_2)^2 + \cdots + (b_m)^2]B$$
$$+ [b_1c_1 + b_2c_2 + \cdots + b_mc_m]C + \cdots + [b_1n_1 + b_2n_2 + \cdots + b_mn_m]N$$
$$- [b_1l_1 + b_2l_2 + \cdots + b_ml_m] = 0$$
$$[c_1a_1 + c_2a_2 + \cdots + c_ma_m^2]A + [c_1b_1 + c_2b_2 + \cdots + c_mb_m]B$$
$$+ [(c_1)^2 + (c_2)^2 + \cdots + (c_m)^2]C + \cdots + [c_1n_1 + c_2n_2 + \cdots + c_mn_m]N$$
$$- [c_1l_1 + c_2l_2 + \cdots + c_ml_m] = 0$$
$$\vdots$$
$$[n_1a_1 + n_2a_2 + \cdots + n_ma_m]A + [n_1b_1 + n_2b_2 + \cdots + n_mb_m]B$$
$$+ [n_1c_1 + n_2c_2 + \cdots + n_mc_m)C + \cdots + [(n_1)^2 + (n_2)^2 + \cdots + (n_m)^2]N$$
$$- [n_1l_1 + n_2l_2 + \cdots + n_ml_m] = 0$$

(10.25)

Generalized equations expressing normal Equations (10.25) can now be written as

$$
\begin{aligned}
(\Sigma\ a^2)A + (\Sigma\ ab)B + (\Sigma\ ac)C + \cdots + (\Sigma\ an)N &= \Sigma\ al \\
(\Sigma\ ba)A + (\Sigma\ b^2)B + (\Sigma\ bc)C + \cdots + (\Sigma\ bn)N &= \Sigma\ bl \\
(\Sigma\ ca)A + (\Sigma\ cb)B + (\Sigma\ c^2)C + \cdots + (\Sigma\ cn)N &= \Sigma\ cl \\
&\vdots \\
(\Sigma\ na)A + (\Sigma\ nb)B + (\Sigma\ nc)C + \cdots + (\Sigma\ n^2)N &= \Sigma\ nl
\end{aligned}
\tag{10.26}
$$

where the a's, b's, c's, ..., n's are the coefficients for the unknowns A, B, C, ..., N; the l values are the observations; and Σ signifies the summation from $i = 1$ to m.

10.7.2 Weighted Case

In a manner similar to that of Section 10.7.1, it can be shown that normal equations may be formed for weighted observation equations systematically in the following manner:

$$
\begin{aligned}
\Sigma(wa^2)A + \Sigma(wab)B + \Sigma(wac)C + \cdots + \Sigma(wan)N &= \Sigma(wal) \\
\Sigma(wba)A + \Sigma(wb^2)B + \Sigma(wbc)C + \cdots + \Sigma(wbn)N &= \Sigma(wbl) \\
\Sigma(wca)A + \Sigma(wcb)B + \Sigma(wc^2)C + \cdots + \Sigma(wcn)N &= \Sigma(wcl) \\
&\vdots \\
\Sigma(wna)A + \Sigma(wnb)B + \Sigma(wnc)C + \cdots + \Sigma(wn^2)N &= \Sigma(wnl)
\end{aligned}
\tag{10.27}
$$

where w are the weights of the observations; and again the a's, b's, c's, ..., n's are the coefficients for the unknowns A, B, C, ..., N; the l values are the observations, and Σ signifies the summation from $i = 1$ to m. Notice that the terms in Equations (10.27) are the same as those in Equations (10.26) except for the addition of the w's, which are the relative weights of the observations. In fact, Equations (10.27) can be thought of as the general set of equations for forming the normal equations, since if the weights are equal, they can all be given a value of 1, in which case they will cancel out of Equations (10.27) to produce the special case given by Equations (10.26).

10.7.3 Advantages of the Systematic Approach

Using the systematic methods just demonstrated, the normal equations can be formed for a set of linear equations without having to write the residual equations, compile their summation equation, or take partial derivatives with respect to the unknowns. Rather, for any set of linear equations, the normal equations for the least-squares solution can be written directly.

10.8 TABULAR FORMATION OF THE NORMAL EQUATIONS

Formulation of normal equations from observation equations can be simplified further by handling Equations (10.26) and (10.27) in a table. In this way, a large quantity of numbers can easily be manipulated.

Tabular formulation of the normal equations for the example in Section 10.6.1 is illustrated below. First, Equations (10.17) are made compatible with the generalized form of Equations (10.21). This is shown in Equations (10.28).

$$
\begin{aligned}
&(7)\quad x + y = 3.0 + v_1 \\
&(8)\quad 2x - y = 1.5 + v_2 \\
&(9)\quad x - y = 0.2 + v_3
\end{aligned}
\tag{10.28}
$$

In Equations (10.28), there are two unknowns, x and y, with different coefficients for each equation. Placing the coefficients and the measurements, l's, for each expression of Equation (10.28) into a table, the normal equations can be formed systematically. Table 10.2 shows the coefficients, the appropriate products, and summations in accordance with Equations (10.26).

Substituting the appropriate terms for $\Sigma(a^2)$, $\Sigma(ab)$, $\Sigma(b^2)$, $\Sigma(al)$, and $\Sigma(bl)$ from Table 10.2 into Equations (10.26), the following normal equations are obtained:

$$
\begin{aligned}
6x - 2y &= 6.2 \\
-2x + 3y &= 1.3
\end{aligned}
\tag{10.29}
$$

Notice that Equations (10.29) are exactly the same as those obtained in Section 10.6.1 using the nontabular least-squares method. That is, Equations (10.29) match Equations (10.20).

10.9 USING MATRICES TO FORM THE NORMAL EQUATIONS

Note that the number of normal equations in a parametric least-squares adjustment is always equal to the number of unknown variables. Often, the

Table 10.2 Tabular formation of normal equations

Eqn.	a	b	l	a^2	ab	b^2	al	bl
(7)	1	1	3.0	1	1	1	3.0	3.0
(8)	2	−1	1.5	4	−2	1	3.0	−1.5
(9)	1	−1	0.2	1	−1	1	0.2	−0.2
				$\Sigma = 6$	$\Sigma = -2$	$\Sigma = 3$	$\Sigma = 6.2$	$\Sigma = 1.3$

system of normal equations becomes quite large. But even when dealing with three unknowns, their solution by hand is time consuming. As a consequence, computers and matrix methods as described in Appendixes A through C are almost always used today. In the following subsections we present the matrix methods used when performing least-squares adjustments.

10.9.1 Equal-Weight Case

To develop the matrix expressions for performing least-squares adjustments, analogy will be made with the systematic procedures demonstrated in Section 10.7. For this development, let a system of observation equations be represented by the matrix notation

$$AX = L + V \tag{10.30}$$

where

$$A = \begin{bmatrix} a_{11} & a_{12} & \cdots & a_{1n} \\ a_{21} & a_{22} & \cdots & a_{2n} \\ \vdots & \vdots & \ddots & \vdots \\ a_{m1} & a_{m2} & \cdots & a_{mn} \end{bmatrix}$$

$$X = \begin{bmatrix} x_1 \\ x_2 \\ \vdots \\ x_n \end{bmatrix} \qquad L = \begin{bmatrix} l_1 \\ l_2 \\ \vdots \\ l_m \end{bmatrix} \qquad V = \begin{bmatrix} v_1 \\ v_2 \\ \vdots \\ v_m \end{bmatrix}$$

Note that the system of observation equations (10.30) is identical to Equations (10.21) except that the unknowns are x_1, x_2, ..., x_n instead of A, B, ..., N; and the coefficients of the unknowns are a_{11}, a_{12}, ..., a_{1n} instead of a_1, b_1, ..., n_1.

Subjecting the matrices above to the manipulations given in the following expression, the normal equations are produced [i.e., matrix Equation (10.31*a*) is equivalent to Equations (10.26)]:

$$A^TAX = A^TL \tag{10.31a}$$

Equation (10.31*a*) can also be expressed as

$$NX = A^TL \tag{10.31b}$$

The correspondence between Equations (10.31) and Equations (10.26) becomes clear if the matrices are multiplied and analyzed as follows:

$$A^TA = \begin{bmatrix} a_{11} & a_{21} & \cdots & a_{m1} \\ a_{12} & a_{22} & \cdots & a_{m2} \\ \vdots & \vdots & \vdots & \vdots \\ a_{1n} & a_{2n} & \cdots & a_{mn} \end{bmatrix} \begin{bmatrix} a_{11} & a_{12} & \cdots & a_{1n} \\ a_{21} & a_{22} & \cdots & a_{2n} \\ \vdots & \vdots & \vdots & \vdots \\ a_{m1} & a_{m2} & \cdots & a_{mn} \end{bmatrix} = \begin{bmatrix} n_{11} & n_{12} & \cdots & n_{1n} \\ n_{21} & n_{22} & \cdots & n_{2n} \\ \vdots & \vdots & \ddots & \vdots \\ n_{n1} & n_{n2} & \cdots & n_{nn} \end{bmatrix}$$

$$A^TL = \begin{bmatrix} a_{11} & a_{21} & \cdots & a_{m1} \\ a_{12} & a_{22} & \cdots & a_{m2} \\ \vdots & \vdots & \vdots & \vdots \\ a_{1n} & a_{2n} & \cdots & a_{mn} \end{bmatrix} \begin{bmatrix} l_1 \\ l_2 \\ \vdots \\ l_m \end{bmatrix} = \begin{bmatrix} \sum_{i=1}^{m} (a_{i1})l_i \\ \sum_{i=1}^{m} (a_{i2})l_i \\ \vdots \\ \sum_{i=1}^{m} (a_{in})l_i \end{bmatrix}$$

The individual elements of the N matrix can be expressed in the following summation forms:

$$\begin{array}{llll} n_{11} = \sum_{i=1}^{m} a_{i1}^2 & n_{12} = \sum_{i=1}^{m} a_{i1}a_{i2} & \cdots & n_{1n} = \sum_{i=1}^{m} a_{i1}a_{in} \\ n_{21} = \sum_{i=1}^{m} a_{i2}a_{i1} & n_{22} = \sum_{i=1}^{m} a_{i2}^2 & \cdots & n_{2n} = \sum_{i=1}^{m} a_{i2}a_{in} \\ \vdots & \vdots & \ddots & \vdots \\ n_{n1} = \sum_{i=1}^{m} a_{in}a_{i1} & n_{n2} = \sum_{i=1}^{m} a_{in}a_{i2} & \cdots & n_{nn} = \sum_{i=1}^{m} a_{in}^2 \end{array}$$

By comparing the summations above with those obtained in Equations (10.26), it should be clear that they are the same. Therefore, it has been demonstrated that Equations (10.31*a*) and (10.31*b*) produce the normal equations of a least-squares adjustment. By inspection, it can also be seen that the N matrix is always symmetric (i.e., $n_{ij} = n_{ji}$).

By employing matrix algebra, the solution of normal equations like Equation (10.31*a*) is

$$X = (A^TA)^{-1}A^TL = N^{-1}A^TL \tag{10.32}$$

Example 10.1 To demonstrate this procedure, the problem of Section 10.6.1 will be solved. Equations (10.28) can be expressed in matrix form as

$$AX = \begin{bmatrix} 1 & 1 \\ 2 & -1 \\ 1 & -1 \end{bmatrix} \begin{bmatrix} x \\ y \end{bmatrix} = \begin{bmatrix} 3.0 \\ 1.5 \\ 0.2 \end{bmatrix} + \begin{bmatrix} v_1 \\ v_2 \\ v_3 \end{bmatrix} = L + V \tag{a}$$

Applying Equation (10.31) to Equation (*a*) yields

$$A^TAX = NX = \begin{bmatrix} 1 & 2 & 1 \\ 1 & -1 & -1 \end{bmatrix} \begin{bmatrix} 1 & 1 \\ 2 & -1 \\ 1 & -1 \end{bmatrix} \begin{bmatrix} x \\ y \end{bmatrix} = \begin{bmatrix} 6 & -2 \\ -2 & 3 \end{bmatrix} \begin{bmatrix} x \\ y \end{bmatrix} \quad (b)$$

and

$$A^TL = \begin{bmatrix} 1 & 2 & 1 \\ 1 & -1 & -1 \end{bmatrix} \begin{bmatrix} 3.0 \\ 1.5 \\ 0.2 \end{bmatrix} = \begin{bmatrix} 6.2 \\ 1.3 \end{bmatrix} \quad (c)$$

Finally, the adjusted unknowns, X matrix, are obtained using the matrix methods of Equation (10.32), which yields

$$X = N^{-1}A^TL = \begin{bmatrix} 6 & -2 \\ -2 & 3 \end{bmatrix}^{-1} \begin{bmatrix} 6.2 \\ 1.3 \end{bmatrix} = \begin{bmatrix} 1.514 \\ 1.442 \end{bmatrix} \quad (d)$$

Notice that the normal equations, and the solution, in this method are the same as those obtained in Section 10.6.1.

10.9.2 Weighted Case

A system of weighted linear observation equations can be expressed in matrix notation as

$$WAX = WL + WV \quad (10.33)$$

Using the same methods as demonstrated in Section 10.9.1, it is possible to show that the normal equations for this weighted system are

$$A^TWAX = A^TWL \quad (10.34a)$$

Equation (10.34a) can also be expressed as

$$NX = A^TWL \quad (10.34b)$$

where $N = A^TWA$.

Using matrix algebra, the least-squares solution of these weighted normal equations is

$$X = (A^TWA)^{-1}A^TWL = N^{-1}A^TWL \quad (10.35)$$

where W is the weight matrix as defined in Chapter 9.

10.10 LEAST SQUARES SOLUTION OF NONLINEAR SYSTEMS

In Appendix C we discuss a method of solving a nonlinear system of equations using a Taylor series approximation. Following this procedure, the least-squares solution for a system of nonlinear equations can be found as follows:

Step 1: Write the first-order Taylor series approximation for each equation.

Step 2: Determine initial approximations for the unknowns in the equations of step 1.

Step 3: Use matrix methods similar to those discussed in Section 10.9 to find the least-squares solution for the equations of step 1 (these are corrections to the initial approximations).

Step 4: Apply the corrections to the initial approximations.

Step 5: Repeat steps 1 through 4 until the corrections become sufficiently small.

A system of nonlinear equations, expressed as functions $F_1, F_2, \ldots, F_m$, that are linearized by a Taylor series approximation can be written as

$$JX = K + V \tag{10.36}$$

where the *Jacobian* matrix J contains the coefficients of the linearized observation equations. The individual matrices in Equation (10.36) are

$$J = \begin{bmatrix} \dfrac{\partial F_1}{\partial x_1} & \dfrac{\partial F_1}{\partial x_2} & \cdots & \dfrac{\partial F_1}{\partial x_n} \\ \dfrac{\partial F_2}{\partial x_1} & \dfrac{\partial F_2}{\partial x_2} & \cdots & \dfrac{\partial F_2}{\partial x_n} \\ \vdots & \vdots & \vdots & \vdots \\ \dfrac{\partial F_m}{\partial x_1} & \dfrac{\partial F_m}{\partial x_2} & \cdots & \dfrac{\partial F_m}{\partial x_n} \end{bmatrix}$$

$$X = \begin{bmatrix} dx_1 \\ dx_2 \\ \vdots \\ dx_n \end{bmatrix} \qquad K = \begin{bmatrix} l_1 - F_1(x_{1_0}, x_{2_0}, \ldots, x_{n_0}) \\ l_2 - F_2(x_{1_0}, x_{2_0}, \ldots, x_{n_0}) \\ \vdots \\ l_m - F_m(x_{1_0}, x_{2_0}, \ldots, x_{n_0}) \end{bmatrix} \qquad V = \begin{bmatrix} v_1 \\ v_2 \\ \vdots \\ v_m \end{bmatrix}$$

The vector of least-squares corrections in the equally weighted system of Equation (10.36) is given by

$$X = (J^T J)^{-1} J^T K = N^{-1} J^T K \tag{10.37}$$

Similarly, the system of weighted equations is

$$WJX = WK \tag{10.38}$$

and its solution is

$$X = (J^TWJ)^{-1}J^TWK = N^{-1}J^TWK \tag{10.39}$$

where W is the weight matrix as defined in Chapter 9.

Example 10.2 Find the least-squares solution for the following system of nonlinear equations:

$$\begin{aligned} F:&\quad x + y - 2y^2 = -4 \\ G:&\quad x^2 + y^2 \quad\;\; = \;\; 8 \\ H:&\quad 3x^2 - y^2 \quad = \;\; 7.7 \end{aligned} \tag{e}$$

SOLUTION

Step 1: Determine the elements of the J matrix by taking partial derivatives of Equations (e) with respect to the unknowns x and y. Then write the first-order Taylor series equations:

$$\frac{\partial F}{\partial x} = 1 \qquad \frac{\partial G}{\partial x} = 2x \qquad \frac{\partial H}{\partial x} = 6x$$

$$\frac{\partial F}{\partial y} = 1 - 4y \qquad \frac{\partial G}{\partial y} = 2y \qquad \frac{\partial H}{\partial y} = -2y$$

$$JX = \begin{bmatrix} 1 & 1 - 4y_0 \\ 2x_0 & 2y_0 \\ 6x_0 & -2y_0 \end{bmatrix} \begin{bmatrix} dx \\ dy \end{bmatrix} = \begin{bmatrix} -4 - F(x_0,y_0) \\ 8 - G(x_0,y_0) \\ 7.7 - H(x_0,y_0) \end{bmatrix} = K \tag{f}$$

Step 2: Determine initial approximations for the solution of the equations. Initial approximations can be derived by solving any two equations for x and y. This was done in Section C.3 for the equations for (F) and (G), and their solution was $x_0 = 2$ and $y_0 = 2$. Using these values, the evaluation of the equations gives:

$$F(x_0,y_0) = -4 \qquad G(x_0,y_0) = 8 \qquad H(x_0,y_0) = 8 \tag{g}$$

Substituting Equations (g) into the K matrix of Equation (f), the K matrix becomes:

$$K = \begin{bmatrix} -4 - (-4) \\ 8 - 8 \\ 7.7 - 8 \end{bmatrix} = \begin{bmatrix} 0 \\ 0 \\ -0.3 \end{bmatrix}$$

It should not be surprising that the first two rows of the K matrix are zero since the initial approximations were determined using these two equations. In successive iterations, these values will change and all terms will become nonzero.

Step 3: Solve the system using Equation (10.37):

$$N = J^TJ = \begin{bmatrix} 1 & 4 & 12 \\ -7 & 4 & -4 \end{bmatrix} \begin{bmatrix} 1 & -7 \\ 4 & 4 \\ 12 & -4 \end{bmatrix} = \begin{bmatrix} 161 & -39 \\ -39 & 81 \end{bmatrix}$$

$$J^TK = \begin{bmatrix} 1 & 4 & 12 \\ -7 & 4 & -4 \end{bmatrix} \begin{bmatrix} 0 \\ 0 \\ -0.3 \end{bmatrix} = \begin{bmatrix} -3.6 \\ 1.2 \end{bmatrix} \qquad (h)$$

Substituting the matrices of Equation (h) into Equation (10.37), the solution for the first iteration is[1]

$$X = N^{-1}J^TK = \begin{bmatrix} -0.02125 \\ 0.00458 \end{bmatrix}$$

Step 4: Apply the corrections to the initial approximations for the first iteration:

$$x_0 = 2.00 - 0.02125 = 1.97875 \qquad y_0 = 2.00 + 0.00458 = 2.00458$$

Step 5: Repeating steps 2 through 4, we get the following results:

$$X = N^{-1}J^TK = \begin{bmatrix} 157.61806 & -38.75082 \\ -38.75082 & 81.40354 \end{bmatrix}^{-1} \begin{bmatrix} -0.12393 \\ 0.75219 \end{bmatrix} = \begin{bmatrix} 0.00168 \\ 0.01004 \end{bmatrix}$$

$$x = 1.97875 + 0.00168 = 1.98043$$

$$y = 2.00458 + 0.01004 = 2.01462$$

Iterating a third time yields extremely small corrections, and thus the final solution, rounded to the hundredths place, is $x = 1.98$ and $y = 2.01$. Notice that N changed by a relatively small amount from the first iteration to the second iteration. This can be expected when the initial approxima-

[1] Note that although the solution represents more significant figures than can be warranted by the observations, it is important to carry more digits than are desired for the final solution. Failure to carry enough digits can result in a system that will never converge; rather, it may *bounce* above and below the solution, or it may take more iterations due to these rounding errors. This mistake has been made by many beginning students. The answer should be rounded only after convergence has been achieved.

tions are close to their final values. Thus if these computations were done by hand, the initial N matrix could be used for each iteration, making it only necessary to recompute J^TK between iterations. However, this procedure should be used with caution, for if the initial approximations vary too much from their final values, it will result in an incorrect solution. One should always perform complete computations when doing the solution with the aid of a computer.

10.11 LEAST-SQUARES FIT OF POINTS TO A LINE OR CURVE

Frequently in engineering work, it is desirable or necessary to fit a straight line or curve to a set of points with known coordinates. In solving this type of problem, it is first necessary to decide on the appropriate mathematical model for the data. The decision on whether to use a straight line, parabola, or some other higher-order curve can generally be made after plotting the data and studying their form, or by checking the size of the residuals after the least-squares solution with the first selected line or curve.

10.11.1 Fitting Data to a Straight Line

Consider the data illustrated in Figure 10.2. The straight line shown in the figure can be represented by the equation

$$y = mx + b \tag{10.40}$$

where x and y are the coordinates of a point, m is the slope of a line, and b is the y intercept at $x = 0$. If the points were truly linear, and there were no measurement or experimental errors, all coordinates would lie on a straight line. However, this is rarely the case as seen in Figure 10.2, and thus it is possible that (1) the points contain errors, (2) the mathematical model is a

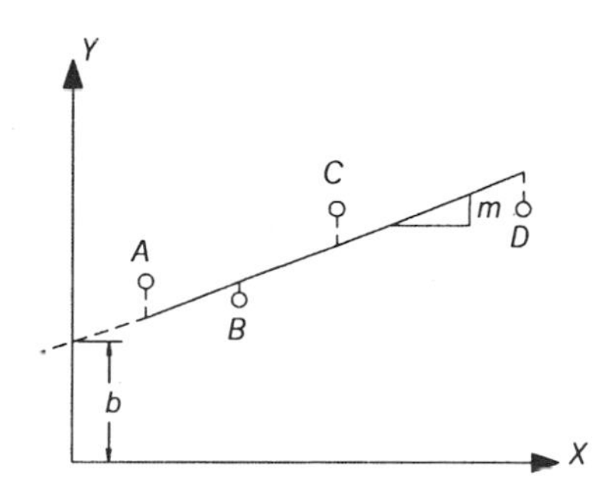

Figure 10.2 Points on a line.

higher-order curve, or both. If a line is selected as the model for the data, the equation of the best-fitting straight line is found by adding residuals to Equations (10.40) which account for the errors shown in the figure. Equations for the four data points A, B, C, and D of Figure 10.2 are rewritten as

$$\begin{aligned} y_A + v_{yA} &= mx_A + b \\ y_B + v_{yB} &= mx_B + b \\ y_C + v_{yC} &= mx_C + b \\ y_D + v_{yD} &= mx_D + b \end{aligned} \tag{10.41}$$

Equations (10.41) contain two unknowns, m and b, with four observations. Their matrix representation is

$$AX = L + V \tag{10.42}$$

where

$$A = \begin{bmatrix} x_a & 1 \\ x_b & 1 \\ x_c & 1 \\ x_d & 1 \end{bmatrix} \qquad X = \begin{bmatrix} m \\ b \end{bmatrix} \qquad L = \begin{bmatrix} y_a \\ y_b \\ y_c \\ y_d \end{bmatrix} \qquad V = \begin{bmatrix} v_{ya} \\ v_{yb} \\ v_{yc} \\ v_{yd} \end{bmatrix}$$

Equation (10.42) is solved by the least-squares method using Equation (10.32). If some data were more reliable than others, relative weights could be introduced and a weighted least-squares solution could be obtained using Equation (10.35).

Example 10.3 Find the best-fit line for the following points, whose x and y coordinates are given in parentheses.

A: (3.00, 4.50) B: (4.25, 4.25) C: (5.50, 5.50) D: (8.00, 5.50)

SOLUTION: Following Equations (10.41), the four observation equations for the coordinate pairs are

$$\begin{aligned} 3.00m + b &= 4.50 + v_a \\ 4.25m + b &= 4.25 + v_b \\ 5.50m + b &= 5.50 + v_c \\ 8.00m + b &= 5.50 + v_d \end{aligned} \tag{i}$$

Rewriting Equations (i) into matrix form yields

$$\begin{bmatrix} 3.00 & 1 \\ 4.25 & 1 \\ 5.50 & 1 \\ 8.00 & 1 \end{bmatrix} \begin{bmatrix} m \\ b \end{bmatrix} = \begin{bmatrix} 4.50 \\ 4.25 \\ 5.50 \\ 5.50 \end{bmatrix} + \begin{bmatrix} v_A \\ v_B \\ v_C \\ v_D \end{bmatrix} \tag{j}$$

To form the normal equations, premultiply matrices A and L of Equation (j) by A^T and get

$$\begin{bmatrix} 121.3125 & 20.75 \\ 20.7500 & 4.00 \end{bmatrix} \begin{bmatrix} m \\ b \end{bmatrix} = \begin{bmatrix} 105.8125 \\ 19.7500 \end{bmatrix} \tag{k}$$

Applying Equation (10.32) the solution of Equation (k) is

$$X = \begin{bmatrix} m \\ b \end{bmatrix} = \begin{bmatrix} 121.3125 & 20.7500 \\ 20.7500 & 4.0000 \end{bmatrix}^{-1} \begin{bmatrix} 105.81250 \\ 19.75000 \end{bmatrix} = \begin{bmatrix} 0.246 \\ 3.663 \end{bmatrix}$$

Thus the most probable values for m and b to the nearest hundredth are 0.25 and 3.66, respectively. To obtain the residuals, matrix equation (10.30) is rearranged and solved:

$$V = AX - L = \begin{bmatrix} 3.00 & 1 \\ 4.25 & 1 \\ 5.50 & 1 \\ 8.00 & 1 \end{bmatrix} \begin{bmatrix} 0.246 \\ 3.663 \end{bmatrix} - \begin{bmatrix} 4.50 \\ 4.25 \\ 5.50 \\ 5.50 \end{bmatrix} = \begin{bmatrix} -0.10 \\ 0.46 \\ -0.48 \\ 0.13 \end{bmatrix}$$

10.11.2 Fitting Data to a Parabola

For certain data sets or in special situations, a parabola will fit the situation best. An example would be fitting a vertical curve to an existing roadbed. The general equation of a parabola is

$$Ax^2 + Bx + C = y \tag{10.43}$$

Again, since the data rarely fit the equation exactly, residuals are introduced. For the data shown in Figure 10.3, the following observation equations can be written:

$$\begin{aligned} Ax_a^2 + Bx_a + C &= y_a + v_a \\ Ax_b^2 + Bx_b + C &= y_b + v_b \\ Ax_c^2 + Bx_c + C &= y_c + v_c \\ Ax_d^2 + Bx_d + C &= y_d + v_d \\ Ax_e^2 + Bx_e + C &= y_e + v_e \end{aligned} \tag{10.44}$$

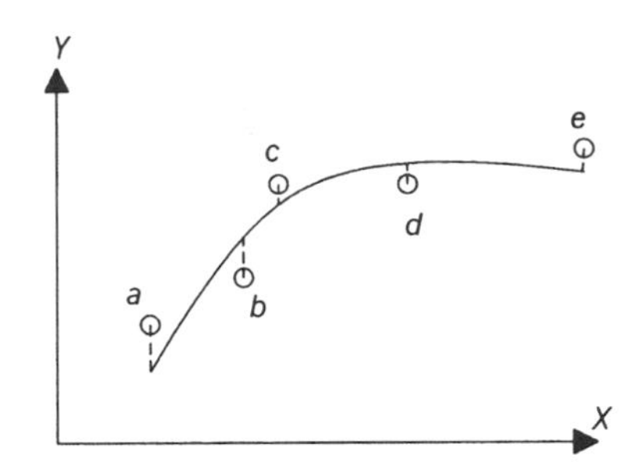

Figure 10.3 Points on a parabolic curve.

Equations (10.44) contain three unknowns, A, B, and C, with five equations. Thus this represents a redundant system that can be solved using least squares. In terms of the unknown coefficients, Equations (10.44) are linear and can be represented in matrix form as

$$AX = L + V \tag{10.45}$$

Since this is a linear system, it is solved using Equation (10.32). If weights were introduced, Equation (10.35) should be used. The steps taken in the solution are similar to those used in Section 10.11.1.

10.12 CALIBRATION OF AN EDM INSTRUMENT

Calibration of an EDM instrument is necessary to ensure confidence in the distances it measures. In calibrating these devices, if they internally make corrections and reductions for atmospheric conditions, earth curvature, and slope, it is first necessary to determine if these corrections are made correctly. Once these corrections are applied properly, the instruments with their reflectors must be checked to determine their constant and scaling corrections. This is often accomplished using a calibration baseline. The observation equation for an electronically measured distance on a calibration baseline is

$$SD_A + C = D_H - D_A + V_{DH} \tag{10.46}$$

where S is a scaling factor for the EDM, C an instrument–reflector constant, D_H the measured horizontal distance with all atmospheric and slope corrections applied, D_A the published horizontal calibrated distance for the baseline, and V_{DH} the residual error for each measurement. This is a linear equation with two unknowns, S and C. Systems of these equations can be solved using Equation (10.32).

Example 10.4 A surveyor wishes to use an instrument–reflector combination that has an unknown constant value. Baseline measurements were carefully made, and following the manufacturer's recommendations, the necessary corrections were applied for the atmospheric conditions, earth curvature, and slope. Use these corrected distances and the published values listed in Table 10.3 to determine the instrument/reflector constant (C) and scaling factor (S) for the system.

SOLUTION: In accordance with Equation (10.46), the matrix equation for this problem is

$$\begin{bmatrix} 149.9975 & 1 \\ 149.9975 & 1 \\ 430.0101 & 1 \\ 430.0101 & 1 \\ 1400.0030 & 1 \\ 1400.0030 & 1 \\ 280.0126 & 1 \\ 280.0126 & 1 \\ 1250.0055 & 1 \\ 1250.0055 & 1 \\ 969.9929 & 1 \\ 969.9929 & 1 \end{bmatrix} \begin{bmatrix} S \\ C \end{bmatrix} = \begin{bmatrix} 150.0175 - 149.9975 \\ 150.0174 - 149.9975 \\ 430.0302 - 430.0101 \\ 430.0304 - 430.0101 \\ 1400.0223 - 1400.0030 \\ 1400.0221 - 1400.0030 \\ 280.0327 - 280.0126 \\ 280.0331 - 280.0126 \\ 1250.0248 - 1250.0055 \\ 1250.0257 - 1250.0055 \\ 970.0119 - 969.9929 \\ 970.0125 - 969.9929 \end{bmatrix} + V$$

Using Equation (10.32), the solution is $S = -0.0000007$ (-0.7 ppm) and $C = 0.0203$ m. Thus the constant value for the instrument/reflector pair is approximately 20 mm.

10.13 LEAST-SQUARES ADJUSTMENT USING CONDITIONAL EQUATIONS

As stated in Section 10.5, observations can also be adjusted using conditional equations. In this section this form of adjustment will be demonstrated by

Table 10.3 EDM instrument–reflector calibration data

Distance	D_A (m)	D_H (m)	Distance	D_A (m)	D_H (m)
0–150	149.9975	150.0175	150–0	149.9975	150.0174
0–430	430.0101	430.0302	430–0	430.0101	430.0304
0–1400	1400.0030	1400.0223	1400–0	1400.0030	1400.0221
150–430	280.0126	280.0327	430–150	280.0126	280.0331
150–1400	1250.0055	1250.0248	1400–150	1250.0055	1250.0257
430–1400	969.9929	970.0119	1400–430	969.9929	970.0125

using the condition that the sum of the angles in the horizon at a point must equal 360°.

Example 10.5 While measuring angles at a station, the horizon was closed. The following measurements and their standard deviations were obtained:

No.	Angle	S
a_1	134°38′56″	±6.7″
a_2	83°17′35″	±9.9″
a_3	142°03′14″	±4.3″

What are the most probable values for the observations above?

SOLUTION: In a conditional adjustment, the most probable set of residuals are found that satisfy a given mathematical condition. In this case the condition is that the sum of the three angles is equal to 360°. Since the three angles measured actually sum to 359°59′45″, the angular misclosure is 15″, and thus errors are present. The following residual equations can be written for the measurements listed above.

$$v_1 + v_2 + v_3 = 360° - (a_1 + a_2 + a_3) = 15'' \qquad (l)$$

where the a's represent the observations and the v are residuals.

Applying the fundamental condition for a weighted least-squares adjustment, the following equation must be minimized:

$$F = w_1v_1^2 + w_2v_2^2 + w_3v_3^2 \qquad (m)$$

where the w's are weights, which are the inverses of the squares of the standard deviations.

Obviously, Equation (l) can be rearranged such that v_3 is expressed as a function of the other two residuals, or

$$v_3 = 15 - (v_1 + v_2) \qquad (n)$$

Substituting Equation (n) into Equation (m),

$$F = w_1v_1^2 + w_2v_2^2 + w_3[15 - (v_1 + v_2)]^2 \qquad (o)$$

Taking the partial derivatives of F with respect to both v_1 and v_2, respectively in Equation (o),

$$\frac{\partial F}{\partial v_1} = 2w_1v_1 + 2w_3[15'' - (v_1 + v_2)](-1) = 0$$
$$\frac{\partial F}{\partial v_2} = 2w_2v_2 + 2w_3[15'' - (v_1 + v_2)](-1) = 0 \tag{p}$$

Rearranging Equations (*p*) and substituting in the appropriate weights yields the following normal equations:

$$\left(\frac{1}{6.7^2} + \frac{1}{4.3^2}\right)v_1 + \frac{1}{4.3^2}v_2 = 15\left(\frac{1}{4.3^2}\right)$$
$$\frac{1}{4.3^2}v_1 + \left(\frac{1}{9.9^2} + \frac{1}{4.3^2}\right)v_2 = 15\left(\frac{1}{4.3^2}\right) \tag{q}$$

Solving Equations (*q*) for v_1 and v_2 gives

$$v_1 = 4.2''$$
$$v_2 = 9.1''$$

By substituting these residual values into Equation (*n*), residual v_3 is determined to be

$$v_3 = 15'' - (4.2'' + 9.1'') = 1.7''$$

Finally, the adjusted observations are obtained by adding the computed residuals to the observations, and thus:

No.	Observed Angle	v	Adjusted Angle
a_1	134°38′56″	4.2″	134°39′00.2″
a_2	83°17′35″	9.1″	83°17′44.1″
a_3	142°03′14″	1.7″	142°03′15.7″
			Σ = 360°00′00.0″

Note that geometric closure has been enforced in the adjusted angles to make their sum exactly 360°. Note also that the angle having the smallest standard deviation received the smallest correction (i.e., its residual is smallest), and the angle having the largest standard deviation received the largest correction.

10.14 EXAMPLE USING OBSERVATION EQUATIONS

Example 10.5 can also be done using observation equations. In this case the three observations are related to their adjusted values and their residuals by writing the following observation equations:

$$\begin{aligned} a_1 &= 134°38'56'' + v_1 \\ a_2 &= 83°17'35'' + v_2 \\ a_3 &= 142°03'14'' + v_3 \end{aligned} \qquad (r)$$

While these equations relate the adjusted observations to their measured values, they cannot be solved in this form. What is needed is the *constraint*,[2] which states that the sum of the three angles equals 360°. This equation is

$$a_1 + a_2 + a_3 = 360° \qquad (s)$$

Rearranging Equation (*s*) to solve for a_3 yields

$$a_3 = 360° - (a_1 + a_2) \qquad (t)$$

Substituting Equation (*t*) into Equations (*r*) gives

$$\begin{aligned} a_1 &= 134°38'56'' + v_1 \\ a_2 &= 83°17'35'' + v_2 \\ 360° - (a_1 + a_2) &= 142°03'14'' + v_3 \end{aligned} \qquad (u)$$

This is a linear problem with two unknowns, a_1 and a_2. The weighted observation equation solution is obtained by solving Equation (10.35). Appropriate matrices for this problem are

$$A = \begin{bmatrix} 1 & 0 \\ 0 & 1 \\ -1 & -1 \end{bmatrix} \quad W = \begin{bmatrix} \frac{1}{6.7^2} & 0 & 0 \\ 0 & \frac{1}{9.9^2} & 0 \\ 0 & 0 & \frac{1}{4.3^2} \end{bmatrix} \quad L = \begin{bmatrix} 134°38'56'' \\ 83°17'35'' \\ 142°03'14'' - 360° \end{bmatrix}$$

Performing matrix manipulations, the coefficients of the normal equations are

[2] Chapter 19 covers the use of constraint equations in least squares adjustment.

$$A^TWA = \begin{bmatrix} 1 & 0 & -1 \\ 0 & 1 & -1 \end{bmatrix} \begin{bmatrix} \frac{1}{6.7^2} & 0 & 0 \\ 0 & \frac{1}{9.9^2} & 0 \\ 0 & 0 & \frac{1}{4.3^2} \end{bmatrix} \begin{bmatrix} 1 & 0 \\ 0 & 1 \\ -1 & -1 \end{bmatrix}$$

$$= \begin{bmatrix} 0.07636 & 0.05408 \\ 0.05408 & 0.06429 \end{bmatrix}$$

$$A^TWL = \begin{bmatrix} 14.7867721 \\ 12.6370848 \end{bmatrix}$$

Finally, X is found from

$$X = (A^TWA)^{-1}A^TWL = \begin{bmatrix} 134°39'00.2'' \\ 83°17'44.1'' \end{bmatrix}$$

Using Equation (t), it can now be determined that a_3 is 360° − 134°39′00.2″ − 83″17′44.1″ = 142°03′15.7″. This is the same result as obtained in Section 10.13. It is important to note that no matter the method of least-squares adjustment, if the procedures are performed properly the same solution will be obtained.

PROBLEMS

10.1 Calculate the most probable values for A and B in the equations below by the method of least squares. Consider the observations to be of equal weight. (Use the tabular method to form normal equations.)

(a) $A + 2B = 10.50 + v_1$
(b) $2A - 3B = 5.55 + v_2$
(c) $2A - B = -10.50 + v_3$

10.2 If observations (a), (b) and (c) in Problem 10.1 have weights of 2, 3, and 1 respectively, solve these equations for the most probable values of A and B using weighted least squares. (Use the tabular method to form normal equations.)

10.3 Repeat Problem 10.1 using matrices.

10.4 Repeat Problem 10.2 using matrices.

10.5 Do Problems 10.1 and 10.2 using the Program MATRIX.

10.6 Three angles were measured to close the horizon at station Red. The measured values and their standard deviations are:

Angle	Value	S
1	114°23′05″	±6.5″
2	138°17′59″	±3.5″
3	107°19′03″	±8.9″

What is the most probable value for each angle?

10.7 The three interior angles of a triangle were measured as:

Angle	Value	S
1	65°23′15″	±7.6″
2	83°15′43″	±5.1″
3	31°21′12″	±8.4″

What is the most probable value for each angle?

10.8 Eight blocks of the Main Street are to be reconstructed. The existing street consists of jogging, short segments as tabulated in the traverse survey data below. Assuming coordinates of X = 1000.0 and Y = 1000.0 at station A, and that the azimuth of AB is 90°, define a new straight alignment for a reconstructed street passing through this area which best conforms to the present alignment. Give the Y intercept and the azimuth of the new alignment.

Course	Length (ft)	Station	Angle to Right
AB	735.7	B	180°17′
BC	464.8	C	179°51′
CD	503.1	D	179°28′
DE	820.0	E	180°33′
EF	917.3	F	179°10′
FG	329.8	G	179°59′
GH	287.4	H	181°02′
HI	345.9		

10.9 Use the Program ADJUST to do Problem 10.8.

10.10 The property corners on a single block with an alley are shown as a straight line with a due east bearing on a recorded plat. During a recent survey, all the lot corners were found, and measurements from the station A to each were obtained. The surveyor wishes to determine the possibility of disturbance of the corners by checking their fit to a straight line. A sketch of the situation is shown in Figure 10.4, and

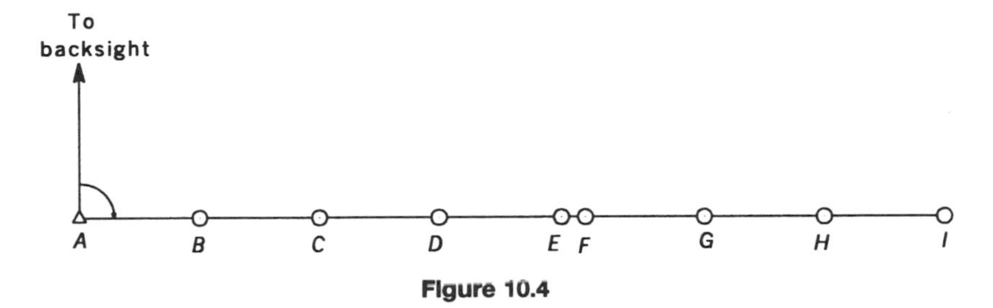

Figure 10.4

the results of the survey are given below. Assuming that station *A* has coordinates of $X = 5000.00$ and $Y = 5000.00$, and that the coordinates of the backsight station are $X = 5000.10$ and $Y = 5200.00$, determine the best-fitting line for the corners. Give the *Y* intercept and the bearing of the best-fit line.

Course	Distance (ft)	Angle at *A*
AB	100.02	90°01′14″
AC	200.03	90°00′11″
AD	300.14	89°57′50″
AE	399.90	90°01′12″
AF	420.02	90°00′54″
AG	519.94	90°00′30″
AH	620.18	89°59′47″
AI	720.08	90°00′26″

10.11 Use the Program ADJUST to do Problem 10.10.

10.12 Calculate a best-fit parabola for the following data obtained on a survey of an existing vertical curve. The curve starts at station 10+00 and ends at station 18+00.

Station	Elevation	Station	Elevation
10+00	51.2	15+00	46.9
11+00	49.5	16+00	47.3
12+00	48.2	17+00	48.3
13+00	47.3	18+00	49.6
14+00	46.8		

10.13 Use the Program ADJUST to do Problem 10.12.

10.14 Using a procedure similar to that in Section 10.7.1, derive Equations (10.27).

10.15 Using a procedure similar to that used in Section 10.9.1, show that the matrix operations in Equation (10.34a) result in the normal equations given by Equation (10.27).

10.16 Discuss the importance of selecting the mathematical model when adjusting data.

10.17 The values for three angles in a plane triangle, measured using the repetition method, are:

Angle	Number of Repetitions	Value
A	2	14°25′20″
B	3	58°16′00″
C	6	107°19′10″

The measured lengths of the courses are:

$$AB = 971.25 \text{ ft} \qquad BC = 253.25 \text{ ft} \qquad CA = 865.28 \text{ ft}$$

The following uncertainties are assumed for each measurement:

$$\sigma_i = \pm 0.004 \text{ ft} \qquad \sigma_t = \pm 0.010 \text{ ft} \qquad \sigma_r = \pm 6.0'' \qquad \sigma_p = \pm 2.3''$$

What are the most probable values for the angles? Use the conditional equation method.

10.18 Do Problem 10.17 using observation equations and a constraint equation as presented in Section 10.14.

10.19 The following data were collected on a calibration baseline. Calibrated and measured distances are D_A and D_H, respectively. Atmospheric refraction and earth curvature corrections were made to the measured distances, which are in units of meters. Determine the instrument–reflector constant and any scaling factor.

Distance	D_A	D_H	Distance	D_A	D_H
0–150	149.9104	149.9447	150–0	149.9104	149.9435
0–430	430.0010	430.0334	430–0	430.0010	430.0340
0–1400	1399.9313	1399.9777	1400–0	1399.9313	1399.9519
150–430	280.0906	280.1238	430–150	280.0906	280.1230
150–1400	1250.0209	1250.0795	1400–150	1250.0209	1250.0664
430–1400	969.9303	969.9546	1400–430	969.9303	969.9630

10.20 A survey of the centerline of a horizontal metric curve is performed to determine the as-built specifications. The coordinates for six points along the curve are:

Point	X	Y
1	10,006.82	10,007.31
2	10,013.12	10,015.07
3	10,024.01	10,031.83
4	10,032.44	10,049.95
5	10,038.26	10,069.04
6	10,041.39	10,088.83

(a) Using Equation (C.10), compute the most probable values for the radius and center of the circle.

(b) If two points located on the tangents have (X, Y) coordinates of (9987.36, 9987.40) and (10,044.09, 10,119.54), what are the coordinates of the PC and PT of the curve?

11

ADJUSTMENT OF LEVEL NETS

11.1 INTRODUCTION

Differential leveling measurements are used to determine differences in elevation between stations. As with all measurements, these observations are subject to random errors which can be adjusted using the method of least squares. In this chapter, the observation equation method for adjusting differential leveling observations by least squares is developed, and several examples are given to illustrate the adjustment procedures.

11.2 OBSERVATION EQUATION

To apply the method of least squares in leveling adjustments, a prototype observation equation is first written for any elevation difference. Figure 11.1 illustrates the functional relationship for the observed elevation difference between two stations *I* and *J*. The equation is expressed as

$$E_j - E_i = \Delta\text{Elev}_{ij} + v_{\Delta\text{Elev}_{ij}} \tag{11.1}$$

This prototype observation equation relates the unknown elevations E_i and E_j of any two stations, *I* and *J*, with the differential leveling observation ΔElev_{ij} and its residual $v_{\Delta\text{Elev}_{ij}}$.

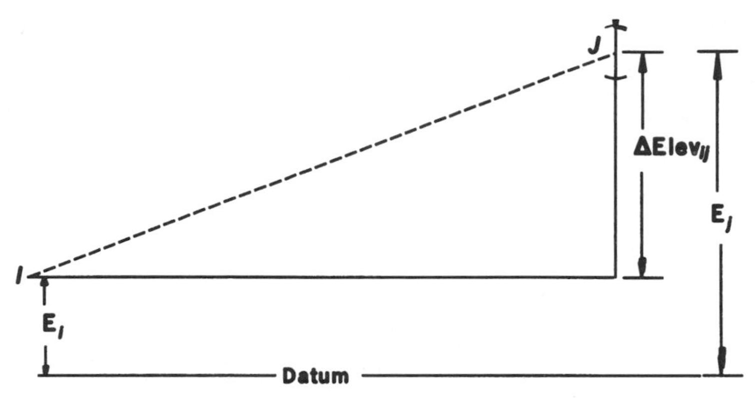

Figure 11.1 Differential leveling observation.

11.3 UNWEIGHTED EXAMPLE

In Figure 11.2, a leveling network and its survey data, given in feet, are shown. Assume that all observations are equal in weight. In this figure, arrows indicate the direction of leveling, and thus for line 1, leveling proceeded from benchmark X to A with an observed elevation difference of +5.10 ft. By substituting into prototype Equation (11.1), an observation equation is written for each measurement in Figure 11.2, which results in the equation set

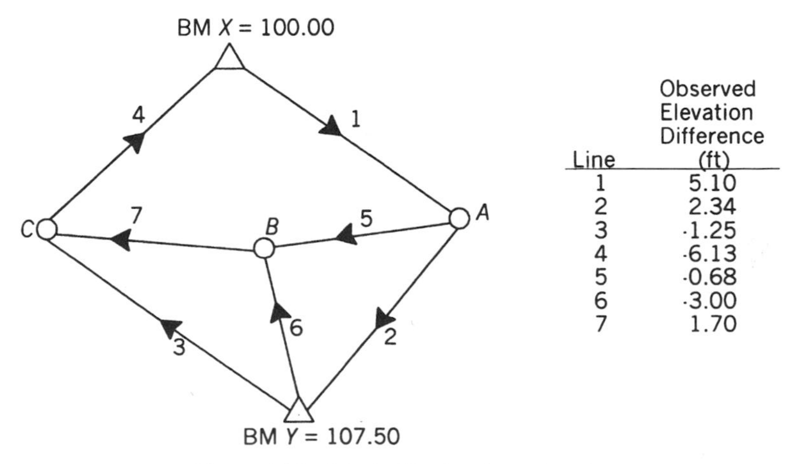

Line	Observed Elevation Difference (ft)
1	5.10
2	2.34
3	-1.25
4	-6.13
5	-0.68
6	-3.00
7	1.70

Figure 11.2 Interlocking leveling network.

$$
\begin{array}{llllll}
A & & & -\text{BM } X = & 5.10 + v_1 & \\
-A & & & +\text{BM } Y = & 2.34 + v_2 & \\
 & & C & -\text{BM } Y = & -1.25 + v_3 & \\
 & & -C & +\text{BM } X = & -6.13 + v_4 & (11.2) \\
-A & B & & = & -0.68 + v_5 & \\
 & B & & -\text{BM } Y = & -3.00 + v_6 & \\
 & -B & C & = & +1.70 + v_7 &
\end{array}
$$

Rearranging so that the known benchmarks are on the right-hand sides of the equations and substituting in their appropriate elevations gives

$$
\begin{array}{lllll}
A & & & = & +105.10 + v_1 \\
-A & & & = & -105.16 + v_2 \\
 & & C & = & +106.25 + v_3 \\
 & & -C & = & -106.13 + v_4 \\
-A & +B & & = & -0.68 + v_5 \\
 & B & & = & +104.50 + v_6 \\
 & -B & +C & = & +1.70 + v_7
\end{array}
\tag{11.3}
$$

In this example there are three unknowns, A, B, and C. Equations (11.3) can be written in the following matrix form:

$$
AX = L + V \tag{11.4}
$$

where

$$
A = \begin{bmatrix} 1 & 0 & 0 \\ -1 & 0 & 0 \\ 0 & 0 & 1 \\ 0 & 0 & -1 \\ -1 & 1 & 0 \\ 0 & 1 & 0 \\ 0 & -1 & 1 \end{bmatrix} \quad X = \begin{bmatrix} A \\ B \\ C \end{bmatrix} \quad L = \begin{bmatrix} 105.10 \\ -105.16 \\ 106.25 \\ -106.13 \\ -0.68 \\ 104.50 \\ 1.70 \end{bmatrix} \quad V = \begin{bmatrix} v_1 \\ v_2 \\ v_3 \\ v_4 \\ v_5 \\ v_6 \\ v_7 \end{bmatrix}
$$

Note in the A matrix that when an unknown does not appear in an equation, its coefficient is zero. Since this is an unweighted example, according to Equation (10.31) the normal equations are

$$A^TAX = NX = \begin{bmatrix} 3 & -1 & 0 \\ -1 & 3 & -1 \\ 0 & -1 & 3 \end{bmatrix} \begin{bmatrix} A \\ B \\ C \end{bmatrix} \quad \text{and} \quad A^TL = \begin{bmatrix} 210.94 \\ 102.12 \\ 214.08 \end{bmatrix} \tag{11.5}$$

The solution of Equation (11.5) can be found using the methods of Equation (10.32):

$$X = N^{-1}A^TL = \begin{bmatrix} 3 & -1 & 0 \\ -1 & 3 & -1 \\ 0 & -1 & 3 \end{bmatrix}^{-1} \begin{bmatrix} 210.94 \\ 102.12 \\ 214.08 \end{bmatrix}$$

$$X = \begin{bmatrix} 0.38095 & 0.14286 & 0.04762 \\ 0.14286 & 0.42857 & 0.14286 \\ 0.04762 & 0.14286 & 0.38095 \end{bmatrix} \begin{bmatrix} 210.94 \\ 102.12 \\ 214.08 \end{bmatrix} = \begin{bmatrix} 105.141 \\ 104.483 \\ 106.188 \end{bmatrix} \tag{11.6}$$

From Equation (11.6), the rounded most probable elevations for A, B, and C are 105.14, 104.48, and 106.19 ft, respectively. To compute the residuals, the following rearranged form of Equations (11.4) is applied:

$$V = AX - L \tag{11.7}$$

From Equation (11.7), the matrix solution for V is

$$V = \begin{bmatrix} 1 & 0 & 0 \\ -1 & 0 & 0 \\ 0 & 0 & 1 \\ 0 & 0 & -1 \\ -1 & 1 & 0 \\ 0 & 1 & 0 \\ 0 & -1 & 1 \end{bmatrix} \begin{bmatrix} 105.141 \\ 104.483 \\ 106.188 \end{bmatrix} - \begin{bmatrix} 105.10 \\ -105.16 \\ 106.25 \\ -106.13 \\ -0.68 \\ 104.50 \\ 1.70 \end{bmatrix} = \begin{bmatrix} 0.041 \\ 0.019 \\ -0.062 \\ -0.058 \\ 0.022 \\ -0.017 \\ 0.005 \end{bmatrix}$$

11.4 WEIGHTED EXAMPLE

In Section 9.6 it was shown that relative weights for adjusting level lines are inversely proportional to the lengths of the lines, or

$$w = \frac{1}{\text{length}} \tag{11.8}$$

The application of weights in the least-squares adjustment of level circuit is illustrated by including the variable line lengths for the unweighted example of Section 11.3. These line lengths for the leveling network of Figure 11.2 and their corresponding relative weights are given in Table 11.1. For convenience, each length is divided into the constant 12, so that integer "relative weights" were obtained. (Note that this is an unnecessary step in the adjustment.) The observation equations are now formed as in Section 11.3, except that in the weighted case, each equation is multiplied by its weight.

$$
\begin{aligned}
w_1(\ A \qquad\qquad\qquad) &= w_1(+105.10) + w_1 v_1 \\
w_2(-A \qquad\qquad\qquad) &= w_2(-105.16) + w_2 v_2 \\
w_3(\qquad\qquad\qquad\ \ C) &= w_3(+106.25) + w_3 v_3 \\
w_4(\qquad\qquad\qquad -C) &= w_4(-106.13) + w_4 v_4 \\
w_5(-A \quad +B \qquad\quad) &= w_5(-0.68) + w_5 v_5 \\
w_6(\qquad\quad B \qquad\quad) &= w_6(+104.50) + w_6 v_6 \\
w_7(\qquad -B \quad +C) &= w_7(1.70) + w_7 v_7
\end{aligned}
\tag{11.9}
$$

After dropping the residual terms in Equation (11.9), they can be written in matrix form as

$$
\begin{bmatrix} 3 & 0 & 0 & 0 & 0 & 0 & 0 \\ 0 & 4 & 0 & 0 & 0 & 0 & 0 \\ 0 & 0 & 6 & 0 & 0 & 0 & 0 \\ 0 & 0 & 0 & 4 & 0 & 0 & 0 \\ 0 & 0 & 0 & 0 & 6 & 0 & 0 \\ 0 & 0 & 0 & 0 & 0 & 6 & 0 \\ 0 & 0 & 0 & 0 & 0 & 0 & 6 \end{bmatrix}
\begin{bmatrix} 1 & 0 & 0 \\ -1 & 0 & 0 \\ 0 & 0 & 1 \\ 0 & 0 & -1 \\ -1 & 1 & 0 \\ 0 & 1 & 0 \\ 0 & -1 & 1 \end{bmatrix}
\begin{bmatrix} A \\ B \\ C \end{bmatrix}
= \begin{bmatrix} 3 & 0 & 0 & 0 & 0 & 0 & 0 \\ 0 & 4 & 0 & 0 & 0 & 0 & 0 \\ 0 & 0 & 6 & 0 & 0 & 0 & 0 \\ 0 & 0 & 0 & 4 & 0 & 0 & 0 \\ 0 & 0 & 0 & 0 & 6 & 0 & 0 \\ 0 & 0 & 0 & 0 & 0 & 6 & 0 \\ 0 & 0 & 0 & 0 & 0 & 0 & 6 \end{bmatrix}
\begin{bmatrix} 105.10 \\ -105.16 \\ 106.25 \\ -106.13 \\ -0.68 \\ 104.50 \\ 1.70 \end{bmatrix}
\tag{11.10}
$$

According to Equation (10.34), the normal equations are

$$(A^T WA)X = NX = A^T WL \tag{11.11}$$

where

Table 11.1 Weights for example in Section 11.2

Line	Length (mi)	Relative Weight
1	4	12/4 = 3
2	3	12/3 = 4
3	2	12/2 = 6
4	3	12/3 = 4
5	2	12/2 = 6
6	2	12/2 = 6
7	2	12/2 = 6

$$N = \begin{bmatrix} 1 & -1 & 0 & 0 & -1 & 0 & 0 \\ 0 & 0 & 0 & 0 & 1 & 1 & -1 \\ 0 & 0 & 1 & -1 & 0 & 0 & 1 \end{bmatrix} \begin{bmatrix} 3 & 0 & & & & & 0 \\ 0 & 4 & & & & & \\ & & 6 & & \text{(zeros)} & & \\ & & & 4 & & & \\ & & \text{(zeros)} & & 6 & & \\ & & & & & 6 & 0 \\ 0 & & & & & 0 & 6 \end{bmatrix}$$

$$\times \begin{bmatrix} 1 & 0 & 0 \\ -1 & 0 & 0 \\ 0 & 0 & 1 \\ 0 & 0 & -1 \\ -1 & 1 & 0 \\ 0 & 1 & 0 \\ 0 & -1 & 1 \end{bmatrix} = \begin{bmatrix} 13 & -6 & 0 \\ -6 & 18 & -6 \\ 0 & -6 & 16 \end{bmatrix}$$

$$A^TWL = \begin{bmatrix} 740.02 \\ 612.72 \\ 1072.22 \end{bmatrix}$$

By using Equation (10.35), the solution for the X matrix is

$$X = N^{-1}(A^TWL) = \begin{bmatrix} 0.0933 & 0.0355 & 0.0133 \\ 0.0355 & 0.0770 & 0.0289 \\ 0.0133 & 0.0289 & 0.0733 \end{bmatrix} \begin{bmatrix} 740.02 \\ 612.72 \\ 1072.22 \end{bmatrix} = \begin{bmatrix} 105.150 \\ 104.489 \\ 106.197 \end{bmatrix} \tag{11.12}$$

Equation (11.7) is now applied to compute the residuals:

$$V = AX - L = \begin{bmatrix} 1 & 0 & 0 \\ -1 & 0 & 0 \\ 0 & 0 & 1 \\ 0 & 0 & -1 \\ -1 & 1 & 0 \\ 0 & 1 & 0 \\ 0 & -1 & 1 \end{bmatrix} \begin{bmatrix} 105.150 \\ 104.489 \\ 106.197 \end{bmatrix} - \begin{bmatrix} 105.10 \\ -105.16 \\ 106.25 \\ -106.13 \\ -0.68 \\ 104.50 \\ 1.70 \end{bmatrix} = \begin{bmatrix} 0.050 \\ 0.010 \\ -0.053 \\ -0.067 \\ 0.019 \\ -0.011 \\ 0.010 \end{bmatrix}$$

It should be noted that these adjusted values (X matrix) and residuals (V matrix) differ slightly from those obtained in the unweighted adjustment of Section 11.3. This illustrates the effect of weights on an adjustment. Although the differences are small, for precise level circuits it is both logical and wise to use a weighted adjustment.

11.5 REFERENCE STANDARD DEVIATION

From Section 9.4.2, the standard deviation of unit weight for a weighted set of observations is

$$S_0 = \sqrt{\frac{\sum wv^2}{n - 1}} \tag{11.13}$$

However, Equation (11.13) applies to a multiple set of observations for a single quantity, where each observation has a different weight. Often, observations are obtained that involve several unknown quantities which are related functionally like those in Equations (11.3) or (11.9). For these types of observations, the standard deviation of unit weight in the unweighted case is

$$S_0 = \sqrt{\frac{\sum v^2}{m - n}} = \sqrt{\frac{\sum v^2}{r}} \quad \text{which in matrix form is} \quad S_0 = \sqrt{\frac{V^TV}{r}} \tag{11.14}$$

where $\Sigma\, v^2$ in matrix form is expressed as V^TV. In Equation (11.14), m is the number of observations and n is the number of unknowns. Thus, there are $r = m - n$ redundant measurements or degrees of freedom.

The standard deviation of unit weight for the weighted case is

$$S_0 = \sqrt{\frac{\sum wv^2}{m - n}} = \sqrt{\frac{\sum wv^2}{r}} \quad \text{which in matrix form is} \quad S_0 = \sqrt{\frac{V^TWV}{r}} \tag{11.15}$$

where $\Sigma\, wv^2$ in matrix form is V^TWV.

Since these standard deviations of unit weight relate to the overall adjustment and not a single quantity, they are referred to as *reference standard deviations*. Computation of the reference standard deviations for both unweighted and weighted examples are illustrated below.

11.5.1 Unweighted Example

In the example of Section 11.3, there were 7 − 3 or 4 degrees of freedom. Using the residuals given in Equation (11.7), the system's reference standard deviation is determined using the algebraic expression of Equation (11.14) as:

$$S_o = \sqrt{\frac{(0.041)^2 + (0.019)^2 + (-0.062)^2 + (-0.058)^2 + (0.022)^2 + (-0.017)^2 + (0.005)^2}{7-3}}$$

$$= \pm 0.05 \text{ ft} \tag{11.16}$$

This can be computed using the matrix expression of Equation (11.14) as

$$S_0 = \sqrt{\frac{V^TV}{r}}$$

$$V^TV = [0.041 \quad 0.019 \quad -0.062 \quad -0.058 \quad 0.022 \quad -0.017 \quad 0.005] \times \begin{bmatrix} 0.041 \\ 0.019 \\ -0.062 \\ -0.058 \\ 0.022 \\ -0.017 \\ 0.005 \end{bmatrix} = 0.010$$

$$S_0 = \sqrt{\frac{0.010}{4}} = \pm 0.05 \text{ ft} \tag{11.17}$$

11.5.2 Weighted Example

Notice that the weights are used when computing the reference standard deviation in Equation (11.15). That is, each residual is squared and multiplied by its weight, and thus the reference standard deviation, computed using non-matrix methods, is

$$S_0 = \sqrt{\frac{3(0.050)^2 + 4(0.010)^2 + 6(-0.053)^2 + 4(-0.067)^2 + 6(0.019)^2 + 6(-0.011)^2 + 6(0.010)^2}{7 - 3}}$$

$$= \sqrt{\frac{0.04598}{4}} = \pm 0.107 \tag{11.18}$$

It is left as an exercise to verify this result by solving the matrix expression of Equation (11.15).

11.6 ANOTHER WEIGHTED ADJUSTMENT

Example 11.1 The level net shown in Figure 11.3 was measured with the following results (the elevation differences and standard deviations are given in meters, and the elevation of A is 437.596 m):

From:	To:	ΔElev	σ	From:	To:	ΔElev	σ
A	B	10.509	0.006	D	A	−7.348	0.003
B	C	5.360	0.004	B	D	−3.167	0.004
C	D	−8.523	0.005	A	C	15.881	0.012

What are the most probable values for the elevations of B, C, and D?

SOLUTION

Step 1: Write the observation equations without their weights:

$$\begin{aligned}
&(1) \quad +B &&= A + 10.509 + v_1 = 448.105 + v_1 \\
&(2) \quad -B + C &&= 5.360 + v_2 \\
&(3) \quad -C + D &&= -8.523 + v_3 \\
&(4) \quad -D &&= -A - 7.348 + v_4 = -444.944 + v_4 \\
&(5) \quad -B + D &&= -3.167 + v_5 \\
&(6) \quad +C &&= A + 15.881 + v_6 = 453.477 + v_6
\end{aligned}$$

Step 2: Rewrite observation equations in matrix form $AX = L + V$ as:

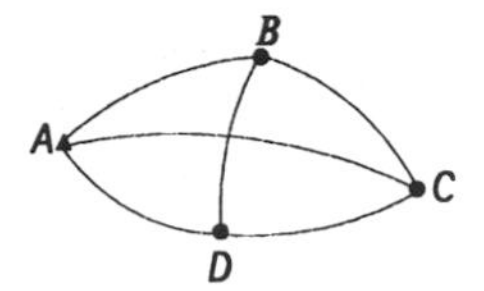

Figure 11.3 Level net.

$$\begin{bmatrix} 1 & 0 & 0 \\ -1 & 1 & 0 \\ 0 & -1 & 1 \\ 0 & 0 & -1 \\ -1 & 0 & 1 \\ 0 & 1 & 0 \end{bmatrix} \begin{bmatrix} B \\ C \\ D \end{bmatrix} = \begin{bmatrix} 448.105 \\ 5.360 \\ -8.523 \\ -444.944 \\ -3.167 \\ 453.477 \end{bmatrix} + \begin{bmatrix} v_1 \\ v_2 \\ v_3 \\ v_4 \\ v_5 \\ v_6 \end{bmatrix} \tag{11.19}$$

Step 3: In accordance with Equations (9.4) and (9.6), form the weight matrix as

$$W = \begin{bmatrix} \frac{1}{0.006^2} & 0 & 0 & 0 & 0 & 0 \\ 0 & \frac{1}{0.004^2} & 0 & 0 & 0 & 0 \\ 0 & 0 & \frac{1}{0.005^2} & 0 & 0 & 0 \\ 0 & 0 & 0 & \frac{1}{0.003^2} & 0 & 0 \\ 0 & 0 & 0 & 0 & \frac{1}{0.004^2} & 0 \\ 0 & 0 & 0 & 0 & 0 & \frac{1}{0.012^2} \end{bmatrix} \tag{11.20}$$

from which

$$W = \begin{bmatrix} 27{,}778 & 0 & 0 & 0 & 0 & 0 \\ 0 & 62{,}500 & 0 & 0 & 0 & 0 \\ 0 & 0 & 40{,}000 & 0 & 0 & 0 \\ 0 & 0 & 0 & 111{,}111 & 0 & 0 \\ 0 & 0 & 0 & 0 & 62{,}500 & 0 \\ 0 & 0 & 0 & 0 & 0 & 6944 \end{bmatrix} \tag{11.21}$$

Step 4: Compute the normal equations using Equation (10.34):

$$(A^T WA)X = NX = A^T WL \tag{11.22}$$

This gives

$$N = \begin{bmatrix} 152{,}778 & -62{,}500 & -62{,}500 \\ -62{,}500 & 109{,}444 & -40{,}000 \\ -62{,}500 & -40{,}000 & 213{,}611 \end{bmatrix} \qquad X = \begin{bmatrix} B \\ C \\ D \end{bmatrix}$$

$$A^T WL = \begin{bmatrix} 12{,}310{,}298.611 \\ 3{,}825{,}065.833 \\ 48{,}899{,}364.722 \end{bmatrix}$$

Step 5: Solve for the X matrix using Equation (10.35), which yields

$$X = \begin{bmatrix} 448.1087 \\ 453.4685 \\ 444.9436 \end{bmatrix} \tag{11.23}$$

Step 6: Compute the residuals using the matrix expression $V = AX - L$.

$$V = \begin{bmatrix} 448.1087 \\ 5.3598 \\ -8.5249 \\ -444.9436 \\ -3.1651 \\ 453.4685 \end{bmatrix} - \begin{bmatrix} 448.105 \\ 5.360 \\ -8.523 \\ -444.944 \\ -3.167 \\ 453.477 \end{bmatrix} = \begin{bmatrix} 0.0037 \\ -0.0002 \\ -0.0019 \\ 0.0004 \\ 0.0019 \\ -0.0085 \end{bmatrix} \tag{11.24}$$

Step 7: Calculate the reference standard deviation for the adjustment using the matrix expression of Equation (11.15).

$$V^T WV = [0.0037 \quad -0.0002 \quad -0.0019 \quad 0.0004 \quad 0.0019 \quad -0.0085]W$$

$$\times \begin{bmatrix} 0.0037 \\ -0.0002 \\ -0.0019 \\ 0.0004 \\ 0.0019 \\ -0.0085 \end{bmatrix} = [1.26976] \tag{11.25}$$

Since the number of system redundancies is the number of observations minus the number of unknowns, $r = 6 - 3 = 3$, and thus S_o is

$$S_0 = \sqrt{\frac{1.26976}{3}} = \pm 0.6575 \text{ m} \tag{11.26}$$

Step 8: Tabulate the results showing the adjusted elevation differences, their residuals, and the final adjusted elevations.

From	To	Adjusted ΔElev	Residual	Station	Adjusted Elevation
A	*B*	10.513	0.004	*A*	437.596
B	*C*	5.360	0.000	*B*	448.109
C	*D*	−8.525	−0.002	*C*	453.468
D	*A*	−7.348	0.000	*D*	444.944
B	*D*	−3.165	0.002		
A	*C*	15.872	−0.009		

PROBLEMS

Note: For problems below requiring least-squares adjustment, if a computer program is not distinctly specified for use in the problem, it is expected that the least-squares algorithm will be solved using the program MATRIX which is included within the diskette supplied with the book.

11.1 For the leveling network shown in Figure 11.4, calculate the most probable elevations for *X* and *Y*. Use an unweighted least-squares adjustment with the observed values given in the accompanying table.

Line	ΔElev
1	+1.00
2	+0.08
3	+0.92
4	−1.00
5	+0.08

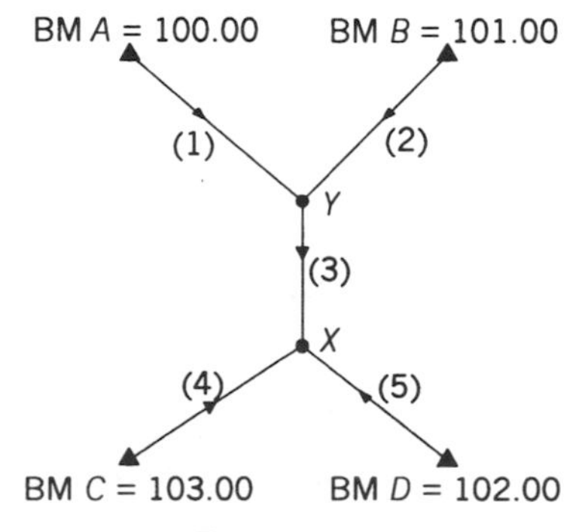

Figure 11.4

11.2 For Problem 11.1, compute the reference standard deviation and tabulate the adjusted observations and their residuals.

11.3 Do Problem 11.1 using Program ADJUST.

11.4 For the leveling network shown in Figure 11.5, calculate the most probable elevations for *X*, *Y*, and *Z*. The observed values and line lengths are given in the table. Apply appropriate weights in the computations.

Line	Length	ΔElev	Line	Length	ΔElev
1	4	+1.05	4	2	−1.95
2	4	−0.95	5	1	+0.10
3	2	+2.10	6	3	+0.05

11.5 For Problem 11.4, compute the reference standard deviation, and tabulate the adjusted observations and their residuals

11.6 Use Program ADJUST to solve Problem 11.4.

11.7 A line of differential levels was run from benchmark Oak (elevation 753.01) to station 13+00 on a proposed road alignment. It continued along the alignment to 19+00. Rod readings were taken on stakes set at each full station. The circuit then closed on benchmark Bridge having an elevation of 764.95 ft. The observed elevation differences, in order, are −3.03, 4.10, 4.03, 7.92, 7.99, −6.00, −6.02, and 2.98 ft. A check measurement between benchmark Rock (elevation = 772.39 ft) and station 16+00 gives an elevation difference of −6.34 ft.

(a) What are the most probable values for the adjusted elevation differences?

(b) What is the reference standard deviation for the adjustment?

(c) Tabulate the adjusted observations and their residuals.

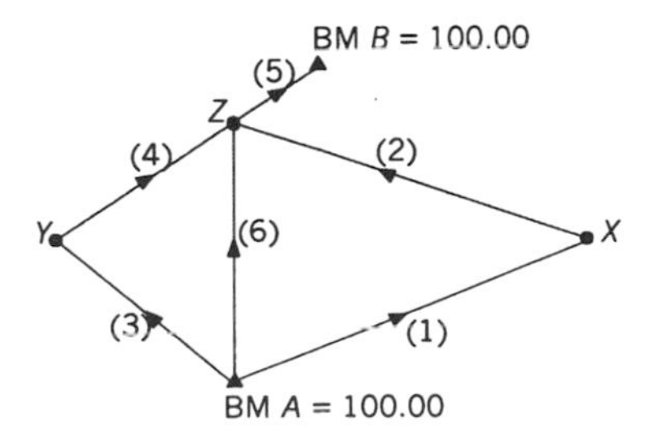

Figure 11.5

11.8 Use Program ADJUST to solve Problem 11.7.

11.9 If the elevation of A is 257.891 m, adjust the following leveling data using the weighted least-squares method.

From	To	ΔElev (m)	Distance (km)
A	B	5.666	1
B	C	48.025	4
C	D	3.021	3
D	E	−13.987	2.5
E	F	20.677	3
F	G	−32.376	5
G	A	−30.973	2
A	C	53.700	5
C	F	9.634	4
F	D	−6.631	3

(a) What are the most probable values for the elevations of the stations?

(b) What is the reference standard deviation?

(c) Tabulate the adjusted observations and their residuals.

11.10 Use Program ADJUST to solve Problem 11.9.

11.11 If the elevation of station 1 is 2395.67 ft, use weighted least squares to adjust the following leveling.

From	To	ΔElev (ft)	Distance (mi)	From	To	ΔElev (ft)	Distance (mi)
1	2	37.17	2.00	4	5	−10.92	2.00
3	4	34.24	1.00	6	1	−28.06	1.00
5	6	−23.12	1.00	7	3	11.21	1.11
1	7	16.99	1.11	7	6	10.89	1.11
2	7	−19.99	1.11	3	8	74.93	1.17
6	3	−0.04	1.00	6	8	74.89	1.17
8	5	−51.96	1.08	8	4	−41.14	1.08
2	3	−9.20	1.00				

(a) What are the most probable values for the elevations for the stations?

(b) What is the adjustment's reference standard deviation?

(c) Tabulate the adjusted observations and their residuals.

11.12 Use Program ADJUST to solve Problem 11.11.

11.13 Precise procedures were applied with a level that can be read to within ±0.4 mm/m to obtain the following listed elevation differences. The line of sight was held to within ±3″ of horizontal, and the sight distances were approximately 50 m in length. Use these specifications and Equation (8.21) to compute standard deviations and hence weights. The elevation of *A* is 100.000 m. Adjust the network by weighted least squares.

From	To	ΔElev (m)	Number of Setups	From	To	ΔElev (m)	Number of Setups
A	*B*	12.378	26	*M*	*D*	−38.238	18
B	*C*	−16.672	42	*C*	*M*	30.338	15
C	*D*	−7.903	32	*M*	*B*	−13.676	40
D	*A*	12.190	37	*A*	*M*	26.051	11

(a) What are the most probable values for the elevations of the stations?

(b) What is the reference standard deviation for the adjustment?

(c) Tabulate the adjusted observations and their residuals.

11.14 Repeat Problem 11.13 using the number of setups for weighting following the procedures discussed in Section 9.6.

11.15 Demonstrate that $\Sigma\, v^2 = V^TV$.

11.16 Demonstrate that $\Sigma\, wv^2 = V^TWV$.

Programming Problems

11.17 Write a program that reads a file of differential leveling observations and writes the matrices *A*, *W*, and *L* in a format suitable for the MATRIX program.

11.18 Write a program that reads matrices *A*, *W*, and *L*, computes the least-squares solution for the unknown station elevations, and writes a file of adjusted elevation differences and their residuals.

11.19 Write a program that reads a file of differential leveling observations, computes the least-squares solution for the adjusted station elevations, and writes a file of adjusted elevation differences and their residuals.

12

PRECISIONS OF INDIRECTLY DETERMINED QUANTITIES

12.1 INTRODUCTION

Following an adjustment, it is important to know the estimated errors in both the adjusted observations and in quantities derived from the adjusted values. For example, after adjusting a level net as described in Chapter 11, the uncertainties in both adjusted elevation differences and computed benchmark elevations should be determined. In Chapter 5, error propagation formulas were developed for indirectly measured quantities which were functionally related to observed values. In this chapter, error propagation formulas are developed for the quantities computed in a least squares solution.

12.2 DEVELOPMENT OF THE COVARIANCE MATRIX

Consider an adjustment involving weighted observation equations like those in the level circuit example of Section 11.4. The matrix form for the system of weighted observation equation is

$$WAX = WL + WV \tag{12.1}$$

and the least-squares solution of the weighted observation equations is given by

$$X = (A^TWA)^{-1}A^TWL \tag{12.2}$$

In this equation, X contains the most probable values for the unknowns, whereas the true values are X_{true}. The true values differ from X by some small amount ΔX, such that

$$X + \Delta X = X_{true} \tag{12.3}$$

where ΔX represents the errors in the adjusted values.

Consider now a small incremental change, ΔL, in the measured values, L, which changes X to its true value, $X + \Delta X$. Then Equation (12.2) becomes

$$X + \Delta X = (A^T WA)^{-1} A^T W(L + \Delta L) \tag{12.4}$$

Expanding Equation (12.4) gives

$$X + \Delta X = (A^T WA)^{-1} A^T WL + (A^T WA)^{-1} A^T W (\Delta L) \tag{12.5}$$

Note in Equation (12.2) that $X = (A^T WA)^{-1} A^T WL$, and thus subtracting this from Equation (12.5) gives

$$\Delta X = (A^T WA)^{-1} A^T W (\Delta L) \tag{12.6}$$

Recognizing ΔL as the errors in the observations, Equation (12.6) can be rewritten as

$$\Delta X = (A^T WA)^{-1} A^T WV \tag{12.7}$$

Now let

$$B = (A^T WA)^{-1} A^T W \tag{12.8}$$

then

$$\Delta X = BV \tag{12.9}$$

Multiplying both sides of Equation (12.9) by their transposes yields

$$\Delta X \, \Delta X^T = (BV)(BV)^T \tag{12.10}$$

Note that in matrix algebra, $(BV)^T = V^T B^T$. Applying this property to Equation (12.10) gives

$$\Delta X \, \Delta X^T = BVV^T B^T \tag{12.11}$$

The expanded left side of Equation (12.11) is

$$
\Delta X(\Delta X^T) = \begin{bmatrix} \Delta x_1^2 & \Delta x_1\,\Delta x_2 & \Delta x_1\,\Delta x_3 & \cdots & \Delta x_1\,\Delta x_n \\ \Delta x_2\,\Delta x_1 & \Delta x_2^2 & \Delta x_2\,\Delta x_3 & \cdots & \Delta x_2\,\Delta x_n \\ \Delta x_3\,\Delta x_1 & \Delta x_3\,\Delta x_2 & \Delta x_3^2 & & \Delta x_3\,\Delta x_n \\ \vdots & \vdots & & \ddots & \vdots \\ \Delta x_n\,\Delta x_1 & \Delta x_n\,\Delta x_2 & \Delta x_n\,\Delta x_3 & \cdots & \Delta x_n^2 \end{bmatrix} \tag{12.12}
$$

Also, the expanded right side of Equation (12.11) is

$$
B \begin{bmatrix} v_1^2 & v_1 v_2 & v_1 v_3 & \cdots & v_1 v_m \\ v_2 v_1 & v_2^2 & v_2 v_3 & \cdots & v_2 v_m \\ v_3 v_1 & v_3 v_2 & v_3^2 & & v_3 v_m \\ \vdots & \vdots & & \ddots & \vdots \\ v_m v_1 & v_m v_2 & v_m v_3 & \cdots & v_m^2 \end{bmatrix} B^T \tag{12.13}
$$

Assume that it is possible to repeat the entire sequence of observations many times, say a times, and that each time a slightly different solution occurs, yielding a different set of X's. Averaging these sets, the left side of Equation (12.11) becomes

$$
\frac{1}{a}\sum (\Delta X)(\Delta X)^T = \begin{bmatrix} \dfrac{\sum \Delta x_1^2}{a} & \dfrac{\sum \Delta x_1\,\Delta x_2}{a} & \cdots & \dfrac{\sum \Delta x_1\,\Delta x_n}{a} \\ \dfrac{\sum \Delta x_2\,\Delta x_1}{a} & \dfrac{\sum \Delta x_2^2}{a} & \cdots & \dfrac{\sum \Delta x_2\,\Delta x_n}{a} \\ \vdots & \vdots & \ddots & \vdots \\ \dfrac{\sum \Delta x_n\,\Delta x_1}{a} & \dfrac{\sum \Delta x_n\,\Delta x_2}{a} & \cdots & \dfrac{\sum \Delta x_n^2}{a} \end{bmatrix} \tag{12.14}
$$

If a is large, the terms in Equation (12.14) are the variances and covariances as defined in Equation (5.7), and Equation (12.14) can be rewritten as

$$
\begin{bmatrix} S_{x_1}^2 & S_{x_1x_2} & \cdots & S_{x_1x_n} \\ S_{x_2x_1} & S_{x_2}^2 & \cdots & S_{x_2x_n} \\ \vdots & \vdots & \ddots & \vdots \\ S_{x_nx_1} & S_{x_nx_2} & \cdots & S_{x_n}^2 \end{bmatrix} = S_x^2 \tag{12.15}
$$

Also, considering a sets of observations, Equation (12.13) becomes

$$B\begin{bmatrix} \frac{\sum v_1^2}{a} & \frac{\sum v_1v_2}{a} & \cdots & \frac{\sum v_1v_m}{a} \\ \frac{\sum v_2v_1}{a} & \frac{\sum v_2^2}{a} & \cdots & \frac{\sum v_2v_m}{a} \\ \vdots & \vdots & \ddots & \vdots \\ \frac{\sum v_mv_1}{a} & \frac{\sum v_mv_2}{a} & \cdots & \frac{\sum v_m^2}{a} \end{bmatrix}B^T \tag{12.16}$$

Recognizing the diagonal terms as variances of the observed quantities, $S^2_{l_i}$, the off-diagonal terms as the covariances $S_{l_il_j}$, and the fact that the matrix is symmetric, Equation (12.16) can be rewritten as

$$B\begin{bmatrix} S^2_{l_1} & S_{l_1l_2} & S_{l_1l_3} & \cdots & S_{l_1l_m} \\ S_{l_1l_2} & S^2_{l_2} & S_{l_2l_3} & \cdots & S_{l_2l_m} \\ S_{l_1l_3} & S_{l_2l_3} & S^2_{l_3} & \cdots & S_{l_3l_m} \\ \vdots & \vdots & \vdots & \ddots & \vdots \\ S_{l_1l_m} & S_{l_2l_m} & S_{l_3l_m} & \cdots & S^2_{l_m} \end{bmatrix}B^T \tag{12.17}$$

In Section 9.1 it was shown that an observation's weight is inversely proportional to its variance. Also, from Equation (9.5), the variance of an observation of weight w can be expressed in terms of the reference variance as

$$S_i^2 = \frac{S_0^2}{w_i} \tag{12.18}$$

Recall from Equation (9.3) that $W = Q^{-1} = \sigma_0^2\Sigma^{-1}$. Therefore, $\Sigma = \sigma_0^2 W^{-1}$, and thus substituting Equation (12.18) into matrix (12.17) and replacing σ_0 with S_0 yields

$$S_0^2 B W_{ll}^{-1} B^T \tag{12.19}$$

Substituting Equation (12.8) into Equation (12.19) gives

$$S_0^2 B W^{-1} B^T = S_0^2 (A^TWA)^{-1} A^T W W^{-1} W^T A ((A^TWA)^{-1})^T \tag{12.20}$$

Since the matrix of normal equations is symmetric, we can show that

$$((A^TWA)^{-1})^T = (A^TWA)^{-1} \tag{12.21}$$

Also, since the weight matrix W is symmetric, $W^T = W$, and thus Equation (12.20) reduces to

$$S_0^2(A^TWA)^{-1}(A^TWA)(A^TWA)^{-1} = S_0^2(A^TWA)^{-1} \tag{12.22}$$

Equation (12.15) is the left side of Equation (12.11), for which Equation (12.22) is the right. That is,

$$S_x^2 = S_0^2(A^TWA)^{-1} = S_0^2N^{-1} = S_0^2Q_{xx} \tag{12.23}$$

In least-squares adjustment, the matrix Q_{xx} of Equation (12.23) is known as the *variance–covariance matrix*, or simply the *covariance matrix*. Diagonal elements of the matrix when multiplied by S_0^2 give variances of the adjusted quantities, and the off-diagonal elements, multiplied by S_0^2, yield covariances. From Equation (12.23), the estimated standard deviation S_i for any unknown parameter, having been computed from a system of observation equations, is expressed as

$$S_i = S_0\sqrt{q_{x_ix_i}} \tag{12.24}$$

where $q_{x_ix_i}$ is the diagonal element (from the ith row and ith column) of the Q_{xx} matrix, which as noted in Equation (12.23), is equal to N^{-1}, the inverse of the matrix of normal equations.

Since the normal equation matrix is symmetric, its inverse is also symmetric, and thus the covariance matrix is a symmetric matrix (i.e., element ij = element ji).[1]

12.3 NUMERICAL EXAMPLES

The results of the level net adjustment in Section 11.3 will be used to illustrate the computation of estimated errors for the adjusted unknowns. From Equation (11.6), the N^{-1} matrix, which is also the Q_{xx} matrix, is

[1] Note that an estimate of the reference variance, σ_0^2, may be computed using either Equation (11.13) or (11.14). However, it should be remembered that this only gives an estimate of the *a priori* (before the adjustment) value for the reference variance. This estimate may be checked using a χ^2 test as discussed in Chapter 4. If it is a valid estimate for σ_0^2, the a priori value for the reference variance should be used. Thus if the a priori value for σ_0^2 is known, it should be used when computing the *a posteriori* (after the adjustment) statistics. When weights are determined as $1/\sigma_i^2$, the implicit assumption made is that the a priori value for $\sigma_0^2 = 1$ [see Equations (9.5) and (9.6)].

$$Q_{xx} = \begin{bmatrix} 0.38095 & 0.14286 & 0.04762 \\ 0.14286 & 0.42857 & 0.14286 \\ 0.04762 & 0.14286 & 0.38095 \end{bmatrix}$$

Also, from Equation (11.17) $S_0 = \pm 0.05$ ft. Now by Equation (12.24), the estimated standard deviations for the unknown benchmark elevations A, B, and C are

$$S_A = S_0 \sqrt{q_{AA}} = \pm 0.05 \sqrt{0.38095} = \pm 0.031 \text{ ft}$$

$$S_B = S_0 \sqrt{q_{BB}} = \pm 0.05 \sqrt{0.42857} = \pm 0.033 \text{ ft}$$

$$S_C = S_0 \sqrt{q_{CC}} = \pm 0.05 \sqrt{0.38095} = \pm 0.031 \text{ ft}$$

In the weighted example of Section 11.4, it should be noted that although this is a weighted adjustment, the a priori value for the reference variance is not known because weights were determined as 1/length, not $1/\sigma_i^2$. From Equation (11.12), the Q_{xx} matrix is

$$Q_{xx} = \begin{bmatrix} 0.0933 & 0.0355 & 0.0133 \\ 0.0355 & 0.0770 & 0.0289 \\ 0.0133 & 0.0289 & 0.0733 \end{bmatrix}$$

Recalling that in Equation (11.18), $S_0 = \pm 0.107$, the estimated errors in the computed elevations of benchmarks A, B, and C are

$$S_A = S_0 \sqrt{q_{AA}} = \pm 0.107 \sqrt{0.0933} = \pm 0.033 \text{ ft}$$

$$S_B = S_0 \sqrt{q_{BB}} = \pm 0.107 \sqrt{0.0770} = \pm 0.030 \text{ ft}$$

$$S_C = S_0 \sqrt{q_{CC}} = \pm 0.107 \sqrt{0.0733} = \pm 0.029 \text{ ft}$$

These standard deviations are at the 68% probability level, and if other percentage errors are desired, these values should be multiplied by their respective t values as discussed in Chapter 3.

12.4 STANDARD DEVIATIONS OF COMPUTED QUANTITIES

In Section 5.1, the generalized law of propagation of variances was developed. Recalled here for convenience, Equation (5.13) was written as

$$\Sigma_{\hat{l}\hat{l}} = A\Sigma_{xx}A^T \qquad (a)$$

where $\hat{l}$ represents the adjusted observations, $\Sigma_{\hat{l}\hat{l}}$ the covariance matrix of the adjusted observations, Σ_{xx} the covariance matrix of the unknown parameters [i.e., $(A^TWA)^{-1}$], and A is the coefficient matrix. Rearranging Equation (9.2), $\Sigma_{xx} = S_0^2 Q_{xx}$. Also from Equation (12.23), $S_x^2 = S_0^2 Q_{xx} = S_0^2(A^TWA)^{-1}$, and thus $\Sigma_{xx} = S_x^2$. Substituting this equality into Equation (a), the estimated standard deviations of the adjusted observations is

$$\Sigma_{\ell\ell} = S_{\ell\ell}^2 = A\Sigma_{xx}A^T = AS_0^2(A^TWA)^{-1}A^T = S_0^2 AQ_{xx}A^T = S_0^2 Q_{\ell\ell} \quad (12.25)$$

where $AQ_{xx}A^T = Q_{\ell\ell}$, which is called the *covariance matrix of the observations.*

Example 12.1 Consider the linear example in Section 11.3. By Equation (12.25), the estimated standard deviations in the adjusted elevation differences are

$$S_0^2 AQ_{xx}A^T = (0.050)^2 \begin{bmatrix} 1 & 0 & 0 \\ -1 & 0 & 0 \\ 0 & 0 & 1 \\ 0 & 0 & -1 \\ -1 & 1 & 0 \\ 0 & 1 & 0 \\ 0 & -1 & 1 \end{bmatrix} \begin{bmatrix} 0.38095 & 0.14286 & 0.04762 \\ 0.14286 & 0.42857 & 0.14286 \\ 0.04762 & 0.14286 & 0.38095 \end{bmatrix}$$

$$\times \begin{bmatrix} 1 & -1 & 0 & 0 & -1 & 0 & 0 \\ 0 & 0 & 0 & 0 & 1 & 1 & -1 \\ 0 & 0 & 1 & -1 & 0 & 0 & 1 \end{bmatrix} \quad (12.26)$$

Performing the required matrix multiplications in Equation (12.26) gives

$$S_{\ell\ell}^2 = 0.050^2$$

$$\times \begin{bmatrix} 0.38095 & -0.38095 & 0.04762 & -0.04762 & -0.23810 & 0.14286 & -0.09524 \\ -0.38095 & 0.38095 & -0.04762 & 0.04762 & 0.23810 & -0.14286 & 0.09524 \\ 0.04762 & -0.04762 & 0.38095 & -0.38095 & 0.09524 & 0.14286 & 0.23810 \\ -0.04762 & 0.04762 & -0.38095 & 0.38095 & -0.09524 & -0.14286 & -0.23810 \\ -0.23810 & 0.23810 & 0.09524 & -0.09524 & 0.52381 & 0.28571 & 0.19048 \\ 0.14286 & -0.14286 & 0.14286 & -0.14286 & 0.28571 & 0.42857 & -0.28571 \\ -0.09524 & 0.09524 & 0.23810 & -0.23810 & -0.19048 & -0.28571 & 0.52381 \end{bmatrix}$$

(12.27)

In accordance with Equation (12.25), the estimated standard deviation of an observation is found by taking the square root of the corresponding diagonal element of the $S_{\ell\ell}^2$ matrix. For instance, for the fifth observation (leveling from A to B), $S_{\ell\ell}(5,5)$ applies, and thus the estimated error in the adjusted elevation difference of that observation is

$$S_{\Delta \text{ELEV}_{AB}} = \pm 0.050\sqrt{0.52381} = \pm 0.036 \text{ ft}$$

An interpretation of the meaning of the value just calculated is that there is a 68% probability that the adjusted elevation difference ($l_5 + v_5 = -0.68 + 0.022 = -0.658$) is within ± 0.036 of its true value. Thus the true value lies between -0.622 and -0.694 ft with 68% probability.

Careful examination of the matrix manipulations involved in solving Equation (12.25) for Example 12.1 reveals that the effort can be reduced significantly. In fact, to obtain the estimated standard deviation in the fifth observation, only the fifth row of the coefficient matrix, A, which represents the elevation difference between A and B need be used in the calculations. That row is $[-1 \quad 1 \quad 0]$. Thus to compute the standard deviation in this observation, the following computations would be made:

$$\begin{aligned} S^2_{\Delta \text{ELEV}_{AB}} &= (0.050)^2[-1 \quad 1 \quad 0] \begin{bmatrix} 0.38095 & 0.14286 & 0.04762 \\ 0.14286 & 0.42857 & 0.14286 \\ 0.04762 & 0.14286 & 0.38095 \end{bmatrix} \begin{bmatrix} -1 \\ 1 \\ 0 \end{bmatrix} \\ &= (0.050)^2[-0.23809 \quad 0.28571 \quad 0.09524] \begin{bmatrix} -1 \\ 1 \\ 0 \end{bmatrix} \\ &= (0.050)^2[0.52380] \end{aligned} \tag{12.28}$$

$$S_{\Delta \text{ELEV}_{AB}} = \pm 0.050\sqrt{0.52380} = \pm 0.036 \text{ ft}$$

Note that this shortcut method produces the same value. Furthermore, because of the zero in the third position of this row from the coefficient matrix, the matrix operations in Equation (12.28) could be further reduced to

$$S^2_{\Delta \text{ELEV}_{AB}} = (0.050)^2[-1 \quad 1] \begin{bmatrix} 0.38095 & 0.14286 \\ 0.14286 & 0.42857 \end{bmatrix} \begin{bmatrix} -1 \\ 1 \end{bmatrix} = (0.050)^2[0.52380]$$

Another use for Equation (12.25) is in the computation of adjusted uncertainties for measurements that were never made. For instance, in the example of Section 11.3, the elevation difference between benchmarks X and B was not observed. But from the results of the adjustment, this elevation difference is $104.48 - 100.00 = 4.48$ ft. The estimated error in this difference can be found by writing an observation equation for it (i.e., $B = X + \Delta\text{ELEV}_{XB}$). This equation does not involve either A or C, and thus in matrix form this difference would be expressed as

$$[0 \quad 1 \quad 0] \tag{12.29}$$

Using this row matrix in the same procedure as in Equation (12.28) gives

$$S^2_{\Delta XB} = (0.050)^2[0 \quad 1 \quad 0] \begin{bmatrix} 0.38095 & 0.14286 & 0.04762 \\ 0.14286 & 0.42857 & 0.14286 \\ 0.04762 & 0.14286 & 0.38095 \end{bmatrix} \begin{bmatrix} 0 \\ 1 \\ 0 \end{bmatrix}$$

$$= (0.050)^2[0.42857]$$

Hence

$$S_{\Delta XB} = \pm 0.050\sqrt{0.42857} = \pm 0.033 \text{ ft}$$

Again, recognizing the presence of the zeros in the row matrix, these computations can be simplified to

$$S^2_{\Delta XB} = (0.050)^2[1][0.42857][1] = (0.050)^2[0.42857]$$

The method illustrated above of eliminating unnecessary matrix computations is formally known as *matrix partitioning*.

Computing uncertainties of quantities that were not actually measured has applications in many other areas. For example, suppose in a triangulation adjustment that the x and y coordinates of stations A and B are calculated and the covariance matrix exists. Equation (12.25) could be applied to determine the estimated error in the length of line AB calculated from the adjusted coordinates of A and B. This is accomplished by relating the length AB to the unknown parameters as

$$\overline{AB} = \sqrt{(X_b - X_a)^2 + (Y_b - Y_a)^2} \tag{12.30}$$

This subject is discussed further in Chapter 14.

An important observation that should be made about the and Q_{ll} and Q_{xx} matrices is that only the coefficient matrix, A, is used in their formation. Since the A matrix contains coefficients that express the relationships of the unknowns to one another, it depends only on the *geometry* of the problem. The only other term in Equation (12.25) is the reference variance, and that depends on the quality of the measurements. These are important concepts that will be revisited in Chapter 20 when the simulation of surveying networks is discussed.

PROBLEMS

For each problem, calculate the estimated errors for the adjusted benchmark elevations. Except where specified otherwise, Program MATRIX should be used.

12.1 Problem 11.1

12.2 Problem 11.4

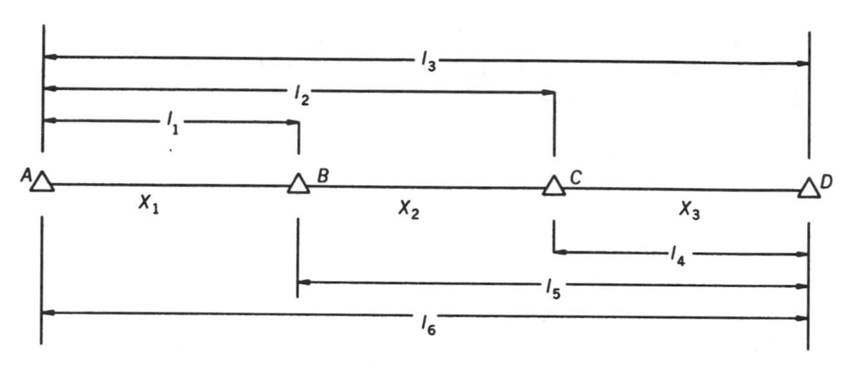

Figure 12.1

12.3 Problem 11.7

12.4 Problem 11.9

12.5 Problem 11.11

12.6 Problem 11.13

For each problem, calculate the estimated errors for the adjusted elevation differences.

12.7 Problem 11.1

12.8 Problem 11.4

12.9 Problem 11.7

12.10 Problem 11.9

12.11 Problem 11.11

12.12 Referring to Figure 12.1 and the measurement data below, calculate adjusted length $\overline{AD}$ and its estimated error. (Assume equal weights.)

$$l_1 = 100.00 \qquad l_2 = 200.00 \qquad l_3 = 300.00 \qquad l_4 = 99.92$$

$$l_5 = 200.04 \qquad l_6 = 299.96$$

12.13 Use Figure 12.2 and the data below to answer the following questions.

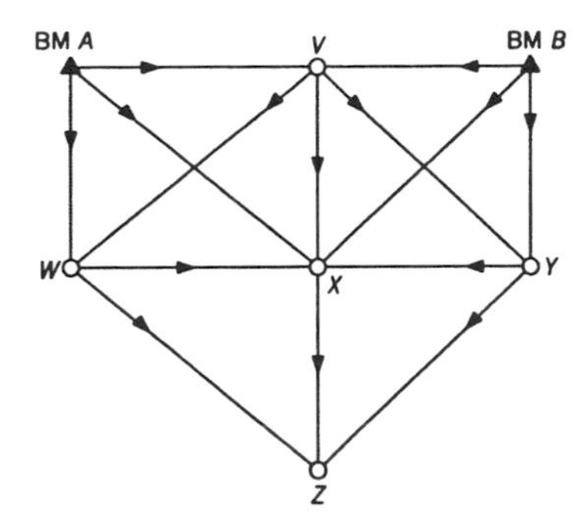

Figure 12.2

Elevation of BM A = 1060.00 ft					Elevation of BM B = 1125.79 ft				
Obs.	From:	To:	ΔElev	σ	Obs.	From:	To:	ΔElev	σ
1	BM A	V	12.33	±0.006	8	Y	Z	3.51	±0.010
2	BM B	V	−53.46	±0.003	9	W	Z	58.84	±0.008
3	V	X	−6.93	±0.007	10	V	W	−28.86	±0.004
4	V	Y	26.51	±0.006	11	BM A	W	−16.52	±0.002
5	BM B	Y	−27.09	±0.009	12	BM B	X	−60.36	±0.009
6	BM A	X	5.38	±0.004	13	W	X	21.86	±0.005
7	Y	X	−33.50	±0.008	14	X	Z	36.90	±0.002

What is the:

(a) most probable elevation for stations V, W, X, Y, and Z?

(b) estimated error in each adjusted elevation?

(c) estimated error in each measurement?

(d) estimated error in the elevation difference from benchmark A to station Z?

12.14 Do Problem 12.13 without including observations 3, 4, and 10.

12.15 Do Problem 12.13 without including observations 4, 8, 9, and 12.

12.16 Use Program ADJUST to do Problems 12.13, 12.14, and 12.15. Discuss how the removal of these observations weakens the network.

Programming Problems

12.17 Adapt the program developed in Problem 11.18 to compute and tabulate:

(a) the adjusted elevations and their estimated errors.

(b) the adjusted elevation differences and their estimated errors.

12.18 Adapt the program developed in Problem 11.19 to compute and tabulate:

(a) the adjusted elevations and their estimated errors.

(b) the adjusted elevation differences and their estimated errors.

13

ADJUSTMENT OF HORIZONTAL SURVEYS: TRILATERATION

13.1 INTRODUCTION

Horizontal surveys are performed for the purpose of determining precise relative horizontal positions of points. They have traditionally been accomplished by trilateration, triangulation, and traverse. These traditional types of surveys involve making distance, direction, and angle measurements. As with all types of surveys, errors will occur in making these measurements, and thus they must be analyzed, and if acceptable, adjusted. In the following three chapters procedures are described for adjusting trilateration, triangulation, and traverse surveys, in that order.

In recent years, the *global positioning system* (GPS) has gradually been replacing these traditional procedures for conducting precise horizontal surveys. In fact, GPS not only yields horizontal positions, but it gives ellipsoid heights as well. Thus GPS provides three-dimensional surveys. Again as with all measurements, GPS surveys contain errors and must be adjusted. In Chapter 16 we discuss the subject of GPS surveying in more detail and illustrate methods for adjusting networks surveyed by this procedure.

Horizontal surveys, especially those covering a large extent, must account for the systematic effects of earth curvature. One way this can be accomplished is to do the computations using coordinates from a mathematically rigorous map projection system such as the state plane system or a local plane coordinate system that rigorously accounts for earth curvature. In the following chapters, methods are developed for adjusting horizontal surveys using parametric equations that are based on plane coordinates. It should be noted that if state plane coordinates are used, the numbers are usually rather large. Consequently, when they are used in mathematical computations, errors due to roundoff and truncation can occur. This can be prevented by translating the origin of the coordinates prior to adjustment, a process that simply involves subtracting a constant value from all coordinates. Then after the adjustment is finished, the true origin is restored by adding the constants to the adjusted values. This procedure is demonstrated with the following example.

Example 13.1 Assume that the NAD83 state plane coordinates of three control stations to be used in a horizontal survey adjustment are as given below. Translate the origin.

Point	Easting (m)	Northing (m)
A	698,257.171	172,068.220
B	698,734.839	171,312.344
C	698,866.717	170,696.617

SOLUTION:

Step 1: Many surveyors prefer to work in feet, and some jobs require it. Thus in this step the eastings and northings, respectively, are converted to X and Y values in feet by multiplying by 3.28083333. This is the factor for converting meters to U.S. survey feet, and is based on there being exactly 39.37 inches per meter. After making the multiplications, the coordinates in feet are:

Point	X (ft)	Y (ft)
A	2,290,865.40	564,527.15
B	2,292,432.55	562,047.25
C	2,292,865.22	560,027.15

Step 2: To reduce the sizes of these numbers, an X constant is subtracted from each X coordinate, and a Y constant is subtracted from each Y coordinate. For convenience, these constants are usually rounded to the nearest thousandth and are normally selected to give the smallest possible coordinates without producing negative values. In this instance, 2,290,000 ft and 560,000 ft are used as the X constant and Y constant, respectively. Subtracting these values from the coordinates above yields:

Point	X' (ft)	Y' (ft)
A	865.40	4527.15
B	2432.55	2047.25
C	2865.22	27.15

These X' and Y' coordinates can then be used in the adjustment. After the adjustment is complete, the coordinates are translated back to their state plane values by reversing the steps described above, that is, by adding 2,290,000 ft to all adjusted X coordinates and adding 560,000 ft to all adjusted Y coordinates. If desired, they could also be converted back to meters. In the horizontal adjustment problems solved later in the book, either translated state plane coordinates or local plane coordinates are used.

In this chapter we concentrate on adjusting trilateration surveys, those involving only horizontal distance measurements. This method of conducting horizontal surveys became common with the introduction of EDM instruments, which enabled accurate distance measurements to be made rapidly and economically. Trilateration is still practiced using today's modern total station instruments, but as noted, the procedure is now giving way to traversing and GPS surveys.

13.2 DISTANCE OBSERVATION EQUATION

In adjusting trilateration surveys by parametric least squares, observation equations are written that relate the observed quantities and their inherent random errors to the most probable values for the x and y coordinates (the parameters) of the stations involved. Referring to Figure 13.1, the following distance equation can be written for any line IJ:

$$l_{ij} + v_{l_{ij}} = \sqrt{(x_j - x_i)^2 + (y_j - y_i)^2} \tag{13.1}$$

where l_{ij} is the measured length of a line between stations I and J, $v_{l_{ij}}$ the residual in the measurement, x_i and y_i the most probable coordinate values for station I, and x_j and y_j are the most probable coordinate values for station J. Equation (13.1) is a nonlinear function involving the unknown variables x_i, y_i, x_j, and y_j, which can be rewritten as

$$F(x_i, y_i, x_j, y_j) = l_{ij} + v_{l_{ij}} \tag{13.2}$$

where $F(x_i, y_i, x_j, y_j) = \sqrt{(x_j - x_i)^2 + (y_j - y_i)^2}$.

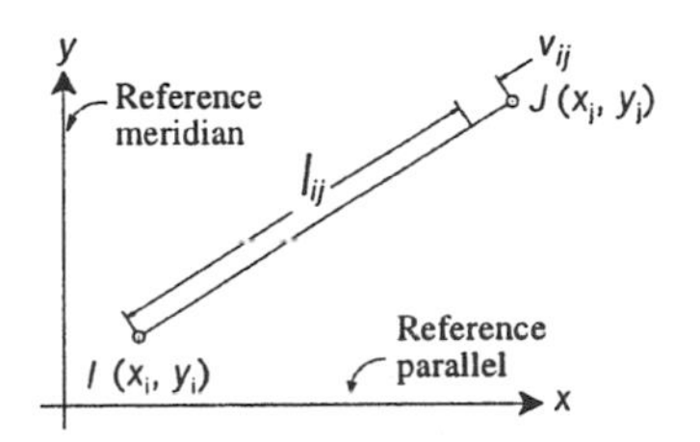

Figure 13.1 Measurement of a distance.

As discussed in Section 10.10, a system of nonlinear equations such as Equation (13.2) can be linearized and solved using a first-order Taylor series approximation. The linearized form of Equation (13.2) is

$$F(x_i, y_i, x_j, y_j) = F(x_{i0}, y_{i0}, x_{j0}, y_{j0}) + \left(\frac{\partial F}{\partial x_i}\right)_0 dx_i + \left(\frac{\partial F}{\partial y_i}\right)_0 dy_i + \left(\frac{\partial F}{\partial x_j}\right)_0 dx_j + \left(\frac{\partial F}{\partial y_j}\right)_0 dy_j \tag{13.3}$$

where x_{i0}, y_{i0}, x_{j0}, and y_{j0} are approximations of the unknowns x_i, y_i, x_j, and y_j; $(\partial F/\partial x_i)_0$, $(\partial F/\partial y_i)_0$, $(\partial F/\partial x_j)_0$, and $(\partial F/\partial y_j)_0$ are the partial derivatives of F with respect to x_i, y_i x_j, and y_j, respectively, evaluated using the approximate coordinate values; and dx_i, dy_i, dx_j, and dy_j are the corrections to the initial approximations such that

$$\begin{aligned} x_i &= x_{i0} + dx_i & y_i &= y_{i0} + dy_i \\ x_j &= x_{j0} + dx_j & y_j &= y_{j0} + dy_j \end{aligned} \tag{13.4}$$

The evaluation of partial derivatives is straightforward and will be illustrated with $\partial F/\partial x_i$. Equation (13.2) can be rewritten as

$$F(x_i, y_i, x_j, y_j) = [(x_j - x_i)^2 + (y_j - y_i)^2]^{1/2} \tag{13.5}$$

Taking the derivative of Equation (13.5) with respect to x_i gives

$$\frac{\partial F}{\partial x_i} = \frac{1}{2}[(x_j - x_i)^2 + (y_j - y_i)^2]^{1/2}[2(x_j - x_i)(-1)] \tag{13.6}$$

and simplifying Equation (13.6) yields

$$\frac{\partial F}{\partial x_i} = \frac{-x_j + x_i}{\sqrt{(x_j - x_i)^2 + (y_j - y_i)^2}} = \frac{x_i - x_j}{IJ} \tag{13.7}$$

Employing the same procedure, the remaining partial derivatives are

$$\frac{\partial F}{\partial y_i} = \frac{y_i - y_j}{IJ} \qquad \frac{\partial F}{\partial x_j} = \frac{x_j - x_i}{IJ} \qquad \frac{\partial F}{\partial y_j} = \frac{y_j - y_i}{IJ} \tag{13.8}$$

If Equations (13.7) and (13.8) are substituted into Equation (13.3) and the results substituted into Equation (13.2), the following prototype linearized distance observation equation is obtained:

$$\frac{x_{i_0} - x_{j_0}}{IJ_0} dx_i + \frac{y_{i_0} - y_{j_0}}{IJ_0} dy_i + \frac{x_{j_0} - x_{i_0}}{IJ_0} dx_j + \frac{y_{j_0} - y_{i_0}}{IJ_0} dy_j = k_{l_{ij}} + v_{l_{ij}} \tag{13.9}$$

where $k_{l_{ij}} = l_{ij} - IJ_0$ and

$$IJ_0 = F(x_{i_0}, y_{i_0}, x_{j_0}, y_{j_0}) = \sqrt{(x_{j_0} - x_{i_0})^2 + (y_{j_0} - y_{i_0})^2}$$

13.3 TRILATERATION ADJUSTMENT EXAMPLE

Even though the geometric figures used in trilateration are many and varied, they are all readily adaptable to the observation equation method in a parametric adjustment. Consider the example shown in Figure 13.2, where the distances are measured from three stations with known coordinates, to a common unknown station U. Since the unknown station has two unknown coordinates and there are three observations, this results in one redundant measurement. That is, the coordinates of station U could be determined using any two of the three measurements. But all three measurements can be used simultaneously and adjusted by the method of least squares to determine the most probable value for the coordinates of station U.

The observation equations are developed by substituting into prototype Equation (13.9). For example, the equation for distance AU is formed by interchanging subscript I with A and subscript J with U in Equation (13.9). In a similar fashion, an equation can be created for each measured line using the following subscript substitutions:

I	J
A	U
B	U
C	U

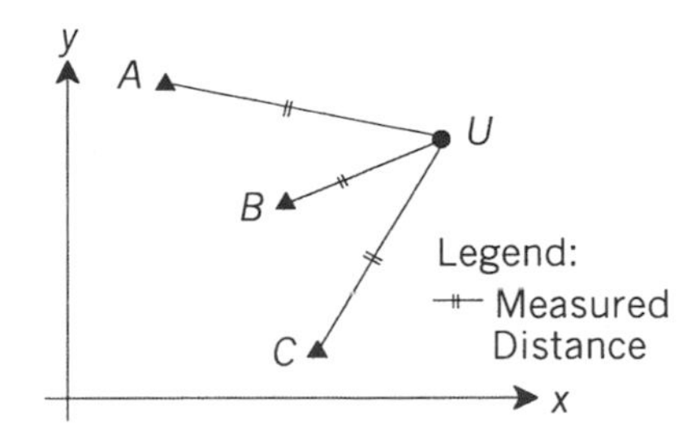

Figure 13.2 Trilateration example.

When one end of the measured line is a control station, its coordinates are fixed, and thus those terms can be dropped in prototype equation (13.9). This can be thought of as setting the dx and dy corrections for the control station equal to zero. In this example, station U always takes the position of J in the prototype equation, and thus only the coefficients corresponding to dx_j and dy_j are used. Using the appropriate substitutions, the following three linearized observation equations result:

$$
\begin{aligned}
\frac{x_{u0} - x_a}{AU_0} dx_u + \frac{y_{u0} - y_a}{AU_0} dy_u &= (l_{au} - AU_0) + v_{l_{au}} \\
\frac{x_{u0} - x_b}{BU_0} dx_u + \frac{y_{u0} - y_b}{BU_0} dy_u &= (l_{bu} - BU_0) + v_{l_{bu}} \\
\frac{x_{u0} - x_c}{CU_0} dx_u + \frac{y_{u0} - y_c}{CU_0} dy_u &= (l_{cu} - CU_0) + v_{l_{cu}}
\end{aligned}
\tag{13.10}
$$

where

$$
\begin{aligned}
AU_0 &= \sqrt{(x_{u_0} - x_a)^2 + (y_{u_0} - y_a)^2}; \\
BU_0 &= \sqrt{(x_{u_0} - x_b)^2 + (y_{u_0} - y_b)^2}; \\
CU_0 &= \sqrt{(x_{u_0} - x_c)^2 + (y_{u_0} - y_c)^2};
\end{aligned}
$$

l_{au}, l_{bu}, and l_{cu} are the measured distances; the v's are residuals; and x_{u_0} and y_{u_0} are initial coordinate approximations for station U. Equations (13.10) can be expressed in matrix form as

$$
JX = K + V \tag{13.11}
$$

where J is the Jacobian matrix of partial derivatives, X the matrix or unknown corrections dx_u and dy_u, K the matrix of constants (i.e., the measured lengths, minus their corresponding lengths computed from the initial approximate coordinates), and V the residual matrix. Equation (13.11) in expanded form is

$$
\begin{bmatrix}
\frac{x_{u0} - x_a}{AU_0} & \frac{y_{u0} - y_a}{AU_0} \\
\frac{x_{u0} - x_b}{BU_0} & \frac{y_{u0} - y_b}{BU_0} \\
\frac{x_{u0} - x_c}{CU_0} & \frac{y_{u0} - y_c}{CU_0}
\end{bmatrix}
\begin{bmatrix} dx_u \\ dy_u \end{bmatrix}
=
\begin{bmatrix} l_{au} - AU_0 \\ l_{bu} - BU_0 \\ l_{cu} - CU_0 \end{bmatrix}
+
\begin{bmatrix} v_{l_{au}} \\ v_{l_{bu}} \\ v_{l_{cu}} \end{bmatrix}
\tag{13.12}
$$

The Jacobian matrix can be formed systematically using the following steps:

Step 1: Head each column with an unknown value.

Step 2: Create a row for every observation.

Step 3: Substitute in the appropriate coefficient corresponding to the column into each row.

If this procedure is followed for this problem, the Jacobian matrix is

$$\begin{bmatrix} & \underline{dx_u} & \underline{dy_u} \\ AU \mid & \dfrac{\partial F}{\partial x_u} & \dfrac{\partial F}{\partial y_u} \\ BU \mid & \dfrac{\partial F}{\partial x_u} & \dfrac{\partial F}{\partial y_u} \\ CU \mid & \dfrac{\partial F}{\partial x_u} & \dfrac{\partial F}{\partial y_u} \end{bmatrix}$$

Once Equation (13.12) is created, the corrections dx_u and dy_u can be computed using Equation (10.37). The corrections are then applied to the initial approximations of x_u and y_u. To obtain the final adjusted values, the solution must be iterated, as discussed in Section 10.10.

Example 13.2 To clarify the computational procedure, a numerical example for Figure 13.2 is presented. Suppose that the measured distances l_{au}, l_{bu}, and l_{cu} are 6049.00, 4736.83, and 5446.49 ft, respectively, and the control stations have coordinates of

$$x_a = 865.40 \qquad x_b = 2432.55 \qquad x_c = 2865.22$$

$$y_a = 4527.15 \qquad y_b = 2047.25 \qquad y_c = 27.15$$

(Note that these are the translated coordinates obtained in Example 13.1). Compute the most probable coordinates for station U.

SOLUTION
First Iteration

Step 1: Calculate approximate coordinates for station U.

a. *Calculate azimuth AB* from the coordinate values of stations A and B.

$$\text{Az}_{AB} = \tan^{-1}\left(\frac{x_b - x_a}{y_b - y_a}\right) + 180°$$

$$= \tan^{-1}\left(\frac{2432.55 - 865.40}{2047.25 - 4527.15}\right) + 180°$$

$$= 147°42'34''$$

b. Calculate the distance between stations A and B from their coordinate values.

$$
\begin{aligned}
AB &= \sqrt{(x_b - x_a)^2 + (y_b - y_a)^2} \\
&= \sqrt{(2432.55 - 865.20)^2 + (2047.25 - 4527.15)^2} \\
&= 2933.58 \text{ ft}
\end{aligned}
$$

c. Calculate azimuth AU_0 using the cosine law in triangle AUB:

$$\cos(\measuredangle UAB) = \frac{(6049.00)^2 + (2933.58)^2 - (4736.83)^2}{2(6049.00)(2933.58)}$$

$$\measuredangle UAB = 50°06'50''$$

$$\text{Az}_{Au0} = 147°42'34'' - 50°06'50'' = 97°35'44''$$

d. Calculate the coordinates for station U.

$$x_{u0} = 865.40 + 6049.00 \sin(97°35'44'') = 6861.325 \text{ ft}$$

$$y_{u0} = 4527.15 + 6049.00 \cos(97°35'44'') = 3727.596 \text{ ft}$$

Step 2: Calculate AU_0, BU_0, and CU_0. For this first iteration, AU_0 and BU_0 are exactly equal to their respective measured distances since x_{u0} and y_{u0} were calculated using these quantities. Thus:

$$AU_0 = 6049.00 \qquad BU_0 = 4736.83$$

$$CU_0 = [(6861.325 - 2865.22)^2 + (3727.596 - 27.15)^2]^{1/2} = 5446.298 \text{ ft}$$

Step 3: Formulate the matrices.

a. The elements of the Jacobian matrix in Equation (13.12) are[1]

$$j_{11} = \frac{6861.325 - 865.40}{6049.00} = 0.991 \qquad j_{12} = \frac{3727.596 - 4527.15}{6049.00} = -0.132$$

$$j_{21} = \frac{6861.325 - 2432.55}{4736.83} = 0.935 \qquad j_{22} = \frac{3727.596 - 2047.25}{4736.83} = 0.355$$

$$j_{31} = \frac{6861.325 - 2865.22}{5446.298} = 0.734 \qquad j_{32} = \frac{3727.596 - 27.15}{5446.298} = 0.679$$

[1] Note that the denominators in the coefficients of step 3a are distances computed from the approximate coordinates. Only the distances computed for the first iteration will match the measured distances exactly. Do not use measured distances for the denominators of these coefficients.

b. The elements of the K matrix in Equation (13.12) are

$$k_1 = 6049.00 - 6049 = 0.000$$

$$k_2 = 4736.83 - 4736.83 = 0.000$$

$$k_3 = 5446.49 - 5446.298 = 0.192$$

Step 4: The matrix solution using Equation (10.37) is

$$X = (J^TJ)^{-1}J^TK$$

$$J^TJ = \begin{bmatrix} 0.991 & 0.935 & 0.734 \\ -0.132 & 0.355 & 0.679 \end{bmatrix} \begin{bmatrix} 0.991 & -0.132 \\ 0.935 & 0.355 \\ 0.735 & 0.679 \end{bmatrix} = \begin{bmatrix} 2.395 & 0.699 \\ 0.699 & 0.605 \end{bmatrix}$$

$$(J^TJ)^{-1} = \frac{1}{0.960} \begin{bmatrix} 0.605 & -0.699 \\ -0.699 & 2.395 \end{bmatrix}$$

$$J^TK = \begin{bmatrix} 0.991 & 0.935 & 0.734 \\ -0.132 & 0.355 & 0.679 \end{bmatrix} \begin{bmatrix} 0.000 \\ 0.000 \\ 0.192 \end{bmatrix} = \begin{bmatrix} 0.141 \\ 0.130 \end{bmatrix}$$

$$X = \frac{1}{0.960} \begin{bmatrix} 0.605 & -0.699 \\ -0.699 & 2.395 \end{bmatrix} \begin{bmatrix} 0.141 \\ 0.130 \end{bmatrix} = \begin{bmatrix} -0.006 \\ 0.222 \end{bmatrix}$$

The revised coordinates of U are

$$x_u = 6861.325 - 0.006 = 6861.319$$

$$y_u = 3727.596 + 0.222 = 3727.818$$

Second Iteration

Step 1: Calculate AU_0, BU_0, and CU_0.

$$AU_0 = \sqrt{(6861.319 - 865.40)^2 + (3727.818 - 4527.15)^2} = 6048.965$$

$$BU_0 = \sqrt{(6861.319 - 2432.55)^2 + (3727.818 - 2047.25)^2} = 4736.909$$

$$CU_0 = \sqrt{(6861.319 - 2865.22)^2 + (3727.818 - 27.15)^2} = 5446.444$$

Notice that these computed distances no longer match their measured counterparts.

Step 2: Formulate the matrices. With these minor changes in the lengths, the J matrix (to three places) does not change, and thus $(J^TJ)^{-1}$ does not change either. However, the K matrix does change, as shown by the following computations:

$$k_1 = 6049.00 - 6048.965 = 0.035$$

$$k_2 = 4736.83 - 4736.909 = -0.079$$

$$k_3 = 5446.49 - 5446.444 = 0.046$$

Step 3: Matrix solution

$$J^T K = \begin{bmatrix} 0.991 & 0.935 & 0.734 \\ -0.132 & 0.355 & 0.679 \end{bmatrix} \begin{bmatrix} 0.035 \\ -0.079 \\ 0.046 \end{bmatrix} = \begin{bmatrix} -0.005 \\ -0.001 \end{bmatrix}$$

$$X = \frac{1}{0.960} \begin{bmatrix} 0.605 & -0.699 \\ -0.699 & 2.395 \end{bmatrix} \begin{bmatrix} -0.005 \\ -0.001 \end{bmatrix} = \begin{bmatrix} -0.002 \\ 0.001 \end{bmatrix}$$

The revised coordinates of U are

$$x_u = 6861.319 - 0.002 = 6861.317$$

$$y_u = 3727.818 + 0.001 = 3727.819$$

Satisfactory convergence is shown by the very small corrections in the second iteration. This problem has also been solved using program ADJUST. Values computed include the most probable coordinates for station U, their standard deviations, the adjusted lengths of the measured distances, their residuals and standard deviations, and the reference variance and standard deviation. These are tabulated as shown below.

```
****************
Adjusted Stations
****************
Station          X            Y           Sx        Sy
--------------------------------------------------------------
        U   6,861.32     3,727.82      0.078     0.154
*****************************
Adjusted Distance Observations
*****************************
  Station     Station
 Occupied     Sighted      Distance          V        S
--------------------------------------------------------------
        A           U      6,048.96     -0.037    0.090
        B           U      4,736.91      0.077    0.060
        C           U      5,446.44     -0.047    0.085
                        ********************
                        Adjustment Statistics
                        ********************
                        S0^2 = 0.00954
                        S0   = ±0.10
```

13.4 FORMULATION OF A GENERALIZED COEFFICIENT MATRIX FOR A MORE COMPLEX NETWORK

In the trilaterated network of Figure 13.3, all lines were measured. Assume that Stations *A* and *C* are control stations. For this network, there are 10 observations and eight unknowns. Stations *A* and *C* can be fixed by giving the terms dx_a, dy_a, dx_c, and dy_c zero coefficients, which effectively drops these terms from the solution. The coefficient matrix formulated from prototype Equation (13.9) has nonzero elements, as indicated in Table 13.1. In this table the appropriate coefficient from Equation (13.9) is indicated by its corresponding unknown terms of dx_i, dy_i, dx_j, or dy_j.

13.5 COMPUTER SOLUTION OF A TRILATERATED QUADRILATERAL

The quadrilateral shown in Figure 13.4 was adjusted using the Program MATRIX. In this problem, points Bucky and Badger are control stations whose coordinates are held fixed. The five observed distances are:

Line	Distance (ft)
Badger–Wisconsin	5870.302
Badger–Campus	7297.588
Wisconsin–Campus	3616.434
Wisconsin–Bucky	5742.878
Campus–Bucky	5123.760

The state plane coordinates for control station Badger are $x = 2{,}410{,}000.000$ and $y = 390{,}000.000$, and for Bucky they are $x = 2{,}411{,}820.000$ and $y = 386{,}881.222$.

Step 1: To solve this problem, approximate coordinates are first computed for stations Wisconsin and Campus. This is done following the same pro-

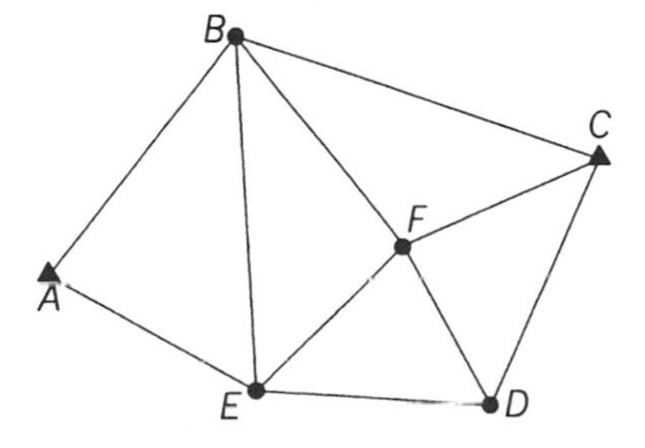

Figure 13.3 Trilateration network.

Table 13.1 Structure of the coefficient matrix for the complex network in Figure 13.3

Distance,	Unknown							
IJ	dx_b	dy_b	dx_d	dy_d	dx_e	dy_e	dx_f	dy_f
AB	dx_j	dy_j	0	0	0	0	0	0
AE	0	0	0	0	dx_j	dy_j	0	0
BC	dx_i	dy_i	0	0	0	0	0	0
BF	dx_i	dy_i	0	0	0	0	dx_j	dy_j
BE	dx_i	dy_i	0	0	dx_j	dy_j	0	0
CD	0	0	dx_j	dy_j	0	0	0	0
CF	0	0	0	0	0	0	dx_j	dy_j
DF	0	0	dx_i	dy	0	0	dx_j	dy_j
DE	0	0	dx_i	dy_i	dx_j	dy_j	0	0
EF	0	0	0	0	dx_i	dy_i	dx_j	dy_j

cedures used in Section 13.3, with the resulting initial approximations being

Wisconsin: $x = 2{,}415{,}776.819$ $y = 391{,}043.461$

Campus: $x = 2{,}416{,}898.227$ $y = 387{,}602.294$

Step 2: Following prototype Equation (13.9) and the procedures outlined in Section 13.4, a table of coefficients is established. For the sake of brevity in Table 13.2, the following station assignments were made: Badger = 1, Bucky = 2, Wisconsin = 3, and Campus = 4.

After forming the J matrix, the K matrix is computed. This is done in a manner similar to *step 3* of the first iteration in Example 13.2. The matrices were entered into a file following the formats as listed in Appendix F for program MATRIX. Following are the input data, matrices for the first and last iterations of this three-iteration solution, and the final results.

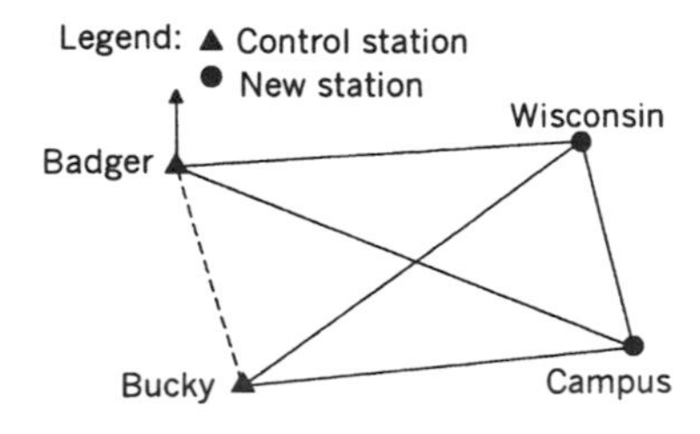

Figure 13.4 Quadrilateral.

Table 13.2 Structure of the coefficient or *J* matrix for example in Figure 13.4

	$dx_{\text{Wisconsin}}$	$dy_{\text{Wisconsin}}$	dx_{Campus}	dy_{Campus}
Badger–Wisconsin 1–3	$\dfrac{x_{3_0} - x_1}{(1\text{–}3)_0}$	$\dfrac{y_{3_0} - y_1}{(1\text{–}3)_0}$	0	0
Badger–Campus 1–4	0	0	$\dfrac{x_{4_0} - x_1}{(1\text{–}4)_0}$	$\dfrac{y_{4_0} - y_1}{(1\text{–}4)_0}$
Wisconsin–Campus 3–4	$\dfrac{x_{3_0} - x_{4_0}}{(3\text{–}4)_0}$	$\dfrac{y_{3_0} - y_{4_0}}{(3\text{–}4)_0}$	$\dfrac{x_{4_0} - x_{3_0}}{(3\text{–}4)_0}$	$\dfrac{y_{4_0} - y_{3_0}}{(3\text{–}4)_0}$
Wisconsin–Bucky 3–2	$\dfrac{x_{3_0} - x_2}{(3\text{–}2)_0}$	$\dfrac{y_{3_0} - y_2}{(3\text{–}2)_0}$	0	0
Campus–Bucky 4–2	0	0	$\dfrac{x_{4_0} - x_2}{(4\text{–}2)_0}$	$\dfrac{y_{4_0} - y_2}{(4\text{–}2)_0}$

Computer Solution of Example Trilaterated Quadrilateral

```
*******************************************
Initial Approximations for Unknown Stations
*******************************************
   Station                  X                         Y
-----------------------------------------------------------------
    Wisconsin          2,415,776.819             391,043.461
       Campus          2,416,898.220             387,602.294

Control Stations
----------------

      Station               X                         Y
-----------------------------------------------------------------
       Badger          2,410,000.000             390,000.000
        Bucky          2,411,820.000             386,881.222


*********************
Distance Observations
*********************
    Station               Station
   Occupied               Sighted                   Distance
-----------------------------------------------------------------
      Badger             Wisconsin                 5,870.302
      Badger                Campus                 7,297.588
   Wisconsin                Campus                 3,616.434
   Wisconsin                 Bucky                 5,742.878
      Campus                 Bucky                 5,123.760
```

First Iteration Matrices

J Dim: 5 × 4

0.98408	0.17775	0.00000	0.00000
0.00000	0.00000	0.94457	-0.32832
-0.30984	0.95079	0.30984	-0.95079
0.68900	0.72477	0.00000	0.00000
0.00000	0.00000	0.99007	0.14058

K Dim: 5 × 1

-0.00026
-5.46135
-2.84579
-0.00021
-5.40507

X Dim 4 × 1

0.084751
-0.165221
-5.531445
0.959315

JtJ Dim: 4 × 4

1.539122	0.379687	-0.096003	0.294595
0.379687	1.460878	0.294595	-0.903997
-0.096003	0.294595	1.968448	-0.465525
0.294595	-0.903997	-0.465525	1.031552

Inv(N) Dim: 4 × 4

1.198436	-1.160169	-0.099979	-1.404084
-1.160169	2.635174	0.194272	2.728324
-0.099979	0.194272	0.583337	0.462054
-1.404084	2.728324	0.462054	3.969873

Final Iteration Matrices

J Dim: 5 × 4

0.98408	0.17772	0.00000	0.00000
0.00000	0.00000	0.94453	-0.32843
-0.30853	0.95121	0.30853	-0.95121
0.68902	0.72474	0.00000	0.00000
0.00000	0.00000	0.99002	0.14092

K Dim: 5×1

-0.05468
0.07901
-0.03675
0.06164
-0.06393

X dim 4×1

0.000627
-0.001286
-0.000040
0.001814

JtJ Dim: 4 × 4

1.538352	0.380777	-0.095191	0.293479
0.380777	1.461648	0.293479	-0.904809
-0.095191	0.293479	1.967465	-0.464182
0.293479	-0.904809	-0.464182	1.032535

Qxx = Inv(N) Dim: 4 × 4

1.198574	-1.160249	-0.099772	-1.402250
-1.160249	2.634937	0.193956	2.725964
-0.099772	0.193956	0.583150	0.460480
-1.402250	2.725964	0.460480	3.962823

```
Q11 = J Qxx Jt Dim: 5 x 5
---------------------------------------------------------------------
 0.838103    0.233921   -0.108806    0.182506   -0.189263
 0.233921    0.662015    0.157210   -0.263698    0.273460
-0.108806    0.157210    0.926875    0.122656   -0.127197
 0.182506   -0.263698    0.122656    0.794261    0.213356
-0.189263    0.273460   -0.127197    0.213356    0.778746
****************
Adjusted Stations
****************
  Station          X               Y           Sx       Sy
---------------------------------------------------------------------
Wisconsin  2,415,776.904   391,043.294    0.1488   0.2206
   Campus  2,416,892.696   387,603.255    0.1038   0.2705


*****************************
Adjusted Distance Observations
*****************************
 Occupied    Sighted      Distance         V          S
---------------------------------------------------------------------
   Badger   Wisconsin     5,870.357     0.055     0.1244
   Badger      Campus     7,297.509    -0.079     0.1106
Wisconsin      Campus     3,616.471     0.037     0.1308
Wisconsin       Bucky     5,742.816    -0.062     0.1211
   Campus       Bucky     5,123.824     0.064     0.1199


-----Reference Standard Deviation = +0.135905 ----
     Iterations » 3
```

Notes

1. As earlier noted, it is important that measured distances not be used in the denominator of the coefficient matrix, J. This is not only theoretically incorrect, it can cause slight differences in the final solution, or even worse, it can cause the system to diverge from any solution! In developing the J matrix, always compute distances based on the current approximate coordinates.
2. The final portion of the output lists the adjusted x and y coordinates of the stations, the reference standard deviation, the standard deviations of the adjusted coordinates, the adjusted line lengths, and their residuals.
3. The Q_{xx} matrix was listed on the last iteration only. It is needed for calculating the estimated errors of the adjusted coordinates using Equation (12.24), and is also necessary for calculating error ellipses. The subject of error ellipses is discussed in Chapter 18.

13.6 ITERATION TERMINATION

When programming a nonlinear least-squares adjustment, some criteria must be established to determine the appropriate point at which to stop the iteration process. Since it is possible to have a set of data that has no solution, it is also important to determine when that condition occurs. In this section, we describe three methods commonly used to indicate the appropriate time stop the iteration process.

13.6.1 Method of Maximum Iterations

The simplest procedure of iteration termination involves limiting the number of iterations to a predetermined maximum. The risk with this method is that if this maximum is too low, a solution may not be reached at the conclusion of the process, and if it is too high, time is wasted on unnecessary iterations. Although this method does not assure convergence, it can prevent the adjustment from continuing indefinitely, which could occur if the solution diverges. When good initial approximations are supplied for the unknown parameters, a limit of 10 iterations should be well beyond what is required for a solution since the least-squares method converges quadratically.

13.6.2 Maximum Correction

This method was used in earlier examples. It involves monitoring the absolute sizes of the corrections. When all corrections become negligibly small, the iteration process is stopped. The term *negligible* is relative. For example, if distances are measured to the nearest foot, it would be foolish to assume that the sizes of the corrections will become less than some small fraction of a foot. Generally, *negligible* is interpreted as a correction that is less than one-half the least count of the smallest unit of measure. For instance, if all distances are measured to the nearest 0.01 ft, it would be appropriate to assume convergence when the absolute size of all corrections is less than 0.005 ft. While the solution may continue to converge with continued iterations, the work to get these corrections is not warranted based on the precision of the measurements.

13.6.3 Monitoring the Adjustment's Reference Variance

The best method for determining convergence involves monitoring the reference variance for its changes between iterations. Since the least-squares method converges quadratically, the iteration process should definitely be stopped if the reference variance increases. An increasing reference variance suggests a diverging solution, which happens when one of two things has occurred: (1) a large blunder exists in the data set and no solution is possible, or (2) the maximum correction size is less than the precision of the measurements. If the second case has occurred, the best solution for the given data set has already been reached, and when another iteration is attempted, the solution will converge, only to diverge on the next iteration. This *apparent bouncing* in the solution is caused by round-off errors, or convergence limits being too stringent for the quality of the data.

By monitoring the reference variance, convergence and divergence can be detected. Convergence is assumed when the change in the reference variance falls below some predefined percentage. Convergence can be generally assumed when the change in the reference variance is less than 1% between iterations. If the size of the reference variance increases, the solution is diverging and the iteration process should be stopped. It should be noted that monitoring changes in the reference variance will always show convergence or divergence in the solution, and thus it is better than the other two methods. However, all methods should be used in concert when doing an adjustment.

PROBLEMS

Note: For problems below requiring least squares adjustment, if a computer program is not distinctly specified for use in the problem, it is expected that the least squares algorithm will be solved using the program MATRIX which is included within the diskette supplied with the book.

13.1 Given the following measured values for the lines in Figure 13.2:

$$AU = 1653.92 \text{ ft} \qquad BU = 1067.24 \text{ ft} \qquad CU = 1298.10 \text{ ft}$$

The control coordinates of stations A, B and C are:

Station	x	y
A	0.00	0.00
B	1000.00	0.00
C	0.00	1000.00

What are the most probable values for the adjusted coordinates of station U?

13.2 Do a weighted least squares adjustment using the data in Problem 13.1 with weights based on the following observational errors.

$$AU = \pm 0.021 \text{ ft} \qquad BU = \pm 0.020 \text{ ft} \qquad CU = \pm 0.021 \text{ ft}$$

(a) What are the most probable values for the adjusted coordinates of station U?

(b) What is the standard deviation of unit weight?

(c) What are the estimated standard deviations in the adjusted coordinates?

(d) Tabulate the adjusted distances, their residuals, and standard deviations.

13.3 Do a least-squares adjustment of the following measured values for the lines in Figure 13.5:

$$DA = 29{,}593.10 \text{ ft}$$

$$BC = 22{,}943.66 \text{ ft}$$

$$AC = 30{,}728.80 \text{ ft}$$

$$CD = 28{,}217.92 \text{ ft}$$

$$BD = 41{,}470.26 \text{ ft}$$

In the adjustment hold the coordinates of

$$x_a = 110{,}000.00 \quad \text{and} \quad y_a = 110{,}000.00$$

$$x_b = 121{,}433.56 \quad \text{and} \quad y_b = 129{,}803.51$$

(a) What are the most probable values for the adjusted coordinates of stations C and D?

(b) What is the standard deviation of unit weight?

(c) What are the estimated standard deviations in the adjusted coordinates?

(d) Tabulate the adjusted distances, their residuals, and standard deviations.

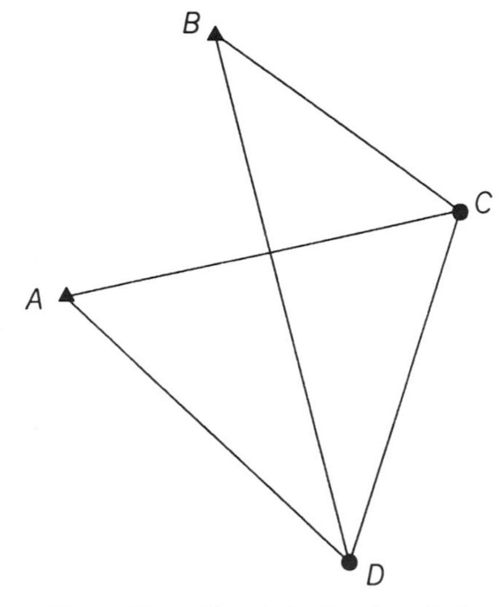

Figure 13.5 Sketch for Problem 13.3.

13.4 Repeat Problem 13.3 using a weighted least-squares adjustment, where the distance standard deviations are

$$DA = \pm 0.297 \text{ ft} \qquad BC = \pm 0.230 \text{ ft} \qquad BD = \pm 0.415 \text{ ft}$$

$$AC = \pm 0.308 \text{ ft} \qquad CD = \pm 0.283 \text{ ft}$$

13.5 Using the trilaterated Figure 13.3 and the data below, compute the most probable station coordinates and their standard deviations.

Initial approximations of stations			Control stations		
Station	Easting	Northing	Station	Easting	Northing
B	10,862.495	11,103.943	*A*	10,000.000	10,000.000
D	11,990.881	9,387.460	*C*	12,487.083	10,528.646
E	10,948.550	9,461.910	*F*	11,595.232	10,131.568

Distance observations

Station Occupied	Station Sighted	Distance	σ
A	*B*	1400.934	0.021
A	*E*	1090.553	0.020
B	*E*	1644.292	0.021
B	*F*	1217.535	0.021
B	*C*	1723.429	0.021
C	*F*	976.254	0.020
C	*D*	1244.399	0.021
D	*F*	842.745	0.020
D	*E*	1044.987	0.020
E	*F*	930.940	0.020

13.6 Use Program ADJUST to do Problems 13.3 and 13.4. Discuss the differences in the two adjustments.

13.7 Use Program ADJUST to do Problem 13.5.

13.8 Describe the methods used to detect convergence in a nonlinear least-squares adjustment, and discuss the advantages and disadvantages of each.

Programming Problems

13.9 Create a spreadsheet that computes the coefficients, and $k_{i_{ij}}$ in Equation (13.9) between stations *I* and *J* given their initial coordinate values. Use this spreadsheet to determine the matrix values necessary for solving Problem 13.5.

13.10 Create a program that reads a data file containing station coordinates and distances and generates the *J*, *W*, and *K* matrices which can be used by program MATRIX. Demonstrate that this program works by using the data of Problem 13.5.

13.11 Create a program that reads a file containing the *J*, *W*, and *K* matrices and finds the most probable value for the station coordinates, the reference standard deviation, and the standard deviations of the station coordinates. Demonstrate that this program works by solving Problem 13.5.

13.12 Create a program that reads a file containing control station coordinates, initial approximations of unknown stations, and distance observations. The program should generate the appropriate matrices for a least-squares adjustment, do the adjustment, and print out the final adjusted coordinates, their standard deviations, the final adjusted distances, their residuals, and the standard deviations in the adjusted distances. Demonstrate that this program works by solving Problem 13.5.

14

ADJUSTMENT OF HORIZONTAL SURVEYS: TRIANGULATION

14.1 INTRODUCTION

Prior to the development of electronic distance measuring equipment and the global positioning system, triangulation was the preferred method for extending horizontal control over long distances. The positions of widely spaced stations were computed from measured angles and a minimal number of measured distances called *baselines*. This method was used extensively by the U.S. Coast and Geodetic Survey in extending much of the national network. Triangulation is still used by many surveyors in establishing horizontal control, although surveys that combine trilateration (distance measurements) with triangulation (angle measurements) are more common. In this chapter, methods are described for adjusting triangulation networks using least squares.

A least-squares triangulation adjustment can use condition equations or observation equations written either in terms of azimuths or angles. In this chapter the observation equation method is presented. The procedure involves a parametric adjustment where the parameters are coordinates in a plane rectangular system such as state plane coordinates. In examples, the specific types of triangulations known as intersections, resections, and quadrilaterals will be adjusted.

14.2 AZIMUTH OBSERVATION EQUATION

The azimuth equation in parametric form is

$$\text{azimuth} = \alpha + C \tag{14.1}$$

where

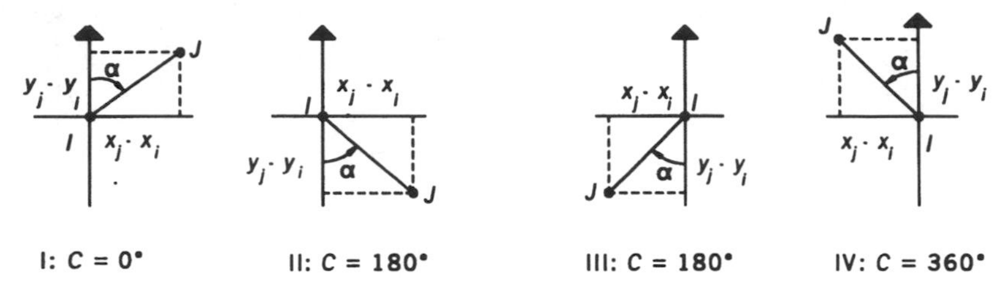

Figure 14.1 Relationship between azimuth and the computed angle α.

$$\alpha = \tan^{-1}\left(\frac{x_j - x_i}{y_j - y_i}\right)$$

and x_i and y_i are the coordinates of the occupied station, I; x_j and y_j are the coordinates of the sighted station, J; and C is a constant that depends on the quadrant in which point J lies, as shown in Figure 14.1. From the figure, Table 14.1 can be constructed, which relates the algebraic sign of the computed angle α in Equation (14.1) to the value of C and the value of the azimuth.

14.2.1 Linearization of the Azimuth Observation Equation

Referring to Equation (14.1), the complete observation equation for the measured azimuth of line IJ in parametric form is

$$\tan^{-1}\left(\frac{x_j - x_i}{y_j - y_i}\right) + C = \text{Az}_{ij} + v_{\text{Az}_{ij}} \tag{14.2}$$

where Az_{ij} is the measured azimuth, $v_{\text{Az}_{ij}}$ the residual in the measured azimuth, x_i and y_i the most probable values for the coordinates of station I, x_j and y_j the most probable values for the coordinates of station J, and C a constant with a value based on Table 14.1. Equation (14.2) is a nonlinear function involving variables x_i, y_i, x_j, and y_j, that can be rewritten as

Table 14.1 Relationships between quadrant, *C*, and azimuth

Quadrant	Sign $(x_j - x_i)$	Sign $(y_j - y_i)$	Sign α	C	Azimuth
I	+	+	+	0°	α
II	+	−	−	180°	α + 180°
III	−	−	+	180°	α + 180°
IV	−	+	−	360°	α + 360°

$$F(x_i, y_i, x_j, y_j) = \text{Az}_{ij} + v_{\text{Az}_{ij}} \tag{14.3}$$

where

$$F(x_i, y_i, x_j, y_j) = \tan^{-1}\left(\frac{x_j - x_i}{y_j - y_i}\right) + C$$

As discussed in Section 10.10, nonlinear equations such as (14.3) can be linearized and solved using a first-order Taylor series approximation. The linearized form of Equation (14.3) is

$$F(x_i, y_i, x_j, y_j) = F(x_{i_0}, y_{i_0}, x_{j_0}, y_{j_0}) + \left(\frac{\partial F}{\partial x_i}\right)_0 dx_i + \left(\frac{\partial F}{\partial y_i}\right)_0 dy_i + \left(\frac{\partial F}{\partial x_j}\right)_0 dx_j + \left(\frac{\partial F}{\partial y_j}\right)_0 dy_j \tag{14.4}$$

where $(\partial F/\partial x_i)_0$, $(\partial F/\partial y_i)_0$, $(\partial F/\partial x_j)_0$, and $(\partial F/\partial y_j)_0$ are the partial derivatives of F with respect to x_i, y_i; x_j, and y_j evaluated at the initial approximations x_{i_0}, y_{i_0}, x_{j_0}, and y_{j_0}; and dx_i, dy_i, dx_j, and dy_j are the corrections applied to the initial approximations after each iteration such that

$$x_i = x_{i_0} + dx_i \qquad y_i = y_{i_0} + dy_i \qquad x_j = x_{j_0} + dx_j \qquad y_j = y_{j_0} + dy_j \tag{14.5}$$

The prototype equation for the derivative of $\tan^{-1}(u)$ with respect to x is

$$\frac{d}{dx}\tan^{-1}(u) = \frac{1}{1 + u^2}\frac{du}{dx} \tag{14.6}$$

Using Equation (14.6), the procedure for determining the $\partial F/dx_i$ will be demonstrated.

$$\begin{aligned}\frac{\partial F}{\partial x_i} &= \frac{1}{1 + [(x_j - x_i)/(y_j - y_i)]^2}\frac{-1}{y_j - y_i}\\ &= \frac{(-1)(y_j - y_i)}{(x_j - x_i)^2 + (y_j - y_i)^2}\\ &= \frac{y_i - y_j}{(IJ)^2}\end{aligned} \tag{14.7}$$

By employing the same procedure, the remaining partial derivatives are

$$\frac{\partial F}{\partial y_i} = \frac{x_j - x_i}{(IJ)^2} \qquad \frac{\partial F}{\partial x_j} = \frac{y_j - y_i}{(IJ)^2} \qquad \frac{\partial F}{\partial y_j} = \frac{x_i - x_j}{(IJ)^2} \tag{14.8}$$

where $(IJ)^2 = (x_j - x_i)^2 + (y_j - y_i)^2$.

If Equations (14.7) and (14.8) are substituted into Equation (14.4) and the result substituted into Equation (14.3), the following prototype azimuth equation is obtained:

$$\frac{y_{i_0} - y_{j_0}}{(IJ_0)^2}\,dx_i + \frac{x_{j_0} - x_{i_0}}{(IJ_0)^2}\,dy_i + \frac{y_{j_0} - y_{i_0}}{(IJ_0)^2}\,dx_j + \frac{x_{i_0} - x_{j_0}}{(IJ_0)^2}\,dy_j$$
$$= k_{\mathrm{Az}_{ij}} + v_{\mathrm{Az}_{ij}} \tag{14.9}$$

where

$$k_{\mathrm{Az}_{ij}} = \mathrm{Az}_{ij} - \tan^{-1}\left(\frac{x_{j_0} - x_{i_0}}{y_{j_0} - y_{i_0}}\right) + C \quad \text{and}$$

$$(IJ_0)^2 = (x_{j_0} - x_{i_0})^2 + (y_{j_0} - y_{i_0})^2$$

14.3 ANGLE OBSERVATION EQUATION

Figure 14.2 illustrates the geometry for an angle observation. In the figure, B is the backsight station, F is the foresight station, and I is the instrument station. As shown in the figure, an angle observation equation can be written as the difference between two azimuth observations, and thus for clockwise angles,

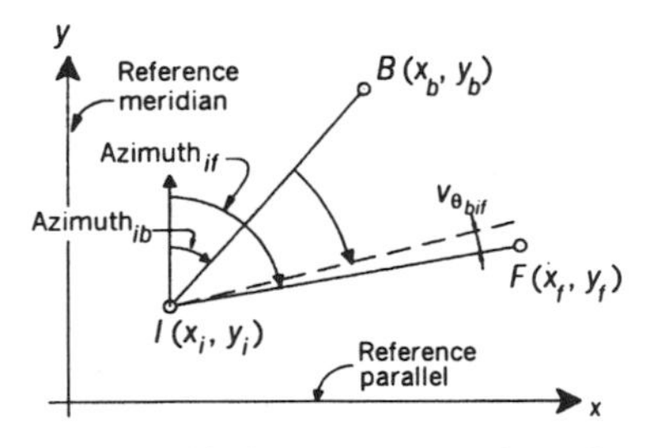

Figure 14.2 Relationship between an angle and two azimuths.

$$\angle BIF = \text{Azimuth}_{if} - \text{Azimuth}_{ib} = \tan^{-1}\left(\frac{x_f - x_i}{y_f - y_i}\right) - \tan^{-1}\left(\frac{x_b - x_i}{y_b - y_i}\right) + D = \theta_{bif} + v_{\theta_{bif}} \tag{14.10}$$

where θ_{bif} is the measured clockwise angle, $v_{\theta_{bif}}$ the residual in the measured angle, x_b and y_b the most probable values for the coordinates of the backsighted station B, x_i and y_i the most probable values for the coordinates of the instrument station I, x_f and y_f the most probable values for the coordinates of the foresighted station F, and D is a constant that depends on the quadrants in which the backsight and foresight occur. This term can be computed as the difference between the C terms from Equation (14.1) as applied to the backsight and foresight azimuths; that is,

$$D = C_{if} - C_{ib}$$

Equation (14.10) is a nonlinear function of x_b, y_b, x_i, y_i, x_f, and y_f which can be rewritten as

$$F(x_b, y_b, x_i, y_i, x_f, y_f) = \theta_{bif} + v_{\theta_{bif}} \tag{14.11}$$

where

$$F(x_b, y_b, x_i, y_i, x_f, y_f) = \tan^{-1}\left(\frac{x_f - x_i}{y_f - y_i}\right) - \tan^{-1}\left(\frac{x_b - x_i}{y_b - y_i}\right) + D$$

Equation (14.11) expressed as a linearized, first-order Taylor series expansion is

$$\begin{aligned} F(x_b, y_b, x_i, y_i, x_f, y_f) &= F(x_{b_0}, y_{b_0}, x_{i_0}, y_{i_0}, x_{f_0}, y_{f_0}) + \left(\frac{\partial F}{\partial x_b}\right)_0 dx_b \\ &+ \left(\frac{\partial F}{\partial y_b}\right)_0 dy_b + \left(\frac{\partial F}{\partial x_i}\right)_0 dx_i + \left(\frac{\partial F}{\partial y_i}\right)_0 dy_i \\ &+ \left(\frac{\partial F}{\partial x_f}\right)_0 dx_f + \left(\frac{\partial F}{\partial y_f}\right)_0 dy_f \end{aligned} \tag{14.12}$$

where $\partial F/dx_b$, $\partial F/dy_b$, $\partial F/dx_i$, $\partial F/dy_i$, $\partial F/dx_f$, and $\partial F/dy_f$ are the partial derivatives of F with respect to x_b, y_b, x_i, y_i, x_f, and y_f, respectively.

Evaluating partial derivatives of the function F and substituting into Equation (14.12), then substituting into Equation (14.11), results in the following prototype angle equation:

$$\frac{y_{i0} - y_{b0}}{(IB_0)^2} dx_b + \frac{x_{b0} - x_{i0}}{(IB_0)^2} dy_b + \left(\frac{y_{b0} - y_{i0}}{(IB_0)^2} - \frac{y_{f0} - y_{i0}}{(IF_0)^2} \right) dx_i$$

$$+ \left(\frac{x_{i0} - x_{b0}}{(IB_0)^2} - \frac{x_{i0} - x_{f0}}{(IF_0)^2} \right)_0 dy_i + \frac{y_{f0} - y_{i0}}{(IF_0)^2} dx_f + \frac{x_{i0} - x_{f0}}{(IF_0)^2} dy_f$$

$$= k_{\theta_{bif}} + v_{\theta_{bif}} \tag{14.13}$$

where

$$k_{\theta_{bif}} = \theta_{bif} - \theta_{bif_0} \qquad \theta_{bif_0} = \tan^{-1}\left(\frac{x_{f0} - x_{i0}}{y_{f0} - y_{i0}}\right) - \tan^{-1}\left(\frac{x_{b0} - x_{i0}}{y_{b0} - y_{i0}}\right) + D$$

$$(IB_0)^2 = (x_{b0} - x_{i0})^2 + (y_{b0} - y_{i0})^2 \quad \text{and} \quad (IF_0)^2 = (x_{f0} - x_{i0})^2 + (y_{f0} - y_{i0})^2$$

In formulating the angle observation equation, remember that I is always assigned to the instrument station, B is the backsight station, and F is the foresight station. This station designation must be strictly followed in employing prototype equation (14.13), as will be demonstrated in the numerical examples that follow.

14.4 ADJUSTMENT OF INTERSECTIONS

When an unknown station is visible from two or more existing control stations, intersection is one of the simplest and sometimes most practical methods for determining the horizontal position of that station. For a unique computation, the method requires the measurement of at least two horizontal angles from two control points. For example, angles θ_1, and θ_2 measured from control stations R and S in Figure 14.3 will enable a unique computation for the position of station U. If additional control is available, the computations for the unknown station's position can be strengthened by measuring redundant angles such as angles θ_3 and θ_4 in Figure 14.3 and applying the method of least squares. In that case, for each of the four independent angles, a linearized observation equation can be written in terms of the two unknown coordinates x_u and y_u.

Example 14.1 Using the method of least squares, compute the most probable coordinates of station U in Figure 14.3 by the least-squares intersection procedure. The following equally weighted horizontal angles were observed from control stations R, S, and T:

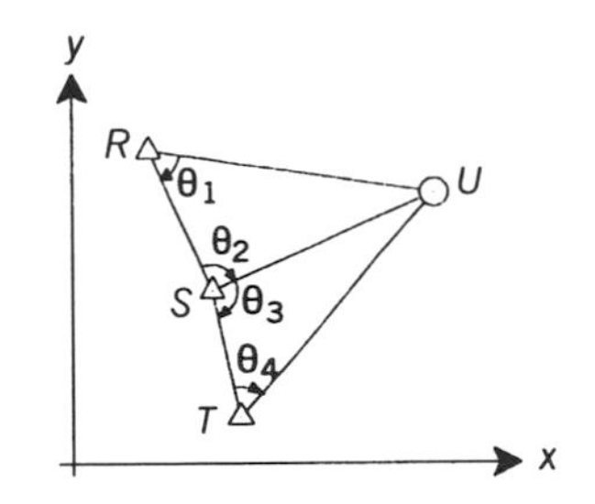

Figure 14.3 Intersection example.

$\theta_1 = 50°06'50''$ $\theta_2 = 101°30'47''$ $\theta_3 = 98°41'17''$ $\theta_4 = 59°17'01''$

The coordinates for the control stations *R*, *S*, and *T* are

$$x_r = 865.40 \qquad x_s = 2432.55 \qquad x_t = 2865.22$$
$$y_r = 4527.15 \qquad y_s = 2047.25 \qquad y_t = 27.15$$

SOLUTION

Step 1: Determine initial approximations for the coordinates of station *U*.

a. Using the coordinates of stations *R* and *S*, distance *RS* is computed as

$$RS = \sqrt{(2432.55 - 865.40)^2 + (4527.15 - 2047.25)^2} = 2933.58 \text{ ft}$$

b. From the coordinates of stations *R* and *S*, the azimuth of the line between *R* and *S* can be determined using Equation (14.2). Then the initial azimuth of line *RU* can be computed by subtracting θ_1 from the azimuth of line *RS*.

$$\text{Az}_{RS} = \tan^{-1}\left(\frac{x_s - x_r}{y_s - y_r}\right) + C = \tan^{-1}\left(\frac{865.40 - 2432.55}{4527.15 - 2047.25}\right) + 180°$$
$$= 147°42'34''$$

$$\text{Azimuth } RU_0 = 147°42'34'' - 50°06'50'' = 97°35'44''$$

c. Using the sine law with ΔRUS, an initial length for RU_0 can be calculated as

$$RU_0 = \frac{RS \sin(\theta_2)}{\sin(180° - \theta_1 - \theta_2)} = \frac{2933.58 \sin(101°30'47'')}{\sin(28°27'23'')} = 6049.00 \text{ ft}$$

d. Using the azimuth and distance for RU_0 computed in steps 1b and 1c, initial coordinates for station U are computed as

$$x_{u_0} = x_r + RU_0 \sin(\text{Az}_{RU_0}) = 865.40 + (6049.00) \sin(97°35'44'')$$
$$= 6861.35 \text{ ft}$$
$$y_{u_0} = y_r + RU_0 \cos(\text{Az}_{RU_0}) = 4527.15 + (6049.00) \cos(97°35'44'')$$
$$= 3727.59 \text{ ft}$$

e. Using the appropriate coordinates, the initial distances for SU and TU are calculated as

$$SU_0 = \sqrt{(6861.35 - 2432.55)^2 + (3727.59 - 2047.25)^2} = 4736.83 \text{ ft}$$
$$TU_0 = \sqrt{(6861.35 - 2865.22)^2 + (3727.59 - 27.15)^2} = 5446.29 \text{ ft}$$

Step 2: Formulate the linearized equations. As in the trilateration adjustment, control station coordinates are held fixed during the adjustment by assigning zeros to their dx and dy values. Thus these terms drop out of prototype equation (14.13). In forming the observation equations, b, i, and f are assigned to the backsight, instrument, and foresight stations, respectively, for each angle. For example, with angle θ_1, B, I, and F are replaced by U, R, and S, respectively. By combining the station substitutions shown in Table 14.2 with prototype equation (14.13), the following observation equations are written for the four observed angles:

$$\frac{y_r - y_{u_0}}{(RU_0)^2} dx_u + \frac{x_{u_0} - x_r}{(RU_0)^2} dy_u$$
$$= \theta_1 - \left\{ \tan^{-1}\left(\frac{x_s - x_r}{y_s - y_r}\right) - \tan^{-1}\left(\frac{x_{u_0} - x_r}{y_{u_0} - y_r}\right) + 0° \right\} + v_1$$

Table 14.2 Substitutions

Angle	B	I	F
θ_1	U	R	S
θ_2	R	S	U
θ_3	U	S	T
θ_4	S	T	U

$$\frac{y_{u_0} - y_s}{(SU_0)^2}\,dx_u + \frac{x_s - x_{u_0}}{(SU_0)^2}\,dy_u$$

$$= \theta_2 - \left\{\tan^{-1}\left(\frac{x_{u_0} - x_s}{y_{u_0} - y_s}\right) - \tan^{-1}\left(\frac{x_r - x_s}{y_r - y_s}\right) + 0^\circ\right\} + v_2$$

$$\frac{y_s - y_{u_0}}{(SU_0)^2}\,dx_u + \frac{x_{u_0} - x_s}{(SU_0)^2}\,dy_u$$

$$= \theta_3 - \left\{\tan^{-1}\left(\frac{x_t - x_s}{y_t - y_s}\right) + \tan^{-1}\left(\frac{x_{u_0} - x_s}{y_{u_0} - y_s}\right) + 180^\circ\right\} + v_3$$

$$\frac{y_{u_0} - y_t}{(TU_0)^2}\,dx_u + \frac{x_t - x_{u_0}}{(TU_0)^2}\,dy_u$$

$$= \theta_4 - \left\{\tan^{-1}\left(\frac{x_{u_0} - x_t}{y_{u_0} - y_t}\right) + \tan^{-1}\left(\frac{x_s - x_t}{y_s - y_t}\right) + 0^\circ\right\} + v_4 \qquad (14.14)$$

Substituting the appropriate values into Equations (14.14) and multiplying the left side of the equations by ρ to achieve unit consistency,[1] the following J and K matrices are formed:

$$J = \rho\begin{bmatrix} \dfrac{4527.15 - 3727.59}{(6049.00)^2} & \dfrac{6861.35 - 865.40}{(6049.00)^2} \\ \dfrac{3727.59 - 2047.25}{(4736.83)^2} & \dfrac{2432.55 - 6861.35}{(4736.83)^2} \\ \dfrac{2047.25 - 3727.59}{(4736.83)^2} & \dfrac{6861.35 - 2432.55}{(4736.83)^2} \\ \dfrac{3727.59 - 27.15}{(5446.29)^2} & \dfrac{2865.22 - 6861.35}{(5446.29)^2} \end{bmatrix} = \begin{bmatrix} 4.507 & 33.800 \\ 15.447 & -40.713 \\ -15.447 & 40.713 \\ 25.732 & -27.788 \end{bmatrix}$$

[1] In order for these observations to be dimensionally consistent, the elements of the K and V matrices must be in radian measure, or in other words, the coefficients of the K and J elements must be in the same units. Since it is most common to work in the sexagesimal system, and since the magnitudes of the angle residuals are generally in the range of seconds, the units of the equations are converted to seconds by (1) multiplying the coefficients in the equation by ρ, which is the number of seconds per radian, or 206,264.8″/rad, and (2) computing the K elements in units of seconds.

$$K = \begin{bmatrix} 50°06'50'' - \left\{ \tan^{-1}\left(\dfrac{2432.55 - 865.40}{2047.25 - 4527.15}\right) - \tan^{-1}\left(\dfrac{6861.35 - 865.40}{3727.59 - 4527.15}\right) + 0° \right\} \\ 101°30'47'' - \left\{ \tan^{-1}\left(\dfrac{6861.35 - 2432.55}{3727.59 - 2047.25}\right) - \tan^{-1}\left(\dfrac{865.40 - 2432.55}{4527.15 - 2047.25}\right) + 0° \right\} \\ 98°41'17'' - \left\{ \tan^{-1}\left(\dfrac{2865.22 - 2432.55}{27.15 - 2047.25}\right) - \tan^{-1}\left(\dfrac{6861.35 - 2432.55}{3727.59 - 2047.25}\right) + 180° \right\} \\ 59°17'01'' - \left\{ \tan^{-1}\left(\dfrac{6861.35 - 2865.22}{3727.59 - 27.15}\right) - \tan^{-1}\left(\dfrac{2432.55 - 2865.22}{2047.25 - 27.15}\right) + 0° \right\} \end{bmatrix}$$

$$= \begin{bmatrix} 0.00'' \\ 0.00'' \\ -0.69'' \\ -20.23'' \end{bmatrix}$$

Notice that the initial coordinates for x_{u_0} and y_{u_0} were calculated using θ_1 and θ_2, and thus their K-matrix values are zero for the first iteration.

Step 3: Matrix solution. The least-squares solution is found by applying Equation (10.37):

$$J^TJ = \begin{bmatrix} 1159.7 & -1820.5 \\ -1820.5 & 5229.7 \end{bmatrix} \qquad Q_{xx} = (J^TJ)^{-1} = \begin{bmatrix} 0.001901 & 0.000662 \\ 0.000662 & 0.000422 \end{bmatrix}$$

$$J^TK = \begin{bmatrix} -509.9 \\ 534.1 \end{bmatrix}$$

$$X = [J^TJ]^{-1}[J^TK] = \begin{bmatrix} 0.001901 & 0.000662 \\ 0.000662 & 0.000422 \end{bmatrix} \begin{bmatrix} -509.9 \\ 534.1 \end{bmatrix} = \begin{bmatrix} dx_u \\ dy_u \end{bmatrix}$$

$$dx_u = -0.62 \text{ ft} \quad \text{and} \quad dy_u = -0.11 \text{ ft}$$

Step 4: Add the corrections to the initial coordinates for station U.

$$\begin{aligned} x_u &= x_{u_0} + dx_u = 6861.35 - 0.62 = 6860.73 \\ y_u &= y_{u_0} + dy_u = 3727.59 - 0.11 = 3727.48 \end{aligned} \tag{14.15}$$

Step 5: Repeat Steps 2 through 4 until negligible corrections occur. The next iteration produced negligible corrections, and thus Equations (14.15) produced the final adjusted coordinates for station U.

Step 6: Compute postadjustment statistics. The residuals for the angles are

$$V = JX - K = \begin{bmatrix} 4.507 & 33.800 \\ 15.447 & -40.713 \\ -15.447 & 40.713 \\ 25.732 & -27.788 \end{bmatrix} \begin{bmatrix} -0.62 \\ -0.11 \end{bmatrix} - \begin{bmatrix} 0.00'' \\ 0.00'' \\ -0.69'' \\ -20.23'' \end{bmatrix} = \begin{bmatrix} -6.5'' \\ -5.1'' \\ 5.8'' \\ 7.3'' \end{bmatrix}$$

The reference variance (standard deviation of unit weight) for the adjustment is computed using Equation (11.14) as

$$V^TV = [-6.5 \quad -5.1 \quad 5.8 \quad 7.3] \begin{bmatrix} -6.5 \\ -5.1 \\ 5.8 \\ 7.3 \end{bmatrix} = [155.2''^2]$$

$$S_0 = \sqrt{\frac{V^TV}{m - n}} = \sqrt{\frac{155.2}{4 - 2}} = \pm 8.8''$$

The estimated errors for the adjusted coordinates of station U, given by Equation (12.24), are

$$S_{x_u} = S_0 \sqrt{Q_{x_u x_u}} = \pm 8.8 \sqrt{0.001901} = \pm 0.38 \text{ ft}$$

$$S_{y_u} = S_0 \sqrt{Q_{y_u y_u}} = \pm 8.8 \sqrt{0.000422} = \pm 0.18 \text{ ft}$$

The estimated error in the position of station U is given by

$$S_u = \sqrt{S_x^2 + S_y^2} = \pm \sqrt{(0.38)^2 + (0.18)^2} = \pm 0.42 \text{ ft}$$

14.5 ADJUSTMENT OF RESECTIONS

Resection is a method used for determining the unknown horizontal position of an occupied station by measuring a minimum of two horizontal angles to a minimum of three stations whose horizontal coordinates are known. If more than three stations are available, redundant observations are obtained and the position of the unknown occupied station can be computed using the least-squares method. Like intersection, resection is suitable for locating an occasional station and is especially well adapted over inaccessible terrain. The method is commonly used for orienting total station instruments in locations favorable for staking projects by radiation using coordinates.

Consider the resection position computation for the occupied station U of Figure 14.4, having observed the three horizontal angles shown between stations P, Q, R, and S whose positions are known. To determine the position of station U, two angles could be measured. The third angle provides a check

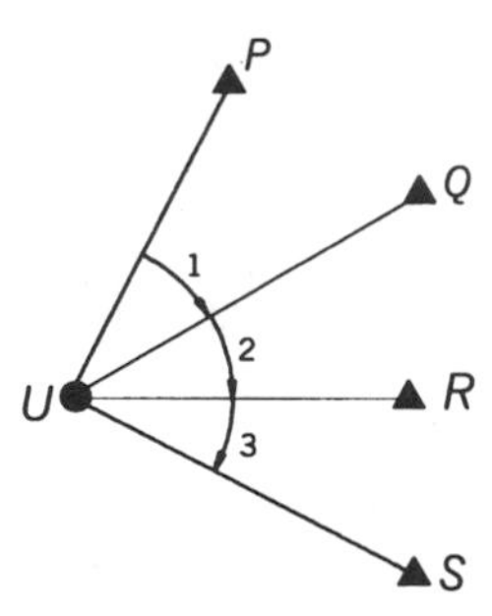

Figure 14.4 Resection.

and allows a least-squares solution for computing the coordinates of station *U*.

Using prototype equation (14.13), a linearized observation equation can be written for each angle. In this problem, the vertex station is occupied and is the only unknown station. Thus all coefficients in the Jacobian matrix follow the form used for the coefficients of dx_i and dy_i in prototype equation (14.13). The method of least squares yields corrections, dx_u and dy_u, which give the most probable coordinate values for station *U*.

14.5.1 Computing Initial Approximations in the Resection Problem

In Figure 14.4 only two angles are necessary to determine the coordinates of station *U*. Using stations *P*, *Q*, *R*, and *U*, a procedure to find the station *U*'s approximate coordinate values is:

Step 1: Let

$$\angle QPU + \angle URQ = G = 360° - (\angle 1 + \angle 2 + \angle RQP) \tag{14.16}$$

Step 2: Using the sine law with triangle *PQU*,

$$\frac{QU}{\sin(\angle QPU)} = \frac{PQ}{\sin(\angle 1)} \tag{a}$$

and with triangle *UQR*,

$$\frac{QU}{\sin(\angle URQ)} = \frac{QR}{\sin(\angle 2)} \tag{b}$$

Step 3: Solving Equations (*a*) and (*b*) for *QU* and setting the resulting equations equal to each other gives

$$\frac{PQ \sin(\angle QPU)}{\sin(\angle 1)} = \frac{QR \sin(\angle URQ)}{\sin(\angle 2)} \tag{c}$$

Step 4: From Equation (*c*), let H be defined as

$$H = \frac{\sin(\angle QPU)}{\sin(\angle URQ)} = \frac{QR \sin(\angle 1)}{PQ \sin(\angle 2)} \tag{14.17}$$

Step 5: From Equation (14.16),

$$\angle QPU = G - \angle URQ \tag{d}$$

Step 6: Solving Equation (14.17) for the $\sin(\angle QPU)$ and substituting Equation (*d*) into the result gives

$$\sin(G - \angle URQ) = H \sin(\angle URQ) \tag{e}$$

Step 7: From trigonometry

$$\sin(\alpha - \beta) = \sin(\alpha)\cos(\beta) - \cos(\alpha)\sin(\beta)$$

Applying this relationship to Equation (*e*) gives us

$$\sin(G - \angle URQ) = \sin(G)\cos(\angle URQ) - \cos(G)\sin(\angle URQ) \tag{f}$$

$$\sin(G)\cos(\angle URQ) - \cos(G)\sin(\angle URQ) = H \sin(\angle URQ) \tag{g}$$

Step 8: Dividing Equation (*g*) by $\cos(\angle URQ)$ and rearranging yields

$$\sin(G) = \tan(\angle URQ)[H + \cos(G)] \tag{h}$$

Step 9: Solving Equation (*h*) for $\angle URQ$, gives

$$\angle URQ = \tan^{-1}\left(\frac{\sin(G)}{H + \cos(G)}\right) \tag{14.18}$$

Step 10: From Figure 14.4,

$$\angle RQU = 180° - (\angle 2 + \angle URQ) \tag{14.19}$$

Step 11: Again applying the sine law, we obtain

$$RU = \frac{QR \sin(\angle RQU)}{\sin(\angle 2)} \tag{14.20}$$

Step 12: Finally, the initial coordinates for station U are

$$
\begin{aligned}
x_u &= x_r + RU \sin(\mathrm{Az}_{RQ} - \angle URQ) \\
y_u &= y_r + RU \cos(\mathrm{Az}_{RQ} - \angle URQ)
\end{aligned}
\tag{14.21}
$$

Example 14.2 The following data are obtained for Figure 14.4. Control stations P, Q, R, and S have the following (x,y) coordinates: P (1303.599, 1458.615), Q (1636.436, 1310.468), R (1503.395, 888.362), and S (1,506.262, 785.061). The measured values for the angles 1, 2, and 3, with standard deviations are:

Backsight	Occupied	Foresight	Angle	S
P	U	Q	30°29′33″	5″
Q	U	R	38°30′31″	6″
R	U	S	10°29′57″	6″

What are the most probable coordinates of station U?

SOLUTION: Using the procedures described in Section 14.5.1, initial approximations for the coordinates of station U are:
(*a*) From Equation (14.10),

$$\angle RQP = \mathrm{Az}_{PQ} - \mathrm{Az}_{QR} = 293°59'38.4'' - 197°29'38.4'' = 96°30'00.0''$$

(*b*) Substituting the appropriate angular values into Equation (14.16), gives

$$G = 360° - (30°29'33'' + 38°30'31'' + 96°30'00.0'') = 194°29'56''$$

(*c*) Substituting the appropriate station coordinates into Equation (13.1) yields

$$PQ = 364.318 \quad \text{and} \quad QR = 442.576$$

(*d*) Substituting the appropriate values into Equation (14.17), H is

$$H = \frac{442.576 \sin(30°29'33'')}{364.318 \sin(38°30'31'')} = 0.99002730$$

(*e*) Substituting previously determined G and H into Equation (14.18), $\angle URQ$ is computed as

$$
\begin{aligned}
\angle URQ &= \tan^{-1}\left(\frac{\sin(194°29'56'')}{0.99002730 + \cos(194°29'56'')}\right) + 180° \\
&= 85°00'22.2'' + 180° = 94°59'36.3''
\end{aligned}
$$

(*f*) Substituting the value of $\angle URQ$ from step 5 into Equation (14.19), $\angle RQU$ is determined to be

$$\angle RQU = 180° - (38°30'31'' + 94°59'36.3'') = 46°29'52.7''$$

(*g*) From Equation (14.20), RU is

$$RU = \frac{442.576 \sin(46°29'52.7'')}{\sin(38°30'31'')} = 515.589$$

(*h*) Using Equation (14.1), the azimuth of RQ is

$$\text{Az}_{RQ} = \tan^{-1}\left(\frac{1636.436 - 1503.395}{1310.468 - 888.362}\right) = 17°29'38.4''$$

(*i*) From Figure (14.4), Az_{RU} is computed as

$$\text{Az}_{RQ} = 197°29'38.4'' - 180° = 17°29'38.4''$$

$$\text{Az}_{RU} = \text{Az}_{RQ} - \angle URQ = 360° + 17°29'38.4'' - 94°59'36.3''$$

$$= 282°30'02.2''$$

(*j*) Using (14.21), the coordinates for station U are

$$x_u = 1503.395 + 515.589 \sin(\text{Az}_{RU}) = 1000.03$$

$$y_u = 888.362 + 515.589 \cos(\text{Az}_{RU}) = 999.96$$

For this problem, using prototype equation (14.13), the J and K matrices are

$$J = \begin{bmatrix} \rho\left(\dfrac{y_p - y_{u_0}}{(UP_0)^2} - \dfrac{y_q - y_{u_0}}{(UQ_0)^2}\right) & \rho\left(\dfrac{x_{u_0} - x_p}{(UP_0)^2} - \dfrac{x_{u_0} - x_q}{(UQ_0)^2}\right) \\ \rho\left(\dfrac{y_q - y_{u_0}}{(UQ_0)^2} - \dfrac{y_r - y_{u_0}}{(UR_0)^2}\right) & \rho\left(\dfrac{x_{u_0} - x_q}{(UQ_0)^2} - \dfrac{x_{u_0} - x_r}{(UR_0)^2}\right) \\ \rho\left(\dfrac{y_r - y_{u_0}}{(UR_0)^2} - \dfrac{y_s - y_{u_0}}{(US_0)^2}\right) & \rho\left(\dfrac{x_{u_0} - x_r}{(UR_0)^2} - \dfrac{x_{u_0} - x_s}{(US_0)^2}\right) \end{bmatrix}$$

$$K = \begin{bmatrix} (\angle 1 - \angle 1_0)'' \\ (\angle 2 - \angle 2_0)'' \\ (\angle 3 - \angle 3_0)'' \end{bmatrix}$$

Also, the weight matrix W is a diagonal matrix composed of the inverses of the variances of the observed angles, or

$$W = \begin{bmatrix} \frac{1}{5^2} & 0 & 0 \\ 0 & \frac{1}{6^2} & 0 \\ 0 & 0 & \frac{1}{6^2} \end{bmatrix}$$

Using the data given for the problem, together with the computed initial approximations, numerical values for the matrices were calculated and the adjustment performed using program ADJUST. The following results were obtained after two iterations. The reader is encouraged to adjust these example problems using both the MATRIX and ADJUST programs supplied.

```
Iteration 1
J MATRIX                              K MATRIX        X MATRIX
---------------------------           ---------       ---------
   184.993596    54.807717            -0.203359       -0.031107
   214.320813   128.785353            -0.159052        0.065296
    59.963802   -45.336838            -6.792817


Iteration 2
J MATRIX                              K MATRIX        X MATRIX
---------------------------           ---------       ---------
   185.018081    54.771738             1.974063        0.000008
   214.329904   128.728773            -1.899346        0.000004
    59.943758   -45.340316            -1.967421


INVERSE MATRIX
---------------------------
 0.00116318   -0.00200050
-0.00200050    0.00500943


*****************
Adjusted Stations
*****************
Station            X              Y            Sx          Sy
------------------------------------------------------------------
      U       999.999      1,000.025       0.0206      0.0427
```

```
***************************
Adjusted Angle Observations
***************************
```

Station Backsighted	Station Occupied	Station Foresighted	Angle	V	S
P	U	Q	30° 29′ 31″	-2.0″	2.3″
Q	U	R	38° 30′ 33″	1.9″	3.1″
R	U	S	10° 29′ 59″	2.0″	3.0″

```
        Redundancies = 1
  Reference Variance = 0.3636
        Reference So = ±0.60
```

14.6 ADJUSTMENT OF TRIANGULATED QUADRILATERALS

The quadrilateral is the basic figure for triangulation. Procedures like those used for adjusting intersections and resections are also used to adjust this type of figure. In fact, the parametric adjustment using the observation equation method can be applied to any triangulated geometric figure, regardless of its shape.

The basic procedure for adjusting a quadrilateral consists in first using a minimum number of the observed angles to solve triangles, and compute initial values for the coordinates. Corrections to these initial coordinates are then calculated by applying the method of least squares. The procedure is iterated until the solution converges. This yields the most probable coordinate values. A statistical analysis of the results is then made. The following example illustrates the procedure.

Example 14.3 The following observations are supplied for Figure 14.5. Adjust this figure by the method of least squares. The observed angles (assume equal weights) are as follows:

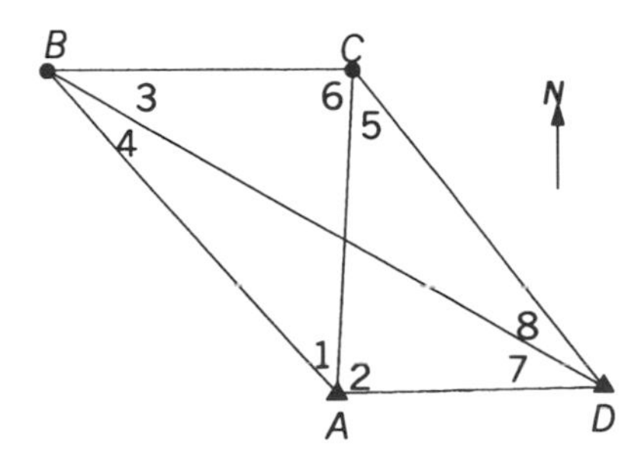

Figure 14.5 Quadrilateral.

$1 = 42°35'29.0''$ $\quad 3 = 79°54'42.1''$ $\quad 5 = 21°29'23.9''$ $\quad 7 = 31°20'45.8''$

$2 = 87°35'10.6''$ $\quad 4 = 18°28'22.4''$ $\quad 6 = 39°01'35.4''$ $\quad 8 = 39°34'27.9''$

The fixed coordinates are

$x_A = 9270.33 \qquad y_A = 8448.90 \qquad x_D = 15{,}610.58 \qquad y_D = 8568.75$

SOLUTION: The coordinates of stations B and C are to be computed in this adjustment. The Jacobian matrix has the form shown in Table 14.3. The subscripts b, i, and f of the dx's and dy's in the table indicate whether stations B and C are the backsight, instrument, or foresight station in Equation (14.13). In developing the coefficient matrix, of course, the appropriate station coordinate substitutions must be made to obtain each coefficient.

A computer program has been used to form the matrices and solve the problem. In the program, the angles were entered in the order of 1 through 8. The X matrix has the form

$$X = \begin{bmatrix} dx_B \\ dy_B \\ dx_C \\ dy_C \end{bmatrix}$$

The following self-explanatory computer listing gives the solution for this example. As shown, one iteration was satisfactory to achieve convergence, since the second iteration produced negligible corrections. Residuals, adjusted coordinates, their estimated errors, and adjusted angles are tabulated at the end of the listing.

Table 14.3 Structure of the coefficient or *J* matrix in Example 14.3

	Unknown			
Angle	dx_B	dy_B	dx_C	dy_C
1	$dx(b)$	$dy(b)$	$dx(f)$	$dy(f)$
2	0	0	$dx(b)$	$dy(b)$
3	$dx(i)$	$dy(i)$	$dx(b)$	$dy(b)$
4	$dx(i)$	$dy(i)$	0	0
5	0	0	$dx(i)$	$dy(i)$
6	$dx(f)$	$dy(f)$	$dx(i)$	$dy(i)$
7	$dx(f)$	$dy(f)$	0	0
8	$dx(b)$	$dy(b)$	$dx(f)$	$dy(f)$

Triangulation Example

```
------------
Example 14.3
------------
Number of Control Stations      » 2
Number of Unknown Stations      » 2
Number of Angle observations    » 8
```

```
******************************************
Initial Approximations for Unknown Stations
******************************************
```

Station	X	Y
B	2,403.600	16,275.400
C	9,649.800	24,803.500

```
Control Stations
----------------
```

Station	X	Y
A	9,270.330	8,448.900
D	15,610.580	8,568.750

```
******************
Angle Observations
******************
```

Station Backsighted	Station Occupied	Station Foresighted	Angle
B	A	C	42° 35′ 29.0″
C	A	D	87° 35′ 10.6″
C	B	D	79° 54′ 42.1″
D	B	A	18° 28′ 22.4″
D	C	A	21° 29′ 23.9″
A	C	B	39° 01′ 35.4″
A	D	B	31° 20′ 45.8″
B	D	C	39° 34′ 27.9″

Iteration 1

```
J MATRIX                                                 K MATRIX     X MATRIX
-----------------------------------------------------   ---------    -----------
  -14.891521 -13.065362  12.605250  -0.292475           -1.811949    1 -0.011149
    0.000000   0.000000 -12.605250   0.292475           -5.801621    2  0.049461
   20.844399  -0.283839 -14.045867  11.934565            3.508571    3  0.061882
    8.092990   1.414636   0.000000   0.000000            1.396963    4  0.036935
    0.000000   0.000000   1.409396  -4.403165           -1.833544
  -14.045867  11.934565   1.440617 -11.642090            5.806415
    6.798531  11.650726   0.000000   0.000000           -5.983393
   -6.798531 -11.650726  11.195854   4.110690            1.818557
```

Iteration 2

```
J MATRIX                                                   K MATRIX    X MATRIX
-------------------------------------------------          ---------   -----------
 -14.891488 -13.065272  12.605219  -0.292521  -2.100998  1  0.000000
   0.000000   0.000000 -12.605219   0.292521  -5.032381  2 -0.000000
  20.844296  -0.283922 -14.045752  11.934605   4.183396  3  0.000000
   8.092944   1.414588   0.000000   0.000000   1.417225  4 -0.000001
   0.000000   0.000000   1.409357  -4.403162  -1.758129
 -14.045752  11.934605   1.440533 -11.642083   5.400377
   6.798544  11.650683   0.000000   0.000000  -6.483846
  -6.798544 -11.650683  11.195862   4.110641   1.474357
```

```
INVERSE MATRIX
-------------------------------------------
 297713237 -211387995  68037326 -460470445
-211387995  324375288  62795552  483966116
  68037326   62795552 160883503   30871391
-460470445  483966116  30871391 1006374249
```

```
*****************
Adjusted Stations
*****************
```

Station	X	Y	Sx	Sy
B	2,403.589	16,275.449	0.4690	0.4895
C	9,649.862	24,803.537	0.3447	0.8622

```
***************************
Adjusted Angle Observations
***************************
```

Station Backsighted	Station Occupied	Station Foresighted	Angle	V	S
B	A	C	42° 35′ 31.1″	2.10″	3.65″
C	A	D	87° 35′ 15.6″	5.03″	4.33″
C	B	D	79° 54′ 37.9″	-4.18″	4.29″
D	B	A	18° 28′ 21.0″	-1.42″	3.36″
D	C	A	21° 29′ 25.7″	1.76″	3.79″
A	C	B	39° 01′ 30.0″	-5.40″	4.37″
A	D	B	31° 20′ 52.3″	6.48″	4.24″
B	D	C	39° 34′ 26.4″	-1.47″	3.54″

```
*********************
Adjustment Statistics
*********************
         Iterations = 2
       Redundancies = 4
 Reference Variance = 31.42936404
       Reference So = ±5.6062
       Convergence!
```

PROBLEMS

Note: For problems requiring least-squares adjustment, if a computer program is not distinctly specified for use in the problem, it is expected that the least-squares algorithm will be solved using the program MATRIX which is included within the diskette supplied with the book.

14.1 Given the following observations and control station coordinates to accompany Figure 14.6, what are the most probable coordinates for station *E* using an unweighted least-squares adjustment?

Control station

Station	X	Y
A	10,000.000	10,000.000
B	11,498.580	10,065.320
C	12,432.170	11,346.190
D	11,490.570	12,468.510

Angle observations

Backsighted	Occupied	Foresighted	Angle	s
E	A	B	90°59′58.5″	5.3″
A	B	E	40°26′02.2″	4.7″
E	B	C	88°08′53.7″	4.9″
B	C	E	52°45′02.2″	4.7″
E	C	D	51°09′58.4″	4.8″
C	D	E	93°13′12.5″	5.0″

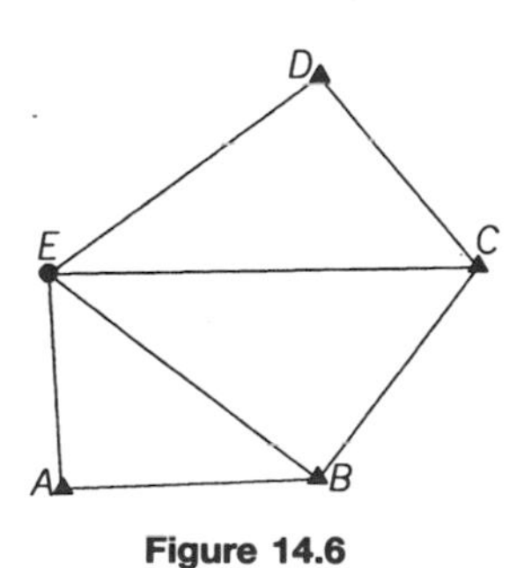

Figure 14.6

14.2 Repeat Problem 14.1 using a weighted least-squares adjustment with weights of $1/s^2$ for each angle. What are the:

(a) most probable coordinates for station *E*?

(b) standard deviation of unit weight?

(c) standard deviations in the adjusted coordinates for station *E*?

(d) adjusted angles, their residuals, and standard deviations?

14.3 Given the following measured angles and control station coordinates for the resection problem of Figure 14.4:

$$1 = 18°26'00'' \qquad 2 = 53°08'00'' \qquad 3 = 18°26'10''$$

Station	X	Y
P	10,000.00	10,000.00
Q	10,000.00	7,500.00
R	10,000.00	2,500.00
S	10,000.00	0.00

Assuming equally weighted angles, what are the most probable coordinates for station *U*?

14.4 If the estimated standard deviations for the angles in Problem 14.3 are $s_1 = \pm 7.5''$, $s_2 = \pm 5.6''$, and $s_3 = \pm 8.6''$, what are the:

(a) most probable coordinates for station *U*?

(b) standard deviation of unit weight?

(c) standard deviations in the adjusted coordinates of station *U*?

(d) adjusted angles, their residuals, and standard deviations?

14.5 The following measured angles and standard deviations apply to the quadrilateral in Figure 14.5. Hold station *D* fixed at $X = 10{,}000$ and $Y = 10{,}000$, and hold the length and azimuth of line *DA* at 2320.13 ft and 271°37′, respectively.

$1 = 51°58'10.0'' \pm 1.0''$ $\quad 2 = 55°42'35.2'' \pm 1.9''$ $\quad 3 = 43°15'15.3'' \pm 1.3''$

$4 = 41°09'19.6'' \pm 1.5''$ $\quad 5 = 55°20'57.6'' \pm 1.0''$ $\quad 6 = 43°37'20.3'' \pm 2.2''$

$7 = 31°09'53.7'' \pm 1.8''$ $\quad 8 = 37°46'30.5'' \pm 0.8''$

Do a weighted adjustment using the standard deviations to calculate weights. (*Hint:* Fixing the azimuth and length of *DA* makes station *A* a control station.) What are the:

(a) most probable coordinates for stations *B* and *C*?

(b) standard deviation of unit weight?

(c) standard deviations in the adjusted coordinates for stations *B* and *C*?

(d) adjusted angles, their residuals, and standard deviations?

14.6 For Figure 14.7 and the following observations, perform a weighted least-squares adjustment. Tabulate the final adjusted:

(a) station coordinate values and their standard deviations.

(b) adjusted angles, their residuals, and standard deviations.

Control stations

Station	X	Y
A	4,270.330	8,448.900
B	5,599.549	16,748.769

Angle observations

Backsighted	Occupied	Foresighted	Angle	s
B	A	C	42°35′22.8″	±2.6″
C	A	D	37°13′42.6″	±2.6″
C	B	D	57°47′41.0″	±2.6″
D	B	A	40°35′23.5″	±2.6″
D	C	A	25°11′16.4″	±2.6″
A	C	B	39°01′31.6″	±2.6″
A	D	B	59°35′29.9″	±2.6″
B	D	C	57°59′32.2″	±2.6″

14.7 Do Problem 14.6 using an unweighted least-squares adjustment. Compare and discuss any differences or similarities between these results and those obtained in Problem 14.6.

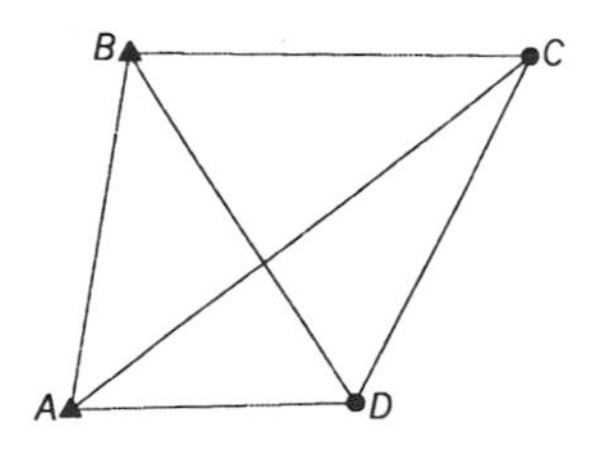

Figure 14.7

14.8 The following measurements and data pertain to the triangulation chain shown in Figure 14.8.

Control stations

Station	X	Y
A	1718.871	632.095
B	2191.570	715.709
G	1590.370	2560.743
H	2076.006	2597.745

Initial approximations for unknown stations

Station	Easting	Northing
C	1668.571	1310.429
D	2139.109	1296.242
E	1617.479	1949.217
F	2028.688	1934.566

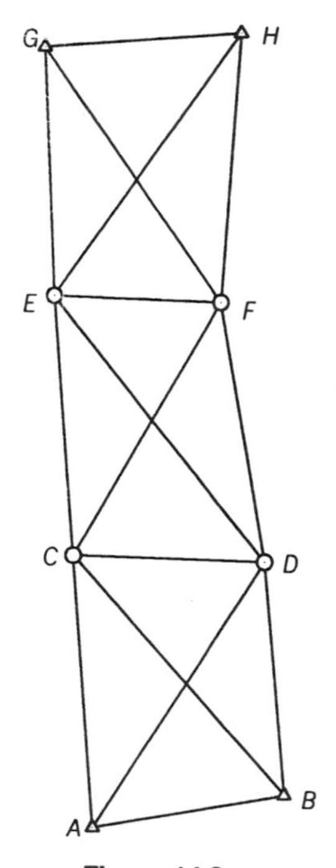

Figure 14.8

Angle observations

Backsight	Occupied	Foresight	Angle	s
C	*A*	*D*	36°33′49.5″	4.6″
D	*A*	*B*	47°38′40.5″	5.7″
A	*B*	*C*	58°42′05.4″	5.7″
C	*B*	*D*	36°09′56.9″	4.9″
B	*C*	*A*	37°05′18.5″	4.6″
D	*C*	*B*	46°56′40.5″	5.7″
B	*D*	*A*	37°29′08.7″	5.0″
A	*D*	*C*	59°24′14.1″	5.8″
E	*C*	*F*	34°33′22.5″	4.8″
F	*C*	*D*	61°44′34.1″	5.9″
C	*D*	*E*	49°39′14.4″	5.7″
E	*D*	*F*	28°48′19.6″	4.6″
F	*E*	*D*	49°20′57.4″	6.2″
D	*E*	*C*	34°02′48.2″	4.6″
D	*F*	*C*	39°47′51.2″	4.8″
C	*F*	*E*	62°03′22.6″	6.5″
G	*E*	*H*	37°47′58.1″	4.8″
H	*E*	*F*	56°46′43.5″	6.3″
E	*F*	*G*	52°58′05.9″	6.4″
G	*F*	*H*	39°04′18.8″	4.7″
F	*G*	*E*	32°27′12.1″	4.8″
H	*G*	*F*	59°21′51.3″	5.7″
F	*H*	*E*	31°10′44.3″	4.6″
E	*H*	*G*	50°22′53.5″	5.6″

Use Program ADJUST to perform a weighted least-squares adjustment. Tabulate the final adjusted:

(a) station coordinates and their standard deviations.

(b) adjusted angles, their residuals, and standard deviations.

14.9 Do Problem 14.8 using an equal-weight least-squares adjustment. Compare and discuss any differences or similarities between these results and those obtained in Problem 14.8. Use Program ADJUST in computing the adjustment.

Use Program ADJUST software to do the problems.

14.10 Problems 14.1 and 14.2

14.11 Problems 14.3 and 14.4

14.12 Problem 14.5

14.13 Problems 14.6 and 14.7

Programming Problems

14.14 Write a spreadsheet that will compute the coefficients for prototype equations (14.9) and (14.13) and their k values, given the coordinates of the appropriate stations. Use this spreadsheet to determine the matrix values necessary to do Problem 14.6.

14.15 Prepare a program that reads a file of station coordinates, observed angles, and their standard deviations, and then:

(a) writes the data to a file in a formatted fashion,

(b) computes the J, K, and W matrices, and

(c) writes the matrices to a file that is compatible with the MATRIX program.

Test this program with Problem 14.6.

14.16 Write a program that reads a file containing the J, K, and W matrices, and then:

(a) writes these matrices in a formatted fashion,

(b) performs one iteration of either a weighted or unweighted least-squares adjustment of Problem 14.6, and

(c) writes the matrices used to compute the solution and the corrections to the station coordinates in a formatted fashion.

14.17 Write a program that reads a file of station coordinates, observed angles, and their standard deviations, and then:

(a) writes the data to a file in a formatted fashion,

(b) computes the J, K, and W matrices,

(c) performs a weighted least-squares adjustment of Problem 14.6, and

(d) writes the matrices used to compute the solution and tabulates the final adjusted station coordinates and their estimated errors, and the adjusted angles together with their residuals and their standard deviations.

14.18 Prepare a spreadsheet that computes initial approximations for the resection problem, and demonstrate its use with Problem 14.3.

15

ADJUSTMENT OF HORIZONTAL SURVEYS: TRAVERSES AND NETWORKS

15.1 INTRODUCTION TO TRAVERSE ADJUSTMENTS

Of the many methods that exist for traverse adjustment, the characteristic that distinguishes the method of least squares from other methods is that distance, angle, and direction observations are adjusted simultaneously. Furthermore, the adjusted observations not only satisfy all geometrical conditions for the traverse, but they provide the most probable values for the given data set. Additionally, the observations can be rigorously weighted based upon their estimated errors, and adjusted accordingly. Given these facts, together with the computational power now provided by computers, it is hard to justify not using least squares for all traverse adjustment work.

In this chapter we describe methods for making traverse adjustments by least squares. As was the case in triangulation adjustments, traverses can be adjusted by least squares using either observation equations or conditional equations. Again, because of the relative ease with which the observation equations can be written and solved, that approach will be discussed here.

15.2 THE OBSERVATION EQUATIONS

When adjusting a traverse using parametric equations, an observation equation is written for each distance, direction, or angle. The necessary linearized observation equations developed previously are recalled in the following equations.

Distance observation equation:

$$\frac{x_{io} - x_{jo}}{IJ_0} dx_i + \frac{y_{io} - y_{jo}}{IJ_0} dy_i + \frac{x_{jo} - x_{io}}{IJ_0} dx_j + \frac{y_{jo} - y_{io}}{IJ_0} dy_j = k_{l_{ij}} + v_{l_{ij}} \tag{15.1}$$

Angle observation equation:

$$\frac{y_{io} - y_{bo}}{(IB_0)^2} dx_b + \frac{x_{bo} - x_{io}}{(IB_0)^2} dy_b + \left(\frac{y_{bo} - y_{io}}{(IB_0)^2} - \frac{y_{fo} - y_{io}}{(IF_0)^2}\right) dx_i + \left(\frac{x_{io} - x_{bo}}{(IB_0)^2} - \frac{x_{io} - x_{fo}}{(IF_0)^2}\right) dy_i + \frac{y_{fo} - y_{io}}{(IF_0)^2} dx_f + \frac{x_{io} - x_{fo}}{(IF_0)^2} dy_f = k_{\theta_{bif}} + v_{\theta_{bif}} \tag{15.2}$$

Azimuth observation equation:

$$\frac{y_{io} - y_{jo}}{(IJ_0)^2} dx_i + \frac{x_{jo} - x_{io}}{(IJ_0)^2} dy_i + \frac{y_{jo} - y_{io}}{(IJ_0)^2} dx_j + \frac{x_{io} - x_{jo}}{(IJ_0)^2} dy_j = k_{\text{Az}_{ij}} + v_{\text{Az}_{ij}} \tag{15.3}$$

The reader should refer to Chapters 13 and 14 to review the specific notation for these equations. As demonstrated with the examples that follow, the azimuth equation may or may not be used in traverse adjustments.

15.3 REDUNDANT EQUATIONS

As noted earlier, one observation equation can be written for each measured angle, distance, or direction in a closed traverse. Thus if there are n sides in the traverse, there are n distances and $n + 1$ angles, assuming that one angle exists for orientation of the traverse. For example, each closed traverse in Figure 15.1 has four sides, four distances, and five angles. Each also has three points whose positions are unknown, and each of these introduces two unknown coordinates into the solution. Thus there are a maximum of $2(n - 1)$ unknowns for any closed traverse. From the foregoing, no matter the number of sides, there will always be a minimum of $r = (n + n + 1) - 2(n - 1) = 3$ redundant equations for any closed traverse that is fixed both positionally and rotationally in space.

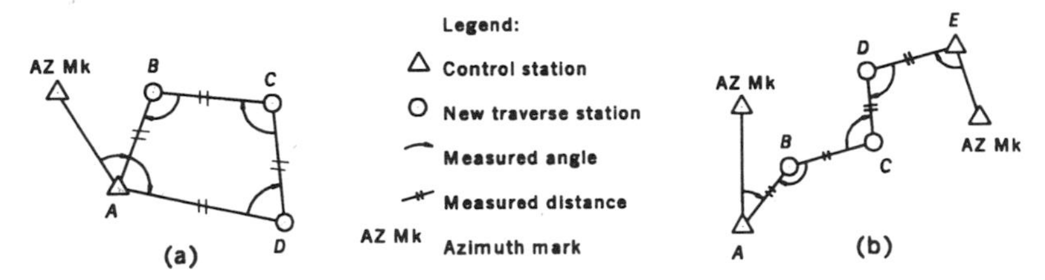

Figure 15.1 (a) Polygon and (b) link traverses.

15.4 NUMERICAL EXAMPLE

Example 15.1 To illustrate a least-squares traverse adjustment, the simple example link traverse shown in Figure 15.2 will be used. The observational data are:

Distances	Angles
$RU = 200.00 \pm 0.05$	$\theta_1 = 240°00' \pm 30''$
$US = 100.00 \pm 0.08$	$\theta_2 = 150°00' \pm 30''$
	$\theta_3 = 240°01' \pm 30''$

SOLUTION

Step 1: Calculate initial approximates for the unknown station coordinates.

$$x_{u_0} = 1000 + 200 \sin(60°) = 1173.20$$

$$y_{u_0} = 1000 + 200 \cos(60°) = 1100.00$$

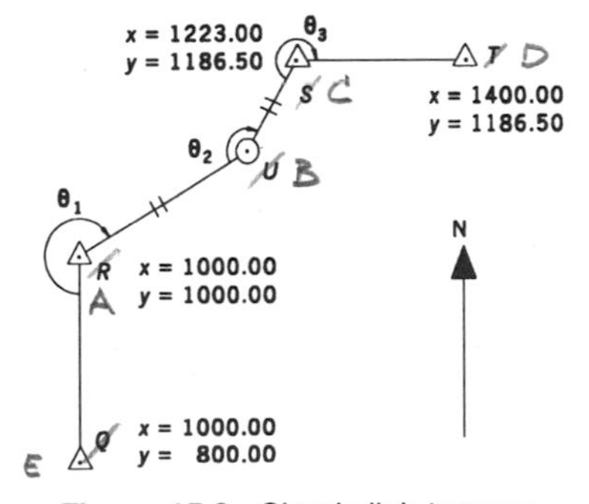

Figure 15.2 Simple link traverse.

Step 2: Formulate the X and K matrices. The traverse in this problem contains only one unknown station with two unknown coordinates. The elements of the X matrix thus consist of the dx_u and dy_u terms. They are the unknown corrections to be applied to the initial approximations for the coordinates of station U. The values in the K matrix are derived by subtracting computed quantities, based on initial coordinates, from their respective observed quantities. Note that since distance RU and angle θ_1 were used to compute initial approximations for station U, their K-matrix values will be zero in the first iteration.

$$X = \begin{bmatrix} dx_u \\ dy_u \end{bmatrix} \qquad K = \begin{bmatrix} k_{l_{ru}} \\ k_{l_{us}} \\ k_{\theta_1} \\ k_{\theta_2} \\ k_{\theta_3} \end{bmatrix} = \begin{bmatrix} 200.00 \text{ ft} - 200.00 \text{ ft} \\ 100.00 \text{ ft} - 99.81 \text{ ft} \\ 240°00'00'' - 240°00'00'' \\ 150°\ 00'00'' - 149°55'51'' \\ 240°01'00'' - 240°04'12'' \end{bmatrix} = \begin{bmatrix} 0.00 \text{ ft} \\ 0.19 \text{ ft} \\ 0'' \\ 249'' \\ -192'' \end{bmatrix}$$

Step 3: Calculate the Jacobian matrix. Since the observation equations are nonlinear, the Jacobian matrix must be formed to obtain the solution. The J matrix is formed using prototype equations (15.1) for distances and (15.2) for angles. As explained in Section 14.4, since the units of the K matrix that relate to the angles are in seconds, the angle coefficients of the J matrix must be multiplied by ρ.

In developing the J matrix using prototype equations (15.1) and (15.2), subscript substitutions were as shown in Table 15.1. Substitutions of numerical values and computation of the J matrix follow.

$$J = \begin{bmatrix} \dfrac{1173.20 - 1000.00}{200.00} & \dfrac{1100.00 - 1000.00}{200.00} \\ \dfrac{1173.20 - 1223.00}{99.81} & \dfrac{1100.00 - 1186.50}{99.81} \\ \left(\dfrac{1100.00 - 1000.00}{(200.00)^2}\right)\rho & \left(\dfrac{1000.00 - 1173.20}{(200.00)^2}\right)\rho \\ \left(\dfrac{1000.00 - 1100.00}{(200.00)^2} - \dfrac{1186.50 - 1100.00}{(99.81)^2}\right)\rho & \left(\dfrac{1173.20 - 1000.00}{(200.00)^2} - \dfrac{1173.20 - 1223.00}{(99.81)^2}\right)\rho \\ \left(\dfrac{1186.50 - 1100.00}{(99.81)^2}\right)\rho & \left(\dfrac{1173.20 - 1223.00}{(99.81)^2}\right)\rho \end{bmatrix}$$

Table 15.1 Subscript substitution

Observation	Subscript Substitution
Length RU	$R = i,\ U = j$
Length US	$U = i,\ S = j$
Angle θ_1	$Q = b,\ R = i,\ U = f$
Angle θ_2	$R = b,\ U = i,\ S = f$
Angle θ_3	$U = b,\ S = i,\ T = f$

$$J = \begin{bmatrix} 0.866 & 0.500 \\ -0.499 & -0.867 \\ 515.7 & -893.2 \\ -2306.6 & 1924.2 \\ 1790.9 & -1031.1 \end{bmatrix}$$

Step 4: Formulate the W matrix. The fact that distance and angle observations have differing measurement units and are combined in an adjustment is resolved by using relative weights that are based on observational variances in accordance with Equation (9.6). This weighting makes the observation equations dimensionally consistent. If weights are not used in traverse adjustments (i.e., equal weights are assumed), the least-squares problem will generally either give unreliable results, or have no solution. Since weights influence the correction size that each observation will receive, it is extremely important to use variances that correspond closely to the measurement errors. The error propagation procedures discussed in Chapter 6 aid in the determination of the estimated errors. Repeating Equation (9.6), the distance and angle weights for this problem are

$$\text{distances: } w_{l_{ij}} = \frac{1}{S_{l_{ij}}^2} \quad \text{and} \quad \text{angles: } w_{\theta_{bif}} = \frac{1}{S_{\theta_{bif}}^2} \tag{15.4}$$

Again the units of the weight matrix must match those of the J and K matrices. That is, the angular weights must be in the same units of measure (seconds) as their counterparts in the other two matrices. Based on the estimated errors in the observations, the W matrix, which is diagonal, is

$$W = \begin{bmatrix} \frac{1}{(0.05)^2} & & & & \\ & \frac{1}{(0.08)^2} & & \text{(zeros)} & \\ & & \frac{1}{(30)^2} & & \\ \text{(zeros)} & & & \frac{1}{(30)^2} & \\ & & & & \frac{1}{(30)^2} \end{bmatrix}$$

$$= \begin{bmatrix} 400.00 & & & & \\ & 156.2 & & \text{(zeros)} & \\ & & 0.0011 & & \\ \text{(zeros)} & & & 0.0011 & \\ & & & & 0.0011 \end{bmatrix}$$

Step 5: Solve the matrix system. This problem is iterative and was solved according to Equation (10.39) using the program MATRIX. (Output from the solution follows.) The first iteration yielded the following corrections to the initial coordinates:

$$dx_u = -0.11 \text{ ft}$$

$$dy_u = -0.01 \text{ ft}$$

Note that a second iteration produced zeros for dx_u and dy_u. A listing of the results from solving this problem using Program ADJUST is given below. The reader is encouraged to use either the MATRIX or ADJUST program to duplicate these results.

Step 6: Compute the a posteriori adjustment statistics. Also from the program MATRIX, the residuals and reference standard deviation are

$$v_{ru} = -0.11 \text{ ft} \qquad v_{us} = -0.12 \text{ ft}$$

$$v_{\theta_1} = -49'' \qquad v_{\theta_2} = -17'' \qquad v_{\theta_3} = 6''$$

$$S_0 = \pm 1.82 \text{ ft}$$

By applying Equation (12.24), and using the appropriate elements from the Q_{xx} matrix of the listing below, the standard deviations in the adjusted coordinates are

$$S_{x_u} = \pm 1.82\sqrt{0.00053} = \pm 0.042 \text{ ft} \quad S_{y_u} = \pm 1.82\sqrt{0.000838} = \pm 0.053 \text{ ft}$$

```
Computer Listing of Example 15.1
*****************************************
Initial Approximations for Unknown Stations
*****************************************
Station           Northing          Easting
_________________________________________________
      U           1,100.00          1,173.20

Control Stations
----------------
Station            Easting          Northing
_________________________________________________
      Q           1,000.00            800.00
      R           1,000.00          1,000.00
      S           1,223.00          1,186.50
      T           1,400.00          1,186.50


Distance Observations
---------------------
 Station      Station
Occupied      Sighted      Distance          s
______________________________________________________
      R            U        200.00        0.050
      U            S        100.00        0.080
```

Angle Observations

Station Backsighted	Station Occupied	Station Foresighted	Angle	s
Q	R	U	240° 00′ 00″	30″
R	U	S	150° 00′ 00″	30″
U	S	T	240° 01′ 00″	30″

First Iteration Matrices

J Dim: 5x2

0.86602	0.50001
-0.49894	-0.86664
515.68471	-893.16591
-2306.62893	1924.25287
1790.94422	-1031.08696

K Dim: 5x1

0.00440
0.18873
2.62001
249.36438
-191.98440

X Dim: 2x1

-0.11
-0.01

W Dim: 5x5

400.00000	0.00000	0.00000	0.00000	0.00000
0.00000	156.25000	0.00000	0.00000	0.00000
0.00000	0.00000	0.00111	0.00000	0.00000
0.00000	0.00000	0.00000	0.00111	0.00000
0.00000	0.00000	0.00000	0.00000	0.00111

N Dim: 2x2

10109.947301	-7254.506002
-7254.506002	6399.173533

Qxx Dim: 2x2

0.000530	0.000601
0.000601	0.000838

Final Iteration Matrices

J Dim: 5x2

0.86591	0.50020
-0.49972	-0.86619
516.14929	-893.51028
-2304.96717	1925.52297
1788.81788	-1032.01269

K Dim: 5x1

0.10723
0.12203
48.62499
17.26820
-5.89319

X Dim: 2x1

0.0000
0.0000

N Dim: 2x2

10093.552221	-7254.153057
-7254.153057	6407.367420

Qxx Dim: 2x2

0.000532	0.000602
0.000602	0.000838

J Qxx Jt Dim: 5x5

0.001130	-0.001195	-0.447052	0.055151	0.391901
-0.001195	0.001282	0.510776	-0.161934	-0.348843
-0.447052	0.510776	255.118765	-235.593233	-19.525532
0.055151	-0.161934	-235.593233	586.956593	-351.363360
0.391901	-0.348843	-19.525532	-351.363360	370.888892

****
Adjusted Stations
****

Station	Northing	Easting	S-N	S-E
U	1,099.99	1,173.09	0.053	0.042

****
Adjusted Distance Observations
****

Station Occupied	Station Sighted	Distance	V	S
R	U	199.89	-0.11	0.061
U	S	99.88	-0.12	0.065

****
Adjusted Angle Observations
****

Station Backsight	Station Occupied	Station Foresight	Angle	V	S
Q	R	U	239° 59′ 11″	-49″	29.0″
R	U	S	149° 59′ 43″	-17″	44.1″
S	S	T	240° 01′ 06″	6″	35.0″

-----Reference Standard Deviation = ±1.82-----
Iterations » 2

15.5 MINIMUM AMOUNT OF CONTROL

All adjustments require some form of control, and failure to supply a sufficient amount will result in an indeterminate solution. A traverse requires a minimum of one control station to fix it in position and one line of known direction to fix it in angular orientation. When a traverse has the minimum amount of control, it is said to be *minimally constrained*. It is not possible to adjust a traverse without this minimum. If minimal constraint is not available, necessary control values can be assumed, and the computational process carried out in arbitrary space. This enables the measured data to be tested for blunders and errors. In Chapter 20 we discuss minimally constrained adjustments.

A *free network adjustment* involves using a *pseudoinverse* to solve systems that have less than the minimum amount of control. This material is beyond the scope of this book. Readers interested in this subject should consult Bjerhammar (1973) or White (1987) in the Bibliography at the end of the book.

15.6 ADJUSTMENT OF NETWORKS

With the introduction of the EDM instrument, and particularly the total station, the speed and reliability of making angle and distance measurements have increased greatly. This has led to systems of horizontal surveys that do not conform to the basic procedures of trilateration, triangulation, or traverse. For example, from any occupied station, angles and distances may be measured to any or all intervisible stations. This creates what is called a *complex network*, more commonly referred to as a *network*. The least-squares solution of a network is readily performed, and is similar to that for a traverse. That is, observation equations are written for each measurement using the prototype equations given in Section 15.2. Coordinate corrections are found using Equation (10.39), and a posteriori error analysis is carried out.

Example 15.2 A network survey was conducted for the project shown in Figure 15.3. Station Q has control coordinates of (1000.00, 1000.00), and the azimuth of line QR is 0°06′24.5″ with an estimated error of ±0.001″. The measurements and their estimated errors are listed in Table 15.2.

Adjust this survey by least squares.

SOLUTION: Using standard traverse coordinate computation methods, the initial (x,y) approximations for station coordinates were determined to be

R: (1003.07, 2640.00) S: (2323.07,2638.46) T: (2661.74, 1096.08)

In Table 15.2, for each measured value the appropriate station subscripts for use in the prototype observation equations are indicated. For example, for the first distance QR, station Q is substituted for i in prototype equation (15.1), and station R replaces j. For the first angle, measured at Q from R to S, station

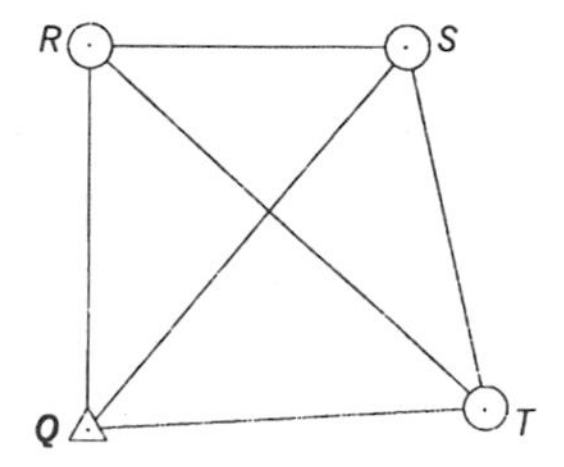

Figure 15.3

Table 15.2 Data for Example 15.2

Distance observations

Occupied, i	Sighted, j	Distance$_{ij}$	S
Q	R	1640.016	0.026
R	S	1320.001	0.024
S	T	1579.123	0.025
T	Q	1664.524	0.026
Q	S	2105.962	0.029
R	T	2266.035	0.030

Angle observations

Backsight, b	Occupied, i	Foresight, f	Angle$_{bif}$	S
R	Q	S	38°48′50.7″	4.0″
S	Q	T	47°46′12.4″	4.0″
T	Q	R	273°24′56.5″	4.4″
Q	R	S	269°57′33.4″	4.7″
R	S	T	257°32′56.8″	4.7″
S	T	Q	279°04′31.2″	4.5″
S	R	T	42°52′51.0″	4.3″
S	R	Q	90°02′26.7″	4.5″
Q	S	R	51°08′45.0″	4.3″
T	S	Q	51°18′16.2″	4.0″
Q	T	R	46°15′02.0″	4.0″
R	T	S	34°40′05.7″	4.0″

Azimuth observations

From, i	To, j	Azimuth	S
Q	R	0°06′24.5″	0.001″

R takes on the subscript b, Q becomes i, and S is substituted for f in prototype equation (15.2).

Table 15.3 shows the structure of the coefficient matrix for this adjustment and indicates by subscripts where the nonzero values occur. In this table, the column headings are the elements of the unknown X matrix dx_r, dy_r, dx_s, dy_s, dx_t, and dy_t. Note that since station Q is a control station, its corrections are set to zero, and thus dx_q and dy_q are not included in the adjustment. Note also in this table that the elements which have been left blank are zeros.

To fix the orientation of the network, the direction of course QR is included as an observation, but with a very small estimated error, ±0.001″. The last row of Table 15.3 shows the inclusion of this constrained observation using prototype equation (15.3). Since for azimuth QR only the foresight station, R, is an unknown, only coefficients for the foresight station j are included in the coefficient matrix.

Table 15.3 Format for coefficient matrix J of Example 15.2

	Unknown					
Observed	dx_r	dy_r	dx_s	dy_s	dx_t	dy_t
QR	j	j				
RS	i	i	j	j		
ST			i	i	j	j
TQ					i	i
QS			j	j		
RT	i	i			j	j
$\angle RQS$	b	b	f	f		
$\angle SQT$			b	b	f	f
$\angle TQR$	f	f			b	b
$\angle QRS$	i	i	f	f		
$\angle RST$	b	b	i	i	f	f
$\angle STQ$			b	b	i	i
$\angle SRT$	i	i	b	b	f	f
$\angle SRQ$	i	i	b	b		
$\angle QSR$	f	f	i	i		
$\angle TSQ$			i	i	b	b
$\angle QTR$	f	f			i	i
$\angle RTS$	b	b	f	f	i	i
Az QR	j	j				

Below are the necessary matrices for the first iteration when doing the weighted least-squares solution of this problem. Note that the numbers have been truncated to five decimal places for publication purposes only. Following these initial matrices, the results of the adjustment are listed as determined with the program ADJUST.

$$J = \begin{bmatrix}
0.00187 & 1.00000 & 0.00000 & 0.00000 & 0.00000 & 0.00000 \\
-1.00000 & 0.00117 & 1.00000 & -0.00117 & 0.00000 & 0.00000 \\
0.00000 & 0.00000 & -0.21447 & 0.97673 & 0.21447 & -0.97673 \\
0.00000 & 0.00000 & 0.00000 & 0.00000 & 0.99833 & 0.05772 \\
0.00000 & 0.00000 & 0.62825 & 0.77801 & 0.00000 & 0.00000 \\
-0.73197 & 0.68133 & 0.00000 & 0.00000 & 0.73197 & -0.68133 \\
-125.77078 & 0.23544 & 76.20105 & -61.53298 & 0.00000 & 0.00000 \\
0.00000 & 0.00000 & -76.20105 & 61.53298 & 7.15291 & -123.71223 \\
125.77078 & -0.23544 & 0.00000 & 0.00000 & -7.15291 & 123.71223 \\
-125.58848 & 156.49644 & -0.18230 & -156.26100 & 0.00000 & 0.00000 \\
-0.18230 & -156.26100 & 127.76269 & 184.27463 & -127.58038 & -28.01362 \\
0.00000 & 0.00000 & -127.58038 & -28.01362 & 134.73329 & -95.69861 \\
61.83602 & -89.63324 & 0.18230 & 156.26100 & -62.01833 & -66.62776 \\
125.58848 & -156.49644 & 0.18230 & 156.26100 & 0.00000 & 0.00000 \\
0.18230 & 156.26100 & -76.38335 & -94.72803 & 0.00000 & 0.00000 \\
0.00000 & 0.00000 & -51.37934 & -89.54660 & 127.58038 & 28.01362 \\
62.01833 & 66.62776 & 0.00000 & 0.00000 & -69.17123 & 57.08446 \\
-62.01833 & -66.62776 & 127.58038 & 28.01362 & -65.56206 & 38.61414 \\
125.76953 & -0.23544 & 0.00000 & 0.00000 & 0.00000 & 0.0000
\end{bmatrix}$$

The weight matrix is

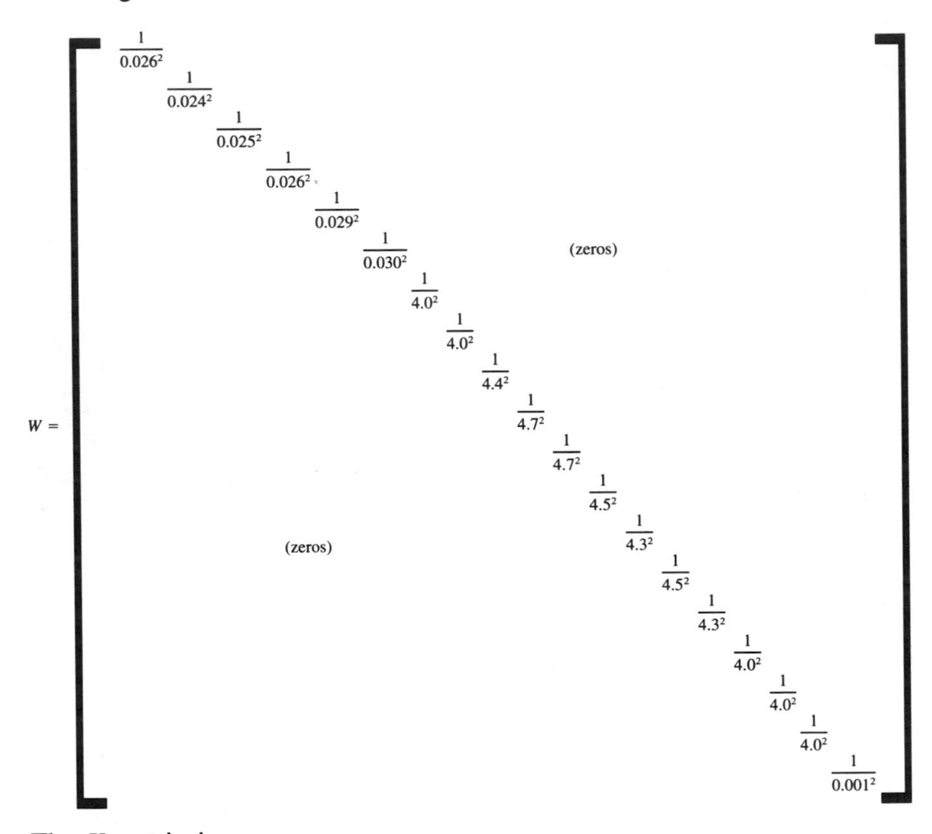

$$W = \operatorname{diag}\left(\frac{1}{0.026^2}, \frac{1}{0.024^2}, \frac{1}{0.025^2}, \frac{1}{0.026^2}, \frac{1}{0.029^2}, \frac{1}{0.030^2}, \frac{1}{4.0^2}, \frac{1}{4.0^2}, \frac{1}{4.4^2}, \frac{1}{4.7^2}, \frac{1}{4.7^2}, \frac{1}{4.5^2}, \frac{1}{4.3^2}, \frac{1}{4.5^2}, \frac{1}{4.3^2}, \frac{1}{4.0^2}, \frac{1}{4.0^2}, \frac{1}{4.0^2}, \frac{1}{0.001^2}\right) \quad \text{(zeros off the diagonal)}$$

The K matrix is:

$$K = \begin{bmatrix} 0.0031 \\ -0.0099 \\ -0.0229 \\ -0.0007 \\ -0.0053 \\ -0.0196 \\ -0.0090 \\ -0.5988 \\ 0.2077 \\ -2.3832 \\ 1.4834 \\ -1.4080 \\ -1.0668 \\ 2.4832 \\ -0.0742 \\ -3.4092 \\ -22.1423 \\ 2.4502 \\ -0.3572 \end{bmatrix}$$

Summary of Results from ADJUST

```
Number of Control Stations       » 1
Number of Unknown Stations       » 3
Number of Distance Observations  » 6
Number of Angle Observations     » 12
Number of Azimuth Observations   » 1
```

****
Initial Approximations for Unknown Stations
****

Station	X	Y
R	1,003.06	2,640.01
S	2,323.07	2,638.47
T	2,661.75	1,096.07

Control Stations

Station	X	Y
Q	1,000.00	1,000.00

****
Distance Observations
****

Station Occupied	Station Sighted	Distance	S
Q	R	1,640.016	0.026
R	S	1,320.001	0.024
S	T	1,579.123	0.025
T	Q	1,664.524	0.026
Q	S	2,105.962	0.029
R	T	2,266.035	0.030

****
Angle Observations
****

Station Backsighted	Station Occupied	Station Foresighted	Angle	S
R	Q	S	38° 48′ 50.7″	4.0″
S	Q	T	47° 46′ 12.4″	4.0″
T	Q	R	273° 24′ 56.5″	4.4″
Q	R	S	269° 57′ 33.4″	4.7″
R	S	T	257° 32′ 56.8″	4.7″
S	T	Q	279° 04′ 31.2″	4.5″
S	R	T	42° 52′ 51.0″	4.3″
S	R	Q	90° 02′ 26.7″	4.5″
Q	S	R	51° 08′ 45.0″	4.3″
T	S	Q	51° 18′ 16.2″	4.0″
Q	T	R	46° 15′ 02.0″	4.0″
R	T	S	34° 40′ 05.7″	4.0″

```
********************
Azimuth Observations
********************
 Station       Station
Occupied       Sighted          Azimuth           S
----------------------------------------------------
   Q              R          0° 06' 24.5"    0.001"

Iteration 1
K MATRIX             X MATRIX
--------             --------
  0.0031             -0.002906
 -0.0099             -0.035262
 -0.0229             -0.021858
 -0.0007              0.004793
 -0.0053              0.003996
 -0.0196             -0.014381
 -0.0090
 -0.5988
  0.2077
 -2.3832
  1.4834
 -1.4080
 -1.0668
  2.4832
 -0.0742
 -3.4092
-22.1423
  2.4502
 -0.3572

Iteration: 2
K MATRIX             X MATRIX
--------             --------
  0.0384              0.000000
 -0.0176             -0.000000
 -0.0155             -0.000000
 -0.0039              0.000000
  0.0087             -0.000000
 -0.0104              0.000000
 -2.0763
 -0.6962
  2.3725
 -0.6444
 -0.5048
 -3.3233
 -1.3435
  0.7444
  3.7319
 -5.2271
-18.5154
  0.7387
  0.0000
```

```
INVERSE MATRIX
-----------------------------------------------------------------------
 0.00000000 0.00000047  0.00000003  0.00000034 0.00000005  0.00000019
 0.00000047 0.00025290  0.00001780  0.00018378 0.00002767  0.00010155
 0.00000003 0.00001780  0.00023696 -0.00004687 0.00006675 -0.00008552
 0.00000034 0.00018378 -0.00004687  0.00032490 0.00010511  0.00022492
 0.00000005 0.00002767  0.00006675  0.00010511 0.00027128  0.00011190
 0.00000019 0.00010155 -0.00008552  0.00022492 0.00011190  0.00038959
```

****
Adjusted Stations
****

Station	X	Y	Sx	Sy
R	1,003.06	2,639.97	0.000	0.016
S	2,323.07	2,638.45	0.015	0.018
T	2,661.75	1,096.06	0.016	0.020

****
Adjusted Distance Observations
****

Station Occupied	Station Sighted	Distance	V	S
Q	R	1,639.978	-0.0384	0.0159
R	S	1,320.019	0.0176	0.0154
S	T	1,579.138	0.0155	0.0158
T	Q	1,664.528	0.0039	0.0169
Q	S	2,105.953	-0.0087	0.0156
R	T	2,266.045	0.0104	0.0163

****
Adjusted Angle Observations
****

Station Backsighted	Station Occupied	Station Foresighted	Angle	V	S
R	Q	S	38° 48′ 52.8″	2.08″	1.75″
S	Q	T	47° 46′ 13.1″	0.70″	1.95″
T	Q	R	273° 24′ 54.1″	-2.37″	2.40″
Q	R	S	269° 57′ 34.0″	0.64″	2.26″
R	S	T	257° 32′ 57.3″	0.50″	2.50″
S	T	Q	279° 04′ 34.5″	3.32″	2.33″
S	R	T	42° 52′ 52.3″	1.34″	1.82″
S	R	Q	90° 02′ 26.0″	-0.74″	2.26″
Q	S	R	51° 08′ 41.3″	-3.73″	1.98″
T	S	Q	51° 18′ 21.4″	5.23″	2.04″
Q	T	R	46° 15′ 20.5″	18.52″	1.82″
R	T	S	34° 40′ 05.0″	-0.74″	1.72″

```
*****************************
Adjusted Azimuth Observations
*****************************
 Station        Station
Occupied        Sighted              Azimuth              V           S"
--------------------------------------------------------------------------
   Q               R             0° 06' 24.5"          0.00"       0.001"

                          ********************
                          Adjustment Statistics
                          ********************
                                 Iterations = 2
                               Redundancies = 13
                         Reference Variance 2.20
                             Reference So ±1.5
                Passed X² test at 99.0% significance level!
                          X² lower value = 3.57
                           X² upper value 29.82
    The a priori value of 1 used in computations involving the
                           reference variance.
                              Convergence!
```

15.7 χ^2 TEST: GOODNESS OF FIT

At the completion of a least-squares adjustment, the significance of the computed reference variance, S_0^2, can be checked statistically. This check is often referred to as a *goodness-of-fit* test since the computation of S_0^2 is based on $\Sigma\, v^2$. That is, as the residuals become larger, so will the computed reference variance, and thus the computed model, deviates more from the observed values. However, the size of the residuals is not the only contributing factor to the size of the reference variance in a weighted adjustment. The stochastic model also plays a role in the size of this value. Thus when the χ^2 test indicates that the null hypothesis should be rejected, it may be due to a blunder in the data or an incorrect decision in selecting the stochastic model for the adjustment. In Chapter 20 these matters are discussed in greater detail. For now, just the reference variance of the adjustment of Example 15.2 will be checked.

In Example 15.2 there are 13 degrees of freedom, and the computed reference variance, S_0^2, is 2.2. In Chapter 9 it was shown that the *a priori* value for the reference variance was 1. A check can now be made to compare the computed value for the reference variance against its a priori value using a two-tailed χ^2 test. For this adjustment, a significance level of 0.01was selected. The procedures for doing the test were outlined in Section 4.10 and the results for this example are shown in Table 15.4. Since $\alpha/2$ is 0.005 and the adjustment had 13 redundant observations, the critical χ^2 value from the table is 29.82. Now it can be seen that the computed χ^2 value is less than the tabular value and thus the test fails to reject the null hypothesis, H_0. Thus the value

Table 15.4 Two-tailed χ^2 test on S_0^2

H_0: $S^2 = 1$
H_a: $S^2 \neq 1$
Test statistic:
$\chi^2 = \dfrac{vS^2}{\sigma^2} = \dfrac{13 \times 2.2}{1} = 28.6$
Rejection region:
$28.6 = \chi^2 > \chi^2_{0.005,13} = 29.82$
$28.6 = \chi^2 < \chi^2_{0.995,13} = 3.565$

of 1 for S_0^2 can be used when computing the standard deviations for the station coordinates and observations since its computed value is only an estimate.

PROBLEMS

Note: For problems below requiring least-squares adjustment, if a computer program is not distinctly specified for use in the problem, it is expected that the least-squares algorithm will be solved using the program MATRIX which is included within the diskette supplied with the book.

15.1 For the link traverse shown in Figure 15.4, assume that the distance and angle standard deviations are ±0.027 ft and ±5″, respectively. Using the control below, adjust the data given in the figure using weighted least squares. The control station coordinates are:

A:	$x = 944.79$	$y = 756.17$	*C:*	$x = 6125.48$	$y = 1032.90$
Mk1:	$x = 991.31$	$y = 667.65$	*Mk2:*	$x = 6225.391$	$y = 1037.109$

(a) What is the reference standard deviation, S_0?

(b) List the adjusted coordinates of station *B* and their standard deviations.

(c) Tabulate the adjusted observations, their residuals, and their standard deviations.

(d) List the inverted normal matrix from the last iteration.

15.2 Adjust by the method of least squares the closed traverse in Figure 15.5. The data are given below.

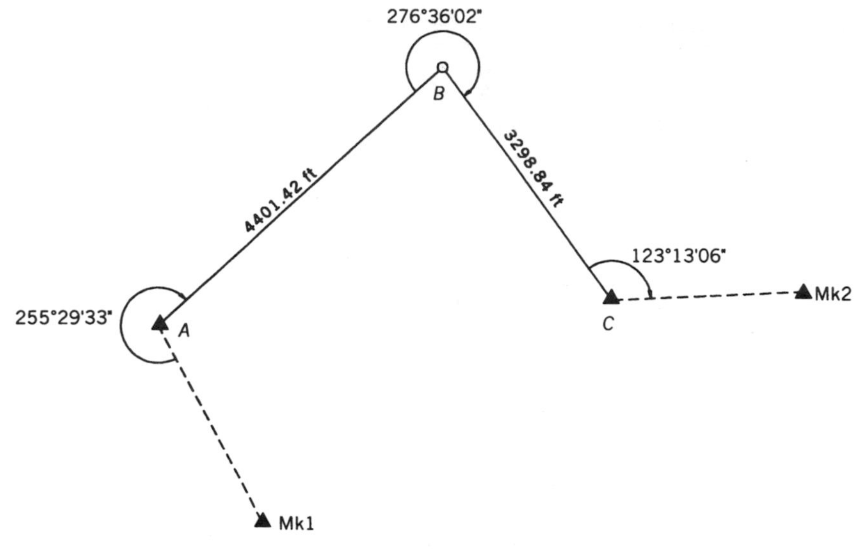

Figure 15.4

(a) What is the reference standard deviation, S_0?

(b) List the adjusted coordinates of the unknown stations and their standard deviations.

(c) Tabulate the adjusted observations, their residuals, and their standard deviations.

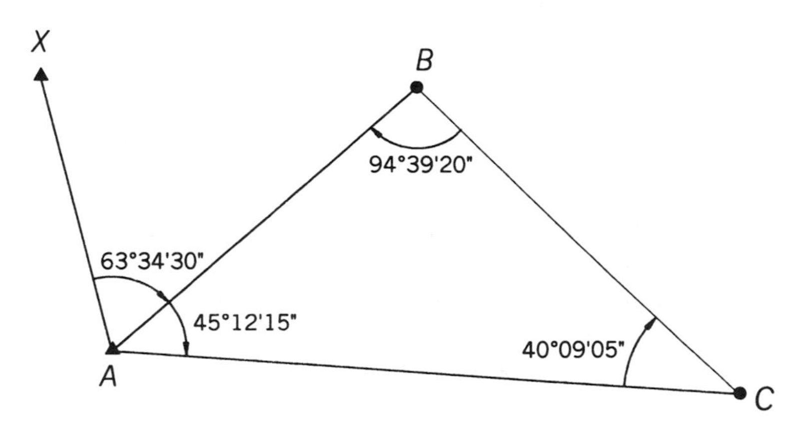

Figure 15.5

(d) List the inverted normal matrix from the last iteration.

Observed angles			Observed distances		
Angle	Value	s	Length	Value	s (ft)
XAB	63°34′30″	5″	*AB*	2731.25	0.29
BAC	45°12′15″	10″	*BC*	3005.81	0.32
CBA	94°39′20″	10″	*CA*	4222.01	0.44
ACB	40°09′05″	10″			

The control station coordinates are

$$X: \quad x = 9974.612 \quad y = 10{,}096.724$$

$$A: \quad x = 10{,}000.000 \quad y = 10{,}000.000$$

15.3 Adjust the network of Figure 15.6 by the method of least squares. The data are listed below.

(a) What is the reference standard deviation, S_o?

(b) List the adjusted coordinates of the unknown stations and their standard deviations.

(c) Tabulate the adjusted observations, their residuals, and their standard deviations.

(d) List the inverted normal matrix from the last iteration.

The coordinates for control station A are

$$x{:}\ 4356.78 \qquad y{:}\ 2340.97$$

The azimuth of line AB is $Az_{AB} = 0°06'23.7''$.

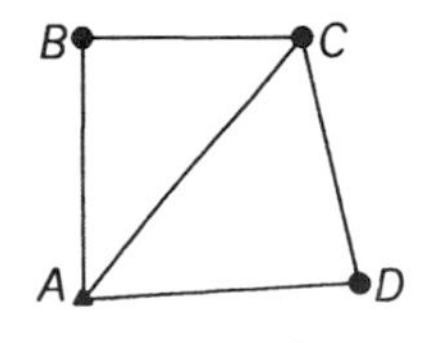

Figure 15.6

Distance observations (ft)

Course	Distance	*s*	Course	Distance	*s*
AB	1639.999	0.023	*BC*	1320.027	0.023
CD	1579.169	0.023	*DA*	1664.497	0.023
AC	2105.995	0.024			

Angle observations

∡	Angle	*s*	∡	Angle	*s*
BAC	38°48′52.5″	2.2″	*CAD*	47°46′13.1″	2.3″
CBA	90°02′25.5″	3.1″	*ACB*	51°08′53.2″	2.6″
DCA	51°18′21.9″	2.3″	*ADB*	46°15′28.8″	2.2″

15.4 Perform a weighted least-squares adjustment using the data given in Problem 14.1, and the additional distances given below.

(a) What is the reference standard deviation, S_0?

(b) List the adjusted coordinates of the unknown stations and their standard deviations.

(c) Tabulate the adjusted observations, their residuals, and their standard deviations.

(d) List the inverted normal matrix from the last iteration.

Occupied	Sighted	Distance	*s*
A	*E*	1297.634	0.021
B	*E*	2000.499	0.022
C	*E*	2511.821	0.023
D	*E*	1959.687	0.022

15.5 Do a weighted least-squares adjustment using the data given in Problem 14.4 and the additional distances given below

(a) What is the reference standard deviation, S_0?

(b) List the adjusted coordinates of the unknown stations and their standard deviations.

(c) Tabulate the adjusted observations, their residuals, and their standard deviations.

(d) List the inverted normal matrix from the last iteration.

Occupied	Sighted	Distance	*s*
U	*P*	7071.073	0.041
U	*Q*	5590.162	0.035
U	*R*	5590.181	0.035
U	*S*	7071.087	0.041

15.6 Using the program ADJUST, do a weighted least-squares adjustment for the data in Problem 14.8 with the additional distances given below,

(a) What is the reference standard deviation, S_0?

(b) List the adjusted coordinates of the unknown stations and their standard deviations.

(c) Tabulate the adjusted observations, their residuals, and their standard deviations.

(d) List the inverted normal matrix from the last iteration.

Occupied	Sighted	Distance	*s*
A	*B*	480.030	0.020
A	*C*	680.184	0.020
C	*D*	470.743	0.020
D	*B*	582.889	0.020
B	*C*	791.958	0.020
A	*D*	785.947	0.020
C	*F*	720.589	0.020
D	*E*	835.748	0.020
C	*E*	640.808	0.020
E	*F*	411.476	0.020
F	*D*	647.791	0.020
E	*H*	794.253	0.020
F	*G*	764.350	0.020
E	*G*	612.127	0.020
F	*H*	664.878	0.020
G	*H*	487.044	0.020

For each problem, does the computed reference variance for the adjustment pass the χ^2 test at a level of significance of 0.05?

15.7 Example 15.1

15.8 Problem 15.1

15.9 Problem 15.2

15.10 Problem 15.3

15.11 Problem 15.4

Programming Problems

15.12 Write a program that reads a file of station coordinates and angle and distance observations and then:

(a) writes the data to a file in a formatted fashion,

(b) computes the J, K, and W matrices, and

(c) writes the matrices to a file that is compatible with the MATRIX program.

(d) Demonstrate this program with Problem 15.6.

15.13 Write a program that reads a file containing the J, K, and W matrices and then:

(a) writes these matrices in a formatted fashion,

(b) performs one iteration of Problem 15.6, and

(c) writes the matrices used to compute the solution, and tabulates the corrections to the station coordinates in a formatted fashion.

15.14 Write a program that reads a file of station coordinates and angle and distance observations and then:

(a) writes the data to a file in a formatted fashion,

(b) computes the J, K, and W matrices,

(c) performs a weighted least-squares adjustment of Problem 15.6,

(d) writes the matrices used in computations in a formatted fashion to a file, and

(e) computes the final adjusted station coordinates, their estimated errors, the adjusted observations, their residuals, and their estimated errors, and writes them to a file in a formatted fashion.

15.15 Develop a spreadsheet that will create the coefficient, weight, and constant matrices for a network. Write the matrices to a file in a format usable by the MATRIX program supplied with this book. Demonstrate its use with Problem 15.6.

16

ADJUSTMENT OF GPS NETWORKS

16.1 INTRODUCTION

For the past three decades, NASA and the U.S. military have been engaged in a space research program to develop a precise positioning and navigation system. The first-generation system, called *TRANSIT*, used six satellites and was based on the *doppler* principle. TRANSIT was made available for commercial use in 1967, and shortly thereafter its use in surveying began. The establishment of a worldwide network of control stations was among its earliest and most valuable applications. Point positioning using TRANSIT required very lengthy observing sessions, and its accuracy was only at about the 1-m level. Thus in surveying it was suitable only for control work on networks consisting of widely spaced points. It was not satisfactory for everyday surveying applications such as traversing or engineering layout.

Encouraged by the success of TRANSIT, a new research program was developed that ultimately led to the creation of the NAVSTAR *global positioning system (GPS)*. This second-generation positioning and navigation system utilizes a constellation of 24 orbiting satellites. The accuracy of GPS was improved substantially over that of the TRANSIT system, and the disadvantage of lengthy observing sessions was also eliminated. Although developed for military applications, civilians, including surveyors, also found uses for the GPS system.

Since its introduction, GPS has been used extensively, and it has been found to be reliable, efficient, and capable of yielding extremely high accuracies. GPS measurements can be taken day or night and in any weather conditions. A significant advantage of GPS is that intervisibility between surveyed points is not necessary. Thus the time-consuming process of clearing lines of sight is avoided. Although most of the earliest applications of GPS were in control work, improvements have now made the system convenient and practical for use in virtually every type of survey, including property surveys, topographic mapping, and construction staking.

This chapter provides a brief introduction to GPS surveying. We explain the basic measurements involved in the system, discuss the errors in those measurements, describe the nature of the adjustments needed to account for those errors, and give procedures for making adjustments of networks surveyed using GPS. An example problem is given to demonstrate the procedures.

16.2 GPS MEASUREMENTS

Fundamentally, the global positioning system operates by measuring distances from receivers located on ground stations having unknown locations to orbiting GPS satellites whose positions are precisely known. Thus, conceptually, GPS surveying is similar to conventional resection, in which distances are measured with an EDM instrument from an unknown station to several control points. (The conventional resection procedure was discussed in Chapter 13 and illustrated in Example 13.2). Of course, there are some differences between GPS position determination and conventional resection. Among them are the manner of measuring the distances and the fact that the control stations used in GPS work are satellites.

All of the satellites in the GPS constellation orbit the earth at nominal altitudes of 20,200 km. Each satellite continuously broadcasts unique electronic signals on two *carrier frequencies*. These carriers are modulated with *pseudorandom noise* (*PRN*) *codes*. The PRN codes consist of unique sequences of binary values (zeros and ones) that are superimposed upon the carriers. These codes appear to be random, but in fact they are generated according to a known mathematical algorithm. The frequencies of the carriers and PRN codes are controlled very precisely at known values.

Distances are determined in GPS surveying by making measurements on these transmitted satellite signals. Two different observational procedures are used: positioning by *pseudoranging* and positioning by *carrier-phase measurements*. Pseudoranging involves determining distances (ranges) between satellites and receivers by measuring precisely the time it takes transmitted signals to travel from satellites to ground receivers. This is done by determining changes in the PRN codes that occur during the time it takes signals to travel from the satellite transmitter to the antenna of the receiver. Then from the known frequency of the PRN codes, very precise travel times are determined. With the velocity and travel times of the signals known, the distances can be computed. Finally, based on the distances, the positions of the ground stations can be calculated. Because pseudoranging is based on measuring PRN codes, this GPS observation technique is also often referred to as the *code measurement* procedure.

In the carrier-phase procedure, the quantities observed are phase changes that occur as a result of the carrier wave traveling from the satellites to the receivers. The principle is similar to the phase-shift method employed by electronic distance-measuring instruments. A major difference, however, is that the satellites are moving, so that the signals cannot be returned to the transmitters for "true" phase-shift measurements. Instead, the phase shifts must be measured at the receivers. But to make true phase-shift observations, the clocks in the satellites and receivers would have to be perfectly synchronized, which of course cannot be achieved. To overcome this timing problem and to eliminate other errors in the system, *differencing* techniques (taking differences between phase measurements) are used. Various differencing procedures can be applied. *Single differencing* by measuring to two satellites simultaneously with one receiver will eliminate receiver clock errors. Single differencing by measuring from two ground receivers simultaneously to one satellite will cancel satellite clock errors. *Double differencing* (subtracting the results of simultaneous measurements from two receivers to two satellites) will eliminate both receiver and satellite clock errors and other system errors.

Another problem in making carrier-phase measurements is that only the phase shift of the last cycle of the carrier wave is measured, and the number of full cycles in the travel distance is unknown. (Recall in EDM work that this problem was overcome by transmitting longer wavelengths and measuring their phase shifts.) Again, because the satellites are moving, this cannot be done in GPS work. However, by extending the differencing technique to what is called *triple differencing*, this ambiguity in the number of cycles cancels out of the solution. Triple differencing consists of differencing the results of two double differences, and therefore it involves making measurements at two different times to two satellites from two stations.

In practice, when surveys are done by measuring carrier phases, four or more satellites are normally observed simultaneously using two or more receivers located on ground stations. Also, the observations are repeated many times. This produces a very large number of redundant observations, from which many difference combinations can be computed.

Of the two GPS observing procedures, pseudoranging yields a somewhat lower order of accuracy, but it is preferred for navigation use because it gives instantaneous point positions of satisfactory accuracy. The carrier-phase technique produces a higher order of accuracy and is therefore the choice for high-precision surveying applications. Adjustment of carrier-phase GPS observations is the subject of this chapter.

The differencing techniques used in carrier-phase measurements described briefly above do not yield positions directly for the points occupied by receivers. Rather, *baselines* (vector distances between stations) are determined. These baselines are actually computed in terms of their coordinate difference components ΔX, ΔY, and ΔZ. These coordinate differences are determined in the reference three-dimensional rectangular coordinate system described in Section 16.4.

To use the GPS carrier-phase procedure in surveying, at least two receivers located on separate stations must be operated simultaneously. Assume, for example, that two stations A and B were occupied for an observing session, that station A is a control point, and that station B is a point of unknown position. The session would yield coordinate differences ΔX_{AB}, ΔY_{AB}, and ΔZ_{AB} between stations A and B. The X, Y, and Z coordinates of station B can then be obtained by adding the baseline components to the coordinates of A, as:

$$\begin{aligned} X_B &= X_A + \Delta X_{AB} \\ Y_B &= Y_A + \Delta Y_{AB} \\ Z_B &= Z_A + \Delta Z_{AB} \end{aligned} \tag{16.1}$$

Because carrier-phase measurements do not yield point positions directly, but rather, give baseline components, this method of GPS surveying is referred to as *relative positioning*. In practice, often more than two receivers are used simultaneously in relative positioning, which enables more than one baseline to be determined during each observing session. Also, after the first observing session, additional points are interconnected in the survey by moving the receivers to other nearby stations. In this procedure, at least one receiver is left at one of the previously occupied stations. By employing this technique, a network of interconnected points can be created. Figure 16.1 illustrates an example of a GPS network. In this figure, stations A and B are control stations, and stations C, D, E, and F are points of unknown position. Creation of such networks is a common procedure employed in GPS relative positioning work.

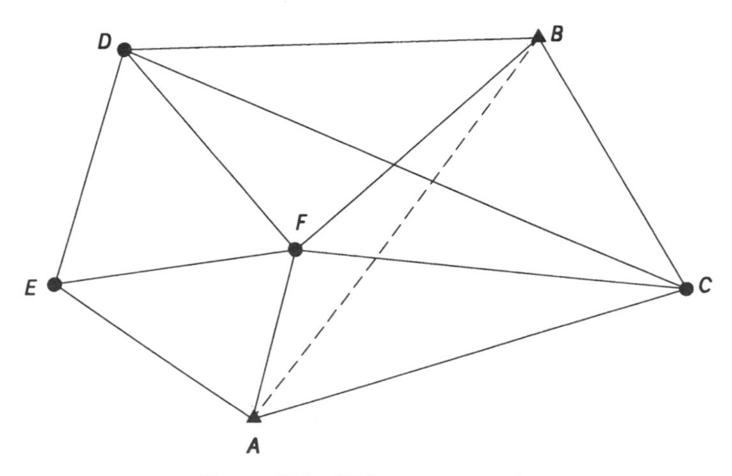

Figure 16.1 GPS survey network.

16.3 GPS ERRORS AND THE NEED FOR ADJUSTMENT

As in all types of surveying measurements, GPS observations contain errors. The principal sources of these errors are (1) errors in the satellite orbits, (2) errors in signal transmission time due to atmospheric conditions, (3) receiver errors, (4) multipath errors (signals being reflected so that they travel indirect routes from satellite to receiver), and (5) errors due to centering the receiver antenna over the ground station and in measuring its height above the station. To account for these and other errors, and to increase the precisions of point position determinations, GPS observations are very carefully made according to strict specifications, and redundant measurements are taken. The fact that errors are present in the observations makes it necessary to analyze the measurements for acceptance or rejection. Also, because redundant observations have been made, they must be adjusted so that all measured values are consistent.

In GPS surveying work where the observations are made using carrier phases, there are two stages where least-squares adjustment is applied. The first is in processing the redundant observations to obtain the adjusted baseline components (ΔX, ΔY, and ΔZ), and the second is in adjusting networks of stations wherein the baseline components have been measured. These adjustments are discussed in more detail later in the chapter.

16.4 REFERENCE COORDINATE SYSTEMS FOR GPS MEASUREMENTS

In GPS surveying, three different reference coordinate systems are involved. First, the satellite positions at the instants of their observation are given in a space-related X_s, Y_s, Z_s three-dimensional rectangular coordinate system. This coordinate system is illustrated in Figure 16.2. In the figure, the elliptical orbit of a satellite is shown. It has one of its two foci at G, the earth's center of gravity. Two points, *perigee* (point where the satellite is closest to *G*) and *apogee* (point where the satellite is farthest from G), define the *line of apsides*. This line, which also passes through the two foci, is the X_s axis of the satellite reference coordinate system. The origin of the system is at *G*, the Y_s axis is in the mean orbital plane, and Z_s is perpendicular to the X_s–Y_s plane. Because a satellite varies only slightly from its mean orbital plane, values of Z_s are small. For each specific instant of time that a given satellite is observed, its coordinates are calculated in its unique X_s, Y_s, Z_s system.

In processing GPS measurements, all X_s, Y_s, Z_s coordinates that were computed for satellite observations are converted to a common earth-related X_e, Y_e, Z_e three-dimensional geocentric coordinate system. This earth-centered, earth-fixed coordinate system, illustrated in Figure 16.3, is also commonly called the *terrestrial geocentric* system, or simply the *geocentric* system. It

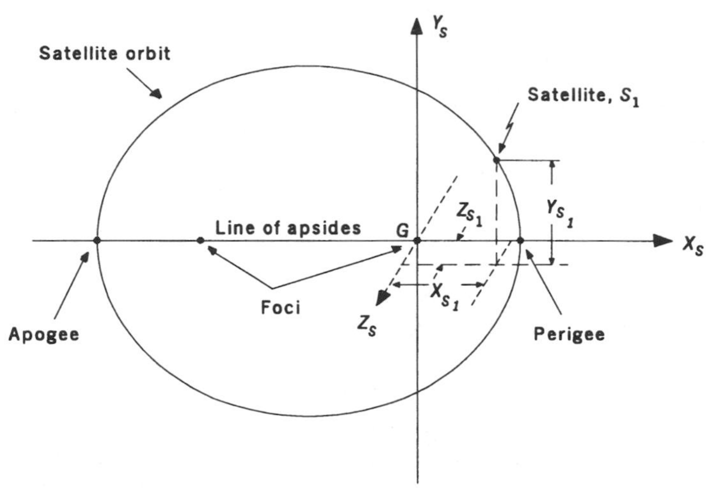

Figure 16.2 Satellite reference coordinate system.

is in this system that the baseline components are computed based on the differencing of observed carrier phase measurements. The origin of this coordinate system is at the earth's gravitational center. The Z_e axis coincides with the earth's conventional international origin (CIO) polar axis, the X_e–Y_e plane is perpendicular to the Z_e axis, and the X_e axis passes through the Greenwich meridian. To convert coordinates from the space-related (X_s, Y_s, Z_s) system to the earth-related (X_e, Y_e, Z_e) geocentric system, four parameters are needed. These are (a) the *inclination angle* i (angle between the orbital plane and the earth's equatorial plane); (b) the *argument of perigee* ω (angle measured in the orbital plane between the equator and the line of apsides); (c) the *right ascension of the ascending node* Ω (angle measured in the plane of the earth's equator from the vernal equinox to the line of intersection between the orbital and equatorial planes); and (d) the *Greenwich hour angle of the vernal equinox* γ (angle measured in the equatorial plane from the Greenwich meridian to the vernal equinox). These four parameters are illustrated in Figure 16.3, and for any satellite at any instant of time they are available. Software provided by GPS equipment manufacturers computes the X_s, Y_s, Z_s coordinates of satellites at their instants of observation, and it also transforms these coordinates into the X_e, Y_e, Z_e geocentric coordinate system used for computing the baseline components.

In order for the results of the baseline computations to be useful to local surveyors, the X_e, Y_e, Z_e coordinates must be converted to *geodetic coordinates* of latitude, longitude, and ellipsoid height. The geodetic coordinate system is illustrated in Figure 16.4, where the parameters are symbolized by ϕ, λ, and h, respectively. Geodetic coordinates are referenced to the World Geodetic

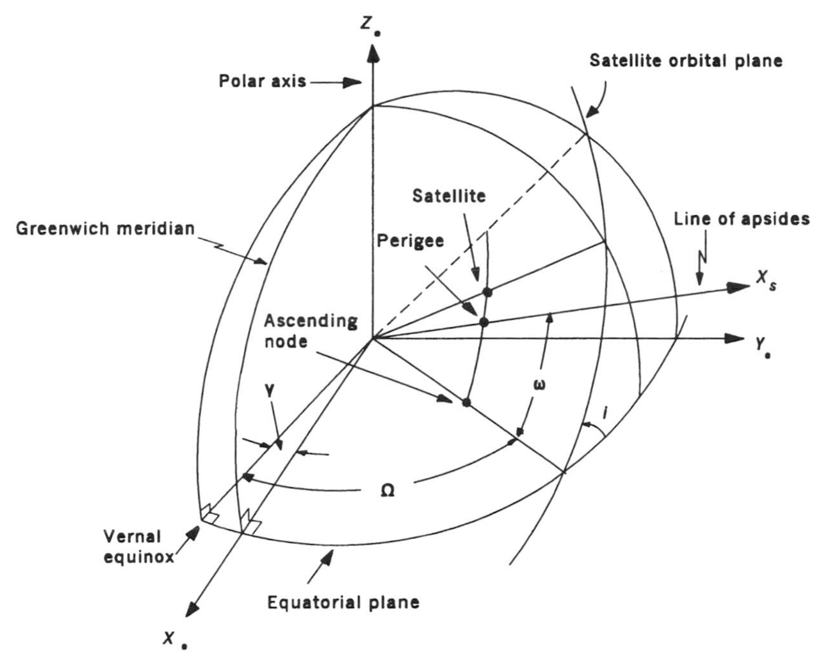

Figure 16.3 Earth-related three-dimensional coordinate system used in GPS carrier-phase differencing computations.

System of 1984, which employs the WGS84 ellipsoid. The center of this ellipsoid is oriented at the earth's gravitational center, and for practical purposes it is the same as the GRS80 ellipsoid used for NAD83. From latitude and longitude, state plane coordinates (which are more convenient for use by local surveyors) can be computed.

It is important to note that ellipsoid heights are not *orthometric heights* (elevations referred to the geoid). To convert ellipsoid heights to elevations, *geoid heights* (vertical distances between the ellipsoid and geoid) must be subtracted from ellipsoid heights.

16.5 CONVERTING BETWEEN THE TERRESTRIAL AND GEODETIC COORDINATE SYSTEMS

GPS networks must include at least one control point, but more are preferable. The geodetic coordinates of these control points will normally be given in the NAD83 reference system. Prior to processing carrier-phase measurements to obtain adjusted baselines for a network, the coordinates of the control

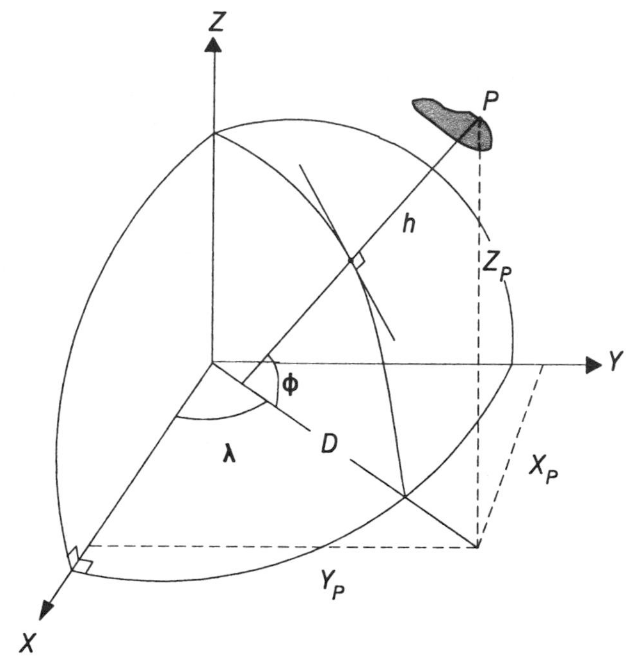

Figure 16.4 Geodetic coordinates (with the earth-related X_e, Y_e, and Z_e geocentric coordinate system superimposed).

stations in the network must be converted from their geodetic values into the earth-related X_e, Y_e, Z_e geocentric system (see Figure 16.4). The equations for making these conversions are

$$X = (N + h)\cos(\phi)\cos(\lambda) \tag{16.2}$$

$$Y = (N + h)\cos(\phi)\sin(\lambda) \tag{16.3}$$

$$Z = [N(1 - e^2) + h]\sin(\phi) \tag{16.4}$$

In the equations above, h is the ellipsoid height of the point, ϕ the geodetic latitude of the point, and λ the geodetic longitude of the point. Also, e^2 is a constant for the ellipsoid, which is computed from

$$e^2 = 2f - f^2 \tag{16.5a}$$

or

$$e^2 = \frac{(a^2 - b^2)}{a^2} \tag{16.5b}$$

where f is the flattening factor of the ellipsoid; a and b are the semimajor and semiminor axes, respectively, of the ellipsoid[1]; and N is the normal to the ellipsoid at the point, which is computed from

$$N = \frac{a}{\sqrt{1 - e^2 \sin^2(\phi)}} \tag{16.6}$$

Example 16.1 Control stations A and B of the GPS network of Figure 16.1 have the following NAD83 geodetic coordinates:

$$\phi_A = 43°15'46.2890'' \qquad \lambda_A = -89°59'42.1640'' \qquad h_A = 1382.618 \text{ m}$$

$$\phi_B = 43°23'46.3626'' \qquad \lambda_B = -89°54'00.7570'' \qquad h_B = 1235.457 \text{ m}$$

Compute their X_e, Y_e, Z_e geocentric coordinates.

SOLUTION: *For station A*: By Equation (16.5*a*),

$$e^2 = \frac{2}{298.25722356} - \left(\frac{1}{298.25722356}\right)^2 = 0.006694379990$$

By Equation (16.6),

$$N = \frac{6{,}378{,}137}{\sqrt{1 - e^2 \sin^2(43°15'46.2890'')}} = 6{,}388{,}188.2524 \text{ m}$$

By Equation (16.2),

$$X_A = (6{,}388{,}188.2524 + 1382.618)\cos(43°15'46.2890'') \times \cos(-89°59'42.1640'') = 402.3509 \text{ m}$$

By Equation (16.3),

$$Y_A = (6{,}388{,}188.2524 + 1382.618)\cos(43°15'46.2890'') \times \sin(-89°59'42.1640'') = -4{,}652{,}995.3011 \text{ m}$$

By Equation (16.4),

[1] The WGS84 ellipsoid is used, whose a, b, and f values are 6,378,137.0000 m, 6,356,752.3142 m, and 1/298.25722356, respectively.

$$Z_A = [6{,}388{,}188.2524(1 - e^2) + 1382.618] \sin(43°15'46.2890'')$$
$$= 4{,}349{,}760.7775 \text{ m}$$

For station B: Following the same procedure as above, the geocentric coordinates for station B are

$$X_B = 8086.0318 \text{ m} \qquad Y_B = -4{,}642{,}712.8474 \text{ m}$$

$$Z_B = 4{,}360{,}439.0833 \text{ m}$$

After completing the network adjustment, it is necessary to convert all X_e, Y_e, Y_e geocentric coordinates into their geodetic values for use by local surveyors. This conversion process follows these steps (refer to Figure 16.4):

Step 1: Determine the longitude, λ, from:

$$\lambda = \tan^{-1}\left(\frac{Y_e}{X_e}\right) \tag{16.7}$$

Step 2: Compute D from

$$D = \sqrt{X_e^2 + Y_e^2} \tag{16.8}$$

Step 3: Calculate an approximate latitude value ϕ_0 from

$$\phi_0 = \tan^{-1}\left[\frac{Z_e}{D\,(1 - e^2)}\right] \tag{16.9}$$

Step 4: Compute an approximate ellipsoid normal value N_0 from

$$N_0 = \frac{a}{\sqrt{1 - e^2 \sin^2(\phi_0)}} \tag{16.10}$$

Step 5: Calculate an improved value for latitude ϕ_0 from

$$\phi_0 = \tan^{-1}\left(\frac{Z_e + e^2 N_0 \sin(\phi_0)}{D}\right) \tag{16.11}$$

Step 6: Use the value of ϕ_0 computed in step 5, and return to step 4. Iterate steps 4 and 5 until there is negligible change in ϕ_0. Using the values from the last iteration for N_0 and ϕ_0, the value for h is now computed[2] as

[2] Equation (16.12) is numerically stable for values of ϕ less than 45°. For values of ϕ greater than 45°, use the equation $h = [Z_e/\sin(\phi_0)] - N_0(1 - e^2)$.

$$h = \frac{D}{\cos(\phi_0)} - N_0 \tag{16.12}$$

Example 16.2 Assume that the final adjusted coordinates for station C of the network of Figure 16.4 were

$$X_C = 12{,}046.5808 \text{ m} \qquad Y_C = -4{,}649{,}394.0826 \text{ m}$$

$$Z_C = 4{,}353{,}160.0634 \text{ m}$$

Compute the NAD83 geodetic coordinates for station C.

SOLUTION: By Equation (16.7),

$$\lambda = \tan^{-1}\left(\frac{-4{,}649{,}394.0826}{12{,}046.5808}\right) = -89°51'05.5691''$$

By Equation (16.8),

$$D = \sqrt{(12{,}046.5808)^2 + (-4{,}649{,}394.0826)^2} = 4{,}649{,}409.6889 \text{ m}$$

Using Equation (16.9), the initial value for ϕ_o is

$$\phi_0 = \tan^{-1}\left[\frac{4{,}353{,}160.0634}{D(1 - e^2)}\right] = 43°18'26.2228''$$

The first iteration for N_0 and ϕ_0 is

$$N_0 = \frac{6{,}378{,}137.0}{1 - e^2 \sin^2(43°18'26.22280'')} = 6{,}388{,}204.8545 \text{ m}$$

$$\phi_0 = \tan^{-1}\left(\frac{4353{,}160.0634 + e^2\, 6{,}388{,}204.8545 \sin(43°18'26.22280'')}{4{,}649{,}409.6889}\right)$$

$$= 43°18'26.1035''$$

The next iteration produced the final values of

$$N_0 = 6{,}388{,}204.8421 \text{ m} \qquad \phi_0 = 43°18'26.1030''$$

Using Equation (16.12), the elevation of the station C is

$$h = \frac{4{,}649{,}409.6889}{\cos(43°18'26.1030'')} - 6{,}388{,}204.8421 = 1103.101 \text{ m}$$

An option within the computer program ADJUST will make these coordinate conversions; both from geodetic to geocentric and from geocentric to geodetic.

16.6 APPLICATION OF LEAST SQUARES IN PROCESSING GPS DATA

The least-squares method is used at two different stages in processing GPS carrier-phase measurements. First, it is applied in the adjustment that yields baseline components between stations from the redundant carrier-phase measurements. Recall that in this procedure, differencing techniques are employed to compensate for errors in the system and to resolve the cycle ambiguities. In the solution, observation equations are written which contain the differences in coordinates between stations as parameters. The reference coordinate system for this adjustment is the X_e, Y_e, Z_e geocentric system. A highly redundant system of equations is obtained because, as described earlier, a minimum of four (and often more) satellites are tracked simultaneously using at least two (and often more) receivers. Furthermore, many repeat observations are taken. This system of equations is solved by least squares to obtain the most probable ΔX, ΔY, and ΔZ components of the baseline vectors. The development of these observation equations is beyond the scope of this book, and thus their solution by least squares is also not covered herein.[3]

Software furnished by manufacturers of GPS receivers will process observed phase changes to form the differencing observation equations, perform the least-squares adjustment, and output the adjusted baseline vector components. The software will also output the covariance matrix, which expresses the correlation between the ΔX, ΔY, and ΔZ components of each baseline. The software is proprietary and thus cannot be included herein.

The second stage where least squares is employed in processing GPS observations is in adjusting baseline vector components in networks. This occurs after the least-squares adjustment of the carrier-phase observations is completed. It is also done in the X_e, Y_e, Z_e geocentric coordinate system. In network adjustments, the goal is to make consistent all X coordinates (and all X-coordinate differences) throughout the figure. The same objective applies for all Y coordinates and for all Z coordinates. As an example, consider the GPS network shown in Figure 16.1. It consists of two control stations and four stations whose coordinates are to be determined. A summary of the baseline measurements obtained from the least-squares adjustment of carrier-

[3] Readers interested in studying these observation equations should consult *GPS Theory and Practice*, by B. Hoffman-Wellenhof et al., or *GPS Satellite Surveying* by A. Leick. Complete references for these publications are given in the Bibliography at the end of this book.

phase observations for this figure is given in Table 16.1. The covariance matrix elements that are listed in the table are used for weighting the observations. These will be discussed in Section 16.8 and for the moment can be ignored.

A network adjustment of Figure 16.1 should yield adjusted X coordinates for the stations (and adjusted coordinate differences between stations) that are all mutually consistent. Specifically for this network, the adjusted X coordinate of station C should be obtained by adding ΔX_{AC} to the X coordinate of station A; and the same value should be obtained by adding ΔX_{BC} to the X coordinate of station B, or by adding ΔX_{DC} to the X coordinate of station D, and so on. Equivalent conditions should exist for the Y and Z coordinates. Note that these conditions do not exist for the data of Table 16.1, which contains the unadjusted baseline measurements. The procedure of adjusting GPS networks is described in detail in Section 16.8 and an example is given.

16.7 NETWORK PREADJUSTMENT DATA ANALYSIS

Prior to adjusting GPS networks, a series of procedures should be followed to analyze the data for internal consistency and to eliminate possible blunders. No control points are needed for these analyses. Depending on the actual measurements taken and the network geometry, these procedures may consist of analyzing (1) differences between fixed and measured baseline components, (2) differences between repeated measurements of the same baseline components, and (3) loop closures. After making these analyses, a "minimally constrained" adjustment is usually performed that will help isolate any blunders that may have escaped the first set of analyses. Procedures for making these analyses are described in the following subsections.

16.7.1 Analysis of Fixed Baseline Measurements

GPS job specifications often require that baseline measurements be taken between fixed control stations. The benefit of making these measurements is to verify the accuracy of both the GPS measurement system and the control being held fixed. Obviously, the smaller the discrepancies between measured and known baseline lengths, the better. If the discrepancies are larger than can be tolerated, the conditions causing them must be investigated. Note that in the data of Table 16.1, one fixed baseline (between control points A and B) was measured. Table 16.2 gives the data for comparing the measured and fixed baseline components. The measured values are listed in column (2), and the fixed components are given in column (3). To compute the fixed values, X_e, Y_e, Z_e geocentric coordinates of the two control stations are first determined from their geodetic coordinates according to procedures discussed in Section 16.5. Then the ΔX, ΔY, and ΔZ differences between the X_e, Y_e, Z_e coordinates for the two control stations are determined. Differ-

Table 16.1 Measured baseline data for the network of Figure 16.1

(1) From:	(2) To:	(3) ΔX	(4) ΔY	(5) ΔZ	(6)	(7)	(8) Covariance matrix elements	(9)	(10)	(11)
A	*C*	11644.2232	3601.2165	3399.2550	9.844E-4	−9.580E-6	9.520E-6	9.377E-4	−9.520E-6	9.827E-4
A	*E*	−5321.7164	3634.0754	3173.6652	2.158E-4	−2.100E-6	2.160E-6	1.919E-4	−2.100E-6	2.005E-4
B	*C*	3960.5442	−6681.2467	−7279.0148	2.305E-4	−2.230E-6	2.070E-6	2.546E-4	−2.230E-6	2.252E-4
B	*D*	−11167.6076	−394.5204	−907.9593	2.700E-4	−2.750E-6	2.850E-6	2.721E-4	−2.720E-6	2.670E-4
D	*C*	15128.1647	−6286.7054	−6371.0583	1.461E-4	−1.430E-6	1.340E-6	1.614E-4	−1.440E-6	1.308E-4
D	*E*	−1837.7459	−6253.8534	−6596.6697	1.231E-4	−1.190E-6	1.220E-6	1.277E-4	−1.210E-6	1.283E-4
F	*A*	−1116.4523	−4596.1610	−4355.9062	7.475E-5	−7.900E-7	8.800E-7	6.593E-5	−8.100E-7	7.616E-5
F	*C*	10527.7852	−994.9377	−956.6246	2.567E-4	−2.250E-6	2.400E-6	2.163E-4	−2.270E-6	2.397E-4
F	*E*	−6438.1364	−962.0694	−1182.2305	9.442E-5	−9.200E-7	1.040E-6	9.959E-5	−8.900E-7	8.826E-5
F	*D*	−4600.3787	5291.7785	5414.4311	9.330E-5	−9.900E-7	9.000E-7	9.875E-5	−9.900E-7	1.204E-4
F	*B*	6567.2311	5686.2926	6322.3917	6.643E-5	−6.500E-7	6.900E-7	7.465E-5	−6.400E-7	6.048E-5
B	*F*	−6567.2310	−5686.3033	−6322.3807	5.512E-5	−6.300E-7	6.100E-7	7.472E-5	−6.300E-7	6.629E-5
A	*F*	1116.4577	4596.1553	4355.9141	6.619E-5	−8.000E-7	9.000E-7	8.108E-5	−8.200E-7	9.376E-5
*A**	*B*	7683.6883	10282.4550	10678.3008	7.240E-4	−7.280E-6	7.520E-6	6.762E-4	−7.290E-6	7.310E-4

*Fixed baseline used only for checking, but not included in adjustment.

Table 16.2 Comparisons of measured and fixed baseline components

(1) Component	(2) Measured (m)	(3) Fixed (m)	(4) Difference	(5) PPM*
ΔX	7,683.6883	7,683.6809	0.0074	0.44
ΔY	10,282.4550	10,282.4537	0.0013	0.08
ΔZ	10,678.3008	10,678.3058	0.0050	0.30

*The total baseline length used in computing these ppm values was 16,697 m, which was derived from the square root of the sum of the squares of ΔX, ΔY, and ΔZ values.

ences (in meters) between the measured and fixed baseline components are given in column (4). Finally the differences, expressed in parts per million (ppm), are listed in column (5). These ppm values are obtained by dividing column (4) differences by their corresponding total baseline lengths and multiplying by 1,000,000.

16.7.2 Analysis of Repeat Baseline Measurements

Another procedure employed in evaluating the consistency of the observed data and in weeding out blunders is to make repeat measurements of certain baselines. These repeat measurements are taken in different observing sessions, and the results compared. In the data of Table 16.1, for example, baselines AF and BF were repeated. Table 16.3 gives comparisons of these measurements using the same procedure that was used in Table 16.2. Again, the ppm values listed in column (5) use the total baseline lengths in the demonimator, which are computed from the square root of the sum of the squares of the measured baseline components.

Table 16.3 Comparisons of repeat baseline measurements

(1) Component	(2) First Observation	(3) Second Observation	(4) Difference	(5) ppm
ΔX_{AF}	1116.4577	−1116.4523	0.0054	0.84
ΔY_{AF}	4596.1553	−4596.1610	0.0057	0.88
ΔZ_{AF}	4355.9141	−4355.9062	0.0079	1.23
ΔX_{BF}	−6567.2310	6567.2311	0.0001	0.01
ΔY_{BF}	−5686.3033	5686.2926	0.0107	1.00
ΔZ_{BF}	−6322.3807	6322.3917	0.0110	1.02

The Federal Geodetic Control Subcommittee (FGCS) has developed a document entitled "Geometric Geodetic Accuracy Standards and Specifications for Using GPS Relative Positioning Techniques." It is intended to serve as a guideline for planning, executing, and classifying geodetic surveys performed by GPS relative positioning methods. This document may be consulted to determine whether or not the ppm values of Tables 16.2 and 16.3 are acceptable for the required order of accuracy for the survey. Besides ppm requirements, the FGCS guidelines specify other criteria that must be met for the different orders of accuracy in connection with repeat baseline measurements.

16.7.3 Analysis of Loop Closures

GPS networks will typically consist of many interconnected closed loops. In the network of Figure 16.1, for example, a closed loop is formed by points *ACBDEA*. Similarly, *ACFA*, *CFBC*, *BDFB*, and so on, are other closed loops. For each closed loop, the algebraic sum of the ΔX components should equal zero. The same condition should exist for the ΔY and ΔZ components. These loop closure conditions are very similar to the latitude and departure closure conditions imposed in closed polygon traverses. An unusually large closure within any loop will indicate that either a blunder or a large error exists in one (or more) of the baselines of the loop.

To compute loop closures, the baseline components are simply added algebraically for that loop. The closure in X components for loop *ACBDEA*, for example, would be computed as

$$cx = \Delta X_{AC} + \Delta X_{CB} + \Delta X_{BD} + \Delta X_{DE} + \Delta X_{EA} \tag{16.13}$$

where cx is the loop closure in X coordinates. Similar equations apply for computing closures in Y and Z coordinates.

Substituting numerical values into Equation (16.13), the closure in X coordinates for loop *ACBDEA* is:

$$\begin{aligned} cx &= 11{,}644.2232 - 3960.5442 - 11{,}167.6076 - 1837.7459 + 5321.7164 \\ &= 0.0419 \text{ m} \end{aligned}$$

Similarly, closures in Y and Z coordinates for that loop are

$$\begin{aligned} cy &= 3601.2165 + 6681.2467 - 394.5204 - 6253.8534 - 3634.0754 \\ &= 0.0140 \text{ m} \end{aligned}$$

$$cz = 3399.2550 + 7279.0148 - 907.9593 - 6596.6697 - 3173.6652$$
$$= -0.0244 \text{ m}$$

For evaluation purposes, loop closures are expressed in terms of the ratios of resultant closures to the total loop lengths. They are given in ppm. For any loop, the resultant closure is the square root of the sum of the squares of its cx, cy, and cz values, and for loop *ACBDEA* the resultant is 0.0505 m. The total length of a loop is computed by summing its legs, each leg being computed from the square root of the sum of the squares of its measured ΔX, ΔY, and ΔZ values. For loop *ACBDEA*, the total loop length is 50,967 m, and the closure ppm ratio is therefore (0.0505/50,967)(1,000,000) = 0.99 ppm. Again these ppm ratios can be compared against values given in the FGCS guidelines to determine if they are acceptable for the order of accuracy of the survey. As was the case with repeat baseline measurements, the FGCS guidelines also specify other criteria that must be met in loop analyses besides the ppm values.

For any network, enough loop closures should be computed so that every baseline is included within at least one loop. This should expose any large blunders that exist. If a blunder does exist, its location can be often be determined through additional loop closure analyses. Assume, for example, that the misclosure of loop ACDEA discloses the presence of a blunder. By also computing the closures of loops *AFCA*, *CFDC*, *DFED*, and *EFAE*, the exact baseline containing the blunder can be detected. In this example, if a large misclosure were found in loop *DFED* and all other loops appeared to be blunder free, the blunder would be in line *DE*.

An option within the computer program ADJUST will make these loop-closure computations. Documentation on the use of this program is given in Appendix F.

16.7.4 Minimally Constrained Adjustment

Prior to making the final adjustment of baseline measurements in a network, a "minimally constrained" least-squares adjustment is usually performed. In this adjustment, also called a "free" adjustment, any one station in the network may be held fixed with arbitrary coordinates. All other stations in the network are therefore free to adjust as necessary to accommodate the baseline measurements. The residuals that result from this adjustment are strictly related to the baseline measurements, not to faulty control coordinates. These residuals are examined, and from them blunders that may have gone undetected through the first set of analyses can be found and eliminated.

16.8 LEAST-SQUARES ADJUSTMENT OF GPS NETWORKS

As noted earlier, because GPS networks contain redundant measurements, they must be adjusted to make all coordinate differences consistent. In applying least squares to the problem of adjusting baselines in GPS networks, observation equations are written that relate station coordinates to the observed coordinate differences and their residual errors. To illustrate the procedure, consider the example of Figure 16.1. For line *AC* of this figure, an observation equation can be written for each measured baseline component as

$$\begin{aligned} X_C &= X_A + \Delta X_{AC} + v_{x_{AC}} \\ Y_C &= Y_A + \Delta Y_{AC} + v_{y_{AC}} \\ Z_C &= Z_A + \Delta Z_{AC} + v_{z_{AC}} \end{aligned} \tag{16.14}$$

Similarly, the observation equations for the baseline components of line *CD* are

$$\begin{aligned} X_D &= X_C + \Delta X_{CD} + v_{x_{CD}} \\ Y_D &= Y_C + \Delta Y_{CD} + v_{y_{CD}} \\ Z_D &= Z_C + \Delta Z_{CD} + v_{z_{CD}} \end{aligned} \tag{16.15}$$

Observation equations of the form above would be written for all measured baselines in any figure. For Figure 16.1 there were a total of 13 measured baselines, so the number of observation equations that can be developed is 39. Also, each of stations *C*, *D*, *E*, and *F* has three unknown coordinates, for a total of 12 unknowns in the problem. Thus there are 39 − 12 = 27 redundant observations in the network. The 39 observation equations can be expressed in matrix form as

$$AX = L + V \tag{16.16}$$

If the observation equations for adjusting the network of Figure 16.1 are written in the same order that the measurements are listed in Table 16.1, the *A*, *X*, *L*, and *V* matrices would be

$$
A = \begin{bmatrix}
1 & 0 & 0 & 0 & 0 & 0 & 0 & 0 & 0 & 0 & 0 & 0 \\
0 & 1 & 0 & 0 & 0 & 0 & 0 & 0 & 0 & 0 & 0 & 0 \\
0 & 0 & 1 & 0 & 0 & 0 & 0 & 0 & 0 & 0 & 0 & 0 \\
0 & 0 & 0 & 1 & 0 & 0 & 0 & 0 & 0 & 0 & 0 & 0 \\
0 & 0 & 0 & 0 & 1 & 0 & 0 & 0 & 0 & 0 & 0 & 0 \\
0 & 0 & 0 & 0 & 0 & 1 & 0 & 0 & 0 & 0 & 0 & 0 \\
1 & 0 & 0 & 0 & 0 & 0 & 0 & 0 & 0 & 0 & 0 & 0 \\
0 & 1 & 0 & 0 & 0 & 0 & 0 & 0 & 0 & 0 & 0 & 0 \\
0 & 0 & 1 & 0 & 0 & 0 & 0 & 0 & 0 & 0 & 0 & 0 \\
0 & 0 & 0 & 0 & 0 & 0 & 1 & 0 & 0 & 0 & 0 & 0 \\
0 & 0 & 0 & 0 & 0 & 0 & 0 & 1 & 0 & 0 & 0 & 0 \\
0 & 0 & 0 & 0 & 0 & 0 & 0 & 0 & 1 & 0 & 0 & 0 \\
1 & 0 & 0 & 0 & 0 & 0 & -1 & 0 & 0 & 0 & 0 & 0 \\
0 & 1 & 0 & 0 & 0 & 0 & 0 & -1 & 0 & 0 & 0 & 0 \\
0 & 0 & 1 & 0 & 0 & 0 & 0 & 0 & -1 & 0 & 0 & 0 \\
0 & 0 & 0 & 1 & 0 & 0 & -1 & 0 & 0 & 0 & 0 & 0 \\
0 & 0 & 0 & 0 & 1 & 0 & 0 & -1 & 0 & 0 & 0 & 0 \\
0 & 0 & 0 & 0 & 0 & 1 & 0 & 0 & -1 & 0 & 0 & 0 \\
0 & 0 & 0 & 0 & 0 & 0 & 0 & 0 & 0 & -1 & 0 & 0 \\
0 & 0 & 0 & 0 & 0 & 0 & 0 & 0 & 0 & 0 & -1 & 0 \\
0 & 0 & 0 & 0 & 0 & 0 & 0 & 0 & 0 & 0 & 0 & -1 \\
1 & 0 & 0 & 0 & 0 & 0 & 0 & 0 & 0 & -1 & 0 & 0 \\
0 & 1 & 0 & 0 & 0 & 0 & 0 & 0 & 0 & 0 & -1 & 0 \\
0 & 0 & 1 & 0 & 0 & 0 & 0 & 0 & 0 & 0 & 0 & -1 \\
0 & 0 & 0 & 1 & 0 & 0 & 0 & 0 & 0 & -1 & 0 & 0 \\
0 & 0 & 0 & 0 & 1 & 0 & 0 & 0 & 0 & 0 & -1 & 0 \\
0 & 0 & 0 & 0 & 0 & 1 & 0 & 0 & 0 & 0 & 0 & -1 \\
0 & 0 & 0 & 0 & 0 & 0 & 1 & 0 & 0 & -1 & 0 & 0 \\
0 & 0 & 0 & 0 & 0 & 0 & 0 & 1 & 0 & 0 & -1 & 0 \\
0 & 0 & 0 & 0 & 0 & 0 & 0 & 0 & 1 & 0 & 0 & -1 \\
0 & 0 & 0 & 0 & 0 & 0 & 0 & 0 & 0 & -1 & 0 & 0 \\
0 & 0 & 0 & 0 & 0 & 0 & 0 & 0 & 0 & 0 & -1 & 0 \\
0 & 0 & 0 & 0 & 0 & 0 & 0 & 0 & 0 & 0 & 0 & -1 \\
0 & 0 & 0 & 0 & 0 & 0 & 0 & 0 & 0 & 1 & 0 & 0 \\
0 & 0 & 0 & 0 & 0 & 0 & 0 & 0 & 0 & 0 & 1 & 0 \\
0 & 0 & 0 & 0 & 0 & 0 & 0 & 0 & 0 & 0 & 0 & 1 \\
0 & 0 & 0 & 0 & 0 & 0 & 0 & 0 & 0 & 1 & 0 & 0 \\
0 & 0 & 0 & 0 & 0 & 0 & 0 & 0 & 0 & 0 & 1 & 0 \\
0 & 0 & 0 & 0 & 0 & 0 & 0 & 0 & 0 & 0 & 0 & 1
\end{bmatrix}
$$

$$
X = \begin{bmatrix} X_C \\ Y_C \\ Z_C \\ X_E \\ Y_E \\ Z_E \\ X_D \\ Y_D \\ Z_D \\ X_F \\ Y_F \\ Z_F \end{bmatrix} \quad
L = \begin{bmatrix} 12046.5741 \\ -4649394.0846 \\ 4353160.0325 \\ -4919.3655 \\ -4649361.2257 \\ 4352934.4427 \\ 12046.5760 \\ -4649394.0941 \\ 4353160.0685 \\ -3081.5758 \\ -4643107.3678 \\ 4359531.1240 \\ 15128.1647 \\ -6286.7054 \\ -6371.0583 \\ -1837.7459 \\ -6253.8534 \\ -6596.6697 \\ -1518.8032 \\ 4648399.1401 \\ -4354116.6837 \\ 10527.7852 \\ -994.9377 \\ -956.6246 \\ -6438.1364 \\ -962.0694 \\ -1182.2305 \\ -4600.3787 \\ 5291.7785 \\ 5414.4311 \\ -1518.8007 \\ 4648399.1400 \\ -4354116.6916 \\ 1518.8008 \\ -4648399.1507 \\ 4354116.7026 \\ 1518.8086 \\ -4648399.1458 \\ 4354116.6916 \end{bmatrix} \quad
V = \begin{bmatrix} v_{x_{AC}} \\ v_{y_{AC}} \\ v_{z_{AC}} \\ v_{x_{AE}} \\ v_{y_{AE}} \\ v_{z_{AE}} \\ v_{x_{BC}} \\ v_{y_{BC}} \\ v_{z_{BC}} \\ v_{x_{BD}} \\ v_{y_{BD}} \\ v_{z_{BD}} \\ v_{x_{DC}} \\ v_{y_{DC}} \\ v_{z_{DC}} \\ v_{x_{DE}} \\ v_{y_{DE}} \\ v_{z_{DE}} \\ v_{x_{FA}} \\ v_{y_{FA}} \\ v_{z_{FA}} \\ v_{x_{FC}} \\ v_{y_{FC}} \\ v_{z_{FC}} \\ v_{x_{FE}} \\ v_{y_{FE}} \\ v_{z_{FE}} \\ \vdots \\ v_{x_{AF}} \\ v_{y_{AF}} \\ v_{z_{AF}} \end{bmatrix}
$$

The numerical values of the elements of the L matrix are determined by rearranging the observation equations. Its first three elements are for the ΔX, ΔY, and ΔZ baseline components of line AC, respectively. Those elements are calculated as follows:

$$
\begin{aligned}
L_X &= X_A + \Delta X_{AC} \\
L_Y &= Y_A + \Delta Y_{AC} \\
L_Z &= Z_A + \Delta Z_{AC}
\end{aligned}
\tag{16.17}
$$

The other elements of the L matrix are formed in the same manner as demonstrated for baseline AC. Before numerical values for the L matrix elements can be obtained, however, the X_e, Y_e, Z_e geocentric coordinates of all control points in the network must be computed. This is done by following the procedures described in Section 16.5 and demonstrated by Example 16.1. That example problem provided the X_e, Y_e, Z_e coordinates of control points A and B of Figure 16.1, which were used together with the data of Table 16.1 to compute the elements of the L matrix given above.

Note that the observation equations for GPS network adjustment are linear and that the only nonzero elements of the A matrix are either 1 and -1. This is the same type of matrix that was developed in adjusting level nets by least squares. In fact, GPS network adjustments are performed in the very same manner as level net adjustments, with the exception of the weights. In GPS relative positioning, the three measured baseline components are correlated. Therefore, a covariance matrix of dimensions 3×3 is derived for each baseline as a product of the least-squares adjustment of the carrier-phase measurements. This covariance matrix is used to weight the observations in the network adjustment in accordance with Equation (9.4). The weight matrix for any GPS network is therefore a block-diagonal type, with an individual 3×3 matrix for each measured baseline on the diagonal. When more than two receivers are used, additional 3×3 matrices are created in the off-diagonal region of the matrix to provide the correlation that exists between the simultaneously measured baselines. All other elements of the matrix are zeros.

The covariances for the observations in the network of Figure 16.1 are given in columns (6) through (11) of Table 16.1. Only the six lower triangular elements of the 3×3 covariance matrix for each observation are listed. This gives complete weighting information, however, because the covariance matrix is symmetrical. Columns 6 through 11 list σ_x^2, σ_{xy}, σ_{xz}, σ_y^2, σ_{yz}, and σ_z^2, respectively. Thus the full 3×3 covariance matrix for baseline AC is

$$
\Sigma_{AC} = \begin{bmatrix} 9.884\text{E-}4 & -9.580\text{E-}6 & 9.520\text{E-}6 \\ -9.580\text{E-}6 & 9.377\text{E-}4 & -9.520\text{E-}6 \\ 9.520\text{E-}6 & -9.520\text{E-}6 & 9.827\text{E-}4 \end{bmatrix}
$$

The complete weight matrix for the example network of Figure 16.1 has dimensions of 39×39. After inverting the full matrix and multiplying by the apriori estimate for the reference variance, S_0^2, in accordance with Equation (9.4), the weight matrix for the network of Figure 16.1 is (note that S_0^2

is taken as 1.0 for this computation, and that no correlation between baselines is included)

$$
W = \begin{bmatrix}
1011.8 & 10.2 & -9.7 & 0 & 0 & 0 & 0 & 0 & 0 & & 0 & 0 & 0 \\
10.2 & 1066.6 & 10.2 & 0 & 0 & 0 & 0 & 0 & 0 & & 0 & 0 & 0 \\
-9.7 & 10.2 & 1017.7 & 0 & 0 & 0 & 0 & 0 & 0 & & 0 & 0 & 0 \\
0 & 0 & 0 & 4634.5 & 50.2 & -49.4 & 0 & 0 & 0 & & 0 & 0 & 0 \\
0 & 0 & 0 & 50.2 & 5209.7 & 54.0 & 0 & 0 & 0 & & 0 & 0 & 0 \\
0 & 0 & 0 & -49.4 & 54.0 & 4988.1 & 0 & 0 & 0 & & 0 & 0 & 0 \\
0 & 0 & 0 & 0 & 0 & 0 & 4339.1 & 37.7 & -39.5 & & 0 & 0 & 0 \\
0 & 0 & 0 & 0 & 0 & 0 & 37.7 & 3927.8 & 38.5 & & 0 & 0 & 0 \\
0 & 0 & 0 & 0 & 0 & 0 & -39.5 & 38.5 & 4441.0 & & 0 & 0 & 0 \\
 & & & & & & & & & \ddots & & & \\
0 & 0 & 0 & 0 & 0 & 0 & 0 & 0 & 0 & & 15111.8 & 147.7 & -143.8 \\
0 & 0 & 0 & 0 & 0 & 0 & 0 & 0 & 0 & & 147.7 & 12336.0 & 106.5 \\
0 & 0 & 0 & 0 & 0 & 0 & 0 & 0 & 0 & & -143.8 & 106.5 & 10667.8
\end{bmatrix}
$$

The system of observation Equations (16.16) is solved by least squares using Equation (10.35). This yields the most probable values for the coordinates of the unknown stations. A complete output for the example of Figure 16.1, obtained using program ADJUST follows.

```
***************
Control Stations
***************
```

Station	X	Y	Z
A	402.3509	−4652995.3011	4349760.7775
B	8086.0318	−4642712.8474	4360439.0833

```
***************
Distance Vectors
***************
```

From	To	ΔX	ΔY	ΔZ	Covariance Matrix Element					
A	C	11644.2232	3601.2165	3399.2550	9.884E-4	−9.580E-6	9.520E-6	9.377E-4	−9.520E-6	9.827E-
A	E	−5321.7164	3634.0754	3173.6652	2.158E-4	−2.100E-6	2.160E-6	1.919E-4	−2.100E-6	2.005E-
B	C	3960.5442	−6681.2467	−7279.0148	2.305E-4	−2.230E-6	2.070E-6	2.546E-4	−2.230E-6	2.252E-
B	D	−11167.6076	−394.5204	−907.9593	2.700E-4	−2.750E-6	2.850E-6	2.721E-4	−2.720E-6	2.670E-
D	C	15128.1647	−6286.7054	−6371.0583	1.461E-4	−1.430E-6	1.340E-6	1.614E-4	−1.440E-6	1.308E-
D	E	−1837.7459	−6253.8534	−6596.6697	1.231E-4	−1.190E-6	1.220E-6	1.277E-4	−1.210E-6	1.283E-
F	A	−1116.4523	−4596.1610	−4355.9062	7.475E-5	−7.900E-7	8.800E-7	6.593E-5	−8.100E-7	7.616E-
F	C	10527.7852	−994.9377	−956.6246	2.567E-4	−2.250E-6	2.400E-6	2.163E-4	−2.270E-6	2.397E-
F	E	−6438.1364	−962.0694	−1182.2305	9.442E-5	−9.200E-7	1.040E-6	9.959E-5	−8.900E-7	8.826E-
F	D	−4600.3787	5291.7785	5414.4311	9.330E-5	−9.900E-7	9.000E-7	9.875E-5	−9.900E-7	1.204E-
F	B	6567.2311	5686.2926	6322.3917	6.643E-5	−6.500E-7	6.900E-7	7.465E-5	−6.400E-7	6.048E-
B	F	−6567.2310	−5686.3033	−6322.3807	5.512E-5	−6.300E-7	6.100E-7	7.472E-5	−6.300E-7	6.629E-
A	F	1116.4577	4596.1553	4355.9141	6.619E-5	−8.000E-7	9.000E-7	8.108E-5	−8.200E-7	9.376E-

NORMAL MATRIX

16093.0	148.0	157.3	0	0	0	−6845.9	−60.0	69.4	−3896.2	−40.1	38.6
148.0	15811.5	159.7	0	0	0	−60.0	−6195.0	−67.6	−40.1	−4622.2	−43.4
−157.3	159.7	17273.4	0	0	0	69.4	−67.6	−7643.2	38.6	−43.4	−4171.5
0	0	0	23352.1	221.9	−249.8	−8124.3	−75.0	76.5	−10593.3	−96.8	123.8
0	0	0	221.9	23084.9	227.3	−75.0	−7832.2	−73.2	−96.8	−10043.0	−100.1
0	0	0	−249.8	227.3	24116.4	76.5	−73.2	−7795.7	123.8	−100.1	−11332.6
−6845.9	−60.0	69.4	−8124.3	−75.0	76.5	29393.8	278.7	−264.4	−10720.0	−106.7	79.2
−60.0	−6195.0	−67.6	−75.0	−7832.2	−73.2	278.7	27831.6	260.2	−106.7	−10128.5	−82.5
69.4	−67.6	−7643.2	76.5	−73.2	−7795.7	−264.4	260.2	27487.5	79.2	−82.5	−8303.5
−3896.2	−40.1	38.6	−10593.3	−96.8	123.8	−10720.0	−106.7	79.2	86904.9	830.9	−874.3
−40.1	−4622.2	−43.4	−96.8	−10043.0	−100.1	−106.7	−10128.5	−82.5	830.9	79004.9	758.1
38.6	−43.4	4171.5	123.8	−100.1	−11332.6	79.2	−82.5	−8303.5	−874.3	758.1	79234.9

CONSTANT MATRIX

−227790228.2336
−23050461170.3104
23480815458.7631
−554038059.5699
−24047087640.5196
21397654262.6187
−491968929.7795
−16764436256.9406
16302821193.7660
−5314817965.0200
−250088821080.1534
238833986827.2840

X MATRIX

12046.5808
−4649394.0826
4353160.0644
−4919.3391
−4649361.2199
4352934.4548
−3081.5831
−4643107.3692
4359531.1233
1518.8012
−4648399.1453
4354116.6914

Degrees of Freedom = 27
Reference Variance = 0.5005
Reference So = ±0.71

```
***************************
Distance Vectors and Residuals
***************************
```

From	To	ΔX	ΔY	ΔZ	Vx	Vy	Vz
A	C	11644.2232	3601.2165	3399.2550	0.00669	0.00203	0.03190
A	E	−5321.7164	3634.0754	3173.6652	0.02645	0.00582	0.01207
B	C	3960.5442	−6681.2467	−7279.0148	0.00478	0.01153	−0.00403
B	D	−11167.6076	−394.5204	−907.9593	−0.00731	−0.00136	−0.00063
D	C	15128.1647	−6286.7054	−6371.0583	−0.00081	−0.00801	−0.00060
D	E	−1837.7459	−6253.8534	−6596.6697	−0.01005	0.00268	0.00117
F	A	−1116.4523	−4596.1610	−4355.9062	0.00198	0.00524	−0.00768
F	C	10527.7852	−994.9377	−956.6246	−0.00563	0.00047	−0.00238
F	E	−6438.1364	−962.0694	−1182.2305	−0.00387	−0.00514	−0.00611
F	D	−4600.3787	5291.7785	5414.4311	−0.00561	−0.00232	0.00082
F	B	6567.2311	5686.2926	6322.3917	−0.00051	0.00534	0.00015
B	F	−6567.2310	−5686.3033	−6322.3807	0.00041	0.00536	−0.01115
A	F	1116.4577	4596.1553	4355.9141	−0.00738	0.00046	−0.00022

```
***************************
Advanced Statistical Values
***************************
```

From	To	±S	Slope Dist	Prec
A	C	0.0105	12,653.538	1,206,000
A	E	0.0090	7,183.255	794,000
B	C	0.0105	10,644.668	1,014,000
B	D	0.0087	11,211.408	1,282,000
D	C	0.0107	17,577.670	1,643,000
D	E	0.0096	9,273.836	961,000
F	A	0.0048	6,430.015	1,344,000
F	C	0.0104	10,617.871	1,020,000
F	E	0.0086	6,616.111	770,000
F	D	0.0083	8,859.035	1,067,000
F	B	0.0048	10,744.075	2,246,000
B	F	0.0048	10,744.075	2,246,000
A	F	0.0048	6,430.015	1,344,000

```
********************
Adjusted Coordinates
********************
```

Station	X	Y	Z	Sx	Sy	Sz
A	402.35087	−4,652,995.30109	4,349,760.77753			
B	8,086.03178	−4,642,712.84739	4,360,439.08326			
C	12,046.58076	−4,649,394.08256	4,353,160.06443	0.0061	0.0061	0.0060
E	−4,919.33908	−4,649,361.21987	4,352,934.45480	0.0052	0.0053	0.0052
D	−3,081.58313	−4,643,107.36915	4,359,531.12333	0.0049	0.0051	0.0051
F	1,518.80119	−4,648,399.14533	4,354,116.69141	0.0027	0.0028	0.0028

PROBLEMS

Note: For problems below requiring least-squares adjustment, if a computer program is not distinctly specified for use in the problem, it is expected that the least-squares algorithm will be solved using the program MATRIX which is included within the diskette supplied with the book.

16.1 Given the following geodetic coordinates, compute geocentric coordinate values for these points.

(a) latitude: 41°18′20.2541″ N
longitude: 76°00′57.0024″ W
height: 364.2408 m

(b) latitude: 43°15′53.0534″ N
longitude: 90°02′36.7203″ W
height: 229.0805 m

(c) latitude: 44°57′45.3603″ N
longitude: 68°12′56.2437″ W
height: 254.3612 m

(d) latitude: 34°58′06.8409″ N
longitude: 123°27′42.0462″ W
height: 564.2408 m

16.2 Given the following geocentric coordinates, compute geodetic coordinate values for these points.

(a) $X = -526{,}125.83563$, $Y = -5{,}572{,}467.69462$, $Z = 3{,}047{,}961.36059$

(b) $X = -1{,}676{,}936.11522$, $Y = -5{,}201{,}821.48801$, $Z = 3{,}279{,}096.22646$

(c) $X = -2{,}764{,}193.49966$, $Y = -3{,}927{,}689.72305$, $Z = 4{,}184{,}084.58852$

(d) $X = 1{,}159{,}583.24626$, $Y = -4{,}656{,}373.27436$, $Z = 4{,}188{,}092.95471$

16.3 Given the following GPS observations and geocentric control station coordinates to accompany Figure 16.5, what are the most probable coordinates for stations B and C using a weighted least-squares adjustment? (All data were collected with only two receivers.)

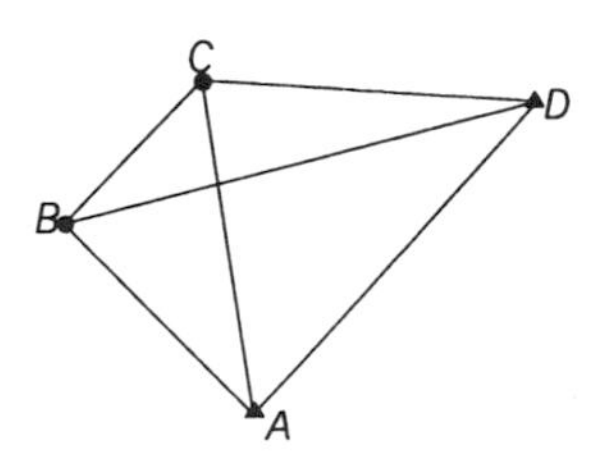

Figure 16.5

Control stations

ID	X	Y	Z
A	1,177,425.88739	−4,674,386.55849	4,162,989.78649
D	1,178,680.69374	−4,673.056.15318	4,164,169.65655

The baseline observations and upper diagonal elements of the covariance matrix for the ΔX, ΔY, and ΔZ values are as follows:

Baseline AB:

$\Delta X = -825.5585$	0.00002199	−0.00000030	0.00000030
$\Delta Y = 492.7369$		0.00002806	−0.00000030
$\Delta Z = 788.9732$			0.00003640

Baseline BC:

$\Delta X = 606.2113$	0.00003096	−0.00000029	0.00000040
$\Delta Y = 558.8905$		0.00002709	−0.00000029
$\Delta Z = 546.7241$			0.00002591

Baseline CD:

$\Delta X = 1474.1569$	0.00004127	−0.00000045	0.00000053
$\Delta Y = 278.7786$		0.00004315	−0.00000045
$\Delta Z = -155.8336$			0.00005811

Baseline AC:

$\Delta X = -219.3510$	0.00002440	−0.00000020	0.00000019
$\Delta Y = 1051.6348$		0.00001700	−0.00000019
$\Delta Z = 1335.6877$			0.00002352

Baseline DB:

$\Delta X = -2080.3644$	0.00003589	−0.00000034	0.00000036
$\Delta Y = -837.6605$		0.00002658	−0.00000033
$\Delta Z = -390.9075$			0.00002982

16.4 Given the following GPS observations and geocentric control station coordinates to accompany Figure 16.6, what are the most probable coordinates for stations B and C using a weighted least-squares adjustment? (All data were collected with only two receivers.)

Control stations

ID	X	Y	Z
A	593,898.8877	4,856,214.5456	4,078,710.7059
D	593,319.2704	−4,855,416.0310	4,079,738.3059

The vector covariance matrices for the ΔX, ΔY, and ΔZ values given are as follows:

Baseline AB:

$\Delta X = 678.034$	5.098E-6	−1.400E-5	6.928E-6
$\Delta Y = 1206.714$		7.440E-5	−3.445E-5
$\Delta Z = 1325.735$			2.018E-5

Baseline BC:

$\Delta X - -579.895$	3.404E-6	2.057E-6	−3.036E-7
$\Delta Y = 145.342$		2.015E-5	−1.147E-5
$\Delta Z = 254.820$			1.873E-5

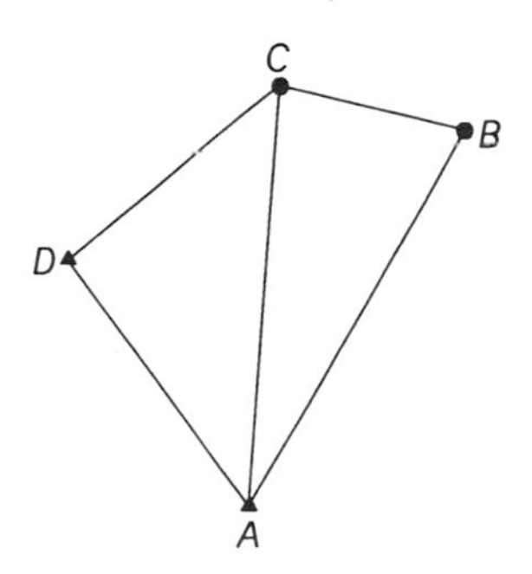

Figure 16.6

Baseline *AC*:

$\Delta X =$ 98.138	6.518E-6	−1.163E-7	−3.811E-7
$\Delta Y =$ 1352.039		3.844E-5	−1.297E-5
$\Delta Z =$ 1580.564			2.925E-5

Baseline *DC*:

$\Delta X =$ 677.758	9.347E-6	−1.427E-5	8.776E-6
$\Delta Y =$ 553.527		2.954E-5	−1.853E-5
$\Delta Z =$ 552.978			1.470E-5

Baseline *DC*:

$\Delta X =$ 677.756	9.170E-6	−1.415E-5	8.570E-6
$\Delta Y =$ 553.533		3.010E-5	−1.862E-5
$\Delta Z =$ 552.975			1.460E-5

16.5 Given the following GPS observations and geocentric control station coordinates to accompany Figure 16.7, what are the most probable coordinates for station *E* using a weighted least-squares adjustment? (All data were collected with only two receivers.)

Control stations

ID	*X*	*Y*	*Z*
A	−1,683,429.825	−4,245,726.842	4,439,764.293
B	−1,524,701.610	−4,230,122.822	4,511,075.501
C	−1,480,308.035	−4,472,815.181	4,287,476.008
D	−1,725,386.928	−4,436,015.964	4,234,036.124

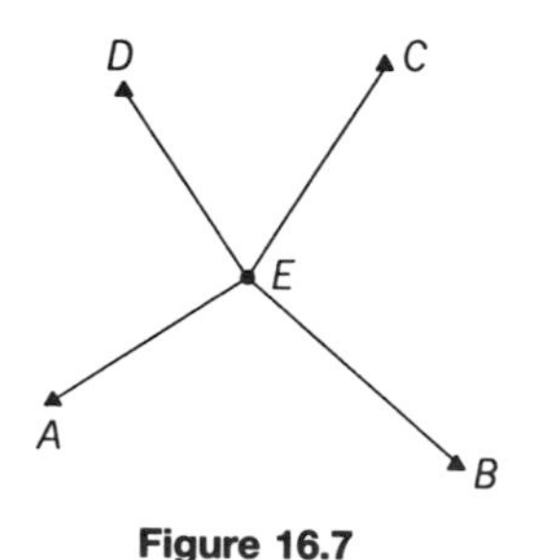

Figure 16.7

The baseline observations and upper diagonal elements of the covariance matrix for the ΔX, ΔY, and ΔZ values are as follows:

Baseline *AE*:

$\Delta X = 94{,}208.555$	0.00001287	−0.00000016	−0.00000019
$\Delta Y = -61{,}902.843$		0.00001621	−0.00000016
$\Delta Z = -24{,}740.272$			0.00001538

Baseline *BE*:

$\Delta X = -64{,}519.667$	0.00003017	−0.00000026	0.00000021
$\Delta Y = -77{,}506.853$		0.00002834	−0.00000025
$\Delta Z = -96{,}051.488$			0.00002561

Baseline *CE*:

$\Delta X = -108{,}913.237$	0.00008656	−0.00000081	−0.00000087
$\Delta Y = 165{,}185.492$		0.00007882	−0.00000080
$\Delta Z = 127{,}548.005$			0.00008647

Baseline *DE*:

$\Delta X = 136{,}165.650$	0.00005893	−0.00000066	−0.00000059
$\Delta Y = 128{,}386.277$		0.00006707	−0.00000064
$\Delta Z = 180{,}987.895$			0.00005225

Baseline *EA*:

$\Delta X = -94{,}208.554$	0.00002284	0.00000036	−0.00000042
$\Delta Y = 61{,}902.851$		0.00003826	−0.00000035
$\Delta Z = 24{,}740.277$			0.00003227

Baseline *EB*:

$\Delta X = 64{,}519.650$	0.00008244	0.00000081	−0.00000077
$\Delta Y = 77{,}506.866$		0.00007737	−0.00000081
$\Delta Z = 96{,}051.486$			0.00008483

Baseline *EC*:

$\Delta X = 108{,}913.236$	0.00002784	−0.00000036	0.00000038
$\Delta Y = -165{,}185.494$		0.00003396	−0.00000035
$\Delta Z = -127{,}547.991$			0.00002621

Baseline *ED*:

$\Delta X = -136{,}165.658$	0.00003024	−0.00000037	0.00000031
$\Delta Y = -128{,}386.282$		0.00003940	−0.00000036
$\Delta Z = -180{,}987.888$			0.00003904

16.6 Given the following GPS observations and geocentric control station coordinates to accompany Figure 16.8, what are the most probable coordinates for stations *B*, *D*, and *E* using a weighted least-squares adjustment? (All data were collected with only two receivers.)

Control stations

ID	*X*	*Y*	*Z*
A	−1,439,383.018	−5,325,949.910	3,190,645.563
C	−1,454,936.177	−5,240,453.494	3,321,529.500

The baseline observations and upper diagonal elements of the covariance matrix for the ΔX, ΔY, and ΔZ values are as follows:

Baseline *AB*:

$\Delta X = -118{,}616.114$	8.145E-4	−7.870E-6	7.810E-6
$\Delta Y = 71{,}775.010$		7.685E-4	−7.820E-6
$\Delta Z = 62{,}170.130$			8.093E-4

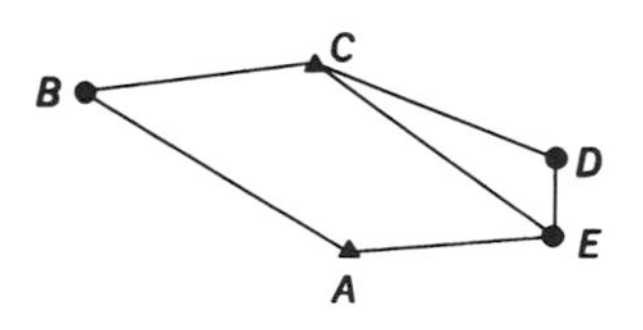

Figure 16.8

Baseline *BC*:

$\Delta X = 103{,}062.915$	8.521E-4	−8.410E-6	8.520E-6
$\Delta Y = 13{,}721.432$		8.040E-4	−8.400E-6
$\Delta Z = 68{,}713.770$			8.214E-4

Baseline *CD*:

$\Delta X = 106{,}488.952$	7.998E-4	−7.850E-6	7.560E-6
$\Delta Y = -41{,}961.364$		8.443E-4	−7.860E-6
$\Delta Z = -21{,}442.604$			7.900E-4

Baseline *DE*:

$\Delta X = -7.715$	3.547E-4	−3.600E-6	3.720E-6
$\Delta Y = -35{,}616.922$		3.570E-4	−3.570E-6
$\Delta Z = -57{,}297.941$			3.512E-4

Baseline *EA*:

$\Delta X = -90{,}928.118$	8.460E-4	−8.380E-6	8.160E-6
$\Delta Y = -7{,}918.120$		8.824E-4	−8.420E-6
$\Delta Z = -52{,}143.439$			8.088E-4

Baseline *CE*:

$\Delta X = 106{,}481.283$	7.341E-4	−7.250E-6	7.320E-6
$\Delta Y = -77{,}578.306$		7.453E-4	−7.290E-6
$\Delta Z = -78{,}740.573$			7.467E-4

16.7 Given the following GPS observations and geocentric control station coordinates to accompany Figure 16.9, what are the most probable coordinates for stations *B*, *D*, *E*, and *F* using a weighted least-squares adjustment? (All data were collected with only two receivers.)

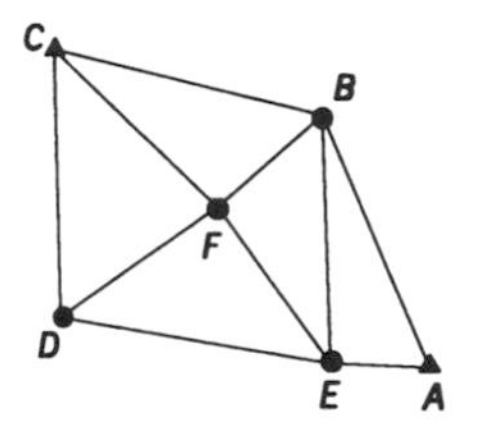

Figure 16.9

Control stations

ID	X	Y	Z
A	−1,612,062.639	−4,384,804.866	4,330,846.142
C	−1,613,505.053	−4,383,572.785	4,331,494.264

The baseline observations and upper diagonal elements of the covariance matrix for the ΔX, ΔY, and ΔZ values are as follows:

Baseline AB:

$\Delta X = -410.891$	7.064E-5	−6.500E-7	6.400E-7
$\Delta Y = 979.896$		6.389E-5	−6.400E-7
$\Delta Z = 915.452$			6.209E-5

Baseline BC:

$\Delta X = -1031.538$	1.287E-5	−1.600E-7	1.900E-7
$\Delta Y = 252.184$		1.621E-5	−1.600E-7
$\Delta Z = -267.337$			1.538E-5

Baseline CD:

$\Delta X = 23.227$	1.220E-5	−9.000E-8	7.000E-8
$\Delta Y = -1035.622$		1.104E-5	−9.000E-8
$\Delta Z = -722.122$			9.370E-6

Baseline *DE*:

$$\begin{matrix} \Delta X = & 1039.772 & 5.335E\text{-}5 & -4.900E\text{-}7 & 5.400E\text{-}7 \\ \Delta Y = & -178.623 & & 4.731E\text{-}5 & -4.800E\text{-}7 \\ \Delta Z = & -3.753 & & & 5.328E\text{-}5 \end{matrix}$$

Baseline *EF*:

$$\begin{matrix} \Delta X = & -434.125 & 7.528E\text{-}5 & -8.300E\text{-}7 & 7.500E\text{-}7 \\ \Delta Y = & 603.781 & & 8.445E\text{-}5 & -8.100E\text{-}7 \\ \Delta Z = & 566.518 & & & 6.771E\text{-}5 \end{matrix}$$

Baseline *EB*:

$$\begin{matrix} \Delta X = & -31.465 & 3.340E\text{-}5 & -4.900E\text{-}7 & 5.600E\text{-}7 \\ \Delta Y = & 962.049 & & 5.163E\text{-}5 & -4.800E\text{-}7 \\ \Delta Z = & 993.212 & & & 4.463E\text{-}5 \end{matrix}$$

Baseline *FA*:

$$\begin{matrix} \Delta X = & 813.523 & 9.490E\text{-}6 & -9.000E\text{-}8 & 8.000E\text{-}8 \\ \Delta Y = & -621.619 & & 7.820E\text{-}6 & -9.000E\text{-}8 \\ \Delta Z = & -488.768 & & & 1.031E\text{-}5 \end{matrix}$$

Baseline *FB*:

$$\begin{matrix} \Delta X = & 402.650 & 1.073E\text{-}5 & -1.600E\text{-}7 & 1.800E\text{-}7 \\ \Delta Y = & 358.278 & & 1.465E\text{-}5 & -1.600E\text{-}7 \\ \Delta Z = & 426.706 & & & 9.730E\text{-}6 \end{matrix}$$

Baseline *FC*:

$$\begin{matrix} \Delta X = & -628.888 & 5.624E\text{-}5 & -6.600E\text{-}7 & 5.700E\text{-}7 \\ \Delta Y = & 610.467 & & 6.850E\text{-}5 & -6.300E\text{-}7 \\ \Delta Z = & 159.360 & & & 6.803E\text{-}5 \end{matrix}$$

Baseline FD:

$$\begin{array}{llll} \Delta X = -605.648 & 8.914\text{E-5} & -8.100\text{E-7} & 8.200\text{E-7} \\ \Delta Y = -425.139 & & 8.164\text{E-5} & -8.100\text{E-7} \\ \Delta Z = -562.763 & & & 7.680\text{E-5} \end{array}$$

16.8 Given the following GPS observations and geocentric control station coordinates to accompany Figure 16.10, what are the most probable coordinates for stations B, D, E, and F using a weighted least-squares adjustment? (All data were collected with only two receivers.)

Control stations

ID	X	Y	Z
A	−2,413,963.823	−4,395,420.994	3,930,059.456
C	−2,413,073.302	−4,393,796.994	3,932,699.132

The baseline observations and the upper diagonal elements of the covariance matrix for the ΔX, ΔY, and ΔZ values are as follows:

Baseline AB:

$$\begin{array}{llll} \Delta X = 535.100 & 4.950\text{E-6} & -9.000\text{E-8} & 7.000\text{E-8} \\ \Delta Y = 974.318 & & 7.690\text{E-6} & -9.000\text{E-8} \\ \Delta Z = 1173.264 & & & 8.090\text{E-6} \end{array}$$

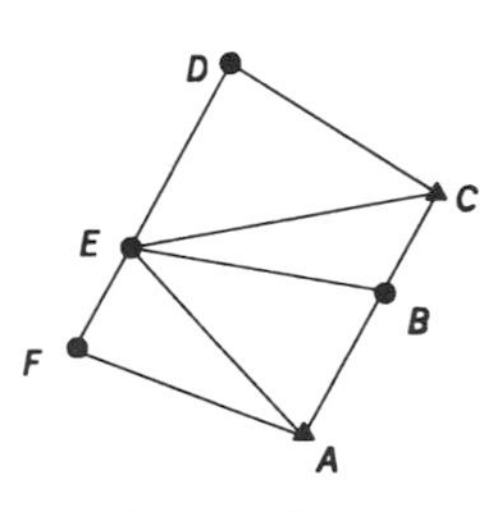

Figure 16.10

Baseline *BC*:

$\Delta X = 355.412$	5.885E-5	−6.300E-7	7.400E-7
$\Delta Y = 649.680$		7.168E-5	−6.500E-7
$\Delta Z = 1466.409$			6.650E-5

Baseline *CD*:

$\Delta X = -1368.545$	6.640E-6	−4.000E-8	7.000E-8
$\Delta Y = 854.284$		4.310E-6	−4.000E-8
$\Delta Z = -71.080$			2.740E-6

Baseline *DE*:

$\Delta X = -671.715$	1.997E-5	−2.500E-7	2.500E-7
$\Delta Y = -1220.263$		2.171E-5	−2.400E-7
$\Delta Z = -951.343$			3.081E-5

Baseline *EF*:

$\Delta X = -374.515$	4.876E-5	−3.600E-7	3.400E-7
$\Delta Y = -679.553$		2.710E-5	−3.700E-7
$\Delta Z = -1439.338$			3.806E-5

Baseline *EA*:

$\Delta X = 1149.724$	8.840E-5	−8.000E-7	8.300E-7
$\Delta Y = -1258.018$		7.925E-5	−8.200E-7
$\Delta Z = -1617.250$			6.486E-5

Baseline *EB*:

$\Delta X = 1684.833$	1.861E-5	−1.600E-7	2.000E-7
$\Delta Y = -283.698$		1.695E-5	−1.600E-7
$\Delta Z = -443.990$			1.048E-5

Baseline *EC*:

$\Delta X = 2040.254$	6.966E-5	−6.400E-7	7.300E-7
$\Delta Y = 365.991$		5.665E-5	−6.300E-7
$\Delta Z = 1022.430$			7.158E-5

Baseline *FA*:

$\Delta X = 1524.252$	2.948E-5	−3.500E-7	3.300E-7
$\Delta Y = -578.473$		3.380E-5	−3.500E-7
$\Delta Z = -177.914$			4.048E-5

Given the data in each problem and using the procedure discussed in Section 16.7.2, analyze any repeat baselines.

16.9 Problem 16.4

16.10 Problem 16.5

Given the data in each problem and using the procedure discussed in Section 16.7.3, analyze the closures of the loops.

16.11 Problem 16.3, loops *ABCDA*, *ABCA*, *ACDA*, and *BCDB*

16.12 Problem 16.4, loops *ABCDA*, *ACBA*, and *ADCA*

16.13 Problem 16.6, loops *ABCDEA* and *CDEC*

16.14 Problem 16.7, loops *ABEA*, *DEFD*, *BFCB*, *CDFC*, and *ABCDEA*

16.15 Problem 16.8, loops *ABCDEFA*, *AECA*, and *BECB*

Use program ADJUST to do each problem.

16.16 Problem 16.1

16.17 Problem 16.2

16.18 Problem 16.3

16.19 Problem 16.4

16.20 Problem 16.5

16.21 Problem 16.6

16.22 Problem 16.7

16.23 Problem 16.8

16.24 Problem 16.11

16.25 Problem 16.12

16.26 Problem 16.13

16.27 Problem 16.14

16.28 Problem 16.15

Programming Problems

16.29 Write a program that reads a file of station coordinates and GPS baseline observations, and then:

(a) writes the data to a file in a formatted fashion,

(b) computes the *A*, *L*, and *W* matrices, and

(c) writes the matrices to a file that is compatible with the MATRIX program.

(d) Demonstrate this program with Problem 16.8.

16.30 Write a program that reads a file containing the *A*, *L*, and *W* matrices and then:

(a) writes these matrices in a formatted fashion,

(b) performs a weighted least-squares adjustment, and

(c) writes the matrices used to compute the solution and tabulates the station coordinates in a formatted fashion.

(d) Demonstrate this program with Problem 16.8.

16.31 Write a program that reads a file of station coordinates and GPS baseline observations, and then:

(a) writes the data to a file in a formatted fashion,

(b) computes the *A*, *L*, and *W* matrices,

(c) performs a weighted least-squares adjustment,

(d) writes the matrices used in computations in a formatted fashion to a file, and

(e) computes the final station coordinates, their estimated errors, the adjusted baseline vectors, their residuals, and their estimated errors, and writes them to a file in a formatted fashion.

(f) Demonstrate this program with Problem 16.8.

17

COORDINATE TRANSFORMATIONS

17.1 INTRODUCTION

The transformation of points from one coordinate system to another is a common problem encountered in surveying and mapping. For instance, a surveyor who works initially in an assumed coordinate system on a project may find it necessary to transfer the coordinates to the state plane coordinate system. In GPS surveying and in the field of photogrammetry, coordinate transformations are used extensively. Since the inception of the North American Datum of 1983 (NAD 83), many land surveyors, management agencies, state departments of transportation, and others have been struggling with the problem of converting their multitudes of stations defined in the 1927 datum (NAD 27) to the 1983 datum. Although several mathematical models have been developed to make these conversions, all involve some form of coordinate transformation. This chapter covers the introductory procedures of using least squares to compute several well-known and often used transformations. More rigorous procedures which employ the general least-squares method are given in Chapter 21.

17.2 TWO-DIMENSIONAL CONFORMAL COORDINATE TRANSFORMATION

The two-dimensional conformal coordinate transformation, also known as the *four-parameter similarity transformation*, has the characteristic that true shape is retained after transformation. It is typically used in surveying when converting separate surveys into a common reference coordinate system. This transformation is a three-step process that involves:

1. *Scaling* to create equal dimensions in the two coordinate systems
2. *Rotation* to make the reference axes of the two systems parallel
3. *Translations* to create a common origin for the two coordinate systems

Scaling and rotation are each defined by one parameter, and translation involves two parameters. Thus there are a total of four parameters in this transformation. The transformation requires a minimum of two points, called *control points*, that are common to both systems. With the minimum of two points, the four parameters of the transformation can be determined uniquely. If more than two control points are available, a least-squares adjustment is possible. After determining the values of the transformation parameters, any points in the original system may be transformed.

17.3 EQUATION DEVELOPMENT

Figure 17.1(*a*) and (*b*) illustrate two independent coordinate systems. In these systems, three common control points, *A*, *B*, and *C* exist (i.e., their coordinates are known in both systems). Points 1 through 4 have their coordinates known only in the *xy* system of Figure 17.1(*b*). The problem is to determine their *X* and *Y* coordinates in the system of Figure 17.1(*a*). The necessary equations are developed as follows:

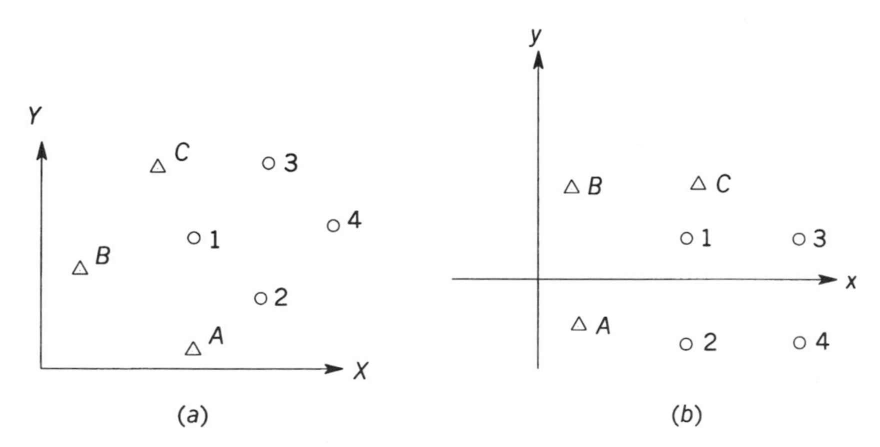

Figure 17.1 Two-dimensional coordinate systems.

Step 1: Scaling. To make line lengths as defined by the (x,y) coordinate system equal to their lengths in the (X,Y) system, it is necessary to multiply (x,y) coordinates by a scale factor, S. Thus scaled coordinates x' and y' are

$$
\begin{aligned}
x' &= Sx \\
y' &= Sy
\end{aligned}
\tag{17.1}
$$

Step 2: Rotation. In Figure 17.2 the (X,Y) coordinate system has been superimposed on the scaled (x',y') system. The rotation angle, θ, is shown between the y' and Y axes. To analyze the effects of this rotation, an (X',Y') system was constructed parallel to the (X,Y) system such that its origin is common with that of the $x'y'$ system. Expressions that give the (X',Y') rotated coordinates for any point (such as point 4 shown) in terms of its (x',y') coordinates are

$$
\begin{aligned}
X' &= x' \cos(\theta) - y' \sin(\theta) \\
Y' &= x' \sin(\theta) + y' \cos(\theta)
\end{aligned}
\tag{17.2}
$$

Step 3: Translation. To finally arrive at X and Y coordinates for a point, it is necessary to translate the origin of the (X',Y') system to the origin of the (X,Y) system. Referring to Figure 17.2, it can be seen that this translation is accomplished by adding translation factors as follows:

$$X = X' + T_X \quad \text{and} \quad Y = Y' + T_Y \tag{17.3}$$

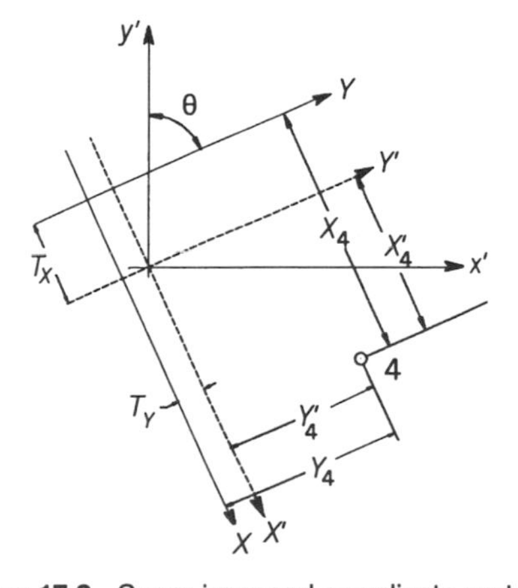

Figure 17.2 Super-imposed coordinate systems.

If Equations (17.1), (17.2), and (17.3) are combined, a single set of equations results that transform the points from (x,y) coordinates of Figure 17.1(b) directly into (X,Y) coordinates of Figure 17.1(a) as

$$
\begin{aligned}
X &= (S\cos\theta)x - (S\sin\theta)y + T_X \\
Y &= (S\sin\theta)x + (S\cos\theta)y + T_Y
\end{aligned}
\tag{17.4}
$$

Now let $S\cos\theta = a$, $S\sin\theta = b$, $T_X = c$, and $T_Y = d$, and add residuals to make redundant equations consistent. Then Equations (17.4) can be written as

$$
\begin{aligned}
ax - by + c &= X + v_X \\
ay + bx + d &= Y + v_Y
\end{aligned}
\tag{17.5}
$$

17.4 APPLICATION OF LEAST SQUARES

Equations (17.5) represent the basic observation equations for a two-dimensional conformal coordinate transformation that have four unknowns: a, b, c, and d. The four unknowns embody the transformation parameters S, θ, T_x, and T_y. Since two equations can be written for every control point, only two control points are needed for a unique solution. When more than two are present, a redundant system exists for which a least-squares solution can be found. As an example, consider the equations that could be written for the situation illustrated in Figure 17.1. There are three control points, A, B, and C, and thus the following six equations can be written:

$$
\begin{aligned}
ax_a - by_a + c &= X_A + v_{X_A} \\
ay_a + bx_a + d &= Y_A + v_{Y_A} \\
ax_b - by_b + c &= X_B + v_{X_B} \\
ay_b + bx_b + d &= Y_B + v_{Y_B} \\
ax_c - by_c + c &= X_C + v_{X_C} \\
ay_c + bx_c + d &= Y_C + v_{Y_C}
\end{aligned}
\tag{17.6}
$$

Equations (17.6) can be expressed in matrix form as

$$AX = L + V \tag{17.7}$$

where

$$A = \begin{bmatrix} x_a & -y_a & 1 & 0 \\ y_a & x_a & 0 & 1 \\ x_b & -y_b & 1 & 0 \\ y_b & x_b & 0 & 1 \\ x_c & -y_c & 1 & 0 \\ y_c & x_c & 0 & 1 \end{bmatrix} \qquad X = \begin{bmatrix} a \\ b \\ c \\ d \end{bmatrix} \qquad L = \begin{bmatrix} X_A \\ Y_A \\ X_B \\ Y_B \\ X_C \\ Y_C \end{bmatrix} \qquad V = \begin{bmatrix} v_{X_A} \\ v_{Y_A} \\ v_{X_B} \\ v_{Y_B} \\ v_{X_C} \\ v_{Y_C} \end{bmatrix}$$

The redundant system can be solved using Equation (10.32). Having obtained the most probable values for the coefficients from the least-squares solution, the (X,Y) coordinates of any additional points whose coordinates are known in the (x,y) system can then be obtained by applying Equations (17.5) (where the residuals are now considered to be zeros).

After the adjustment, the scale factor S and rotation angle θ can be computed with the following equations.

$$\begin{aligned} \theta &= \tan^{-1}\left(\frac{b}{a}\right) \\ S &= \frac{a}{\cos(\theta)} \end{aligned} \tag{17.8}$$

Example 17.1 A survey conducted in an arbitrary (x,y) coordinate system produced station coordinates for A, B, and C as well as for stations 1 through 4. Stations A, B, and C also have known state plane easting and northing coordinates. It is required to derive the state plane coordinates of stations 1 through 4. Table 17.1 has a tabulation of the arbitrary coordinates and state plane coordinates.

Table 17.1 Data for Example 17.1

Point	Easting	Northing	x	y
A	1,049,422.40	51,089.20	121.622	−128.066
B	1,049,413.95	49,659.30	141.228	187.718
C	1,049,244.95	49,884.95	175.802	135.728
1			174.148	−120.262
2			513.520	−192.130
3			754.444	−67.706
4			972.788	120.994

SOLUTION: A computer listing from program ADJUST is given below for the problem. The output includes the input data, the coordinates of transformed points, the transformation coefficients, and their standard deviations. Note that the program formed the A and L matrices in accordance with Equation (17.7). After obtaining the solution using Equation (10.32), the program solved Equation (17.8) to obtain the rotation angle and scale factor of the transformation.

```
Two Dimensional Conformal Coordinate Transformation of File >
-------------------------------------------------------------
ax - by + Tx = X + VX
bx + ay + Ty = Y + VY

A MATRIX                                                    L MATRIX
-------------------------------------------------          -----------
 121.622        128.066        1.000        0.000        1049422.400
-128.066        121.622        0.000        1.000          51089.200
 141.228       -187.718        1.000        0.000        1049413.950
 187.718        141.228        0.000        1.000          49659.300
 175.802       -135.728        1.000        0.000        1049244.950
 135.728        175.802        0.000        1.000          49884.950

Control Points
     Point            X                Y           VX         VY
----------------------------------------------------------------
         A   1,049,422.400   51,089.200   -0.004    0.029
         B   1,049,413.950   49,659.300   -0.101    0.077
         C   1,049,244.950   49,884.950    0.105   -0.106

Transformation Parameters:
  a =      -4.51249 ±       0.00058
  b =      -0.25371 ±       0.00058
 Tx =   1050003.715 ±         0.123
 Ty =     50542.131 ±         0.123

Rotation = 183° 13′ 05.0″
   Scale = 4.51962
Adjustment's Reference Variance = 0.0195

Transformed Points
     Point            X                Y           ±Sx        ±Sy
-----------------------------------------------------------------
         1   1,049,187.361   51,040.629   0.173    0.173
         2   1,047,637.713   51,278.829   0.339    0.339
         3   1,046,582.113   50,656.241   0.453    0.453
         4   1,045,644.713   49,749.336   0.578    0.578

Qxx MATRIX
-----------------------------------------------
 0.00001699  0.00000000 -0.00248493 -0.00110681
 0.00000000  0.00001699  0.00110681 -0.00248493
-0.00248493  0.00110681  0.76875593 -0.00000000
-0.00110681 -0.00248493 -0.00000000  0.76875593
```

17.5 TWO-DIMENSIONAL AFFINE COORDINATE TRANSFORMATION

The two-dimensional affine coordinate transformation is also known as the *six-parameter transformation.* It is a slight variation from the two-dimensional conformal transformation. In the affine transformation there is the additional allowance for two different scale factors, one in the x direction and the other in the y direction. This transformation is commonly used in photogrammetry to transform photo coordinates from an arbitrary measurement photo coordinate system to the camera fiducial system, and to account for differential shrinkages that occur in the x and y directions. As in the conformal transformation, the affine transformation also applies two translations of the origin, and a rotation about the origin, plus a small nonorthogonality correction between the x and y axes. This results in a total of six unknowns. The mathematical model for the affine transformation is

$$\begin{aligned} ax + by + c &= X + V_X \\ dx + ey + f &= Y + V_Y \end{aligned} \tag{17.9}$$

These equations are linear and can be solved uniquely when three control points exist (i.e., points whose coordinates are known in the both systems). This is because for each point, an equation set in the form of Equations (17.9) can be written and three points yields 6 equations involving the 6 unknowns. If more than three control points are available, a least-squares solution can be obtained. Assume, for example, that four common points (1, 2, 3, 4) exist. Then the equation system would be

$$\begin{aligned} ax_1 + by_1 + c &= X_1 + V_{X1} \\ dx_1 + ey_1 + f &= Y_1 + V_{Y1} \\ ax_2 + by_2 + c &= X_2 + V_{X2} \\ dx_2 + ey_2 + f &= Y_2 + V_{Y2} \\ ax_3 + by_3 + c &= X_3 + V_{X3} \\ dx_3 + ey_3 + f &= Y_3 + V_{Y3} \\ ax_4 + by_4 + c &= X_4 + V_{X4} \\ dx_4 + ey_4 + f &= Y_4 + V_{Y4} \end{aligned} \tag{17.10}$$

In matrix notation, Equations (17.10) are expressed as $AX = L + V$, where

$$
\begin{bmatrix} x_1 & y_1 & 1 & 0 & 0 & 0 \\ 0 & 0 & 0 & x_1 & y_1 & 1 \\ x_2 & y_2 & 1 & 0 & 0 & 0 \\ 0 & 0 & 0 & x_2 & y_2 & 1 \\ x_3 & y_3 & 1 & 0 & 0 & 0 \\ 0 & 0 & 0 & x_3 & y_3 & 1 \\ x_4 & y_4 & 1 & 0 & 0 & 0 \\ 0 & 0 & 0 & x_4 & y_4 & 1 \end{bmatrix} \begin{bmatrix} a \\ b \\ c \\ d \\ e \\ f \end{bmatrix} = \begin{bmatrix} X_1 \\ Y_1 \\ X_2 \\ Y_2 \\ X_3 \\ Y_3 \\ X_4 \\ Y_4 \end{bmatrix} + \begin{bmatrix} v_{X_1} \\ v_{Y_1} \\ v_{X_2} \\ v_{Y_2} \\ v_{X_3} \\ v_{Y_3} \\ v_{X_4} \\ v_{Y_4} \end{bmatrix} \tag{17.11}
$$

The most probable values for the unknown parameters are computed using least squares Equation (10.32). They are then used to transfer the remaining points from the (x,y) coordinate system to the (X,Y) coordinate system.

Example 17.2 Photo coordinates, which have been measured using a digitizer, must be transformed into the camera's fiducial coordinate system. The four fiducial points and two additional points were measured in the digitizer's (x,y) coordinate system and are listed in Table 17.2 together with the known camera fiducial coordinates, (X,Y).

SOLUTION: A self-explanatory computer solution from the program ADJUST which gives the least-squares solution for an affine transformation is shown below.

```
Two Dimensional Affine Coordinate Transformation of File > EX17.2.DAT
---------------------------------------------------------------
ax + by + c = X + VX
dx + ey + f = Y + VY
                        A matrix                                     L matrix
---------------------------------------------------------------    -------------
    0.764      5.960     1.000     0.000      0.000     0.000          -113.000
    0.000      0.000     0.000     0.764      5.960     1.000             0.003
    5.062     10.541     1.000     0.000      0.000     0.000             0.001
    0.000      0.000     0.000     5.062     10.541     1.000           112.993
    9.663      6.243     1.000     0.000      0.000     0.000           112.998
    0.000      0.000     0.000     9.663      6.243     1.000             0.003
    5.350      1.654     1.000     0.000      0.000     0.000             0.001
    0.000      0.000     0.000     5.350      1.654     1.000          -112.999
```

Table 17.2 Coordinates of points for Example 17.2

Point	X	Y	x	y	Sx	Sy
1	−113.000	0.003	0.764	5.960	0.026	0.028
3	0.001	112.993	5.062	10.541	0.024	0.030
5	112.998	0.003	9.663	6.243	0.028	0.022
7	0.001	−112.999	5.350	1.654	0.024	0.026
306			1.746	9.354		
307			5.329	9.463		

```
Control Points
      Point          X              Y            VX           VY
------------------------------------------------------------------
          1      -113.000         0.003        0.101        0.049
          3         0.001       112.993       -0.086       -0.057
          5       112.998         0.003        0.117        0.030
          7         0.001      -112.999       -0.086       -0.043

Transformation Parameters:
 a =       25.37152 ± 0.02532
 b =        0.82220 ± 0.02256
 c =       -137.183 ± 0.203
 d =       -0.80994 ± 0.02335
 e =       25.40166 ± 0.02622
 f =       -150.723 ± 0.216

Adjustment's Reference Variance = 34.9248

Transformed Points
      Point          X               Y            ±Sx          ±Sy
------------------------------------------------------------------
        306       -85.193         85.470        0.296        0.330
        307         5.803         85.337        0.324        0.352

Qxx MATRIX
------------------------------------------------------------------
 0.00001836  0.00000013 -0.00009377  0.00000000  0.00000000  0.00000000
 0.00000013  0.00001458 -0.00008950  0.00000000  0.00000000  0.00000000
-0.00009377 -0.00008950  0.00118089  0.00000000  0.00000000  0.00000000
 0.00000000  0.00000000  0.00000000  0.00001561 -0.00000042 -0.00008828
 0.00000000  0.00000000  0.00000000 -0.00000042  0.00001968 -0.00011256
 0.00000000  0.00000000  0.00000000 -0.00008828 -0.00011256  0.00133931
```

17.6 TWO-DIMENSIONAL PROJECTIVE COORDINATE TRANSFORMATION

The two-dimensional projective coordinate transformation is also known as the *eight-parameter transformation.* It is appropriate to use when one two-dimensional coordinate system is projected onto another nonparallel system. This transformation is commonly used in photogrammetry, and it can also be used to transform NAD 27 coordinates into the NAD 83 system. In their final form, the two-dimensional projective coordinate transformation equations are

$$X = \frac{a_1x + b_1y + c_1}{a_3x + b_3y + 1}$$
$$Y = \frac{a_2x + b_2y + c_2}{a_3x + b_3y + 1} \tag{17.12}$$

Upon inspection, it can be seen that these equations are similar to the affine transformation. In fact, if a_3 and b_3 were equal to zero, these equations are the affine transformation. With eight unknowns, this transformation requires a minimum of four control points (points having coordinates known in both systems). If there are more than four control points, the least-squares solution can be used. Since these are nonlinear equations, they must be linearized and solved using Equation (10.37) or (10.39). The linearized form of these equations is

$$\begin{bmatrix} \left(\frac{\partial X}{\partial a_1}\right)_0 & \left(\frac{\partial X}{\partial b_1}\right)_0 & \left(\frac{\partial X}{\partial c_1}\right)_0 & 0 & 0 & 0 & \left(\frac{\partial X}{\partial a_3}\right)_0 & \left(\frac{\partial X}{\partial b_3}\right)_0 \\ 0 & 0 & 0 & \left(\frac{\partial Y}{\partial a_2}\right)_0 & \left(\frac{\partial Y}{\partial b_2}\right)_0 & \left(\frac{\partial Y}{\partial c_2}\right)_0 & \left(\frac{\partial Y}{\partial a_3}\right)_0 & \left(\frac{\partial Y}{\partial b_3}\right)_0 \end{bmatrix} \begin{bmatrix} da_1 \\ db_1 \\ dc_1 \\ da_2 \\ db_2 \\ dc_2 \\ da_3 \\ db_3 \end{bmatrix} = \begin{bmatrix} X - X_0 \\ Y - Y_0 \end{bmatrix} \tag{17.13}$$

where

$$\frac{\partial X}{\partial a_1} = \frac{x}{a_3 x + b_3 y + 1} \quad \frac{\partial X}{\partial b_1} = \frac{y}{a_3 x + b_3 y + 1} \quad \frac{\partial X}{\partial c_1} = \frac{1}{a_3 x + b_3 y + 1}$$

$$\frac{\partial Y}{\partial a_2} = \frac{x}{a_3 x + b_3 y + 1} \quad \frac{\partial Y}{\partial b_2} = \frac{y}{a_3 x + b_3 y + 1} \quad \frac{\partial Y}{\partial c_2} = \frac{1}{a_3 x + b_3 y + 1}$$

$$\frac{\partial X}{\partial a_3} = -\frac{a_1 x + b_1 y + c_1}{(a_3 x + b_3 y + 1)^2} x \qquad \frac{\partial X}{\partial b_3} = -\frac{a_1 x + b_1 y + c_1}{(a_3 x + b_3 y + 1)^2} y$$

$$\frac{\partial Y}{\partial a_3} = -\frac{a_2 x + b_2 y + c_2}{(a_3 x + b_3 y + 1)^2} y \qquad \frac{\partial Y}{\partial a_3} = -\frac{a_2 x + b_2 y + c_2}{(a_3 x + b_3 y + 1)^2} y$$

For each control point, a set of equations of the form of Equation (17.13) can be written. A redundant system of equations can be solved by least squares to yield the eight unknown parameters. With these values known, the remaining points in the (x,y) coordinate system are transformed into the (X,Y) system.

Example 17.3 Given the data in Table 17.3, determine the best-fit projective transformation parameters and use them to transform points 7 and 8 into the (X,Y) coordinate system.

Table 17.3 Data for Example 17.3

Point	*X*	*Y*	*x*	*y*	*Sx*	*Sy*
1	1420.407	895.362	90.0	90.0	0.3	0.3
2	895.887	351.398	50.0	40.0	0.3	0.3
3	−944.926	641.434	−30.0	20.0	0.3	0.3
4	968.084	−1384.138	50.0	−40.0	0.3	0.3
5	1993.262	−2367.511	110.0	−80.0	0.3	0.3
6	−3382.284	3487.762	−100.0	80.0	0.3	0.3
7			−60.0	20.0	0.3	0.3
8			−10.0	−10.0	0.3	0.3

SOLUTION: The program ADJUST was used to solve this problem and the results follow.

```
Two Dimensional Projective Coordinate Transformation of File
-------------------------------------------------------------
a1x + b1y + c1
---------------  = X + VX
a3x + b3y + 1

a2x + b2y + c2
---------------  = Y + VY
a3x + b3y + 1
```

Iteration 1

```
J MATRIX                                                                  K MATRIX
-----------------------------------------------------------------         --------
  58.445  58.445 0.649    0.000    0.000 0.000  -83007.064  -83007.064       0.158
   0.000   0.000 0.000   58.445   58.445 0.649  -52334.927  -52334.927      -0.088
  39.064  31.251 0.781    0.000    0.000 0.000  -35012.162  -28009.729      -0.388
   0.000   0.000 0.000   39.064   31.251 0.781  -13719.422  -10975.538       0.195
 -32.608  21.739 1.087    0.000    0.000 0.000  -30791.646   20527.764      -0.636
   0.000   0.000 0.000  -32.608   21.739 1.087   20924.176  -13949.451      -0.250
  44.644 -35.715 0.893    0.000    0.000 0.000  -43186.235   34548.988       0.729
   0.000   0.000 0.000   44.644  -35.715 0.893   61787.575  -49430.060      -0.120
  85.941 -62.502 0.781    0.000    0.000 0.000 -171318.968  124595.613      -0.198
   0.000   0.000 0.000   85.941  -62.502 0.781  203472.866 -147980.266       0.090
-131.572 105.258 1.316    0.000    0.000 0.000 -445062.971  356050.376       0.362
   0.000   0.000 0.000 -131.572  105.258 1.316  458870.304 -367096.243       0.175

X MATRIX
------------
 0.0027039323
 0.0035789799
-0.0156917169
 0.0014774310
 0.0031559533
 0.0018094254
 0.0000022203
 0.0000030887
```

Iteration 2

J MATRIX

58.427	58.427	0.649	0.000	0.000	0.000	-82976.429	-82976.429
0.000	0.000	0.000	58.427	58.427	0.649	-52318.350	-52318.350
39.057	31.246	0.781	0.000	0.000	0.000	-35007.346	-28005.877
0.000	0.000	0.000	39.057	31.246	0.781	-13720.555	-10976.444
-32.608	21.739	1.087	0.000	0.000	0.000	-30792.864	20528.576
0.000	0.000	0.000	-32.608	21.739	1.087	20925.127	-13950.084
44.644	-35.715	0.893	0.000	0.000	0.000	-43186.259	34549.007
0.000	0.000	0.000	44.644	-35.715	0.893	61790.974	-49432.779
85.941	-62.502	0.781	0.000	0.000	0.000	-171319.428	124595.948
0.000	0.000	0.000	85.941	-62.502	0.781	203479.697	-147985.234
-131.568	105.254	1.316	0.000	0.000	0.000	-445033.570	356026.856
0.000	0.000	0.000	-131.568	105.254	1.316	458858.476	-367086.781

K MATRIX

0.242
-0.082
-0.429
0.102
-0.603
-0.276
0.739
-0.059
-0.199
0.165
0.250
0.150

X MATRIX

0.0000030015
0.0000036563
0.0000215690
0.0000021650
0.0000030314
-0.0000339827
0.0000000027
0.0000000033

```
Transformation Parameters:
a1 =     25.00274 ± 0.01538
b1 =      0.80064 ± 0.01896
c1 =    -134.715 ± 0.377
a2 =    -8.00771 ± 0.00954
b2 =    24.99811 ± 0.01350
c2 =    -149.815 ± 0.398
a3 =     0.00400 ± 0.00001
b3 =     0.00200 ± 0.00001

Adjustment's Reference Variance = 3.8888
            Number of Iterations = 2
```

Control Points

Point	X	Y	VX	VY
1	1,420.165	895.444	-0.242	0.082
2	896.316	351.296	0.429	-0.102
3	-944.323	641.710	0.603	0.276
4	967.345	-1,384.079	-0.739	0.059
5	1,993.461	-2,367.676	0.199	-0.165
6	-3,382.534	3,487.612	-0.250	-0.150

Transformed Points

Point	X	Y	$\pm Sx$	$\pm Sy$
7	-2,023.678	1,038.310	2.273	1.329
8	-417.837	-340.142	0.484	0.463

```
Qxx MATRIX
----------------------------------------------------------------------------------------------------------
 0.0000608536  0.0000722729 -0.0002616333  0.0000291975  0.0000443578 -0.0001538108  0.0000000412  0.0000000534
 0.0000722729  0.0000924235 -0.0003130703  0.0000371049  0.0000551414 -0.0003199965  0.0000000512  0.0000000653
-0.0002616333 -0.0003130703  0.0365393286  0.0001159112  0.0000320019 -0.0152030788  0.0000000263 -0.0000000852
 0.0000291975  0.0000371049  0.0001159112  0.0000233979  0.0000303896 -0.0004082364  0.0000000238  0.0000000291
 0.0000443578  0.0000551414  0.0000320019  0.0000303896  0.0000468536 -0.0005388005  0.0000000346  0.0000000434
-0.0001538108 -0.0003199965 -0.0152030788 -0.0004082364 -0.0005388005  0.0407119017 -0.0000002846 -0.0000002320
 0.0000000412  0.0000000512  0.0000000263  0.0000000238  0.0000000346 -0.0000002846  0.0000000000  0.0000000000
 0.0000000534  0.0000000653 -0.0000000852  0.0000000291  0.0000000434 -0.0000002320  0.0000000000  0.0000000001
```

17.7 THREE-DIMENSIONAL CONFORMAL COORDINATE TRANSFORMATION

The three-dimensional conformal coordinate transformation is also known as the *seven-parameter similarity transformation.* In notation used herein, it transfers points from one three-dimensional coordinate system (xyz), to another (XYZ). It is applied in the process of reducing data from GPS surveys, and is also used extensively in photogrammetry. This transformation involves seven parameters—three rotations, three translations, and one scale factor. To develop the equations, imagine an $x'y'z'$ coordinate system that is parallel with XYZ, but whose origin is common with the origin of the xyz system. The three sequential two-dimensional rotations that follow—ω, ϕ, and κ, convert coordinates from $x'y'z'$ to xyz.

In Figure 17.3 the rotation ω, about the x' axis, expressed in matrix form is

$$X_1 = M_1 X' \qquad (a)$$

where:

$$X_1 = \begin{bmatrix} x_1 \\ y_1 \\ z_1 \end{bmatrix} \qquad M_1 = \begin{bmatrix} 1 & 0 & 0 \\ 0 & \cos(\omega) & \sin(\omega) \\ 0 & -\sin(\omega) & \cos(\omega) \end{bmatrix} \qquad X' = \begin{bmatrix} x' \\ y' \\ z' \end{bmatrix}$$

In Figure 17.4 the rotation ϕ, about the y_1 axis, expressed in matrix form is

$$X_2 = M_2 X_1 \qquad (b)$$

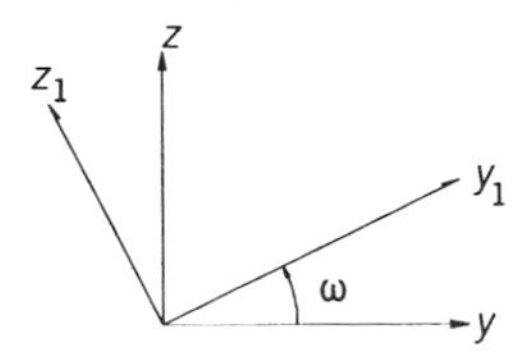

Figure 17.3 ω rotation.

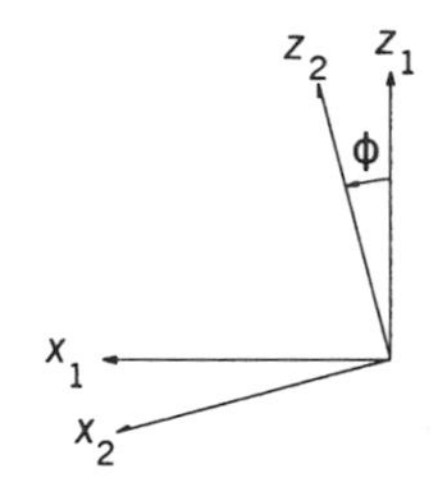

Figure 17.4 φ rotation.

where

$$X_2 = \begin{bmatrix} x_2 \\ y_2 \\ z_2 \end{bmatrix} \quad \text{and} \quad M_2 = \begin{bmatrix} \cos(\phi) & 0 & -\sin(\phi) \\ 0 & 1 & 0 \\ \sin(\phi) & 0 & \cos(\phi) \end{bmatrix}$$

In Figure 17.5 the rotation κ, about the z_2 axis, expressed in matrix form is

$$\overline{X} = M_3X_2 \tag{c}$$

where

$$\overline{X} = \begin{bmatrix} x \\ y \\ z \end{bmatrix} \quad \text{and} \quad M_3 = \begin{bmatrix} \cos(\kappa) & \sin(\kappa) & 0 \\ -\sin(\kappa) & \cos(\kappa) & 0 \\ 0 & 0 & 1 \end{bmatrix}$$

Substituting (*a*) into (*b*), and in turn substituting in to (*c*) gives

$$\overline{X} = M_3M_2M_1X' = MX' \tag{d}$$

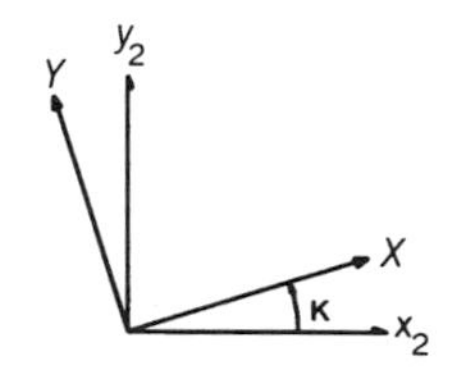

Figure 17.5 κ rotation.

The three matrices M_3, M_2, and M_1 in (*d*), when multiplied together, develop a single *rotation matrix* M for the transformation whose individual elements are

$$M = \begin{bmatrix} m_{11} & m_{12} & m_{13} \\ m_{21} & m_{22} & m_{23} \\ m_{31} & m_{32} & m_{33} \end{bmatrix}$$

where

$$\begin{aligned}
m_{11} &= \cos(\phi)\cos(\kappa) \\
m_{12} &= \sin(\omega)\sin(\phi)\cos(\kappa) + \cos(\omega)\sin(\kappa) \\
m_{13} &= -\cos(\omega)\sin(\phi)\cos(\kappa) + \sin(\omega)\sin(\kappa) \\
m_{21} &= -\cos(\phi)\sin(\kappa) \\
m_{22} &= -\sin(\omega)\sin(\phi)\sin(\kappa) + \cos(\omega)\cos(\kappa) \\
m_{23} &= \cos(\omega)\sin(\phi)\sin(\kappa) + \sin(\omega)\cos(\kappa) \\
m_{31} &= \sin(\phi) \\
m_{32} &= -\sin(\omega)\cos(\phi) \\
m_{33} &= \cos(\omega)\cos(\phi)
\end{aligned}$$

Rotation matrix M is *orthogonal*, and thus has the property that its transpose is its inverse, i.e., $M^T = M^{-1}$. Using that property, the following can be written:

$$X' = M^T X \tag{e}$$

Multiplying the expanded terms of the matrix X' of Equation (*e*) by a scale factor, S, and adding translations factors T_x, T_y, and T_z yields the following mathematical model for the transformation:

$$\begin{aligned}
X &= S(m_{11}\,x + m_{21}\,y + m_{31}\,z) + T_x \\
Y &= S(m_{12}\,x + m_{22}\,y + m_{32}\,z) + T_y \\
Z &= S(m_{13}\,x + m_{23}\,y + m_{33}\,z) + T_z
\end{aligned} \tag{17.14}$$

Equations (17.14) involve seven unknowns (S, ω, ϕ, κ, T_x, T_y, T_z). For a unique solution, seven equations must be written. This requires a minimum of two control stations with known X–Y and x–y coordinates, plus three stations with known Z and z coordinates. If there are more than the minimum number of control points, a least-squares solution can be used. Equations (17.14) are nonlinear in their unknowns and thus must be linearized for a solution. The following linearized equations can be written for each control point:

$$
\begin{bmatrix}
\left(\frac{\partial X}{\partial S}\right)_0 & 0 & \left(\frac{\partial X}{\partial \phi}\right)_0 & \left(\frac{\partial X}{\partial \kappa}\right)_0 & 1 & 0 & 0 \\
\left(\frac{\partial Y}{\partial S}\right)_0 & \left(\frac{\partial Y}{\partial \omega}\right)_0 & \left(\frac{\partial Y}{\partial \phi}\right)_0 & \left(\frac{\partial Y}{\partial \kappa}\right)_0 & 0 & 1 & 0 \\
\left(\frac{\partial Z}{\partial S}\right)_0 & \left(\frac{\partial Z}{\partial \omega}\right)_0 & \left(\frac{\partial Z}{\partial \phi}\right)_0 & \left(\frac{\partial Z}{\partial \kappa}\right)_0 & 0 & 0 & 1
\end{bmatrix}
\begin{bmatrix} dS \\ d\omega \\ d\phi \\ d\kappa \\ dTx \\ dTy \\ dTz \end{bmatrix}
= \begin{bmatrix} X - X_0 \\ Y - Y_0 \\ Z - Z_0 \end{bmatrix} \tag{17.15}
$$

where

$$\frac{\partial X}{\partial S} = m_{11}x + m_{21}y + m_{31}z$$

$$\frac{\partial Y}{\partial S} = m_{12}x + m_{22}y + m_{32}z \qquad \frac{\partial Z}{\partial S} = m_{31}x + m_{32}y + m_{33}z$$

$$\frac{\partial Y}{\partial \omega} = -S[m_{13}x + m_{23}y + m_{33}z] \qquad \frac{\partial Z}{\partial \omega} = S[m_{12}x + m_{22}y + m_{32}z]$$

$$\frac{\partial X}{\partial \phi} = S[-\sin(\phi)\cos(\kappa)\,x + \sin(\phi)\sin(\kappa)\,y + \cos(\phi)\,z]$$

$$\frac{\partial Y}{\partial \phi} = S[\sin(\omega)\cos(\phi)\cos(\kappa)\,x - \sin(\omega)\cos(\phi)\sin(\kappa)\,y + \sin(\omega)\sin(\phi)\,z]$$

$$\frac{\partial Z}{\partial \phi} = S[-\cos(\omega)\cos(\phi)\cos(\kappa)\,x + \cos(\omega)\cos(\phi)\sin(\kappa)\,y - \cos(\omega)\sin(\phi)\,z]$$

$$\frac{\partial X}{\partial \kappa} = S[m_{21}x - m_{11}y]$$

$$\frac{\partial Y}{\partial \kappa} = S[m_{22}x - m_{12}\,y] \qquad \frac{\partial Z}{\partial \kappa}\; S[m_{23}x - m_{13}y]$$

Example 17.4 Three-dimensional coordinates x, y and z were measured for six points. Four of these points (1, 2, 3, and 4) were control points whose coordinates were also known in the XYZ control system. The data are shown in Table 17.4. Compute the parameters of a three-dimensional conformal coordinate transformation and use them to transform points 5 and 6 into the XYZ system.

SOLUTION: The results from the program ADJUST are presented below.

Table 17.4 Data for three-dimensional conformal coordinate transformation

Point	X	Y	Z	$x \pm S_x$	$y \pm S_y$	$z \pm S_z$
1	10037.81	5262.09	772.04	1094.883 ± 0.007	820.085 ± 0.008	109.821 ± 0.005
2	10956.68	5128.17	783.00	503.891 ± 0.011	1598.698 ± 0.008	117.685 ± 0.009
3	8780.08	4840.29	782.62	2349.343 ± 0.006	207.658 ± 0.005	151.387 ± 0.007
4	10185.80	4700.21	851.32	1395.320 ± 0.005	1348.853 ± 0.008	215.261 ± 0.009
5				265.346 ± 0.005	1003.470 ± 0.007	78.609 ± 0.003
6				784.081 ± 0.006	512.683 ± 0.008	139.551 ± 0.008

```
3D Coordinate Transformation of File: EX17-4.DAT
Iteration 1
--------------------------------------------------

J MATRIX                                                                      K MATRIX
-----------------------------------------------------------------             --------
     0.000   102.452   1284.788 1.000 0.000 0.000   -206.164                   -0.000
   -51.103    -7.815   -195.197 0.000 1.000 0.000  -1355.718                    0.000
 -1287.912   195.697      4.553 0.000 0.000 1.000     53.794                    0.000
     0.000   118.747   1418.158 1.000 0.000 0.000    761.082                   -0.000
   -62.063    28.850    723.004 0.000 1.000 0.000  -1496.689                   -0.000
 -1421.832  -722.441     42.501 0.000 0.000 1.000     65.331                   -0.000
     0.000   129.863   1706.020 1.000 0.000 0.000  -1530.174                    0.060
   -61.683   -58.003  -1451.826 0.000 1.000 0.000  -1799.945                    0.209
 -1709.922  1452.485    -41.580 0.000 0.000 1.000     64.931                    0.000
     0.000   204.044   1842.981 1.000 0.000 0.000    -50.417                    0.033
  -130.341    -1.911    -46.604 0.000 1.000 0.000  -1947.124                   -0.053
 -1849.740    47.857     15.851 0.000 0.000 1.000    137.203                    0.043

X MATRIX
-------------
-0.0000347107
-0.0000103312
-0.0001056763
 0.1953458986
-0.0209088384
-0.0400969773
-0.0000257795
Iteration 2
J MATRIX                                                                      K MATRIX
-----------------------------------------------------------------             --------
     0.000   102.446   1284.733 1.000 0.000 0.000   -206.308                   -0.064
   -51.144    -7.813   -195.328 0.000 1.000 0.000  -1355.694                   -0.037
 -1287.855   195.829      4.567 0.000 0.000 1.000     53.838                   -0.001
     0.000   118.750   1418.196 1.000 0.000 0.000    760.923                   -0.025
   -62.114    28.818    722.835 0.000 1.000 0.000  -1496.767                    0.057
 -1421.868  -722.272     42.484 0.000 0.000 1.000     65.386                   -0.011
     0.000   129.843   1705.820 1.000 0.000 0.000  -1530.365                    0.007
   -61.730   -57.958  -1451.969 0.000 1.000 0.000  -1799.781                    0.028
 -1709.719  1452.630    -41.520 0.000 0.000 1.000     64.982                   -0.007
     0.000   204.036   1842.926 1.000 0.000 0.000    -50.624                    0.033
  -130.399    -1.917    -46.797 0.000 1.000 0.000  -1947.114                   -0.091
 -1849.680    48.053     15.864 0.000 0.000 1.000    137.268                    0.024
```

```
X MATRIX
-------------
 0.0000000007
 0.0000000046
-0.0000000033
-0.0000000588
-0.0000008524
 0.0000030138
 0.0000000053
```

Measured Points

Point	x	y	z	Sx	Sy	Sz
1	1094.883	820.085	109.821	0.007	0.008	0.005
2	503.891	1598.698	117.685	0.011	0.008	0.009
3	2349.343	207.658	151.387	0.006	0.005	0.007
4	1395.320	1348.853	215.261	0.005	0.008	0.009

Control Points

Point	X	VX	Y	VY	Z	VZ
1	10037.810	0.064	5262.090	0.037	772.040	0.001
2	10956.680	0.025	5128.170	-0.057	783.000	0.011
3	8780.080	-0.007	4840.290	-0.028	782.620	0.007
4	10185.800	-0.033	4700.210	0.091	851.320	-0.024

```
Transformation Coefficients
---------------------------
Scale =         0.94996 +/- 0.00004
Omega =    2° 17' 05.3" +/- 0° 00' 30.1"
  Phi =   -0° 33' 02.8" +/- 0° 00' 09.7"
Kappa =  224° 32' 10.9" +/- 0° 00' 06.9"
   Tx =       10233.858 +/-      0.065
   Ty =        6549.981 +/-      0.071
   Tz =         720.897 +/-      0.213
```

Reference Standard Deviation: 8.663
Degrees of Freedom: 5
Iterations: 2

Transformed Coordinates

Point	X	Sx	Y	Sy	Z	Sz
5	10722.020	0.073	5691.221	0.080	766.068	0.248
6	10043.246	0.072	5675.898	0.080	816.867	0.248

```
Qxx MATRIX
-----------------------------------------------------------------------
 0.0000000003  0.0000000000  0.0000000000 -0.0000000068 0.0000000210  0.0000004081 -0.0000000000
 0.0000000000  0.0000000000 -0.0000000000 -0.0000000039 0.0000000025  0.0000000487 -0.0000000000
 0.0000000000 -0.0000000000  0.0000000000 -0.0000000247 0.0000000064  0.0000000008 -0.0000000000
-0.0000000068 -0.0000000039 -0.0000000247  0.0000561837 0.0000020823 -0.0000095524  0.0000000101
 0.0000000210  0.0000000025  0.0000000064  0.0000020823 0.0000669290  0.0000283700  0.0000000296
 0.0000004081  0.0000000487  0.0000000008 -0.0000095524 0.0000283700  0.0006032234 -0.0000000016
-0.0000000000 -0.0000000000 -0.0000000000  0.0000000101 0.0000000296 -0.0000000016  0.0000000000
```

Note that in this adjustment, with four control points available having *X*, *Y*, and *Z* coordinates, 12 equations could be written, three for each point. With seven unknown parameters, this gave $12 - 7 = 5$ degrees of freedom in the solution.

17.8 STATISTICALLY VALID PARAMETERS

Besides the coordinate transformations described in preceding sections, it is possible to develop numerous others. For example, polynomial equations of various degrees could be used to transform data. As additional terms are added to a polynomial, the resulting equation will force better fits on any given data set. However, caution should be exercised when doing this since the resulting transformation parameters may not be statistically significant.

As an example, when using a two-dimensional conformal coordinate transformation with a data set having four control points, nonzero residuals would be expected. However, if a projective transformation were used, this data set would yield a unique solution and thus the residuals would be zero. Is the projective a more appropriate transformation for this data set? Is this truly a better fit? Guidance in the answers to these questions can be obtained by checking the statistical validity of the parameters.

The adjusted parameters divided by their standard deviations represent a t statistic with v degrees of freedom. If a parameter is to be judged as statistically different from zero, and thus significant, the computed t value (the test statistic) must be greater than $t_{\alpha/2,v}$. Simply stated, the test statistic is

$$t = \frac{|\text{parameter}|}{S} \tag{17.16}$$

For example, in the adjustment in of Example 17.2, the following computed t values are found:

Parameter	S	t Value
$a =$ 25.37152	±0.02532	1002
$b =$ 0.82220	±0.02256	36.4
$c =$ −137.183	±0.203	675.8
$d =$ −0.80994	±0.02335	34.7
$e =$ 25.40166	±0.02622	968.8
$f =$ −150.723	±0.216	697.8

In this problem there were eight equations involving six unknowns and thus 2 degrees of freedom. From the t distribution table (Table D-3), $t_{0.025,2} = 4.303$. Because all computed t values are greater than 4.303, each parameter is significantly different from zero at a 95% level of confidence.

From the adjustment results of Example 17.3, the computed t values are listed below.

Parameter	S	t Value
$a1 =$ 25.00274	±0.01538	1626
$b1 =$ 0.80064	±0.01896	42.3
$c1 =$ −134.715	±0.377	357.3
$a2 =$ −8.00771	±0.00954	839.4
$b2 =$ 24.99811	±0.01350	1851.7
$c2 =$ −149.815	±0.398	376.4
$a3 =$ 0.00400	±0.00001	400
$b3 =$ 0.00200	±0.00002	100

This adjustment has eight unknown parameters and 12 observations. From the t distribution table (Table D.3), $t_{0.025,4} = 2.776$. By comparing the tabular t value against each computed value, all parameters are again significantly different from zero at a 95% confidence level. This is true for a_3 and b_3 even though they seem relatively small at 0.004 and 0.002, respectively. Using this statistical technique, a check can be made to determine when the projective transformation is not appropriate, since it defaults to an affine transformation when a_3 and b_3 are both statistically equivalent to zero.

PROBLEMS

Note: For problems below requiring least-squares adjustment, if a computer program is not distinctly specified for use in the problem, it is expected that the least-squares algorithm will be solved using the program MATRIX which is included within the diskette supplied with the book.

17.1 Points A, B, and C have their coordinates known in both an (X,Y) system and an (x,y) system. Points D, E, F, and G have their coordinates known only in the (x,y) system. These coordinates are shown in the table below. Using a two-dimensional conformal coordinate transformation, calculate the most probable coordinates for D, E, F, and G in the (X,Y) system.

Point	X	Y	x	y
A	3000.00	2000.00	1767.77	1060.66
B	1000.00	3000.00	707.11	707.11
C	3000.00	4000.00	1060.66	1767.77
D			1414.21	1414.21
E			353.52	1060.66
F			1414.21	−353.52
G			353.52	−353.52

17.2 For Problem 17.1, what are the values of the:

(a) rotation angle?

(b) scale factor?

(c) translation in X?

(d) translation in Y?

17.3 For the given data, use a two-dimensional conformal coordinate transformation to determine (X,Y) coordinates for the points having only (x,y) coordinates. (All units are mm.)

	Measured		Control	
Point	x	y	X	Y
1	−21.420	0.930	−113.000	0.003
2	−12.490	102.585	−103.982	103.963
3	90.590	112.660	0.001	112.993
7	90.605	−113.115	0.001	−112.999
8	−12.470	−104.040	−103.948	−103.961
32	16.990	−91.715		
22	−6.520	−1.995		
12	6.095	90.365		
13	85.635	97.005		
23	92.225	−2.035		
33	90.510	−95.620		

17.4 Use a two-dimensional affine coordinate transformation to do Problem 17.3.

17.5 Use a two-dimensional projective coordinate transformation to do Problem 17.3.

17.6 For the data of Problem 17.3, which two-dimensional transformation is most appropriate and why? Use a 0.01 level of significance.

17.7 Determine the appropriate two-dimensional transformation for the following data set at a 0.01 level of significance.

Point	*X* (m)	*Y* (m)	*x* (mm)	*y* (mm)	*Sx*	*Sy*
1	2181.578	2053.274	89.748	91.009	0.019	0.020
2	1145.486	809.022	49.942	39.960	0.016	0.021
3	−855.426	383.977	−29.467	20.415	0.028	0.028
4	1087.225	−1193.347	50.164	−40.127	0.028	0.028
5	2540.778	−2245.477	109.599	−80.310	0.018	0.021
6	−2595.242	1926.548	−100.971	79.824	0.026	0.022

17.8 Using a weighted three-dimensional conformal coordinate transformation, determine the transformation parameters for the following data set.

Point	*X* (m)	*Y* (m)	*Z* (m)	*x* (mm)	*y* (mm)	*z* (mm)	*Sx* (mm)	*Sy* (mm)	*Sz* (mm)
1	8948.16	6678.50	756.51	1094.97	810.09	804.73	0.080	0.084	0.153
2	8813.93	5755.23	831.67	508.31	1595.68	901.78	0.080	0.060	0.069
3	8512.60	7937.11	803.11	2356.23	197.07	834.47	0.097	0.177	0.202
4	8351.02	6483.62	863.24	1395.18	1397.64	925.96	0.043	0.161	0.120

17.9 Do Problem 17.8 using an equal-weight least-squares adjustment.

17.10 Do Problem 10.19, and determine whether the derived constant and scale factor are statistically significant at a 0.01 level of significance.

Use program ADJUST to do each problem.

17.11 Problem 17.6

17.12 Problem 17.7

17.13 Problem 17.8

Programming Problems

Develop a spreadsheet that calculates the coefficient and constants matrices for each transformation.

17.14 A two-dimensional conformal coordinate transformation

17.15 A two-dimensional affine coordinate transformation

17.16 A two-dimensional projective coordinate transformation

17.17 A three-dimensional conformal coordinate transformation

18

ERROR ELLIPSE

18.1 INTRODUCTION

As discussed previously, after completing a least-squares adjustment, the estimated standard deviations in the coordinates of an adjusted station can be calculated from covariance matrix elements. These standard deviations provide error estimates in the reference axes directions. In graphical representation, they are half the dimensions of a *standard error rectangle* centered on each point. The standard error rectangle has dimensions of $2S_x$ by $2S_y$, as illustrated for point B in Figure 18.1, but this is not a true representation of the error present at the station.

Simple deductive reasoning can be used to show the basic problem. Assume in Figure 18.1 that the (x,y) coordinates of station A have been computed from the measurement of distance AB and azimuth α_{AB}, which is approximately 30°. Further assume that the measured azimuth has no error at all, but that the distance has a very large error, say ± 2 ft. From Figure 18.1 it should then be readily apparent that the largest uncertainty in the station's position would not lie in either cardinal direction. That is, neither S_x nor S_y represents the largest positional uncertainty for the station. Rather, the largest uncertainty would be expected to exist collinear with line AB, and be approximately equal to the estimated error in the distance. This is, in fact, what happens.

In the usual case, the position of a station is uncertain in both direction and distance, and the estimated error of the adjusted station therefore involves the errors of two jointly distributed variables, the x and y coordinates. Thus the positional error at a station follows a *bivariate normal distribution*. The general shape of this distribution for a station is shown in Figure 18.2. In this figure the three-dimensional view or *surface plot* [Figure 18.2(a)] of the bivariate normal distribution curve is shown, along with its *contour plot* [Figure

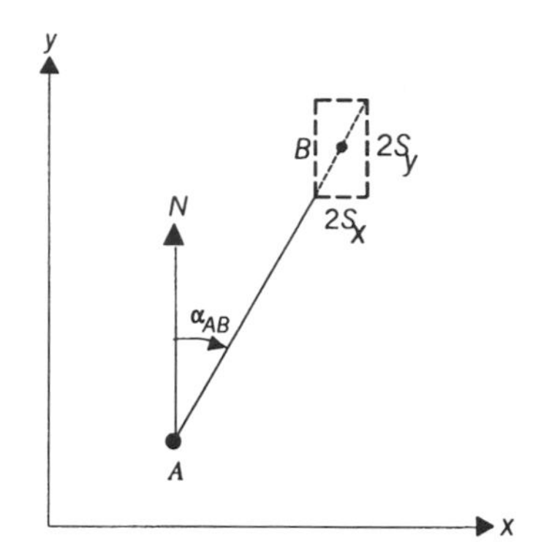

Figure 18.1 Standard error rectangle at station *B*.

18.2(*b*)]. Note that the ellipses shown in the (x,y) plane of figure (b) can be obtained by passing planes at varying levels through figure (a) parallel to the (x,y) plane. The volume of the region inside the intersection of any plane with the surface of figure (a) represents the probability level of the ellipse. The orthogonal projection of the surface plot of Figure (a) onto the xz plane would give the normal distribution curve of the x coordinate, from which S_x is obtained. Likewise, its orthogonal projection onto the yz plane would give the normal distribution in the y coordinate, from which S_y is obtained.

To describe the estimated error of a station fully, it is only necessary to show the orientation and lengths of the semiaxes of the error ellipse. A detailed diagram of an error ellipse is shown in Figure 18.3. In this figure, the *standard error ellipse* of a station is shown (i.e., one whose arcs are tangent to the sides of the standard error rectangle). The orientation of the ellipse

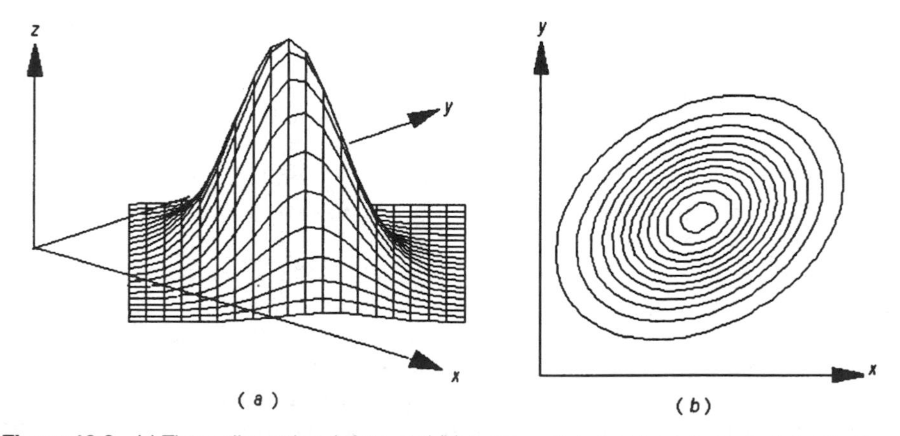

Figure 18.2 (a) Three-dimensional view, and (b) contour plot of a bivariate normal distribution.

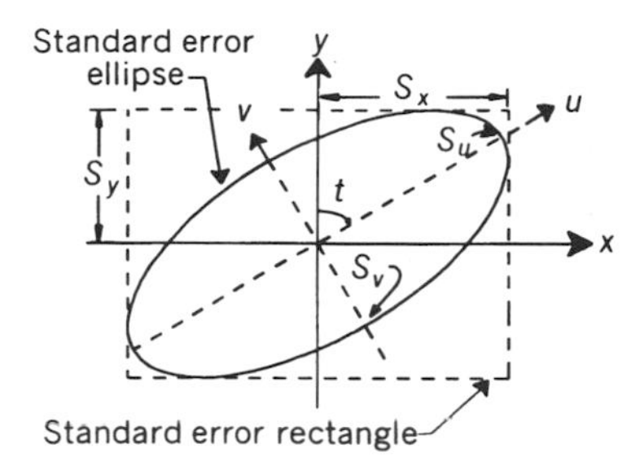

Figure 18.3 Standard error ellipse.

depends upon the t angle, which fixes the directions of the auxiliary, orthogonal (u,v) axes along which the ellipse axes lie. The u axis defines the weakest direction in which the station's adjusted position is known. In other words, it lies in the direction of maximum expected error in the station's coordinates. The v axis is orthogonal to u, and defines the strongest direction in which the station's position is known, or the direction of minimum error. For any station, the value of t that orients the ellipse to provide these maximum and minimum values can be determined after the adjustment from the elements of the covariance matrix.

The exact probability level of the standard error ellipse is dependent on the number of degrees of freedom in the adjustment. This standard error ellipse can be modified in dimensions to represent any percent probability through the use of the critical values tabulated for the F distribution. It will be shown later that the percent probability within the boundary of the standard error ellipse for a simple closed traverse is only 35%.

18.2 COMPUTATION OF ELLIPSE ORIENTATION AND SEMIAXES

As shown in Figure 18.4, the method for calculating the orientation angle, t, that yields maximum and minimum semiaxes involves a two-dimensional axes rotation. Notice that the t angle is defined as a clockwise angle from the y

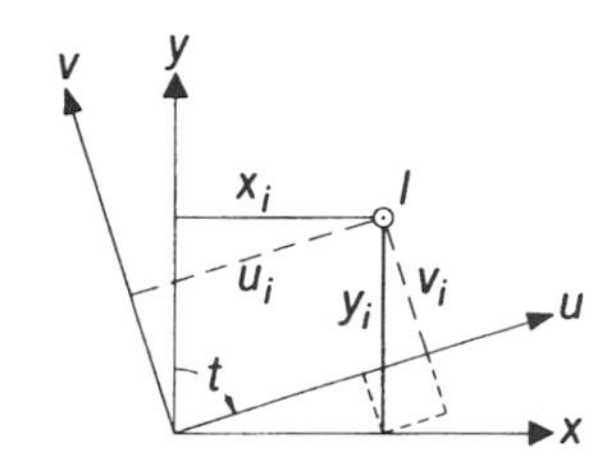

Figure 18.4 Two-dimensional axes rotation.

axis to the u axis. To propagate the errors in a point I from the (x,y) system into an orthogonal (u,v) system, the generalized law of the propagation of variances, discussed in Chapter 5, is used. The specific value for t that yields the maximum error along the u axis must be determined. The following steps accomplish this task.

Step 1: Any point I in the (u,v) system can be represented with respect to its (x,y) coordinates as

$$\begin{aligned} u_i &= x_i \sin(t) + y_i \cos(t) \\ v_i &= -x_i \cos(t) + y_i \sin(t) \end{aligned} \tag{18.1}$$

Rewriting Equations (18.1) in matrix form yields

$$\begin{bmatrix} u_i \\ v_i \end{bmatrix} = \begin{bmatrix} \sin(t) & \cos(t) \\ -\cos(t) & \sin(t) \end{bmatrix} \begin{bmatrix} x_i \\ y_i \end{bmatrix} \tag{18.2}$$

or in simplified matrix notation,

$$Z = RX \tag{18.3}$$

Step 2: Assume that for the adjustment problem in which I appears, there is a Q_{xx} matrix for the (x,y) system. The problem is to develop, from the Q_{xx} matrix, a new covariance matrix Q_{zz} for the (u,v) system. This can be done using the generalized law for the propagation of variances, given in Chapter 5 as

$$\Sigma_{zz} = S_0^2 R Q_{xx} R^T \tag{18.4}$$

where:

$$R = \begin{bmatrix} \sin(t) & \cos(t) \\ -\cos(t) & \sin(t) \end{bmatrix}$$

Since S_0^2 is a scalar, it can temporarily be dropped and recalled again after the derivation. Thus,

$$Q_{ZZ} = R Q_{xx} R^T = \begin{bmatrix} q_{uu} & q_{uv} \\ q_{uv} & q_{vv} \end{bmatrix} \tag{18.5}$$

where

$$Q_{xx} = \begin{bmatrix} q_{xx} & q_{xy} \\ q_{xy} & q_{yy} \end{bmatrix}$$

Step 3: Expanding Equation (18.5), the elements of the Q_{ZZ} matrix are

$$Q_{zz} =$$

$$\begin{bmatrix} \begin{pmatrix} \sin^2(t)\ q_{xx} + \cos(t)\sin(t)\ q_{xy} \\ +\sin(t)\cos(t)\ q_{xy} + \cos^2(t)\ q_{yy} \end{pmatrix} & \begin{pmatrix} -\sin(t)\cos(t)\ q_{xx} - \cos^2(t)\ q_{xy} \\ +\sin^2(t)\ q_{xy} + \cos(t)\sin(t)\ q_{yy} \end{pmatrix} \\ \begin{pmatrix} -\cos(t)\sin(t)\ q_{xx} + \sin^2(t)\ q_{xy} \\ -\cos^2(t)\ q_{xy} + \sin(t)\cos(t)\ q_{yy} \end{pmatrix} & \begin{pmatrix} \cos^2(t)\ q_{xx} - \sin(t)\cos(t)\ q_{xy} \\ -\cos(t)\sin(t)\ q_{xy} + \sin^2(t)\ q_{yy} \end{pmatrix} \end{bmatrix}$$

$$= \begin{bmatrix} q_{uu} & q_{uv} \\ q_{uv} & q_{vv} \end{bmatrix} \tag{18.6}$$

Step 4: The q_{uu} element of Equation (18.6) can be rewritten as

$$q_{uu} = \sin^2(t)q_{xx} + 2\cos(t)\sin(t)q_{xy} + \cos^2(t)q_{yy} \tag{18.7}$$

[The following trigonometric identities are useful in developing the equation for t]

$$\sin(2t) = 2\sin(t)\cos(t) \tag{a}$$

$$\cos(2t) = \cos^2(t) - \sin^2(t) \tag{b}$$

$$\cos^2(t) + \sin^2(t) = 1 \tag{c}$$

Substituting identity (a) into Equation (18.7) gives

$$q_{uu} = \sin^2(t)\ q_{xx} + \cos^2(t)\ q_{yy} + 2\frac{\sin(2t)}{2}\ q_{xy} \tag{18.8}$$

Regrouping the first two terms and adding the necessary terms to Equation (18.8) to maintain equality yields

$$\begin{aligned} q_{uu} = {} & \frac{q_{xx} + q_{yy}}{2}[\sin^2(t) + \cos^2(t)] + \frac{q_{xx}\sin^2(t)}{2} + \frac{q_{yy}\cos^2(t)}{2} \\ & - \frac{q_{yy}\sin^2(t)}{2} - \frac{q_{xx}\cos^2(t)}{2} + \sin(2t)\ q_{xy} \end{aligned} \tag{18.9}$$

Substituting identity (c) and regrouping Equation (18.9) gives

$$\begin{aligned} q_{uu} = {} & \frac{q_{xx} + q_{yy}}{2} + \frac{q_{yy}}{2}[\cos^2(t) - \sin^2(t)] \\ & - \frac{q_{xx}}{2}[\cos^2(t) - \sin^2(t)] + q_{xy}\sin(2t) \end{aligned} \tag{18.10}$$

Finally, substituting identity (b) into Equation (18.10) yields

$$q_{uu} = \frac{q_{xx} + q_{yy}}{2} + \frac{q_{yy} - q_{xx}}{2} \cos(2t) + q_{xy} \sin(2t) \tag{18.11}$$

To find the value of t that maximizes q_{uu}, differentiate q_{uu} in Equation (18.11) with respect to t and set the results equal to zero. This gives

$$\frac{dq_{uu}}{dt} = \frac{q_{yy} - q_{xx}}{2} 2[-\sin(2t)] + q_{xy}\, 2 \cos(2t) = 0 \tag{18.12}$$

Reducing Equation (18.12) gives

$$(q_{yy} - q_{xx}) \sin(2t) = 2q_{xy} \cos(2t) \tag{18.13}$$

Finally, dividing Equation (18.13) by cos(2t) yields

$$\frac{\sin(2t)}{\cos(2t)} = \frac{2q_{xy}}{q_{yy} - q_{xx}} = \tan(2t) \tag{18.14}$$

Equation (18.14) is used to compute 2t, and hence the desired angle t which yields the maximum value of q_{uu}. Note that the correct quadrant of 2t must be determined by examination of the sign of the numerator and denominator in Equation (18.14) before dividing by two to obtain t. Table 18.1 shows the proper quadrant for the different possible sign combinations of the numerator and denominator.

Correlation between the latitude and departure of a station was discussed in Chapter 7. Similarly, the adjusted coordinates of a station are also correlated. Computing the value of t that yields the maximum and minimum values for the semiaxes is equivalent to rotating the covariance matrix until the off-diagonals are nonzero. Thus the u and v coordinate values will be uncorrelated, which is equivalent to setting the q_{uv} element of Equation (18.6) equal to zero. Using the trigonometric identities noted previously, the element q_{uv} from Equation (18.6) can be written as

Table 18.1 Selection of the proper quadrant for 2*t*[a] in Equation (18.4)

Algebraic Sign of:		
Numerator	Denominator	Quadrant of 2t
+	+	1
+	−	2
−	−	3
−	+	4

[a]When calculating for t, always remember to select the proper quadrant of 2t before dividing by 2.

$$q_{uv} = \frac{q_{xx} - q_{yy}}{2} \sin(2t) + q_{xy} \cos(2t) \tag{18.15}$$

Setting q_{uv} equal to zero and solving for t gives

$$\frac{\sin(2t)}{\cos(2t)} = \tan(2t) = \frac{2q_{xy}}{q_{yy} - q_{xx}} \tag{18.16}$$

Note that this yields the same result as Equation (18.14), which verifies the absence of correlation.

Step 5: In a fashion similar to that demonstrated in step 4, the q_{vv} element of Equation (18.6) can be rewritten as

$$q_{vv} = q_{xx} \cos^2(t) - 2q_{xy} \cos(t) \sin(t) + q_{yy} \sin^2(t) \tag{18.17}$$

In summary, the t angle, semimajor ellipse axis (q_{uu}), and semiminor axis (q_{vv}) are calculated using Equations (18.14), (18.7) and (18.17), respectively. These equations are repeated here in order for convenience. Note that the equations use only elements from the covariance matrix.

$$\tan(2t) = \frac{2q_{xy}}{q_{yy} - q_{xx}} \tag{18.18}$$

$$q_{uu} = q_{xx} \sin^2(t) + 2q_{xy} \cos(t) \sin(t) + q_{yy} \cos^2(t) \tag{18.19}$$

$$q_{vv} = q_{xx} \cos^2(t) - 2q_{xy} \cos(t) \sin(t) + q_{yy} \sin^2(t) \tag{18.20}$$

Equation (18.18) gives the t angle that the u axis makes with the y axis. Equation (18.19) gives the numerical value for q_{uu}, which when multiplied by the reference variance S_0^2 gives the variance along the u axis. The square root of the variance is the *semimajor axis* of the standard error ellipse. Equation (18.20) gives the numerical value for q_{vv}, which when multiplied by S_0^2 gives the variance along the v axis. The square root of this variance is the *semiminor axis* of the standard error ellipse. Thus the semimajor and semiminor axes are

$$S_u = S_0 \sqrt{q_{uu}} \quad \text{and} \quad S_v = S_0 \sqrt{q_{vv}} \tag{18.21}$$

18.3 EXAMPLE OF STANDARD ERROR ELLIPSE CALCULATIONS

In this section the error ellipse data for the trilateration example in Section 13.5 will be calculated. From the computer listing given for the solution of that problem, the following values are recalled:

1. $S_0 = \pm 0.136$ ft
2. The unknown X and Q_{xx} matrices were

$$X = \begin{bmatrix} dX_{\text{Wis.}} \\ dY_{\text{Wis.}} \\ dX_{\text{Campus}} \\ dY_{\text{Campus}} \end{bmatrix}$$

$$Q_{xx} = \begin{bmatrix} 1.198574 & -1.160249 & -0.099772 & -1.402250 \\ -1.160249 & 2.634937 & 0.193956 & 2.725964 \\ -0.099772 & 0.193956 & 0.583150 & 0.460480 \\ -1.402250 & 2.725964 & 0.460480 & 3.962823 \end{bmatrix}$$

18.3.1 Error Ellipse for Station Wisconsin

The tangent of $2t$ is

$$\tan(2t) = \frac{2(-1.160249)}{2.634937 - 1.198574} = -1.6155$$

Note that the sign of the numerator is negative and the denominator is positive. Thus from Table 18.1, angle $2t$ is in the fourth quadrant. Therefore, 360° must be added to the computed angle. Hence

$$2t = \tan^{-1}(-1.6155) + 360° = -58°14.5' + 360° = 301°45.5'$$

and

$$t = 150°53'$$

Substituting the appropriate values into Equation (18.21), S_u is

$$\begin{aligned} S_u &= \pm 0.136 \\ &\quad \times \sqrt{1.198574 \sin^2(t) + 2(-1.160249)\cos(t)\sin(t) + 2.634937\cos^2(t)} \\ &= \pm 0.25 \text{ ft} \end{aligned}$$

Similarly, substituting the appropriate values into Equation (18.21), S_v is

$$\begin{aligned} S_v &= \pm 0.136 \\ &\quad \times \sqrt{1.198574 \cos^2(t) - 2(-1.160249)\cos(t)\sin(t) + 2.634937\sin^2(t)} \\ &= \pm 0.10 \text{ ft} \end{aligned}$$

Note that the standard deviations in the x and y coordinates as computed by Equation (12.24) are

$$S_x = S_0 \sqrt{q_{xx}} = \pm 0.136 \sqrt{1.198574} = \pm 0.15 \text{ ft}$$

$$S_y = S_0 \sqrt{q_{yy}} = \pm 0.136 \sqrt{2.634937} = \pm 0.22 \text{ ft}$$

18.3.2 Error Ellipse for Station Campus

Using the same procedures as in Section 18.3.1, the error ellipse data for the station Campus are

$$2t = \tan^{-1}\left(\frac{2 \times 0.460480}{3.962823 - 0.583150}\right) = 15°14'$$

$$t = 7°37'$$

$$\begin{aligned} S_u &= \pm 0.136 \\ &\quad \times \sqrt{0.583150 \sin^2(t) + 2(0.460480)\cos(t)\sin(t) + 3.962823 \cos^2(t)} \\ &= \pm 0.27 \text{ ft} \end{aligned}$$

$$\begin{aligned} S_v &= \pm 0.136 \\ &\quad \times \sqrt{0.583150 \cos^2(t) - 2(0.460480)\cos(t)\sin(t) + 3.962823 \sin^2(t)} \\ &= \pm 0.10 \text{ ft} \end{aligned}$$

$$S_x = S_0 \sqrt{q_{xx}} = \pm 0.136 \sqrt{0.583150} = \pm 0.10 \text{ ft}$$

$$S_y = S_0 \sqrt{q_{yy}} = \pm 0.136 \sqrt{3.962823} = \pm 0.27 \text{ ft}$$

18.3.3 Drawing the Standard Error Ellipse

To draw the error ellipses for Stations Wisconsin and Campus of Figure 18.5, the error rectangle is first constructed by laying out the values of S_x and S_y using a convenient scale along the x and y axes, respectively. For this example, an ellipse scale of 4800 times the map scale was selected. The t angle is laid off clockwise from the positive y axis to construct the u axis. The v axis is drawn 90° counterclockwise from u. The values of S_u and S_v are laid off along the u and v axes, respectively, to locate the semiaxis points. Finally, the ellipse is constructed so that it is tangent to the error rectangle and passes through its semiaxes points (refer to Figure 18.3).

18.4 ANOTHER EXAMPLE

In this section the standard error ellipse for station U in Example 15.1 is calculated. For the adjustment, $S_0 = \pm 1.82$ ft, and the X and Q_{xx} matrices are

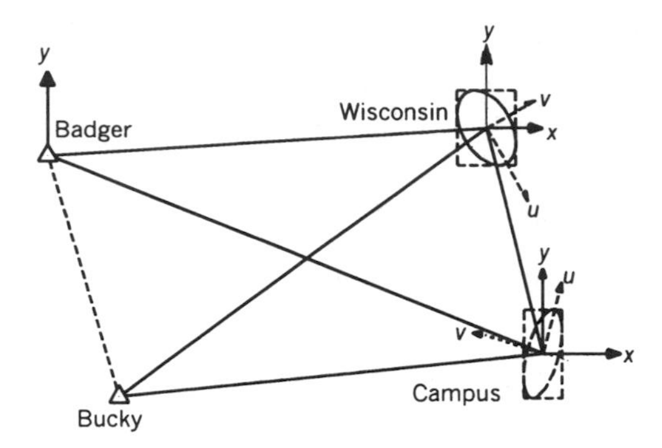

Figure 18.5 Error ellipses for the example problem of Section 13.5.

$$X = \begin{bmatrix} dx_u \\ dy_u \end{bmatrix}$$

$$Q_{xx} = \begin{bmatrix} q_{xx} & q_{xy} \\ q_{xy} & q_{yy} \end{bmatrix} = \begin{bmatrix} 0.000532 & 0.000602 \\ 0.000602 & 0.000838 \end{bmatrix}$$

Error ellipse calculations:

$$\tan(2t) = \frac{2(0.000602)}{0.000838 - 0.000532} = 3.9346$$

$$2t = 75°44'$$

$$t = 37°52'$$

$$\begin{aligned} S_u &= \pm 1.82 \\ &\quad \times \sqrt{0.000532 \sin^2(t) + 2(0.000602) \cos(t) \sin(t) + 0.000838 \cos^2(t)} \\ &= \pm 0.066 \text{ ft} \end{aligned}$$

$$\begin{aligned} S_v &= \pm 1.82 \\ &\quad \times \sqrt{0.000532 \cos^2(t) - 2(0.000602) \cos(t) \sin(t) + 0.000838 \sin^2(t)} \\ &= \pm 0.015 \text{ ft} \end{aligned}$$

$$S_x = \pm 1.82 \sqrt{0.000532} = \pm 0.042 \text{ ft}$$

$$S_y = \pm 1.82 \sqrt{0.000838} = \pm 0.053 \text{ ft}$$

Figure 18.6 shows the plotted standard error ellipse and its error rectangle.

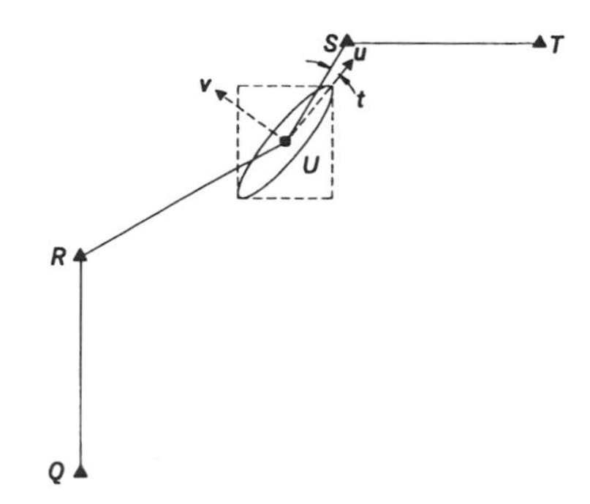

Figure 18.6 Error ellipse for Example 15.1.

18.5 ERROR ELLIPSE CONFIDENCE LEVEL

The calculations in Sections 18.3 and 18.4 produce *standard error ellipses.* These ellipses can be modified to produce error ellipses at any confidence level α by using the F statistic with 2 numerator degrees of freedom and the degrees of freedom for the adjustment in the denominator. Since the F statistic represents variance ratios for varying degrees of freedom, it can be expected that with increases in the number of degrees of freedom, there will be corresponding increases in precision. The $F_{(\alpha,2,\text{degrees of freedom})}$ statistic modifiers for various confidence levels are listed in Table 18.2. Notice that as the degrees of freedom increase, the F statistic modifiers decrease rapidly and begin to stabilize for larger degrees of freedom. The confidence level of an error ellipse can be increased to any level by using the multiplier

Table 18.2 $F_{\alpha,2,\text{degrees of freedom}}$ **statistics for selected probability levels**

	Probability		
Degrees of Freedom	90%	95%	99%
1	49.50	199.50	4999.50
2	9.00	19.00	99.00
3	5.46	9.55	30.82
4	4.32	6.94	18.00
5	3.78	5.79	13.27
10	2.92	4.10	7.56
15	2.70	3.68	6.36
20	2.59	3.49	5.85
30	2.49	3.32	5.39
60	2.39	3.15	4.98

$$c = \sqrt{2F_{(\alpha,2,\text{degrees of freedom})}} \tag{18.22}$$

To calculate percent probability for the semimajor and semiminor axes, the following equations are used:

$$\begin{aligned} S_{u\%} &= S_u \sqrt{2F_{(\alpha,2,\text{degrees of freedom})}} = S_u c \\ S_{v\%} &= S_v \sqrt{2F_{(\alpha,2,\text{degrees of freedom})}} = S_v c \end{aligned} \tag{18.23}$$

From the foregoing, it should be apparent that as the number of degrees of freedom increase, error ellipse sizes decrease. Using the techniques discussed in Chapter 4, the values listed in Table 18.2 are for the F distribution at 90% ($\alpha = 0.10$), 95% ($\alpha = 0.05$), and 99% ($\alpha = 0.01$) percent probability. These probabilities are most commonly used. Values from this table are substituted into Equation (18.22) to determine the value of c for the probability selected and the given number of degrees of freedom in the adjustment. This table is for convenience only and does not contain the values necessary for all situations that might arise.

Example 18.1 Calculate the 95% error ellipse for station Wisconsin of Section 18.3.1. (Note that the degrees of freedom in this example is 1.)

SOLUTION: Using Equations (18.23) yields

$$S_{u_{95\%}} = \pm 0.25 \sqrt{2(199.50)} = \pm 4.99 \text{ ft}$$

$$S_{v_{95\%}} = \pm 0.10 \sqrt{2(199.5)} = \pm 2.00 \text{ ft}$$

$$S_{x_{95\%}} = \pm 0.15 \sqrt{2 \times 199.5} = \pm 3.00 \text{ ft}$$

$$S_{y_{95\%}} = \pm 0.22 \sqrt{2 \times 199.5} = \pm 4.39 \text{ ft}$$

The probability of the standard error ellipse can be found by setting the multiplier $2F_{(\alpha,2,\text{degrees of freedom})}$ equal to 1 so that $F_{(\alpha,2,\text{degrees of freedom})} = 0.5$. For a simple closed traverse with 3 degrees of freedom, this means that $F_{\alpha,2,3} = 0.5$. The value of α that satisfies this condition is 0.65, which was found by trial-and-error procedures using the program STATS. Thus the percent probability of the standard error ellipse in a simple closed traverse is $(1 - 0.65) \times 100\%$, or 35%. The reader is encouraged to verify this result using Program STATS. It is also left as an exercise for the reader to show that the percent probability for the standard error ellipse ranges from 35% to only 39% for horizontal surveys that have less than 100 degrees of freedom.

18.6 ERROR ELLIPSE ADVANTAGES

Besides providing critical information regarding the precision of an adjusted station position, a major advantage of error ellipses is that they offer a method of making a visual comparison of the relative precisions between any two stations. By viewing the shapes, sizes, and orientations of error ellipses, various surveys can be compared rapidly and meaningfully.

18.6.1 Survey Network Design

The sizes, shapes, and orientations of error ellipses are dependent on (1) the control used to constrain the adjustment, (2) the observational precisions, and (3) the geometry of the survey. The last two of these elements are variables that can be readily altered in the design of a survey in order to produce optimal results. In designing surveys that involve the traditional measurements of distances and angles, estimated precisions can be computed for observations made with differing combinations of equipment and field procedures. Also, trial variations in station placement, (station locations create the network geometry), can be made. Then these varying combinations of measurements and geometry can be processed through least-squares adjustments and the resulting station error ellipses computed, plotted, and checked against the desired results. Once acceptable goals are achieved in this process, the measurement equipment and procedures, and network geometry that provide these results can be adopted. This overall process is called *network design.* It enables surveyors to select the equipment and field techniques, and to decide on the number and locations of stations that provide the highest precision at lowest cost. It can be done in the office before entering the field.

In designing networks to be surveyed using the traditional measurements of distance, angle, direction and azimuth, it is important to understand the relationships of those measurements to the resulting positional uncertainties of the stations involved. The following relationships apply:

1. Distance measurements strengthen the positions of stations in directions collinear with the measured lines.
2. Angle, direction, and azimuth observations strengthen the positions of stations in directions perpendicular to the lines of sight.

A simple analysis made with reference to Figure 18.1 should clarify the two relationships above. Assume first that the length of line *AB* was measured precisely, but its direction was not observed. Then the positional uncertainty of station *B* would be held within close tolerances by the measured distance in the direction collinear with *AB*. However, the distance measurement would do nothing to keep line *AB* from rotating, and in fact the position of *B* would be very weak perpendicular to *AB*. On the other hand, if the direction of *AB* had been observed precisely but its length had not been measured, the positional strength of station *B* would be strong in the direction perpendicular to *AB*. But an angle measurement alone does nothing to fix distances between observed stations, and thus the position of station *B* would be very weak along line *AB*. If both the length and direction *AB* were measured with equal precision, a positional uncertainty for station *B* that is more uniform in all directions could be expected. In a survey network that consists of many interconnected stations, analyzing the effects of measurements is not quite as simple as was just demonstrated for the single line *AB*. Nevertheless, the two basic relationships stated above still apply.

Uniform positional strength in all directions for all stations is the desired goal in survey network design. This would be achieved if, following least-squares adjustment, all error ellipses were circular in shape and of equal size. Although this goal is seldom possible, by diligently analyzing various combinations of geometric figures together with different combinations of measurements and precisions, this goal can be approached. Sometimes, however, other overriding factors, such as station accessibility, terrain, and vegetation, preclude the actual use of an optimal design.

The network design process discussed above can be aided significantly with the use of aerial photos and/or topographic maps. These enable layout of trial station locations, and they also permit an analysis of the accessibility of these stations and their intervisibility to be investigated. A field reconnaissance should be made, however, before adopting the final design.

The global positioning system (GPS) has brought about dramatic changes in all areas of surveying, and network design is not an exception. Although GPS does require overhead visibility at each receiver station for tracking satellites, problems of intervisibility between ground stations have been overcome. Thus networks having uniform geometry can normally be laid out. Also, each station in the network is occupied in a GPS survey, and the *XYZ* coordinates of the stations are determined directly. This simplifies the problem of designing networks to attain error ellipses of uniform shapes and sizes. The geometric configuration of observed satellites is an important consideration, however, which affects station precisions. For more discussion on designing GPS surveys, readers are referred to books devoted to the subject of GPS surveying.

18.6.2 Example Network

Figure 18.7 shows error ellipses for two surveys of a network. Figure 18.7(*a*) illustrates the error ellipses from a trilateration survey with the nine stations, two of which (Red and Bug) were control stations. The survey includes 19 distance observations and 5 degrees of freedom. Figure 18.7(*b*) shows the error ellipses of the same network that was measured using triangulation and

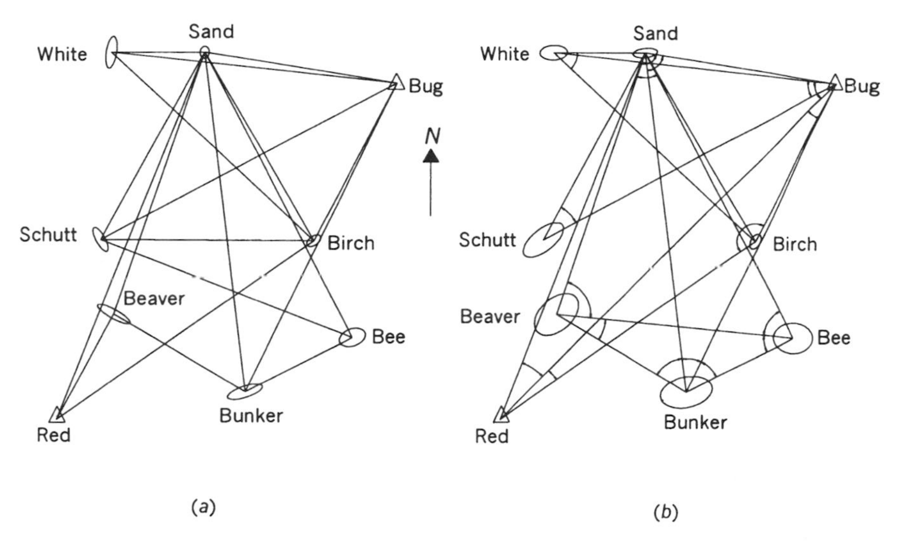

Figure 18.7 Network analysis using error ellipses: (*a*) trilateration with nineteen distances; (*b*) triangulation with nineteen angles.

a baseline from stations Red to Bug. This survey includes 19 observed angles and thus also has 5 degrees of freedom.

With respect to these two figures, and keeping in mind that the smaller the ellipse, the higher the precision, the following general observations can be made:

1. In both figures, Stations Sand and Birch have the highest precisions. This, of course, is expected due to their proximity to control station Bug, and because of the density of observations made to these stations, which include direct measurements from both control stations.
2. The large size of error ellipses at stations Schutt, Beaver, Bunker, and Bee of Figure 18.7(*b*) show that they have lower precision. This, too, is expected because there are fewer observations made to those stations. Also, neither Beaver or Bee are connected directly by observations to either of the control stations.
3. Stations White and Schutt of Figure 18.7(*a*) have relatively high east–west precisions and relatively low north–south precisions. Examination of the network geometry reveals that this could be expected. Distance measurements to those two points from station Red, if they are possible, plus an observed distance between White and Schutt would greatly improve the north–south precision.

4. Stations Beaver and Bunker of Figure 18.7(*a*) have relatively low precisions east–west and relatively high precisions north–south. Again, this is expected when examining the network geometry.
5. The smaller error ellipses of Figure 18.7(*a*) suggest that the trilateration survey will yield superior precision to the triangulation survey of Figure 18.7(*b*). This is expected since the EDM measurements had a stated uncertainty of ±(5mm + 5 ppm). In a 5000 ft distance this yields an uncertainty of ±0.030 ft. To achieve the same precision, the comparable angle uncertainty would need to be

$$\theta'' = \frac{S}{R}\rho = \frac{\pm 0.030}{5000}\, 206{,}264.8''/\text{rad} = \pm 1.2''$$

   The proposed theodolite and field procedures for the project that yielded the error ellipses of Figure 18.7(*b*) had an expected uncertainty of only ±6″. Very probably, the ultimate design would include both measured distances and angles.

These examples serve to illustrate the value of computing station error ellipses in an *a priori* analysis. The comparisons were easily and quickly made by examining the ellipses in the two figures. Similar information would have been difficult, if not impossible, to determine from standard deviations. By varying the survey, it is possible ultimately to find a design that provides optimal results in terms of meeting a uniformly acceptable precision and survey economy.

PROBLEMS

18.1 Calculate the semiminor and semimajor axes of the standard error ellipse for the adjusted position of station *U* in the trilateration Example 13.2. Plot the figure using a scale of 1:12,000 and the error ellipse using an appropriate scale.

18.2 Calculate the semiminor and semimajor axes of the 95% confidence error ellipse for Problem 18.1. Plot this ellipse superimposed over the ellipse of Problem 18.1.

18.3 Repeat Problem 18.1 but for Example 14.1.

18.4 Repeat Problem 18.2 but for Problem 18.3.

18.5 Same as Problem 18.1 except for the adjusted position of station *U* of example 14.2. Use a scale of 1:24,000 for the figure and a scale of 1:12 for the error ellipse.

Calculate the error ellipse data for the unknown stations in each problem.

18.6 Problem 14.2

18.7 Problem 14.5

18.8 Problem 14.6

18.9 Problem 14.7

18.10 Problem 15.1

18.11 Problem 15.2

18.12 Problem 15.3

18.13 Problem 15.5

18.14 Problem 15.6

Using a level of significance of 0.05, compute 95% probable error ellipses for the stations in each problem.

18.15 Problem 15.3

18.16 Problem 15.6

18.17 Using the program STATS, determine the percent probability of the standard error ellipse for a horizontal survey with:
- **(a)** 5 degrees of freedom.
- **(b)** 10 degrees of freedom.
- **(c)** 25 degrees of freedom.
- **(d)** 50 degrees of freedom.
- **(e)** 100 degrees of freedom.

Programming Problems

18.18 Develop a program that takes the Q_{xx} matrix and S_0^2 from a horizontal adjustment and computes error ellipse data for the unknown stations.

18.19 Develop a spreadsheet that does the same as described for Problems 18.15 and 18.16.

19

CONSTRAINT EQUATIONS

19.1 INTRODUCTION

When doing an adjustment, it is sometimes necessary to fix an observation to a specific value. For instance, in Chapter 13, it was shown that the coordinates of a control station can be fixed by setting its *dx* and *dy* corrections to zero, and thus the corrections and their corresponding coefficients in the *J* matrix were removed from the least-squares solution. This is called a *constrained adjustment.* Another constrained adjustment occurs when the direction or length of a line is held to a specific value or when an elevation difference between two stations is fixed in differential leveling. In this chapter methods for developing observational constraints are discussed. However, before discussing constraints, the procedure for including control station coordinates in an adjustment is described.

19.2 ADJUSTMENT OF CONTROL STATION COORDINATES

In examples in preceding chapters, when control station coordinates were excluded from the adjustments, and hence their values held fixed, constrained adjustments were being performed. That is, the measurements were being forced to fit the control coordinates. Control is not perfect, however, and not all control is of equal reliability. This is evidenced by the fact that different orders of accuracy are used to classify control.

When more than minimal control is held fixed in an adjustment, the measurements are forced to fit this control. For example, if the coordinates of two control stations are held fixed, but their actual positions are not in agreement with the values given by their held coordinates, the measurements will be adjusted to match the erroneous control. Simply stated, precise measurements may be forced to fit less precise control. This was not a major problem in the days of transits and tapes, but does happen with modern instrumentation. This topic is discussed in more detail in Chapter 20.

To clarify the problem further, suppose that a new survey is tied to two existing control stations set from two previous surveys. Assume also that the precision of the existing control stations, listed as a distance precision between stations, is 1:10,000. If it is intended that the new survey have a precision of 1:50,000 it is planned at a higher accuracy level than the control to which it must fit. If both existing control stations are fixed in the adjustment, the new measurements would have to adjust to fit the errors of the existing control stations, and after the adjustment, their residuals would probably show a *poor fit*. In this case it would be better to allow the control coordinates to adjust according to their assigned quality.

The observation equations for control station coordinates are

$$
\begin{aligned}
x' &= x + v_x \\
y' &= y + v_y
\end{aligned}
\tag{19.1}
$$

where x' and y' are the adjusted coordinate values of the control station, x and y the published coordinate values of the control station, and v_x and v_y the residuals for the respective published coordinate values.

To allow the control to adjust, Equations (19.1) must be included in the adjustment for each control station. To *fix* a control station in this scheme, high weights are assigned to the station's coordinates. Low weights will allow a control station's coordinates to adjust. With proper weighting, all control stations can be allowed to adjust in accordance with their expected levels of accuracy. In Chapter 20 it will be shown that when the control is included as observations, *poor* measurements and control stations can be isolated in the adjustment by using weights.

Example 19.1 A trilateration survey was completed for the network shown in Figure 19.1. The following data applies:

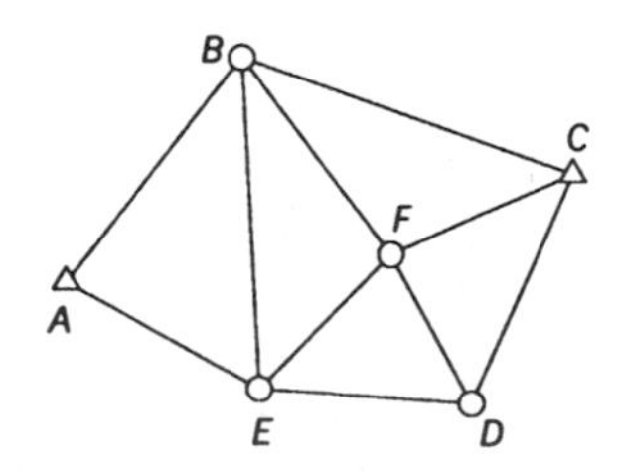

Figure 19.1 Trilateration network.

Control stations

Station	X	Y
A	10,000.00	10,000.00
C	12,487.08	10,528.65

Distance observations

From	To	Distance	S	From	To	Distance	S
A	B	1400.91	0.023	B	E	1644.29	0.023
A	E	1090.55	0.022	B	F	1217.54	0.022
B	C	1723.45	0.023	D	F	842.75	0.022
C	F	976.26	0.022	D	E	1044.99	0.022
C	D	1244.40	0.023	E	F	930.93	0.022

Perform a least-squares adjustment of this survey, holding the control coordinates of stations A and C by appropriate weights. (Assume that these control stations have a precision of 1:10,000.)

SOLUTION: The J, X, and K matrices formed in this adjustment are

$$J = \begin{bmatrix}
\frac{\partial D_{AB}}{\partial x_A} & \frac{\partial D_{AB}}{\partial y_A} & \frac{\partial D_{AB}}{\partial x_B} & \frac{\partial D_{AB}}{\partial y_B} & 0 & 0 & 0 & 0 & 0 & 0 & 0 & 0 \\
\frac{\partial D_{AE}}{\partial x_A} & \frac{\partial D_{AE}}{\partial y_A} & 0 & 0 & 0 & 0 & 0 & 0 & \frac{\partial D_{AE}}{\partial x_E} & \frac{\partial D_{AE}}{\partial y_E} & 0 & 0 \\
0 & 0 & \frac{\partial D_{BE}}{\partial x_B} & \frac{\partial D_{BE}}{\partial y_B} & 0 & 0 & 0 & 0 & \frac{\partial D_{BE}}{\partial x_E} & \frac{\partial D_{BE}}{\partial y_E} & 0 & 0 \\
0 & 0 & \frac{\partial D_{BF}}{\partial x_B} & \frac{\partial D_{BF}}{\partial y_B} & 0 & 0 & 0 & 0 & 0 & 0 & \frac{\partial D_{BF}}{\partial x_F} & \frac{\partial D_{BF}}{\partial y_F} \\
0 & 0 & \frac{\partial D_{BC}}{\partial x_B} & \frac{\partial D_{BC}}{\partial y_B} & \frac{\partial D_{BC}}{\partial x_C} & \frac{\partial D_{BC}}{\partial y_C} & 0 & 0 & 0 & 0 & 0 & 0 \\
0 & 0 & 0 & 0 & \frac{\partial D_{CF}}{\partial x_C} & \frac{\partial D_{CF}}{\partial y_C} & 0 & 0 & 0 & 0 & \frac{\partial D_{CF}}{\partial x_F} & \frac{\partial D_{CF}}{\partial y_F} \\
0 & 0 & 0 & 0 & \frac{\partial D_{CD}}{\partial x_C} & \frac{\partial D_{CD}}{\partial y_C} & \frac{\partial D_{CD}}{\partial x_D} & \frac{\partial D_{CD}}{\partial y_D} & 0 & 0 & 0 & 0 \\
0 & 0 & 0 & 0 & 0 & 0 & \frac{\partial D_{DF}}{\partial x_D} & \frac{\partial D_{DF}}{\partial y_D} & 0 & 0 & \frac{\partial D_{DF}}{\partial x_F} & \frac{\partial D_{DF}}{\partial y_F} \\
0 & 0 & 0 & 0 & 0 & 0 & \frac{\partial D_{DE}}{\partial x_D} & \frac{\partial D_{DE}}{\partial y_D} & \frac{\partial D_{DE}}{\partial x_E} & \frac{\partial D_{DE}}{\partial y_E} & 0 & 0 \\
0 & 0 & 0 & 0 & 0 & 0 & 0 & 0 & \frac{\partial D_{EF}}{\partial x_E} & \frac{\partial D_{EF}}{\partial y_E} & \frac{\partial D_{EF}}{\partial x_F} & \frac{\partial D_{EF}}{\partial y_F} \\
1 & 0 & 0 & 0 & 0 & 0 & 0 & 0 & 0 & 0 & 0 & 0 \\
0 & 1 & 0 & 0 & 0 & 0 & 0 & 0 & 0 & 0 & 0 & 0 \\
0 & 0 & 0 & 0 & 1 & 0 & 0 & 0 & 0 & 0 & 0 & 0 \\
0 & 0 & 0 & 0 & 0 & 1 & 0 & 0 & 0 & 0 & 0 & 0
\end{bmatrix}$$

$$X = \begin{bmatrix} dx_A \\ dy_A \\ dx_B \\ dy_B \\ dx_C \\ dy_C \\ dx_D \\ dy_D \\ dx_E \\ dy_E \\ dx_F \\ dy_F \end{bmatrix} \qquad K = \begin{bmatrix} L_{AB} - AB_0 \\ L_{AE} - AE_0 \\ L_{BE} - BE_0 \\ L_{BF} - BF_0 \\ L_{BC} - BC_0 \\ L_{CF} - CF_0 \\ L_{CD} - CD_0 \\ L_{DF} - DF_0 \\ L_{DE} - DE_0 \\ L_{EF} - EF_0 \\ x_A - x_{A0} \\ y_A - y_{A0} \\ x_C - x_{C0} \\ y_C - y_{C0} \end{bmatrix}$$

Notice that the last four rows of the J matrix correspond to observation equations (19.1) for the coordinates of control stations A and C. Each coordinate has a row with *one* in the column corresponding to its correction. Obviously, by including the control station coordinates, four unknowns have been added to the adjustment: dx_A, dy_A, dx_C, and dy_C. However, four observations have also been added, and thus the number of redundancies is unaffected by adding the coordinate observation equations. That is, the adjustment has the same number of degrees of freedom with or without the control equations.

It is possible to weight a control station according to the precision of its coordinates. Unfortunately, control stations are published with distance precisions rather than covariance matrix elements, which are required in weighting. However, estimates of the standard deviations of the coordinates can be computed from the published distance precisions. That is, if the distance precision between stations A and C is 1:10,000, their coordinates should have estimated errors that yield a distance precision of 1:10,000 or better between the stations. To find the estimated errors in the coordinates that yield the appropriate distance precision between the stations, the error propagation equation can be applied to the distance formula. The resulting equation is

$$S_{Dij}^2 = \left(\frac{\partial D_{ij}}{\partial x_i}\right)^2 S_{x_i}^2 + \left(\frac{\partial D_{ij}}{\partial y_i}\right)^2 S_{y_i}^2 + \left(\frac{\partial D_{ij}}{\partial x_j}\right)^2 S_{x_j}^2 + \left(\frac{\partial D_{ij}}{\partial y_j}\right)^2 S_{y_j}^2 \quad (19.2)$$

where $S_{D_{ij}}^2$ is the variance in distance D_{ij}, and $S_{x_i}^2$, $S_{y_i}^2$, $S_{x_j}^2$, and $S_{y_j}^2$ are variances in the coordinates of the endpoints of the line. Assuming that the estimated errors for the coordinates of Equation (19.2) are equal, and substituting in the appropriate partial derivatives, we obtain

$$S_{Dij}^2 = 2\left(\frac{\partial D_{ij}}{\partial x}\right)^2 S_x^2 + 2\left(\frac{\partial D_{ij}}{\partial y}\right)^2 S_y^2 = 2\left(\frac{\Delta x}{IJ} S_c\right)^2 + 2\left(\frac{\Delta y}{IJ} S_c\right)^2 \quad (19.3)$$

In Equation (19.3), S_c is the standard deviation in the x and y coordinates. [Note that the partial derivatives in Equation (19.3) were described in Section 13.2.] Factoring $2S_c^2$ from Equation (19.3) gives

$$S_{D_{ij}}^2 = 2S_c^2 \frac{x^2 + y^2}{IJ^2} = 2S_c^2 \tag{19.4}$$

where $S_c^2 = S_x^2 = S_y^2$.

From the coordinates of A and C, distance AC is 2542.65 ft. For a distance precision of 1:10,000, the maximum distance uncertainty $S_{D_{ij}}$ is $\pm 2542.65/10{,}000 = \pm 0.25$ ft. Assuming equal coordinate uncertainties, then

$$\pm 0.25 \text{ ft} = S_x \sqrt{2} = S_y \sqrt{2}$$

$$\text{Thus } S_x = S_y = \pm 0.25/\sqrt{2} = \pm 0.18 \text{ ft}$$

These standard deviations are used to weight the control in the adjustment. The entire weight matrix used in this adjustment follows.

$W =$

$$\begin{bmatrix}
\frac{1}{0.023^2} & 0 & 0 & 0 & 0 & 0 & 0 & 0 & 0 & 0 & 0 & 0 & 0 & 0 \\
0 & \frac{1}{0.022^2} & 0 & 0 & 0 & 0 & 0 & 0 & 0 & 0 & 0 & 0 & 0 & 0 \\
0 & 0 & \frac{1}{0.023^2} & 0 & 0 & 0 & 0 & 0 & 0 & 0 & 0 & 0 & 0 & 0 \\
0 & 0 & 0 & \frac{1}{0.022^2} & 0 & 0 & 0 & 0 & 0 & 0 & 0 & 0 & 0 & 0 \\
0 & 0 & 0 & 0 & \frac{1}{0.023^2} & 0 & 0 & 0 & 0 & 0 & 0 & 0 & 0 & 0 \\
0 & 0 & 0 & 0 & 0 & \frac{1}{0.022^2} & 0 & 0 & 0 & 0 & 0 & 0 & 0 & 0 \\
0 & 0 & 0 & 0 & 0 & 0 & \frac{1}{0.023^2} & 0 & 0 & 0 & 0 & 0 & 0 & 0 \\
0 & 0 & 0 & 0 & 0 & 0 & 0 & \frac{1}{0.022^2} & 0 & 0 & 0 & 0 & 0 & 0 \\
0 & 0 & 0 & 0 & 0 & 0 & 0 & 0 & \frac{1}{0.022^2} & 0 & 0 & 0 & 0 & 0 \\
0 & 0 & 0 & 0 & 0 & 0 & 0 & 0 & 0 & \frac{1}{0.022^2} & 0 & 0 & 0 & 0 \\
0 & 0 & 0 & 0 & 0 & 0 & 0 & 0 & 0 & 0 & \frac{1}{0.18^2} & 0 & 0 & 0 \\
0 & 0 & 0 & 0 & 0 & 0 & 0 & 0 & 0 & 0 & 0 & \frac{1}{0.18^2} & 0 & 0 \\
0 & 0 & 0 & 0 & 0 & 0 & 0 & 0 & 0 & 0 & 0 & 0 & \frac{1}{0.18^2} & 0 \\
0 & 0 & 0 & 0 & 0 & 0 & 0 & 0 & 0 & 0 & 0 & 0 & 0 & \frac{1}{0.18^2}
\end{bmatrix}$$

The adjustment results, obtained using the program ADJUST, are shown below.

```
******************************
Adjusted Distance Observations
******************************
```

No.	From	To	Distance	Residual
1	A	B	1,400.910	−0.000
2	A	E	1,090.550	−0.000
3	B	E	1,644.288	−0.002
4	B	F	1,217.544	0.004
5	B	C	1,723.447	−0.003
6	C	F	976.263	0.003
7	C	D	1,244.397	−0.003
8	D	E	1,044.988	−0.002
9	E	F	930.933	0.003
10	D	F	842.753	0.003

```
*************************
Adjusted Control Stations
*************************
```

No.	Sta.	Northing	Easting	N Res	E Res
1	C	10,528.650	12,487.080	0.000	0.002
2	A	10,000.000	10,000.000	−0.000	−0.002

Reference Standard Deviation = 0.25
Degrees of Freedom = 2

```
*****************
Adjusted Unknowns
*****************
```

Station	Northing	Easting	σ North	σ East	t ang°	A axis	B axis
A	10,000.000	9,999.998	0.033	0.045	168.000	0.046	0.033
B	11,103.933	10,862.483	0.039	0.034	65.522	0.040	0.033
C	10,528.650	12,487.082	0.033	0.045	168.000	0.046	0.033
D	9,387.462	11,990.882	0.040	0.038	49.910	0.044	0.033
E	9,461.900	10,948.549	0.039	0.034	110.409	0.039	0.033
F	10,131.563	11,595.223	0.033	0.034	17.967	0.034	0.033

Notice that the control stations were adjusted slightly, as evidenced by their residuals. Also note the error ellipse data computed for each control station.

19.3 HOLDING CONTROL STATION COORDINATES AND DIRECTIONS OF LINES FIXED IN A TRILATERATION ADJUSTMENT

As demonstrated in Example 13.1, the coordinates of a control station are easily fixed during an adjustment. This is accomplished by assigning values of zero to the coefficients of the dx and dy correction terms. This method removes their corrections from the equations. In that particular example, each observation equation had only two unknowns, since one end of each measured line was a control station which was held fixed during the adjustment. This was a special case of a method known as *solution by elimination of constraints*.

This method can be shown in matrix notation as

$$A_1X_1 + A_2X_2 = L_1 + V \tag{19.5}$$

$$C_1X_1 + C_2X_2 = L_2 \tag{19.6}$$

where A_1, A_2, X_1, X_2, L_1, and L_2 are the A, X, and L matrices partitioned by the constraint equations, as shown in Figure 19.2; C_1 and C_2 are the partitions of the matrix C consisting of the coefficients of the constraint equations; and V is the residual matrix. In this method, matrices A, C, and X are partitioned into two matrix equations that separate the constrained and unconstrained observations. Careful consideration should be given to the partition of C_1 since this matrix cannot be singular. If singularity exists, a new set of constraint equations that are independent must be determined. Also, since each constraint equation will remove one parameter from the adjustment, the number of constraints must not be so large that the remaining A_1 and X_1 have no independent equations or are themselves singular.

From Equation (19.6), solve for X_1 in terms of C_1, C_2, X_2, and L_2 as

$$X_1 = C_1^{-1}(L_2 - C_2X_2) \tag{19.7}$$

Substituting Equation (19.7) into Equation (19.5) yields

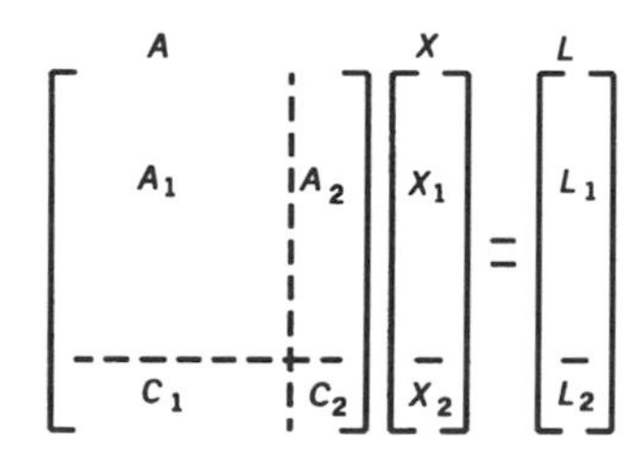

Figure 19.2 *A*, *X* and *L* matrices partitioned by constraint equations.

$$A_1[C_1^{-1}(L_2 - C_2X_2)] + A_2X_2 = L_1 + V \tag{19.8}$$

Rearranging Equation (19.8), regrouping, and dropping V for the time being gives

$$(-A_1C_1^{-1}C_2 + A_2)X_2 = L_1 - A_1C_1^{-1}L_2 \tag{19.9}$$

Letting $A' = -A_1C_1^{-1}C_2 + A_2$, Equation (19.9) can be rewritten as

$$A'X_2 = L_1 - A_1C_1^{-1}L_2 \tag{19.10}$$

Now Equation (19.10) can be solved for X_2, which in turn is substituted into Equation (19.7) to solve for X_1.

It can be seen that in the solution by elimination of constraints, the constraint equations are used to eliminate unknown parameters from the adjustment, thereby fixing certain geometric conditions during the adjustment. This method was used when the coordinates of the control stations were *removed* from the adjustments in previous chapters. In the following subsection, this method is used to hold the azimuth of a line during an adjustment.

19.3.1 Holding the Direction of a Line Fixed by Elimination of Constraints

Using this method, constraint equations are written and then functionally substituted into the observation equations to eliminate unknown parameters. To illustrate, consider that in Figure 19.3, the direction of a line *IJ* is to be held fixed during an adjustment. Thus the position of *J* is constrained to move linearly along *IJ* during the adjustment. If *J* moves to *J'* after adjustment, the relationship between the direction of *IJ* and dx_j and dy_j is

$$dx_j = dy_j \tan(\alpha) \tag{19.11}$$

Suppose, for example, that in Figure 19.4, the direction of line *AB* is to be held fixed during a trilateration adjustment. Noting that station *A* is to be held

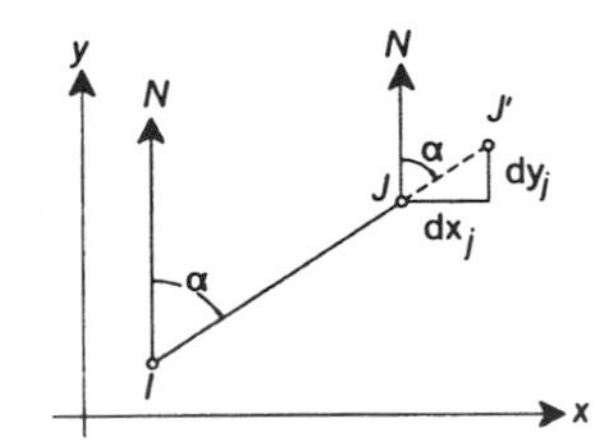

Figure 19.3 Holding direction *IJ* fixed.

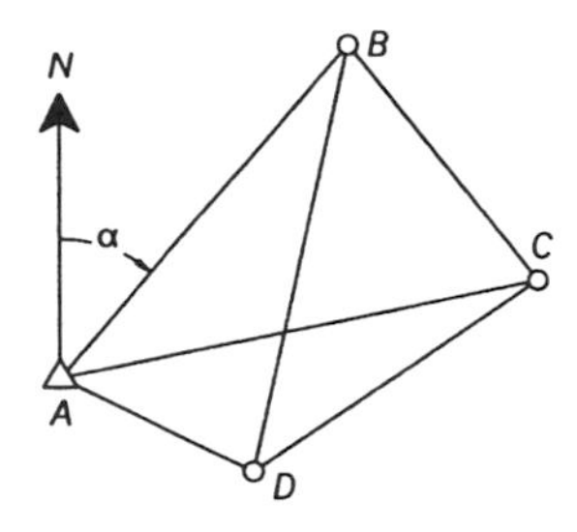

Figure 19.4 Holding direction *AB* fixed in a trilateration adjustment.

fixed and using prototype equation (13.9), the following observation equation results for measured distance *AB*:

$$k_{l_{ab}} + v_{l_{ab}} = \frac{x_{b0} - x_a}{AB_0} dx_b + \frac{y_{b0} - y_a}{AB_0} dy_b \tag{19.12}$$

Now based on Equation (19.11), the following relationship is written for line *AB*:

$$dx_b = dy_b \tan(\alpha) \tag{19.13}$$

Substituting Equation (19.13) into Equation (19.12) gives

$$k_{l_{ab}} + v_{l_{ab}} = \left[\frac{x_{b0} - x_a}{AB_0} \tan(\alpha)\right] dy_b + \frac{y_{b0} - y_a}{AB_0} dy_b \tag{19.14}$$

Factoring dy_b in Equation (19.14), the following observation equation is obtained:

$$k_{l_{ab}} + v_{l_{ab}} = \frac{(x_{b_0} - x_a)\tan(\alpha) + (y_{b_0} - y_a)}{AB_0} dy_b \tag{19.15}$$

Using this same method, the coefficients of dy_b for measured lines *BC* and *BD* are also determined, resulting in the *J* matrix shown in Table 19.1. For this example the *K*, *X*, and *V* matrices are

$$K = \begin{bmatrix} AB - AB_0 \\ AC - AC_0 \\ AD - AD_0 \\ BC - BC_0 \\ BD - BD_0 \\ CD - CD_0 \end{bmatrix} \qquad X = \begin{bmatrix} dy_b \\ dy_c \\ dx_c \\ dy_d \\ dx_d \end{bmatrix} \qquad V = \begin{bmatrix} v_{l_{ab}} \\ v_{l_{ac}} \\ v_{l_{ad}} \\ v_{l_{bc}} \\ v_{l_{bd}} \\ v_{l_{cd}} \end{bmatrix}$$

Table 19.1 *J* matrix for Figure 19.3

Distance	Unknown				
	dy_b	dy_c	dx_c	dx_d	dy_d
AB	$[(x_b - x_a)\tan(\alpha) + (y_b - y_a)]/AB$	0	0	0	0
AC	0	$(y_c - y_a)/AC$	$(x_c - x_a)/AC$	0	0
AD	0	0	0	$(y_d - y_a)/AD$	$(x_d - x_a)/AD$
BC	$[(x_b - x_c)\tan(\alpha) + (y_c - y_b)/BC$	$y_c - y_b/BC$	$(x_c - x_b)/BC$	0	0
BD	$[(x_b - x_d)\tan(\alpha) + (y_b - y_d)/BD$	0	0	$(y_d - y_b)/BD$	$(x_d - x_b)/BD$
CD	0	$(y_c - y_d)/CD$	$(x_c - x_d)/CD$	$(y_d - y_c)/CD$	$(x_d - x_c)/CD$

(Note: In the equations above, initial approximations must be used for all unknowns)

19.4 HELMERT'S METHOD

Another method of introducing constraints was originally presented by F. R. Helmert in 1872. In this procedure, the constraint equations border the reduced normal equations as

$$\begin{bmatrix} A^TWA & \vdots & C^T \\ \text{----} & \text{----} & \text{----} \\ C & \vdots & 0 \end{bmatrix} \begin{bmatrix} X_1 \\ \text{----} \\ X_2 \end{bmatrix} = \begin{bmatrix} A^TWL_1 \\ \text{----} \\ L_2 \end{bmatrix} \tag{19.16}$$

To establish this matrix expression, the normal matrix and its matching constants matrix are formed, as has been done in Chapters 12 through 18. Following this, the observation equations for the constraints are formed. These observation equations are then included in the normal matrix as additional rows $[C]$ and columns $[C^T]$ in Equation (19.16), and their constants are added to the constants matrix as additional rows [L_2 in Equation (19.16)]. The inverse of this bordered normal matrix is computed, and thus the matrix solution of the Equation (19.16) is

$$\begin{bmatrix} X_1 \\ \text{----} \\ X_2 \end{bmatrix} = \begin{bmatrix} A^TWA & \vdots & C^T \\ \text{----} & \text{----} & \text{----} \\ C & \vdots & 0 \end{bmatrix}^{-1} \begin{bmatrix} A^TWL_1 \\ \text{----} \\ L_2 \end{bmatrix} \tag{19.17}$$

where X_2 is not used in the subsequent solution for the unknowns. This procedure is illustrated below using several examples.

Example 19.2 *Constrained Differential Leveling Adjustment.* In the leveling network of Figure 19.5, the elevation difference between stations B and E is to be held at -17.60 ft. The elevation of A is 1300.62 ft. Measured elevation differences for each line are shown below.

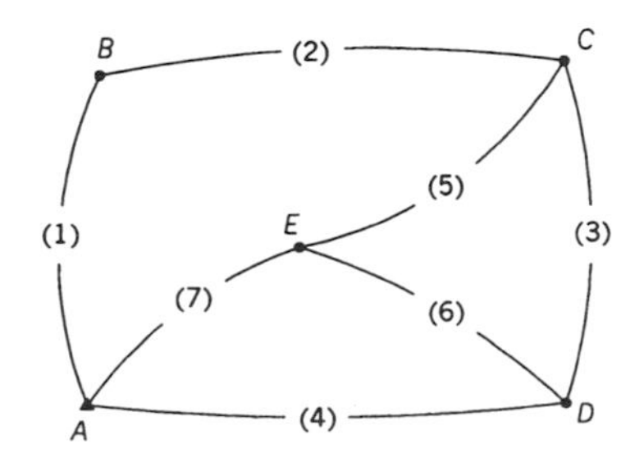

Figure 19.5 Differential leveling network.

Line	From	To	Elevation	S
1	A	B	25.15	0.07
2	B	C	−10.57	0.05
3	C	D	−1.76	0.03
4	D	A	−12.65	0.08
5	C	E	−7.06	0.03
6	E	D	5.37	0.05
7	E	A	−7.47	0.05

Perform a least-squares adjustment of this level net, constraining the required elevation difference using Helmert's method.

SOLUTION: The A, X, and L matrices are

$$A = \begin{bmatrix} 1 & 0 & 0 & 0 \\ -1 & 1 & 0 & 0 \\ 0 & -1 & 1 & 0 \\ 0 & 0 & -1 & 0 \\ 0 & -1 & 0 & 1 \\ 0 & 0 & 1 & -1 \\ 0 & 0 & 0 & -1 \end{bmatrix} \qquad X = \begin{bmatrix} B \\ C \\ D \\ E \end{bmatrix} \qquad L = \begin{bmatrix} 1325.77 \\ -10.55 \\ -1.76 \\ -1313.27 \\ -7.06 \\ 5.37 \\ 1308.09 \end{bmatrix}$$

The weight matrix (W) is

$$W = \begin{bmatrix} 204.08 & 0 & 0 & 0 & 0 & 0 & 0 \\ 0 & 400 & 0 & 0 & 0 & 0 & 0 \\ 0 & 0 & 1111.11 & 0 & 0 & 0 & 0 \\ 0 & 0 & 0 & 156.25 & 0 & 0 & 0 \\ 0 & 0 & 0 & 0 & 1111.11 & 0 & 0 \\ 0 & 0 & 0 & 0 & 0 & 400 & 0 \\ 0 & 0 & 0 & 0 & 0 & 0 & 400 \end{bmatrix}$$

The reduced normal equations are

$$\begin{bmatrix} 604.08 & -400.00 & 0.00 & 0.00 \\ -400.00 & 2622.22 & -1111.11 & -1111.11 \\ 0.00 & -1111.11 & 1667.36 & -400.00 \\ 0.00 & -1111.11 & -400.00 & 1911.11 \end{bmatrix} \begin{bmatrix} B \\ C \\ D \\ E \end{bmatrix} = \begin{bmatrix} 274{,}793.30 \\ 5{,}572.00 \\ 205{,}390.90 \\ 513{,}243.60 \end{bmatrix} \tag{a}$$

The reduced normal matrix is now bordered by the constraint equation

$$E - B = -17.60$$

which has an observation equation form of

$$\begin{bmatrix} -1 & 0 & 0 & 1 \end{bmatrix} \begin{bmatrix} B \\ C \\ D \\ E \end{bmatrix} = [-17.60] \qquad (b)$$

The left side of Equation (*b*) is now included as an additional row and column to the reduced normal matrix in Equation (*a*). The lower-right corner diagonal element of the newly bordered normal matrix is assigned a value of 0. Similarly, the right-hand side of Equation (*b*) is included as an additional row in the constants matrix of Equation (*a*). Thus the bordered-normal equations are

$$\begin{bmatrix} 604.08 & -400.00 & 0.00 & 0.00 & -1 \\ -400.00 & 2622.22 & -1111.11 & -1111.11 & 0 \\ 0.00 & -1111.11 & 1667.36 & -400.00 & 0 \\ 0.00 & -1111.11 & -400.00 & 1911.11 & 1 \\ -1 & 0 & 0 & 1 & 0 \end{bmatrix} \begin{bmatrix} B \\ C \\ D \\ E \\ X_2 \end{bmatrix}$$

$$= \begin{bmatrix} 274{,}793.30 \\ 5{,}572.00 \\ 205{,}390.90 \\ 513{,}243.60 \\ -17.60 \end{bmatrix} \qquad (c)$$

Notice in Equation (*c*) that an additional unknown, X_2, is added at the bottom of the X matrix to make it dimensionally consistent with the bordered-normal matrix. Solving Equation (*c*), gives the following X matrix of unknowns:

$$X = \begin{bmatrix} 1325.686 \\ 1315.143 \\ 1313.390 \\ 1308.086 \\ -28.003 \end{bmatrix} \qquad (d)$$

From the X matrix in Equation (*d*), the elevation of station B is 1325.686, and that for station E is 1308.086. Thus the elevation difference between stations B and E is exactly -17.60, which was enforced by the constraint condition.

Example 19.3 *Constraining the Azimuth of a Line.* Helmert's method can also be used to constrain the direction of a line. In Figure 19.6 the bearing of line *AB* is to remain at its record value of N 0°04′ E. The data for this trilaterated network are:

Control stations

Station	x	y
A	1000.000	1000.000

Station initial approximations

Station	x	y
B	1003.07	3640.00
C	2323.07	3638.46
D	2496.08	1061.74

Distance observations

Occupied	Sighted	Distance	S	Occupied	Sighted	Distance	S
A	*C*	2951.604	0.025	*C*	*D*	2582.534	0.024
A	*B*	2640.017	0.024	*D*	*A*	1497.360	0.021
B	*C*	1320.016	0.021	*B*	*D*	2979.325	0.025

Adjust this figure by least squares, holding the direction of the line *AB* using Helmert's method.

SOLUTION: Using methods discussed in Chapter 13, the reduced normal equations for the trilaterated system are

$$\begin{bmatrix} 2690.728 & -706.157 & -2284.890 & 0.000 & -405.837 & 706.157 \\ -706.157 & 2988.234 & 0.000 & 0.000 & 706.157 & -1228.714 \\ -2284.890 & 0.000 & 2624.565 & 529.707 & -8.737 & 124.814 \\ 0.000 & 0.000 & 529.707 & 3077.557 & 124.814 & -1783.060 \\ -405.837 & 706.157 & -8.737 & 124.814 & 2636.054 & -742.112 \\ 706.157 & -1228.714 & 124.814 & -1783.060 & -742.112 & 3015.328 \end{bmatrix}$$

$$\times \begin{bmatrix} dx_b \\ dy_b \\ dx_c \\ dy_c \\ dx_d \\ dy_d \end{bmatrix} = \begin{bmatrix} -6.615 \\ 6.944 \\ 21.601 \\ 17.831 \\ 7.229 \\ 11.304 \end{bmatrix} \qquad (e)$$

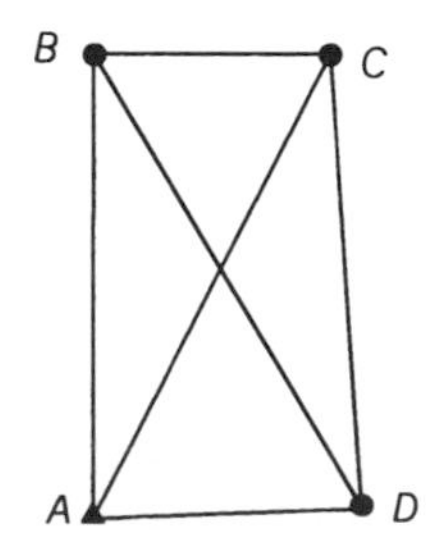

Figure 19.6 Network for Example 19.3.

Following prototype equation (14.9), the linearized equation for the azimuth of line *AB* is

$$\begin{bmatrix} 78.13 & -0.09 & 0 & 0 & 0 & 0 \end{bmatrix} \begin{bmatrix} dx_b \\ dy_b \\ dx_c \\ dy_c \\ dx_d \\ dy_d \end{bmatrix} = [0.139] \qquad (f)$$

The observation equation for the constrained direction [Equation (f)] is then added to border the matrix of reduced normal equations (e), which yields the following matrices:

$$\begin{bmatrix} 2690.728 & -706.157 & -2284.890 & 0.000 & -405.837 & 706.157 & 78.13 \\ -706.157 & 2988.234 & 0.000 & 0.000 & 706.157 & -1228.714 & -0.09 \\ -2284.890 & 0.000 & 2624.565 & 529.707 & -8.737 & 124.814 & 0 \\ 0.000 & 0.000 & 529.707 & 3077.557 & 124.814 & -1783.060 & 0 \\ -405.837 & 706.157 & -8.737 & 124.814 & 2636.054 & -742.112 & 0 \\ 706.157 & -1228.714 & 124.814 & -1783.060 & -742.112 & 3015.328 & 0 \\ 78.13 & -0.09 & 0 & 0 & 0 & 0 & 0 \end{bmatrix}$$

$$\times \begin{bmatrix} dx_b \\ dy_b \\ dx_c \\ dy_c \\ dx_d \\ dy_d \\ dx_2 \end{bmatrix} = \begin{bmatrix} -6.615 \\ 6.944 \\ 21.601 \\ 17.831 \\ 7.229 \\ 11.304 \\ 0.139 \end{bmatrix} \qquad (g)$$

This is a nonlinear problem, and thus the solution must be iterated until convergence. The first two iterations yielded the X matrices listed as X_1 and X_2 below. The third iteration resulted in negligible corrections to the unknowns. The total of these corrections in shown as X_T below.

$$X_1 = \begin{bmatrix} 0.00179 \\ 0.00799 \\ 0.00636 \\ 0.01343 \\ 0.00460 \\ 0.01540 \\ -0.00337 \end{bmatrix} \qquad X_2 = \begin{bmatrix} -0.00001 \\ -0.00553 \\ 0.00477 \\ -0.00508 \\ -0.00342 \\ -0.00719 \\ -0.00359 \end{bmatrix} \qquad X_T = \begin{bmatrix} 0.00178 \\ 0.00247 \\ 0.01113 \\ 0.00835 \\ 0.00117 \\ 0.00821 \\ -0.00696 \end{bmatrix}$$

Adding the coordinate corrections of X_T to the initial approximations gives the final coordinates for stations *B*, *C*, and *D* as

B: (1003.072, 3640.003) *C*: (2323.081, 3638.468)

$$D: (2496.081, 1061.748)$$

Checking the solution: Using Equation (14.1), check to see that the direction of line AB was held to the value of the constraint.

$$\text{Az}_{AB} = \tan^{-1}\left(\frac{3.072}{2640.003}\right) = 0°04'00'' \text{ (check!)}$$

19.5 REDUNDANCIES IN A CONSTRAINED ADJUSTMENT

The number of redundancies in an adjustment increases by one for each parameter that is removed by a constraint equation. An expression for determining the number of redundancies is

$$r = m - n + c \tag{19.18}$$

where r is the number of redundancies (degrees of freedom) in the system, m the number of observations in the system, n the number of unknown parameters in the system, and c the number of mathematically independent constraints applied to the system. In Example 19.2 there were seven observations in a differential leveling network that had four stations with unknown elevations. One constraint was added to the system of equations that fixed the elevation difference between B and E as -17.60. In this way the elevation of B and E became mathematically dependent. By applying Equation (19.18), it can be seen that the number of redundancies in the system is $r = 7 - 4 + 1 = 4$. Without the aforementioned constraint, this adjustment would have only $7 - 4 = 3$ redundancies. Thus the constraint making the elevations of B and E mathematically dependent, added 1 degree of freedom to the adjustment.

Care must be used when adding constraints to an adjustment. It would be possible to add as many mathematically independent constraint equations as there are unknown parameters. If that were done, all unknowns would be constrained or fixed and it would be a waste of time if an adjustment were done. Furthermore, it is also possible to add constraints that are not mathematically independent equations. Under these circumstances, even if the system of equations has a solution, two mathematically dependent constraints would remove only one unknown parameter, and thus the redundancies in the system would increase by only one.

19.6 ENFORCING CONSTRAINTS THROUGH WEIGHTING

The methods described above for handling constraint equations can often be avoided simply by heavily weighting the measurements to be constrained in an otherwise ordinary weighted least-squares adjustment. This was done in Example 15.2 to fix the direction of a line. As a further demonstration of the procedure of enforcing constraints by weighting, Example 19.3 will be adjusted by writing observation equations for azimuth AB and the control station coordinates X_A and Y_A. These observations will be fixed by assigning a 0.001″ standard deviation to the azimuth of line AB and standard deviations of 0.001 ft to the coordinates of station A.

The J, K, and W matrices for the first iteration of this problem are listed below.

$$J = \begin{bmatrix} -0.448254 & -0.893906 & 0.000000 & 0.000000 & 0.448254 & 0.893906 & 0.000000 & 0.000000 \\ -0.001163 & -0.999999 & 0.001163 & 0.999999 & 0.000000 & 0.000000 & 0.000000 & 0.000000 \\ 0.000000 & 0.000000 & -0.999999 & 0.001167 & 0.999999 & -0.001167 & 0.000000 & 0.000000 \\ 0.000000 & 0.000000 & 0.000000 & 0.000000 & -0.066993 & 0.997753 & 0.066993 & -0.997753 \\ -0.999150 & -0.041233 & 0.000000 & 0.000000 & 0.000000 & 0.000000 & 0.999150 & 0.041233 \\ 0.000000 & 0.000000 & -0.501120 & 0.865378 & 0.000000 & 0.000000 & 0.501120 & -0.865378 \\ -78.130503 & 0.090856 & 78.130503 & -0.090856 & 0.000000 & 0.000000 & 0.000000 & 0.000000 \\ 1.000000 & 0.000000 & 0.000000 & 0.000000 & 0.000000 & 0.000000 & 0.000000 & 0.000000 \\ 0.000000 & 1.000000 & 0.000000 & 0.000000 & 0.000000 & 0.000000 & 0.000000 & 0.000000 \end{bmatrix}$$

$$K = \begin{bmatrix} -0.003256 \\ 0.015215 \\ 0.015102 \\ 0.012283 \\ 0.006604 \\ -0.021151 \\ 0.139140 \\ 0.000000 \\ 0.000000 \end{bmatrix}$$

$$W = \begin{bmatrix} \frac{1}{0.025^2} & 0 & 0 & 0 & 0 & 0 & 0 & 0 & 0 \\ 0 & \frac{1}{0.024^2} & 0 & 0 & 0 & 0 & 0 & 0 & 0 \\ 0 & 0 & \frac{1}{0.021^2} & 0 & 0 & 0 & 0 & 0 & 0 \\ 0 & 0 & 0 & \frac{1}{0.024^2} & 0 & 0 & 0 & 0 & 0 \\ 0 & 0 & 0 & 0 & \frac{1}{0.021^2} & 0 & 0 & 0 & 0 \\ 0 & 0 & 0 & 0 & 0 & \frac{1}{0.025^2} & 0 & 0 & 0 \\ 0 & 0 & 0 & 0 & 0 & 0 & \frac{\rho^2}{0.001^2} & 0 & 0 \\ 0 & 0 & 0 & 0 & 0 & 0 & 0 & \frac{1}{0.001^2} & 0 \\ 0 & 0 & 0 & 0 & 0 & 0 & 0 & 0 & \frac{1}{0.001^2} \end{bmatrix}$$

The results of the adjustment from program ADJUST are presented below.

```
*****************
Adjusted Stations
*****************
```

				Standard Error Ellipses			Computed
Station	X	Y	Sx	Sy	Su	Sv	t
A	1,000.000	1,000.000	0.0010	0.0010	0.0010	0.0010	135.00°
B	1,003.072	3,640.003	0.0010	0.0217	0.0217	0.0010	0.07°
C	2,323.081	3,638.468	0.0205	0.0248	0.0263	0.0186	152.10°
D	2,496.081	1,061.748	0.0204	0.0275	0.0281	0.0196	16.37°

```
******************************
Adjusted Distance Observations
******************************
```

Station Occupied	Station Sighted	Distance	V	S
A	C	2,951.620	0.0157	0.0215
A	B	2,640.004	−0.0127	0.0217
B	C	1,320.010	−0.0056	0.0205
C	D	2,582.521	−0.0130	0.0215
D	A	1,497.355	−0.0050	0.0206
B	D	2,979.341	0.0159	0.0214

```
*****************************
Adjusted Azimuth Observations
*****************************
```

Station Occupied	Station Sighted	Azimuth	V	S
A	B	0°04′00″	0.0″	0.0″

```
*********************
Adjustment Statistics
*********************
```

Iterations = 2

Redundancies = 1

Reference Variance = 1.499

Reference So = ±1.2

```
Passed X² test at 95.0% significance level!
X² lower value = 0.00
X² upper value = 5.02
A priori value of 1 used for reference variance
in computations of statistics.
Convergence!
```

Notice in the adjustment above that the coordinates of control station *A* remained fixed and the residual of the azimuth of line *AB* is zero. Thus these values were held fixed without the inclusion of constraint equation simply by heavily weighting these observations. Also note that the final adjusted coordinates of stations *B*, *C*, and *D* match the solution in Example 19.3.

PROBLEMS

Note: For problems below requiring least-squares adjustment, if a computer program is not distinctly specified for use in the problem, it is expected that the least-squares algorithm will be solved using the program MATRIX which is included within the diskette supplied with the book.

19.1 Given the following measured lengths in a trilateration survey:

$AB = 22{,}867.12 \pm 0.116$ ft $DA = 29{,}593.60 \pm 0.149$ ft

$BC = 22{,}943.74 \pm 0.116$ ft $AC = 30{,}728.64 \pm 0.155$ ft

$CD = 28{,}218.26 \pm 0.142$ ft $BD = 41{,}470.07 \pm 0.208$ ft

Adjust this survey by least squares using the elimination of constraints method to hold the coordinates of *A* at $x_a = 30{,}000.00$ and $y_a = 30{,}000.00$ and the azimuth of line *AB* to 30°00′00.00″ ±0.01″ from north. Assume that initial approximations for the coordinate values of stations *B*, *C*, and D are

B: (41,433.56; 49,803.51) *C*: (60,054.84; 36,399.65)

D: (51,386.93; 9,545.64)

Give the adjusted coordinates of stations *B*, *C*, and *D* and the standard deviations of those coordinates.

19.2 Do Problem 19.1 using Helmert's method.

19.3 For the following traverse data, use Helmert's method and perform an adjustment holding the coordinates of station *A* fixed and the azimuth of line AB fixed at 73°28′11.0″. Assume initial approximations for the x-y coordinate values of stations B and C as

$$B: (1{,}654.793;\ 1{,}204.559) \quad \text{and} \quad C: (3{,}186.351;\ 2{,}403.474)$$

Control station

Station	x	y
A	983.322	1005.274

Distance observations

Occupied	Sighted	Distance	S
A	B	700.42	0.020
B	C	1945.01	0.022
C	A	2609.24	0.024

Angle observations

Backsight	Occupied	Foresight	Angle	S
A	B	C	158°28′34.2″	4.5″
B	C	A	5°39′06.0″	2.9″
C	A	B	15°52′22.1″	4.1″

19.4 Given the following differential leveling data, adjust it using Helmert's method. Hold the elevation of A to 100.00 ft and the elevation difference ΔElev_{AD} to −5.00 ft.

From	To	ΔElev	S	From	To	ΔElev	S
A	B	19.997	0.005	D	E	9.990	0.006
B	C	−19.984	0.008	E	A	−5.000	0.004
C	D	−4.998	0.003				

19.5 Using the method of weighting discussed in Section 19.6, adjust the data in:

(a) Problem 19.1.

(b) Problem 19.3.

(c) Problem 19.4.

(d) Compare the results with those of Problem 19.1, 19.3, or 19.4 as appropriate.

19.6 Do Problem 12.12 holding distance BC to 100.00 ft, using:

(a) the elimination of constraints method.

(b) Helmert's method.

(c) Compare the results of the adjustments from the different methods.

19.7 Do Problem 12.13 holding the elevation difference between V and Z to 30.00 ft using:

(a) the elimination of constraints method.
(b) Helmert's method.
(c) the weighting technique discussed in Section 19.6.
(d) Compare the results of the adjustments from the different methods.

19.8 Assuming that the distance precision between *A* and *C* is 1:50,000, do Problem 13.6 by including the control in the adjustment. Compare the answer with that obtained in the solution of Problem 13.6.

Programming Problems

19.9 Develop a spreadsheet that computes the coefficients for the *J* matrix in a trilateration adjustment with a constrained azimuth. Use this spreadsheet to compute the coefficients in Problem 19.1.

19.10 Develop a program that computes a constrained least-squares adjustment of a trilateration network using Helmert's method. Use this program to solve Problem 19.2.

20

BLUNDER DETECTION IN HORIZONTAL SURVEY NETWORKS

20.1 INTRODUCTION

Up to this point, data sets have always been used that were assumed to be free of blunders. However, when adjusting real measurements, the data sets often contain blunders. Not all blunders are large, but no matter their sizes, it is desirable to remove them from the data set. In this chapter some methods are discussed that are used to detect blunders before and after an adjustment.

Many examples can be cited that illustrate mishaps that have resulted from undetected blunders in survey data. Few could have been more costly and embarrassing, however, than a blunder of about 1 mile that occurred in an early nineteenth-century survey of the border between the United States and Canada near the north end of Lake Champlain. Following the survey, construction of a U.S. military fort was begun. The project was abandoned two years later when the blunder was detected and a resurvey showed that the fort was actually located on Canadian soil. The abandoned facility was subsequently named "Fort Blunder"!

As discussed in previous chapters, measurements should conform to the theory of the normal distribution, which means that occasionally large random errors will occur. However, according to the theory, this should seldom happen, and thus many large errors in data sets are actually blunders. Common blunders in data sets include number transposition, entry and recording mistakes, station misidentifications, and others. When blunders are present in a data set, a least-squares adjustment may not be possible or will, at a minimum, produce poor or invalid results. To be safe, the results of an adjustment should never be accepted without an analysis of the postadjustment statistics.

20.2 A PRIORI METHODS FOR DETECTING BLUNDERS IN MEASUREMENTS

In making adjustments, it should always be assumed that there are possible measurement blunders in the data, and appropriate methods should be used to isolate and remove them. It is especially important to eliminate blunders when the adjustment is nonlinear because they can cause the solution to diverge. In this chapter, several methods are discussed that can be used to isolate blunders in a horizontal adjustment.

20.2.1 Use of the *K* Matrix

In horizontal surveys, the easiest method available for detecting blunders is to use the redundant measurements. When initial approximations for station coordinates are computed using standard surveying methods, they should be *close* to their final adjusted values. Thus the difference between observations computed from the initial approximations and their measured values (*K* matrix) are expected to be small in size. If an observational blunder is present, there are two possible situations that could occur with regard to the *K*-matrix values. If the measurement containing the blunder is not used to compute initial coordinates, its corresponding *K*-matrix value will be relatively large. However, if an observation with a blunder is used in the computation of the initial coordinates of a station, the remaining redundant measurements to that station should have relatively large values.

Figure 20.1 shows the two possible situations. In Figure 20.1(*a*), a distance blunder is present in line *BP* and is shown by the length *PP*′. However, this distance was not used in computing the coordinates of station *P*, and thus the *K*-matrix value for $BP' - BP_0$ will suggest the presence of a blunder by its relatively large size. In Figure 20.1(*b*), the blundered distance *BP* was used

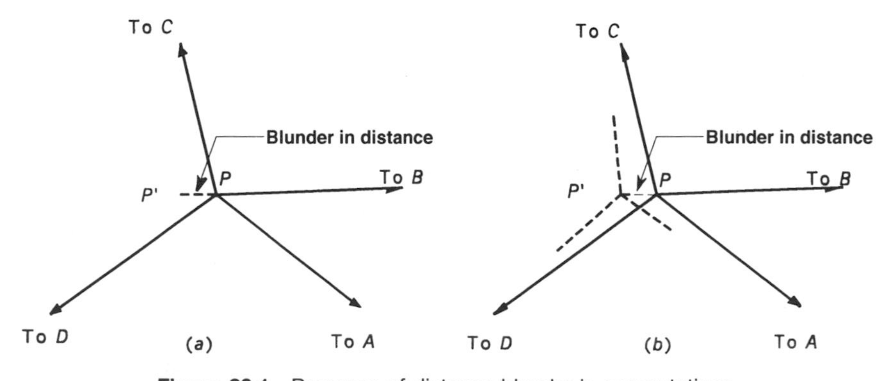

Figure 20.1 Presence of distance blunder in computations.

to compute the coordinates of station P'. In this case, the measurements of angles and distances connecting P with A, C, and D may show large discrepancies in the K matrix when compared with other values. It is possible, however, that some redundant measurements may agree reasonably with their computed values since a shift in a station's position can occur along a sight line for an angle or along a radius for a distance. Still most redundant observations will have large K-matrix values and thus raise suspicions that a blunder exists in one of the measurements used to compute the coordinates of station P.

20.2.2 Traverse Closure Checks

As mentioned in Chapter 8, errors can be propagated throughout a traverse to determine the anticipated closure. Large, complex networks can be broken into smaller link and loop traverses to check estimated closures against their actual values, as was done in Chapter 8. When a loop fails to meet its estimated closure, observations included in the computations are suspected to contain blunders.

Figure 20.2(*a*) and (*b*) show a graphical technique to isolate a traverse distance blunder and a traverse angular blunder, respectively. In Figure 20.2(*a*), a blunder in distance *CD* is shown. Notice that the remaining courses, *DE* and *EA*, are translated by the blunder in the direction of course *CD*. Thus the length of closure line ($A'A$) will be nearly equal to the length of the blunder in *CD*, and have a direction that is consistent with the azimuth of *CD*. Since other measurements contain small random errors, the length and direction of closure line, $A'A$, will not exactly match the blunder. However, when one blunder is present in a traverse, the misclosure and the blunder will be close in both length and direction.

In the traverse of Figure 20.2(*b*), the effect of an angular blunder at traverse station *D* is illustrated. As shown, the courses *DE*, *EF*, and FA' will be rotated

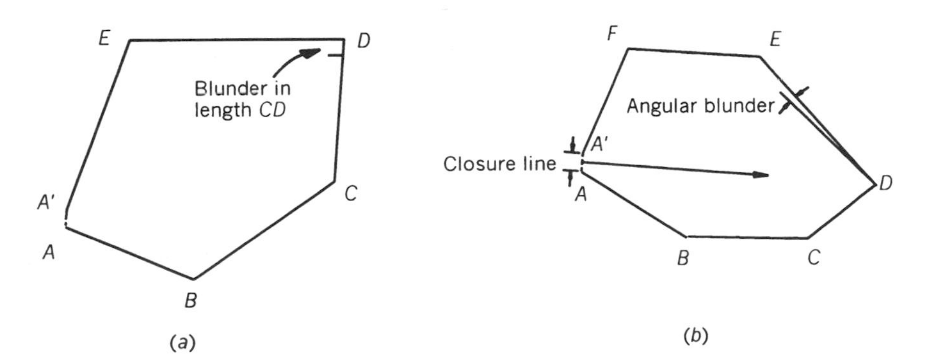

Figure 20.2 Effects of a single blunder on traverse closure.

about station D. Thus the perpendicular bisector of the closure line AA' will point to station D. Again due to random errors in other measurements, the perpendicular bisector may not precisely intersect the blunder, but it should be close enough to identify the angle with the blunder. Since the angle at the initial station is not used in traverse computations, it is possible to isolate a single angular blunder by beginning traverse computations at the station with the suspected blunder. In this case, when the blunder is not used in the computations, estimated misclosure errors will be met and the blunder can be isolated to the single unused angle. Thus if the traverse computations were started at station D in Figure 20.2(b), the traverse misclosure would be within anticipated tolerance since the angle at D would not be used in the traverse computations.

20.3 A POSTERIORI BLUNDER DETECTION

When doing a least-squares adjustment involving more than the minimum amount of control, both a *minimally-* and *fully-constrained* adjustment should always be made. In a *minimally-constrained adjustment*, the data need only satisfy the appropriate geometric closures, and thus it will not be influenced by control errors. After the adjustment, a χ^2 test[1] can be used to check the a priori value of the reference variance against its a posteriori estimate. However, this test is not a good indicator of the presence of a blunder since it is sensitive to poor relative weighting. Thus the a posteriori residuals should also be checked for the presence of large discrepancies. If no large discrepancies are present, the observational weights should be altered and the adjustment rerun. Since this test is sensitive to weights, the procedures described in Chapters 6 through 9 should be used for building the stochastic model of the adjustment.

Besides the sizes of the residuals, the *signs of the residuals* may also indicate a problem in the data. From normal probability theory, residuals are expected to be small and randomly distributed. In Figure 20.3 a small section of a larger network is shown. Notice that the distance residuals between stations A and B are all positive, which is not expected from normally distributed data. Thus it is possible that either a blunder or systematic error is present in some or all of the survey. Part of the problem could stem from control coordinate discrepancies if both A and B are control stations. This possibility can be isolated by doing a minimally constrained adjustment.

Although residuals sizes can suggest measurement errors, they do not necessarily identify the measurements that contain blunders. This is due to the fact that the least-squares method generally spreads a large observational error or blunder out radially from its source. However, this condition is not unique

[1] Statistical testing was discussed in Chapter 4.

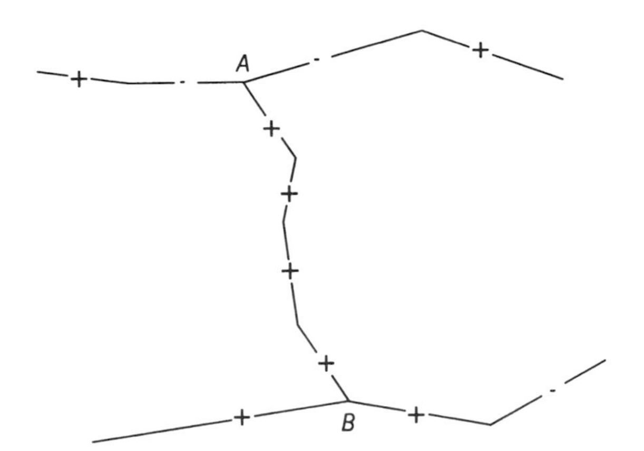

Figure 20.3 Distribution of residuals by sign.

to least-squares adjustments since any arbitrary adjustment method, including the compass rule for traverse adjustment, will also spread a single measurement error or blunder throughout the entire measurement set.

While an abnormally large residual may suggest the presence of a blunder in that measurement, this is not always true. One reason for this could be poor relative weighting in the observations. For example, suppose that angle *GAH* in the Figure 20.4 has a small blunder but has been given a relatively high weight. In this case the largest residual may well appear in a length between stations *G* and *H*, *B* and *H*, *C* and *F*, and most noticeably between *D* and *E* due to their distance from station *A*. This is because the angular

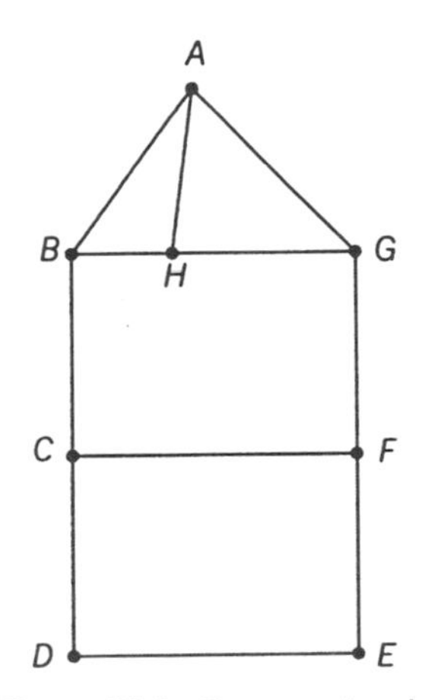

Figure 20.4 Survey network.

blunder will cause the network to spread or compress. When this happens, the sign of the distance residuals between B and H, G and H, C and F, and D and E may all be the same and thus indicate the problem. Again this situation can be minimized by using proper methods to determine observational variances so they are truly reflective of the estimated errors in the observations.

20.4 DEVELOPMENT OF THE COVARIANCE MATRIX FOR THE RESIDUALS

In Chapter 4 it was shown how a sample data set could be tested at any confidence level to isolate observational residuals that were too large. Statistical blunder detection utilizes the residual cofactor matrix. To develop this matrix, from previous work, the adjustment of a linear problem can be expressed in matrix form as

$$L + V = AX + C \tag{20.1}$$

where C is a constants vector, A is the coefficient matrix, X is the estimated parameter matrix, L is the observation matrix, and V is the residual vector. Equation (20.1) can be rewritten in the more familiar solution form of:

$$V = AX - T \tag{20.2}$$

where $T = L - C$, which has a covariance matrix of $W^{-1} = S^2Q_{ll}$.

The solution of Equation (20.2) results in the expression

$$X = (A^TWA)^{-1}A^TWT \tag{20.3}$$

Letting ϵ represent a vector of true errors for the observations, we have

$$L - \epsilon = A\overline{X} + C \tag{20.4}$$

where $\overline{X}$ is the matrix of true values for the unknown parameters X, and thus

$$T = L - C = A\overline{X} + \epsilon \tag{20.5}$$

Substituting Equations (20.3) and (20.5) into Equation (20.2) gives

$$V = A(A^TWA)^{-1}A^TW(A\overline{X} + \epsilon) - (A\overline{X} + \epsilon) \tag{20.6}$$

Expanding Equation (20.6) yields

$$V = A(A^TWA)^{-1}A^TW\epsilon - \epsilon + A(A^TWA)^{-1}A^TWA\overline{X} - A\overline{X} \tag{20.7}$$

Since $(A^TWA)^{-1}(A^TWA)$ is the identity matrix, Equation (20.7) can be simplified to

$$V = A(A^TWA)^{-1}A^TW\epsilon - \epsilon + A\overline{X} - A\overline{X} \tag{20.8}$$

Factoring $W\epsilon$ from Equation (20.8), yields

$$V = -(W^{-1} - A(A^TWA)^{-1}A^T)W\epsilon \tag{20.9}$$

Recognizing $(A^TWA)^{-1} = Q_{xx}$ and defining $Q_{vv} = W^{-1} - AQ_{xx}A^T$, Equation (20.9) can be rewritten as

$$V = -Q_{vv}W\epsilon \tag{20.10}$$

where $Q_{vv} = W^{-1} - AQ_{xx}A^T = W^{-1} - Q_{ll}$.

The Q_{vv} matrix is the cofactor matrix for the residuals, and is both singular and *idempotent*. Being singular, it has no inverse. When a matrix is idempotent, the following properties exist for the matrix: (a) The square of the matrix is equal to the original matrix (i.e., $Q_{vv}\,Q_{vv} = Q_{vv}$), (b) every diagonal element is between zero and 1, and (c) the sum of the diagonal elements equals the degrees of freedom for the adjustment. This sum is known as the *trace* of a matrix, and the relationship is expressed mathematically as

$$q_{11} + q_{22} + \cdots + q_{mm} = \text{degrees of freedom} \tag{20.11}$$

Another property of an idempotent matrix is that the sum of the squares of the elements in any single row or column equals the diagonal element; that is,

$$q_{ii} = q_{i1}^2 + q_{i2}^2 + \cdots + q_{im}^2 = q_{1i}^2 + q_{2i}^2 + \cdots + q_{mi}^2 \tag{20.12}$$

Now consider the case when all measurements have zero errors except for a particular observation l_i which contains a blunder of size Δl_i. A vector of the true errors, $\Delta\epsilon$, can be expressed as

$$\Delta\epsilon = \begin{bmatrix} 0 \\ 0 \\ \vdots \\ 0 \\ \Delta l_i \\ 0 \\ \vdots \\ 0 \end{bmatrix} = \Delta l_i \begin{bmatrix} 0 \\ 0 \\ \vdots \\ 0 \\ 1 \\ 0 \\ \vdots \\ 0 \end{bmatrix} \tag{20.13}$$

If the original measurements are uncorrelated, the specific correction for Δv_i can be expressed as

$$\Delta v_i = -q_{ii}w_{ii}\,\Delta l_i = -r_i\,\Delta l_i \tag{20.14}$$

where q_{ii} is the ith diagonal element of the Q_{vv} matrix, w_{ii} is the ith diagonal term of the weight matrix, W, and $r_i = q_{ii}w_{ii}$ is the observational *redundancy number*.

Notice that when the system has a unique solution, r_i will equal zero, and if the measurement is fully constrained, r_i would equal 1. The redundancy numbers provide insight into the geometric strength of the adjustment. An adjustment that in general has low redundancy numbers will have measurements that lack sufficient checks to isolate blunders, and thus the chance for undetected blunders to exist in the measurements is high. Conversely, a high overall redundancy number enables a high level of internal measurement checking and thus there is a lower chance of accepting measurements that contain blunders. The quotient of r/m is called the *relative redundancy* of the adjustment, where r is the total number of redundant measurements in the system and m is the number of observations.

20.5 DETECTION OF OUTLIERS IN OBSERVATIONS

Equation (20.14) gives the expected correction, v_i to an observation which can be used to isolate measurement blunders. *Standardized residuals* are computed using the diagonal elements of the Q_{vv} matrix as

$$\bar{v}_i = \frac{v_i}{\sqrt{q_{ii}}} \tag{20.15}$$

where $\bar{v}_i$ is the standardized residual, v_i the computed residual, and q_{ii} the diagonal element of the Q_{vv} matrix. Using the Q_{vv} matrix, the standard deviation in the residual is $S_0\sqrt{q_{ii}}$. Thus if the denominator of Equation (20.15) is multiplied by S_0, a t statistic is defined. If the residual is significantly different from zero, the observation used to derive the statistic is considered to be a blunder. The *test statistic* for this hypothesis test is

$$t_i = \frac{v_i}{S_0\sqrt{q_{ii}}} = \frac{v_i}{S_v} = \frac{\bar{v}_i}{S_0} \tag{20.16}$$

Baarda (1968) (see the Bibliography) computed rejection criteria for various significance levels (see Table 20.1), determining the α and β levels for type I and type II errors. The interpretation of these criteria is shown in Figure 20.5. When a blunder is present in the data set, the F distribution is shifted, and a statistical test for this shift may be performed. As with any other statistical test, two types of errors can occur. A type I error occurs when data are rejected that do not contain blunders, and a type II error occurs when a

Table 20.1 Rejection criteria with corresponding significance levels

α	$1 - \alpha$	β	$1 - \beta$	Rejection criteria
0.05	0.95	0.80	0.20	2.8
0.001	0.999	0.80	0.20	4.1
0.001	0.999	0.999	0.001	6.6

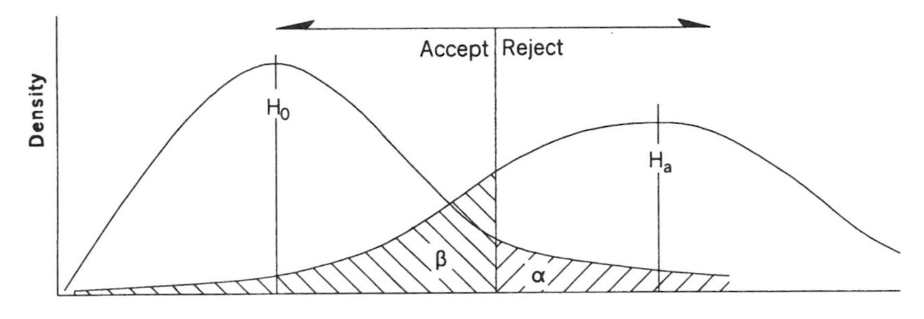

Figure 20.5 Effects of a blunder on the *t* distribution.

blunder is not detected in a data set where one is actually present. The rejection criteria are represented by the vertical line in Figure 20.5, and their corresponding significance levels are shown in Table 20.1. In practice, authors[2] have reported that 3.29 also works well as a criterion for rejection of blunders.

Thus the approach is to use a rejection level given by a t distribution test with $r - 1$ degrees of freedom. In either approach, the observation with the largest absolute value of t_i as given by Equation (20.16) is rejected when it is greater than the rejection level. That is, the observation is rejected when

$$\frac{|v_i|}{S_0\sqrt{q_{ii}}} > \text{rejection level} \tag{20.17}$$

Since the existence of any blunder in the data set will affect the remaining observations, and since Equation (20.17) is dependent on S_0, whose value was computed from data containing blunders, all observations that are detected as blunders should not be removed in a single pass. Instead, only the largest or

[2] References relating to the use of 3.29 as the rejection criterion are made in Amer (1979) and Harvey (1994) (see the Bibliography).

largest unrelated group of measurements should be deleted. Furthermore, since Equation (20.17) is dependent on S_0, it is possible to rewrite the equation so that it can be computed during the final iteration of a nonlinear adjustment. This equation is

$$\bar{v}_i = \frac{|v_i|}{\sqrt{q_{ii}}} > S_0 \times \text{rejection level} \tag{20.18}$$

A summary of procedures for this manner of blunder detection is as follows:

Step 1: Locate all standardized residuals that meet the rejection criteria of Equation (20.17) or (20.18).

Step 2: Remove the largest detected blunder or unrelated blunder groups.

Step 3: Rerun the adjustment.

Step 4: Continue steps 1 through 3 until all detected blunders are removed.

Step 5: If more than one observation is removed in steps 1 through 4, reenter the observations in the adjustment in a *one-at-a-time* fashion. Check the observation after each adjustment to see if it is again detected as a blunder. If it is, remove it from the adjustment or have that observation remeasured.

Again it should be noted that this form of blunder detection is sensitive to improper relative weighting in observations, and thus it is important to use weights that are reflective of the observational uncertainties. Proper methods of computing estimated errors in observations, and weighting, were discussed in Chapters 6 through 9.

20.6 TECHNIQUES USED IN ADJUSTING CONTROL

As discussed in Chapter 19, some control is necessary in each adjustment. However, since control itself is not perfect, this raises the question of how control should be managed. If control stations that contain errors are heavily weighted, the adjustment will improperly associate the control errors with the measurements. This effect can be removed by using only the minimum amount of control required to fix the project both positionally and rotationally in space. That is, if only the coordinates of one station and the direction of one line are held fixed during the adjustment, the observations will not be constricted by the control. Thus the observations will only be adjusted to satisfy the internal geometric conditions of the network. If more than minimum control is used, these additional constraints will be factored into the adjustment.

Statistical blunder detection can help isolate poor control or the presence of systematic errors in measurements. Using a minimally constrained adjustment, the data set can be screened for blunders. After becoming confident that the blunders are removed from the data set, a fully constrained adjustment can be performed. Following the fully constrained adjustment, an F test is used to compare the ratio of the minimally and fully constrained reference variances. The ratio should be 1.[3] When the two reference variances are found to be statistically different, two possible causes may exist. The first is that some of the control may contain errors, and these must be isolated and removed. The second is that some observations contain systematic errors. Since systematic errors are not compensating in nature, they will appear as blunders in the fully constrained adjustment. If systematic errors are suspected, they should be identified and removed from the original data set and the entire adjustment procedure redone. If no systematic errors are identified,[4] different combinations of control stations should be used in the constrained adjustments until the problem is isolated. By following this type of systematic approach, a single control station that has questionable coordinates can be isolated.

Extreme caution should always be used when dealing with control stations. Although it is possible that a control station was disturbed or that the original published coordinates contained errors, that is probably not the case. A prudent surveyor will check for physical evidence of disturbance and talk with other surveyors before deciding to discard a station as bad control. If the station was set by a local, state, or federal agency, the surveyor should contact the proper authorities and report the problem.

20.7 DATA SET WITH BLUNDERS

Example 20.1 The network shown in Figure 20.6 was established to provide control for mapping in the area of stations 1 through 6. It began from two National Geodetic Survey, Second Order, Class II (1:20,000 precision) control stations, 2000 and 2001. The data for the job was gathered by five different field crews in a class environment. The procedures discussed in Chapter 6 were used to estimate the observational errors. The problem is to check for blunders in the data set using a rejection level of $3.29S_0$.

[3] The ratio of the reference variances from the minimally and fully constrained adjustments should be 1 since both reference variances should be statistically equal; that is, $S^2_{\text{minimally constrained}} = S^2_{\text{fully constrained}}$.

[4] When adjusting data that covers a large region, it is essential that geodetic corrections (e.g., spherical excess, reduction to the ellipsoid, etc.) to the data be considered and applied where necessary. These corrections are systematic in nature and can cause errors when fitting to more than minimal control.

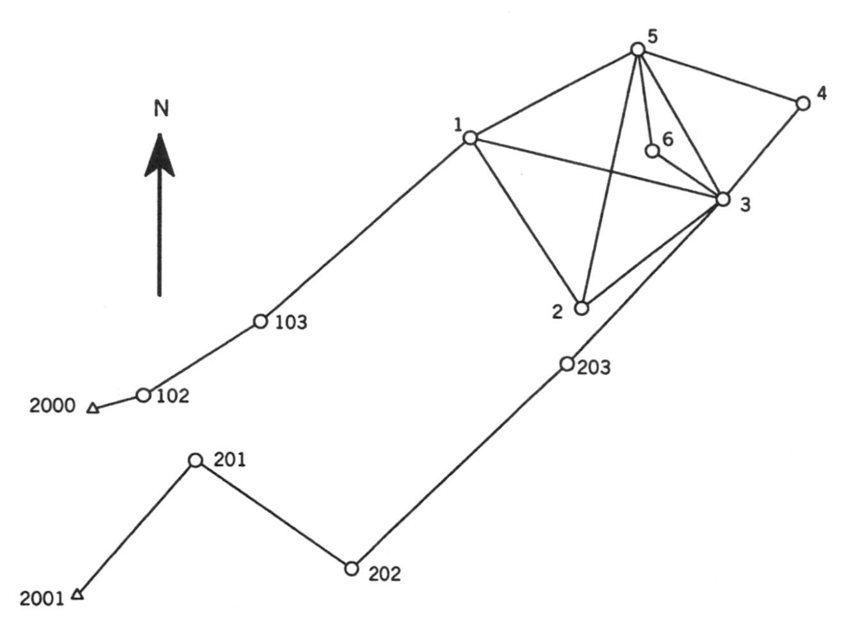

Figure 20.6 Data set with blunders.

Control stations

Station	Northing	Easting
2000	419,710.09	2,476,334.60
2001	419,266.82	2,476,297.98

Abstract of angles

Backsight	Occupied	Foresight	Angle	*S*
102	2000	2001	109°10′54.0″	±25.5″
2000	102	103	162°58′16.0″	±28.9″
102	103	1	172°01′43.0″	±11.8″
2000	2001	201	36°04′26.2″	±7.4″
2001	201	202	263°54′18.7″	±9.7″
201	202	203	101°49′55.0″	±8.1″
202	203	3	176°49′10.0″	±8.4″
203	3	2	8°59′56.0″	±6.5″
2	1	3	316°48′00.5″	±6.3″
3	5	4	324°17′44.0″	±8.1″
6	5	3	338°36′38.5″	±10.7″
1	5	3	268°49′32.5″	±9.8″
2	5	3	318°20′54.5″	±7.0″
2	3	1	51°07′11.0″	±7.2″
2	3	5	98°09′36.5″	±10.3″
2	3	6	71°42′51.5″	±15.1″
2	3	4	167°32′28.0″	±14.5″

Abstract of distances

Occupied	Sighted	Distance (ft)	S (ft)
2001	201	425.90	±0.022
201	202	453.10	±0.022
202	203	709.78	±0.022
203	3	537.18	±0.022
5	3	410.46	±0.022
5	4	397.89	±0.022
5	6	246.61	±0.022
5	1	450.67	±0.022
5	2	629.58	±0.022
3	2	422.70	±0.022
3	1	615.74	±0.022
3	5	410.44	±0.022
3	6	201.98	±0.022
3	4	298.10	±0.022
1	2	480.71	±0.022
1	3	615.74	±0.022
2000	102	125.24	±0.022
102	103	327.37	±0.022
103	1	665.79	±0.022

Initial approximations were computed for the stations as follows:

Station	Northing	Easting
1	420,353.62	2,477,233.88
2	419,951.98	2,477,497.99
3	420,210.17	2,477,832.67
4	420,438.88	2,478,023.86
5	420,567.44	2,477,630.64
6	420,323.31	2,477,665.36
102	419,743.39	2,476,454.17
103	419,919.69	2,476,728.88
201	419,589.24	2,476,576.25
202	419,331.29	2,476,948.76
203	419,819.56	2,477,463.90

SOLUTION: Do the a priori check of the computed observations versus their K-matrix values. In this check, only one angle is detected as having a difference great enough to suspect that it contained a blunder. This is angle 3–5–4, which was measured as 324°17′44.0″ but was computed as 317°35′31.2″. Since this difference should not create a problem with convergence during the adjustment, the angle remained in the data set and the adjustment was attempted. The results of the first trial adjustment are shown below. The software used the rejection criteria procedure based on Equation (20.18) for its blunder detection. A rejection level $3.29S_0$ was used for comparison against the standardized residuals. The column headed Std. Res. represents the standardized residual of the observation as defined by Equation (20.15), and the column headed Red. Num. represents the redundancy number of the observation as defined by Equation (20.14).

```
*****************************
Adjusted Distance Observations
*****************************
```

No.	From	To	Distance	Residual	Std. Res.	Red. Num.
1	1	3	616.234	0.494	26.148	0.7458
2	1	2	480.943	0.233	12.926	0.6871
3	1	3	616.234	0.494	26.148	0.7458
4	**3**	**4**	**267.044**	**−31.056**	**−1821.579**	**0.6169**
5	3	6	203.746	1.766	107.428	0.5748
6	3	5	413.726	3.286	171.934	0.7719
7	3	2	422.765	0.065	3.500	0.7312
8	5	2	630.949	1.369	75.909	0.6791
9	5	1	449.398	−1.272	−79.651	0.5377
10	5	6	247.822	1.212	75.418	0.5488
11	5	4	407.125	9.235	631.032	0.4529
12	5	3	413.726	3.266	170.888	0.7719
13	102	103	327.250	−0.120	−17.338	0.1018
14	103	1	665.702	−0.088	−12.395	0.1050
15	201	202	453.362	0.262	91.903	0.0172
16	202	203	709.856	0.076	10.737	0.1048
17	203	3	537.241	0.061	8.775	0.1026
18	2000	102	125.056	−0.184	−28.821	0.0868
19	2001	201	425.949	0.049	7.074	0.1008

```
***************************
Adjusted Angle Observations
***************************
```

No.	From	Occ	To	Angle	Residual	Std. Res.	Red. Num.
1	2	1	3	316°49′55.1″	114.6″	28.041	0.4164
2	2	3	4	167°36′00.2″	212.2″	25.577	0.3260
3	2	3	6	71°43′01.5″	10.0″	1.054	0.3990
4	2	3	5	97°55′09.3″	−867.2″	−101.159	0.6876
5	2	3	1	51°06′14.6″	−56.4″	−11.156	0.4985
6	203	3	2	8°59′36.3″	−19.7″	−13.003	0.0550
7	2	5	3	318°25′14.4″	259.9″	44.471	0.6949
8	1	5	3	268°58′49.8″	557.3″	78.590	0.5288

9	6	5	3	338°42′53.4″	374.9″	63.507	0.3058
10	**3**	**5**	**4**	**322°02′24.7″**	**−8119.3″**	**−1781.060**	**0.3197**
11	2000	102	103	162°23′50.9″	−2065.1″	−110.371	0.4194
12	102	103	1	171°57′46.9″	−236.1″	−112.246	0.0317
13	2001	201	202	263°58′31.6″	252.9″	104.430	0.0619
14	201	202	203	101°52′56.4″	181.4″	57.971	0.1493
15	202	203	3	176°50′15.9″	65.9″	23.278	0.1138
16	102	2000	2001	109°40′18.6″	1764.6″	106.331	0.4234
17	2000	2001	201	36°07′56.4″	210.2″	104.450	0.0731

```
                    ********************
                    Adjustment Statistics
                    ********************
      Adjustment's Reference Standard Deviation = 487.79
                  Rejection Level = 1604.82
```

The proper procedure for removing blunders is to remove the single observation which is greater in magnitude than the rejection level selected for the adjustment and is also greater in magnitude than the value of any other standardized residual in the adjustment. This procedure prevents removing observations that are connected to blunders and thus are inherently affected by their presence. By comparing the values of the standardized residuals against the rejection level of the adjustment, it can seen that both a single distance (3–4) and angle (3–5–4) are possible blunders since their standardized residuals are greater than the rejection level chosen. However, upon inspection of Figure 20.6 it can be seen that a blunder in distance 3–4 will directly affect angle 3–5–4, and distance 3–4 has the standardized residual that is greatest in magnitude. This explains the previous a priori rejection of this angle measurement. Thus only distance 3–4 should be removed from the observations. After removing this distance from the observations the adjustment was rerun, with the results shown below.

```
*****************************
Adjusted Distance Observations
*****************************
```

No.	From	To	Distance	Residual	Std. Res.	Red. Num.
1	1	3	615.693	−0.047	−2.495	0.7457
2	1	2	480.644	−0.066	−3.647	0.6868
3	1	3	615.693	−0.047	−2.495	0.7457
4	2001	201	425.902	0.002	0.265	0.1009
5	3	6	201.963	−0.017	−1.032	0.5765
6	3	5	410.439	−0.001	−0.032	0.7661
7	3	2	422.684	−0.016	−0.858	0.7314
8	5	2	629.557	−0.023	−1.280	0.6784

9	5	1	450.656	−0.014	−0.858	0.5389
10	5	6	246.590	−0.020	−1.241	0.5519
11	5	4	397.885	−0.005	−0.380	0.4313
12	5	3	410.439	−0.021	−1.082	0.7661
13	102	103	327.298	−0.072	−10.380	0.1018
14	103	1	665.751	−0.039	−5.506	0.1049
15	201	202	453.346	0.246	86.073	0.0172
16	202	203	709.807	0.027	3.857	0.1049
17	203	3	537.193	0.013	1.922	0.1027
18	2000	102	125.101	−0.139	−21.759	0.0868

****
Adjusted Angle Observations
****

No.	From	Occ	To	Angle	Residual	Std. Res.	Red. Num.
1	2	1	3	316°47′54.2″	−6.3″	−1.551	0.4160
2	2	3	4	167°32′31.0″	3.0″	0.380	0.2988
3	2	3	6	71°42′46.0″	−5.5″	−0.576	0.3953
4	2	3	5	98°09′18.6″	−17.9″	−2.088	0.6839
5	2	3	1	51°07′04.1″	−6.9″	−1.360	0.4978
6	203	3	2	8°59′26.7″	−29.3″	−19.340	0.0550
7	2	5	3	318°20′51.4″	−3.1″	−0.532	0.6933
8	1	5	3	268°50′03.4″	30.9″	4.353	0.5282
9	6	5	3	338°36′37.1″	−1.4″	−0.238	0.3049
10	3	5	4	324°17′43.6″	−0.4″	−0.381	0.0160
11	**2000**	**102**	**103**	**162°24′10.2″**	**−2045.8″**	**−109.353**	**0.4193**
12	**102**	**103**	**1**	**171°57′51.2″**	**−231.8″**	**−110.360**	**0.0316**
13	2001	201	202	263°58′20.3″	241.6″	99.714	0.0619
14	201	202	203	101°52′34.7″	159.7″	51.023	0.1494
15	202	203	3	176°49′56.1″	46.1″	16.273	0.1138
16	**102**	**2000**	**2001**	**109°40′17.7″**	**1763.7″**	**106.280**	**0.4233**
17	2000	2001	201	36°07′46.9″	200.7″	99.688	0.0732

****
Adjustment Statistics
****

Adjustment's Reference Standard Deviation = 30.62
Rejection Level = 100.73

After this adjustment, analysis of standardized residuals indicates that the most likely angles to yet contain blunders are observations 11, 12, and 16. Of these, observation 12 displays the highest standardized residual. Looking at Figure 20.6, it is seen that this angle attaches the northern traverse leg to control station 2000. This is a crucial observation in the network if any hopes of redundancy in the orientation of the network are to be maintained. Since this is a flat angle (i.e., nearly 180°), it is possible that the backsight and foresight stations were reported incorrectly, which can be checked by reversing stations 102 and 1. However, without further field checking it cannot be guaranteed that this occurred. A decision must ultimately be made about whether this angle should be remeasured. For now, however, this measurement will be discarded and another trial adjustment made. In this stepwise blunder detection process, it is always wise to remove as few observations as possible. In no case should observations that are blunder free be deleted. This can and does happen, however, in trial blunder detection adjustments. But through persistent and careful processing, ultimately only those observations that contain blunders will be identified and eliminated. The results of the adjustment after removing the angle 12 are shown below.

```
*****************
Adjusted Stations
*****************
```

			Standard Error Ellipses Computed				
Station	X	Y	Sx	Sy	Su	Sv	t
1	2,477,233.72	420,353.59	0.071	0.069	0.092	0.036	133.47°
2	2,477,497.89	419,951.98	0.050	0.083	0.090	0.037	156.01°
3	2,477,832.55	420,210.21	0.062	0.107	0.119	0.034	152.80°
4	2,477,991.64	420,400.58	0.077	0.121	0.138	0.039	149.71°
5	2,477,630.43	420,567.45	0.088	0.093	0.123	0.036	136.74°
6	2,477,665.22	420,323.32	0.071	0.096	0.114	0.036	145.44°
102	2,476,455.89	419,741.38	0.024	0.018	0.024	0.017	80.86°
103	2,476,735.05	419,912.42	0.051	0.070	0.081	0.031	147.25°
201	2,476,576.23	419,589.23	0.020	0.022	0.024	0.017	37.73°
202	2,476,948.74	419,331.29	0.029	0.041	0.042	0.029	14.24°
203	2,477,463.84	419,819.58	0.040	0.077	0.081	0.032	160.84°

```
*****************************
Adjusted Distance Observations
*****************************
```

Station Occupied	Station Sighted	Distance	V	Std.Res.	Red. Num.
2001	201	425.88	−0.023	−3.25	0.102
201	202	453.09	−0.005	−3.25	0.006
202	203	709.76	−0.023	−3.25	0.104
203	3	537.16	−0.023	−3.25	0.103
5	3	410.45	−0.011	−0.60	0.767
5	4	397.89	−0.003	−0.19	0.436
5	6	246.60	−0.014	−0.83	0.556

5	1	450.68	0.013	0.80	0.542
5	2	629.58	0.003	0.15	0.678
3	2	422.70	0.003	0.16	0.736
3	1	615.75	0.008	0.40	0.745
3	5	410.45	0.009	0.44	0.767
3	6	201.97	−0.013	−0.78	0.580
1	2	480.71	−0.003	−0.19	0.688
1	3	615.75	0.008	0.40	0.745
2000	102	125.26	0.020	3.25	0.082
102	103	327.39	0.023	3.25	0.101
103	1	665.81	0.023	3.25	0.104

```
**************************
Adjusted Angle Observations
**************************
```

Station Backsighted	Station Occupied	Station Foresighted	Angle	V	Std.Res.	Red. Num.
102	2000	2001	109°11′11.1″	17.06″	3.25	0.042
2000	102	103	162°58′05.1″	−10.95″	−3.25	0.014
2000	2001	201	36°04′23.8″	−2.45″	−3.25	0.010
2001	201	202	263°54′15.7″	−2.97″	−3.25	0.009
201	202	203	101°49′46.3″	−8.72″	−3.25	0.110
202	203	3	176°49′01.0″	−8.98″	−3.25	0.109
203	3	2	8°59′51.1″	−4.91″	−3.25	0.054
2	1	3	316°48′02.8″	2.29″	0.57	0.410
3	5	4	324°17′43.8″	−0.19″	−0.19	0.016
6	5	3	338°36′37.0″	−1.51″	−0.26	0.302
1	5	3	268°49′43.7″	11.20″	1.57	0.528
2	5	3	318°20′51.1″	−3.44″	−0.59	0.691
2	3	1	51°07′14.4″	3.45″	0.68	0.497
2	3	5	98°09′22.0″	−14.55″	−1.71	0.680
2	3	6	71°42′48.5″	−2.97″	−0.31	0.392
2	3	4	167°32′29.5″	1.48″	0.19	0.294

```
*********************
Adjustment Statistics
*********************

          Iterations = 4
        Redundancies = 12
  Reference Variance = 1.316
        Reference So = ±1.1

Possible blunder in observations with Std. Res. > 4
Convergence!
```

From analysis of the results above, all measurements containing blunders appear to have been removed. However, it should also be noted that several remaining distance and angle observations have very low redundancy numbers. This identifies them as unchecked observations, which is an undesirable situation. Thus, good judgment dictates remeasurement of the observations deleted. This weakness can also be seen in the size of the standard error ellipses for the stations shown in Figure 20.7. Note especially the rotation of the error ellipses, which show that the uncertainty is primarily in a direction perpendicular to the line to stations 1 and 102. This condition is predictable since the angle 102–103–1 has been removed from the data set. Furthermore, the crew on the northern leg never measured an angle at station 1 which would tie into station 103, and thus the position of station 103 was found by the intersection of two distances that nearly form a straight line. This results in a larger error in the direction perpendicular to the lines at this station.

This example demonstrates the process used statistically to detect and remove measurements with blunders. Whether the observations should be remeasured depends on the intended use of the survey. Obviously, they would add to the strength to the network and reduce error ellipse sizes.

Measurements between stations 102 and 201 would also contribute to the overall strength in the network. However, because a building obstructs that line, these observations could not be obtained. This is a common problem in network design (i.e., it is sometimes physically impossible to gather measurements that would contribute to the total network strength. Thus a

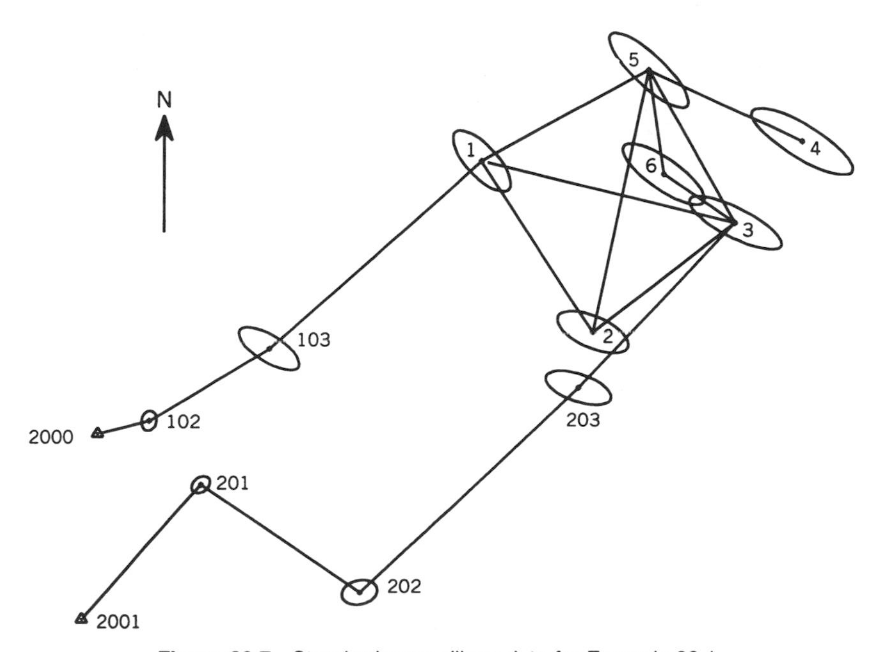

Figure 20.7 Standard error ellipse data for Example 20.1.

compromise must be made between the *ideal* network and what is physically obtainable. Balancing these aspects requires careful planning before the observations are collected. Of course, nowadays, line obstructions that occur due to terrain, vegetation, or buildings can usually be overcome by using GPS.

20.8 SOME FURTHER CONSIDERATIONS

Equation (20.14) shows the relationship between blunders and their effects on residuals as $\Delta v_i = -r_i \, \Delta l_i$. From this relationship note that the effect of the blunder, Δl_i, on the residual, Δv_i, is directly proportional to the redundancy number, r_i. Therefore:

1. If r_i is large ($\approx$ 1), the blunder greatly affects the residual and may be easy to find.
2. If r_i is small ($\approx$ 0), then the blunder has little affect on the residual and will be hard to find.
3. If $r_i = 0$, the blunder is undetectable and the parameters will be incorrect since the error has not been detected.

Since redundancy numbers can range from 0 to 1, it is possible to compute the minimum detectable error for a single blunder. For example, suppose that instead of using Baarda's value of 4.1, or 3.29 as suggested by other researchers, a value of 4.0 is used to isolate measurement blunders. Then if the reference variance of the adjustment is 6, all measurements that have standardized residuals greater than 24.0, (4.0 $\times$ 6), are possible blunders. However, from Equation (20.14), it can be seen that for an observation with a redundancy number of 0.2, ($r_i = 0.2$), and a standardized residual of $\Delta v_i = 24.0$, the minimum detectable error is 24.0/0.2, or 120! Thus a blunder, Δl_i, in this observation as large as five times the desired level can go undetected, due to its low redundancy number. This situation can be extended to observations that have no observational checks, that is, r_i is 0. In this case, Equation (20.14) shows that it is impossible to detect any blunder, Δl_i, in the observation since $\Delta v_i / r_i$ is indeterminate.

With this taken into consideration, it has been shown (Amer, 1983; Caspary, 1987) that a marginally detectable blunder in an individual observation is

$$\Delta l_i = S \sqrt{\frac{\lambda_0}{q_{ii} w_{ii}^2}} \tag{20.19}$$

where λ_0 is the mean of the noncentral normal distribution shown in Figure 20.5, known as the *noncentrality parameter*. This parameter is the translation of the normal distribution that leads to rejection of the null hypothesis and whose values can be taken from nomograms developed by Baarda (1968). The sizes of the values obtained from Equation (20.19) provide a clear insight into weak areas of the network.

20.8.1 Internal Reliability

Internal reliability is found by examining how well measurements check geometrically with each other. As mentioned previously, if a station is uniquely determined, q_{ii} will be zero in Equation (20.19), and thus the value of Δl_i will become infinity, indicating the lack of measurement self-checking. Since Equation (20.19) is independent of the actual observations, it provides a method of detecting weak areas in networks. To minimize the sizes of the undetected blunders in a network, the redundancy numbers of the individual measurements should approach their maximum value of 1. Furthermore, for uniform network strength, the individual redundancy numbers, r_i, should be close to the global relative redundancy of r/m, where r is the number of redundant measurements and m is the number of measurements in the network. Weak areas in the network are located by finding regions where the redundancy numbers become small in comparison to relative redundancy.

20.8.2 External Reliability

An undetected blunder of size Δl_i has a direct effect on the adjusted parameters. External reliability is the effect of the undetected blunders on these parameters. As Δl_i a blunder in Equation (20.13) increases, so will its effect on ΔX. The value of ΔX is given by

$$\Delta X = (A^T W A)^{-1} A^T W \, \Delta\epsilon \tag{20.20}$$

Again, this equation is datum independent. From Equation (20.20) it can be seen that to minimize the value of ΔX_i, the size of redundancy numbers must be increased. Baarda suggested using average coordinate values in determining the effect of an undetected blunder with the following equation:

$$\lambda_0 = \Delta X^T \, Q_{xx} \, \Delta X \tag{20.21}$$

where λ_0 represents the noncentrality parameter.

The noncentrality parameter should remain as small as possible to minimize the effects of undetected blunders on the coordinates. Note that as the redundancy numbers on the measurements become small, the effects of undetected blunders become large. Thus the effect on the coordinates of a station from a blunder in a measurement with a low redundancy number is greater than the effect of a similar blunder in a measurement with a high redundancy number. In fact, a measurement with a high redundancy number and blunder is likely to be detected as a blunder.

A traverse sideshot can be used to explain this phenomenon. Since the angle and distance to the station are unchecked in a sideshot, the coordinates of the station will change according to the size of the measurement blunders. These two measurements will have redundancy numbers of zero since a sideshot consists of geometrically unchecked measurements. This situation is neither good nor acceptable in a well-designed measurement system. In network design one should always check the redundancy numbers of the anticipated observations and strive to achieve a uniformly high value for all observations. Redundancy numbers above 0.5 are generally sufficient to provide well-checked measurements.

20.9 SURVEY DESIGN

In Chapters 7 and 18, the topic of measurement system design was discussed. Redundancy numbers can now be added to this discussion, as well-designed networks will allow measurement blunders to be detected and removed. In Section 20.8.1 it was stated that if blunder removal is to occur uniformly throughout the system, the redundancy numbers should be close to the system's global relative redundancy. Furthermore, in Section 20.8.2 it was noted that redundancy numbers should be greater than 0.5. By combining these two additional concepts with that of error ellipse size and shape, and also including stochastic model planning, an overall methodology for designing measurement systems can be obtained.

To begin the design process, the approximate positions for stations to be included in the survey must be determined. These locations can be determined from topographic maps, photo measurements, or previous survey data. The approximate locations of the control stations should be dictated by their desired locations, the surrounding terrain, vegetation, soils, sight line obstructions, and so on. Field reconnaissance at this phase of the design process is generally worthwhile to verify sight lines and accessibility of stations. That is, often, moving a station only a small distance from the original design location will vastly enhance visibility to and from other stations. By using topographic maps in this process, sight-line ground clearances can be checked by constructing profiles between stations.

When approximate station coordinates are determined, a stochastic model for the measurement system can be designed following the procedures discussed in Chapter 6. In this design process, considerations should be given to the abilities of the field personnel, quality of the equipment, and measurement procedures. After the design is completed, specifications for field crews can be written based on these design parameters. These specifications should include the type of instrument used, number of turnings for angle measurements, accuracy of instrument leveling and centering, horizon closure requirements, as well as many other items.

Once the stochastic model is designed, simulated observations are computed from the station coordinates, and a least-squares adjustment of the measurements is be done. Since actual observations were not made but were computed from the station coordinates, the adjustment will converge in a single iteration with no measurement residuals. Thus the reference variance used to compute the error ellipse axes and coordinate standard deviations must be given an a priori value of 1. Having completed the adjustment, the network can be checked for geometrically weak areas, unacceptable error ellipse sizes or shapes, and so on. This inspection may dictate the need for any or all of the following: (1) more measurements, (2) different measurement procedures, (3) different equipment, (4) more stations, (5) different network geometry, and so on. In any event, a clear picture of results obtainable from the measurement system will be provided by the simulated adjustment.

It should be noted that what is expected from the design may not actually occur due to numerous and varied reasons. Thus systems are generally overdesigned. However, this tendency to overdesign should be tempered with the knowledge that it will raise the costs of the survey. Thus a balance should be found between the design and costs. Experienced surveyors know what level of accuracy is necessary for each job, and they design the measurement system to achieve that accuracy. It would be foolish always to design a system for maximum accuracy regardless of the survey's intended use.

For convenience, the steps involved in network design are summarized below.

Step 1: Using a topographic map or aerial photos, lay out possible positions for stations.

Step 2: Use the topographic map together with air photos to check sight lines for possible obstructions and ground clearance.

Step 3: Do field reconnaissance, checking sight lines for obstructions not shown on the map or photos, and adjust positions of stations as necessary.

Step 4: Determine approximate coordinates for the stations from the map or photos.

Step 5: Compute values of observations using the coordinates from step 4.

Step 6: Using methods discussed in Chapter 6, compute the standard deviation of each observation based on available equipment, and field measuring procedures.

Step 7: Perform a least-squares adjustment, to compute observational redundancy numbers, standard deviations of station coordinates, and error ellipses at a specified percent probability.

Step 8: Inspect the solution for weak areas based on redundancy numbers and ellipse shapes. Add or remove observations as necessary, or reevaluate measurement procedures and equipment.

Step 9: Evaluate the costs of the survey, and determine if some other method of measurement (GPS, for example) may be more cost-effective.

Step 10: Write specifications for field crews.

PROBLEMS

Note: For problems below requiring least-squares adjustment, if a computer program is not distinctly specified for use in the problem, it is expected that the least-squares algorithm will be solved using the program MATRIX which is included within the diskette supplied with the book.

20.1 Discuss the effects of a distance blunder on a traverse closure and explain how it can be identified.

20.2 Discuss the effects of an angle blunder on a traverse closure and explain how it can be identified.

20.3 Explain why it is it is important not to have an observational redundancy number of zero.

20.4 Describe the methods used to isolate control quality problems.

20.5 Summarize the general procedures used in isolating measurement blunders.

20.6 Outline the procedures used in survey network design.

20.7 The following data for Figure 20.8 were gathered. Assuming that the control stations have a published precision of 1:20,000, apply the procedures discussed in this chapter to isolate and remove any apparent blunders in the data.

Unknown stations (approx.)

Station	Easting	Northing
B	2507.7	2500.6
C	4999.9	998.6
E	1597.6	200.0
F	2501.0	1009.6

Control stations

Station	Easting	Northing
A	982.083	1000.204
D	2686.270	58.096

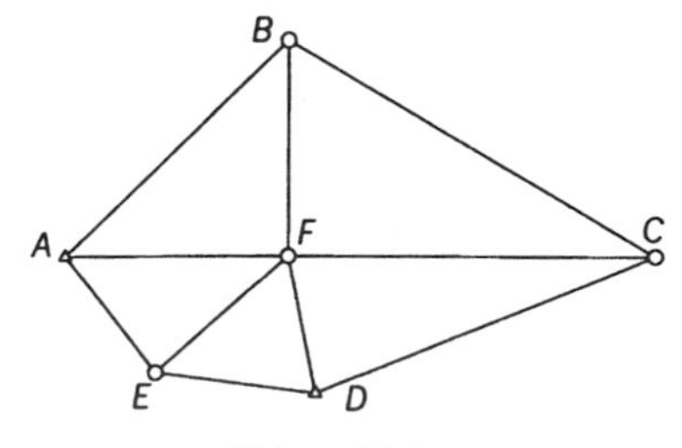

Figure 20.8

Distance observations

Occupied	Sighted	Distance	S	Occupied	Sighted	Distance	S
A	*B*	2139.769	0.023	*E*	*F*	1231.086	0.021
A	*F*	1518.945	0.021	*E*	*A*	1009.552	0.020
B	*C*	2909.771	0.025	*D*	*F*	969.386	0.020
B	*F*	1491.007	0.021	*D*	*E*	1097.873	0.021
C	*D*	2497.459	0.023	*C*	*F*	2498.887	0.023

Angle observations

Backsight	Occupied	Foresight	Angle	S
F	*A*	*E*	52°47′12.3″	3.4″
B	*A*	*F*	44°10′04.6″	2.8″
F	*B*	*A*	45°13′12.5″	2.8″
C	*B*	*F*	59°10′54.5″	2.7″
F	*C*	*B*	27°30′00.0″	2.4″
D	*C*	*F*	22°22′28.2″	2.5″
F	*D*	*C*	78°53′43.8″	3.3″
E	*D*	*F*	71°33′18.7″	3.8″
A	*E*	*F*	85°42′04.9″	3.7″
F	*E*	*D*	49°17′32.8″	3.4″
A	*F*	*B*	90°36′40.5″	3.1″
B	*F*	*C*	89°59′37.6″	2.8″
C	*F*	*D*	78°43′47.8″	3.3″
D	*F*	*E*	59°09′10.5″	3.6″
E	*F*	*A*	41°30′43.3″	3.1″

20.8 Using the data set *Blunder.dat* on the diskette that accompanies this book, isolate any blunders that are detectable with a rejection criteria of 3.29.

Practical Problems

20.9 Design a 6 mile by 6 mile control network having a minimum of eight control stations using a topographic map of your local area. Design a traditional measurement network made up of angles, azimuths, and distances so that the largest ellipse axis at a 95% confidence level is less than 0.20 ft and so that all observations have redundancy numbers greater than 0.5. In the design, specify the shortest permissible sight distance, the largest permissible errors in pointing, reading, and instrument and target setup errors, the number of repetitions necessary for each angle measurement; and the necessary quality of angle and distance measuring instruments. Use realistic values for the instruments. Plot profiles of sight lines for each observation.

20.10 Design a 6 mile by 6 mile GPS control network to be established by differential GPS which has a minimum of eight control stations. Use a topographic map of your local area to select station locations. Design the survey so that all baseline observations included in the network have redundancy numbers greater than 0.5. In the design, use a unit matrix for the covariance matrix of the baselines.

21

GENERAL LEAST-SQUARES METHOD AND ITS APPLICATION TO CURVE FITTING AND COORDINATE TRANSFORMATIONS

21.1 INTRODUCTION TO GENERAL LEAST SQUARES

When fitting points to a straight line it must be recognized that both the x and y coordinates contain errors. Yet in the mathematical model presented in Section 10.11.1, the residuals, as illustrated in Figure 10.2, are only applied to the y coordinate. Because both coordinates contain errors, this mathematical model fails to account for the x coordinate being a measurement. In this chapter the *general least-squares method* is presented and its use in performing adjustments where the observation equations involve more than a single measurement is demonstrated.

21.2 GENERAL LEAST-SQUARES EQUATIONS FOR FITTING A STRAIGHT LINE

Consider the data illustrated in Figure 10.2. To account properly for both the x and y coordinates being measurements, the observation equation must contain residuals for both x and y. That is, Equation (10.40) must be rewritten as

$$F(x,y) = (y + v_y) - m(x + v_x) - b = 0 \tag{21.1}$$

where x and y are a point's coordinate pair with residuals v_x and v_y, respectively, m is the slope of the line, and b is the y intercept. Equation (21.1) contains v_x, v_y, m, and b as unknowns and is nonlinear. Thus its solution is obtained by using the methods outlined in Section 10.10. The resulting linearized form of Equation (21.1) is

$$\frac{\partial F}{\partial x} v_x + \frac{\partial F}{\partial y} v_y + \frac{\partial F}{\partial m} dm + \frac{\partial F}{\partial b} db = m_0 x + b_0 - y \tag{21.2}$$

where the partial derivatives are

$$\frac{\partial F}{\partial x} = -m \qquad \frac{\partial F}{\partial y} = 1 \qquad \frac{\partial F}{\partial m} = -x \qquad \frac{\partial F}{\partial b} = -1 \tag{21.3}$$

Using the four data points A, B, C, and D of Figure 10.2, and substituting Equations (21.3) into (21.2), the following four equations can be written

$$\begin{aligned} -m_0 v_{x_A} + v_{y_A} - x_A\, dm - db &= m_0 x_A + b_0 - y_A \\ -m_0 v_{x_B} + v_{y_B} - x_B\, dm - db &= m_0 x_B + b_0 - y_B \\ -m_0 v_{x_C} + v_{y_C} - x_C\, dm - db &= m_0 x_C + b_0 - y_C \\ -m_0 v_{x_D} + v_{y_D} - x_D\, dm - db &= m_0 x_D + b_0 - y_D \end{aligned} \tag{21.4}$$

In matrix form, Equations (21.4) can be written as

$$BV + JX = K \tag{21.5}$$

where

$$B = \begin{bmatrix} -m_0 & 1 & 0 & 0 & 0 & 0 & 0 & 0 \\ 0 & 0 & -m_0 & 1 & 0 & 0 & 0 & 0 \\ 0 & 0 & 0 & 0 & -m_0 & 1 & 0 & 0 \\ 0 & 0 & 0 & 0 & 0 & 0 & -m_0 & 1 \end{bmatrix} \tag{21.6}$$

$$V = \begin{bmatrix} v_{xA} \\ v_{yA} \\ v_{xB} \\ v_{yB} \\ v_{xC} \\ v_{yC} \\ v_{xD} \\ v_{yD} \end{bmatrix} \qquad J = \begin{bmatrix} -x_A & -1 \\ -x_B & -1 \\ -x_C & -1 \\ -x_D & -1 \end{bmatrix}$$

$$X = \begin{bmatrix} dm \\ db \end{bmatrix} \qquad K = \begin{bmatrix} m_0 x_A + b_0 - y_A \\ m_0 x_B + b_0 - y_B \\ m_0 x_C + b_0 - y_C \\ m_0 x_D + b_0 - y_D \end{bmatrix}$$

Now since both x and y are measured coordinates, they each have individual standard errors. Assuming that the coordinates are from independent measurements, the four points will have eight measured coordinates and a cofactor matrix of

$$Q = \frac{1}{\sigma_0^2}\begin{bmatrix} \sigma_{x_A}^2 & 0 & 0 & 0 & 0 & 0 & 0 & 0 \\ 0 & \sigma_{y_A}^2 & 0 & 0 & 0 & 0 & 0 & 0 \\ 0 & 0 & \sigma_{x_B}^2 & 0 & 0 & 0 & 0 & 0 \\ 0 & 0 & 0 & \sigma_{y_B}^2 & 0 & 0 & 0 & 0 \\ 0 & 0 & 0 & 0 & \sigma_{x_C}^2 & 0 & 0 & 0 \\ 0 & 0 & 0 & 0 & 0 & \sigma_{y_C}^2 & 0 & 0 \\ 0 & 0 & 0 & 0 & 0 & 0 & \sigma_{x_D}^2 & 0 \\ 0 & 0 & 0 & 0 & 0 & 0 & 0 & \sigma_{y_D}^2 \end{bmatrix}$$

21.3 GENERAL LEAST-SQUARES SOLUTION

In solving the general least-squares problem a so-called *equivalent* solution is achieved. For this solution, the following *equivalent weight matrix* is created for the system:

$$W_e = (BQB^T)^{-1} \tag{21.7}$$

where B is as defined in Equation (21.6). Using the equivalent weight matrix in Equation (21.7), the *equivalent matrix system* is

$$J^TW_eJX = J^TW_eK \tag{21.8}$$

Equation (21.8) has the solution

$$X = (J^TW_eJ)^{-1}J^TW_eK \tag{21.9}$$

Since this is a nonlinear equation system, the corrections in matrix X are applied to the initial approximations, and the method is repeated until the system converges. The *equivalent residuals* vector V_e is found following the usual procedure of

$$V_e = JX - K \tag{21.10}$$

Using Equation (21.10), the observational residuals are

$$V = QB^TW_eV_e \tag{21.11}$$

Also, since this is a nonlinear problem and the observations are being adjusted, the observations should also be updated according to their residuals. Thus the updated observations for the second iteration are

$$l_i' = l_i + v_i \tag{21.12}$$

where l_i' are the observations for the second iteration, l_i the original observations, and v_i the corresponding residuals of the observations. Generally, in practice, the original observations are "close" to their final adjusted values, and thus Equation (21.12) is not actually used since the second iteration is only a check for convergence.

Finally, the reference variance for the adjustment can be computed using the equivalent residuals and weight matrix employing the equation

$$S_0^2 = \frac{V_e^T W_e V_e}{r} \tag{21.13}$$

where r is the number of redundancies in the system.

It should be noted that the same results can be obtained using the observational residuals and the following expression:

$$S_0^2 = \frac{V^T W V}{r} \tag{21.14}$$

Example 21.1: Numerical Solution of the Straight-Line-Fit Problem Recall the least-squares fit of points to a line in Section 10.11.1. In that example, the measured coordinate pairs were

A: (3.00, 4.50) B: (4.25, 4.25) C: (5.50, 5.50) D: (8.00, 5.50)

and the solution for the slope of the line, m, and y intercept, b, were

$$\begin{aligned} m &= 0.246 \\ b &= 3.663 \end{aligned} \tag{a}$$

Also, the residuals were

$$V = AX - L = \begin{bmatrix} -0.10 \\ 0.46 \\ -0.48 \\ 0.13 \end{bmatrix}$$

Now this problem will be solved using the general least-squares method. Assume that the following Q matrix is given:

$$Q = \begin{bmatrix} 0.020^2 & 0 & 0 & 0 & 0 & 0 & 0 & 0 \\ 0 & 0.015^2 & 0 & 0 & 0 & 0 & 0 & 0 \\ 0 & 0 & 0.023^2 & 0 & 0 & 0 & 0 & 0 \\ 0 & 0 & 0 & 0.036^2 & 0 & 0 & 0 & 0 \\ 0 & 0 & 0 & 0 & 0.033^2 & 0 & 0 & 0 \\ 0 & 0 & 0 & 0 & 0 & 0.028^2 & 0 & 0 \\ 0 & 0 & 0 & 0 & 0 & 0 & 0.016^2 & 0 \\ 0 & 0 & 0 & 0 & 0 & 0 & 0 & 0.019^2 \end{bmatrix}$$

SOLUTION: The step-by-step procedure for solving this problem using general least squares follows.

Step 1: Compute the initial approximations. Initial approximations for both m and b are found by using two points and solving the unique system. For this example the values from the solution of Section 10.11.1 as given above will be used.

Step 2: Develop the appropriate matrices. In accordance with Equation (21.6), the B matrix is

$$B = \begin{bmatrix} -0.246 & 1 & 0 & 0 & 0 & 0 & 0 & 0 \\ 0 & 0 & -0.246 & 1 & 0 & 0 & 0 & 0 \\ 0 & 0 & 0 & 0 & -0.246 & 1 & 0 & 0 \\ 0 & 0 & 0 & 0 & 0 & 0 & -0.246 & 1 \end{bmatrix}$$

Using Equation (21.7), the equivalent weight matrix is

$$W_e = (BQB^T)^{-1} = \frac{1}{10{,}000}\begin{bmatrix} 2.5 & 0.0 & 0.0 & 0.0 \\ 0.0 & 13.3 & 0.0 & 0.0 \\ 0.0 & 0.0 & 8.5 & 0.0 \\ 0.0 & 0.0 & 0.0 & 3.8 \end{bmatrix}^{-1}$$

$$= \begin{bmatrix} 4012.7 & 0 & 0 & 0 \\ 0 & 753.0 & 0 & 0 \\ 0 & 0 & 1176.6 & 0 \\ 0 & 0 & 0 & 2656.1 \end{bmatrix}$$

Step 3: Solve the system. The first iteration corrections are found using Equation (21.9) as

$$X = (J^TW_eJ)^{-1}J^TW_eK = \begin{bmatrix} -0.0318 \\ 0.1907 \end{bmatrix}$$

where the J, K, and J^TW_eJ matrices for the first iteration were

$$J = \begin{bmatrix} -3.00 & -1 \\ -4.25 & -1 \\ -5.50 & -1 \\ -8.00 & -1 \end{bmatrix} \qquad K = \begin{bmatrix} 0.246(3.00) + 3.663 - 4.50 \\ 0.246(4.25) + 3.663 - 4.25 \\ 0.246(5.50) + 3.663 - 5.50 \\ 0.246(8.00) + 3.663 - 5.50 \end{bmatrix} = \begin{bmatrix} -0.0990 \\ 0.4585 \\ -0.4840 \\ 0.1310 \end{bmatrix}$$

$$J^T W_e J = \begin{bmatrix} 255{,}297.91 & 42{,}958.45 \\ 42{,}958.45 & 8{,}598.40 \end{bmatrix} \qquad J^T W_e K = \begin{bmatrix} 72.9737 \\ 273.5321 \end{bmatrix}$$

Step 4: Apply the corrections. Applying the corrections to m_0 and b_0, to update the initial approximations for the second iteration,

$$m = 0.246 - 0.0318 = 0.2142$$

$$b = 3.663 + 0.1907 = 3.8537$$

Second Iteration

On the second iteration, only the unknown parameters are updated, and thus only B, W_e, and K matrices differ from their first iteration counterparts. Their second iteration values are

$$B = \begin{bmatrix} -0.2142 & 1.0000 & 0.0000 & 0.0000 & 0.0000 & 0.0000 & 0.0000 & 0.0000 \\ 0.0000 & 0.0000 & -0.2142 & 1.0000 & 0.0000 & 0.0000 & 0.0000 & 0.0000 \\ 0.0000 & 0.0000 & 0.0000 & 0.0000 & -0.2142 & 1.0000 & 0.0000 & 0.0000 \\ 0.0000 & 0.0000 & 0.0000 & 0.0000 & 0.0000 & 0.0000 & -0.2142 & 1.0000 \end{bmatrix}$$

$$W_e = \begin{bmatrix} 4109.3 & 0.0 & 0.0 & 0.0 \\ 0.0 & 757.4 & 0.0 & 0.0 \\ 0.0 & 0.0 & 199.1 & 0.0 \\ 0.0 & 0.0 & 0.0 & 2682.8 \end{bmatrix} \qquad K = \begin{bmatrix} -0.00370 \\ 0.51405 \\ -0.46820 \\ 0.06730 \end{bmatrix}$$

The corrections after this iteration are

$$X = \begin{bmatrix} 0.00002 \\ 0.00068 \end{bmatrix}$$

Thus m and b are 0.214 and 3.854 to three decimal places. Using Equation (21.10), the equivalent residual vector is

$$V_e = \begin{bmatrix} -0.0030 \\ 0.5148 \\ -0.4674 \\ 0.0681 \end{bmatrix}$$

Using Equation (21.11), the observation residual vector is

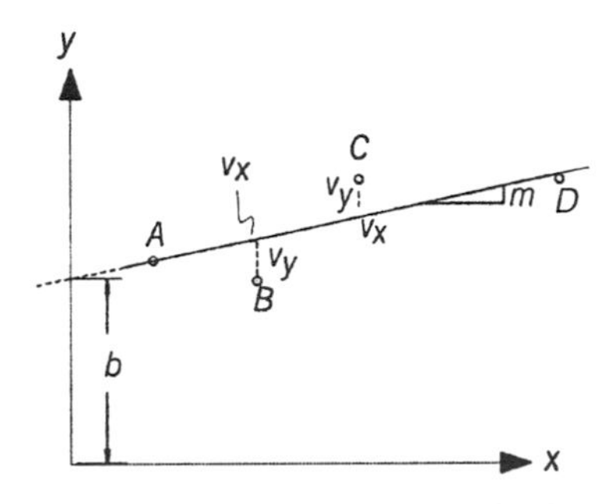

Figure 21.1 General least-squares fit of points to a line.

$$V = QB^TW_eV_e = \begin{bmatrix} 0.0010 \\ -0.0027 \\ -0.0442 \\ 0.5053 \\ 0.1307 \\ -0.4394 \\ -0.0100 \\ 0.0660 \end{bmatrix}$$

A graphical interpretation of the residuals is shown in Figure 21.1. Notice how the equivalent residuals are aligned with the y-axis, and that the observational residuals exist in the primary x and y axes directions. These residuals are shown more clearly in Figure 21.2, which is an enlarged view of the portion of Figure 21.1 that surrounds data point C. The equivalent residual of C is -0.4674 from the line, and the observational residuals $v_{x_c} = 0.1307$ and $v_{y_c} = -0.4394$ are parallel to the x and y axes, respectively. This general solution is more appropriate for adjusting the actual coordinate measurements. Note that this solution is somewhat different from that determined previously in Example 10.3.

Using Equation (21.13), the reference standard deviation is

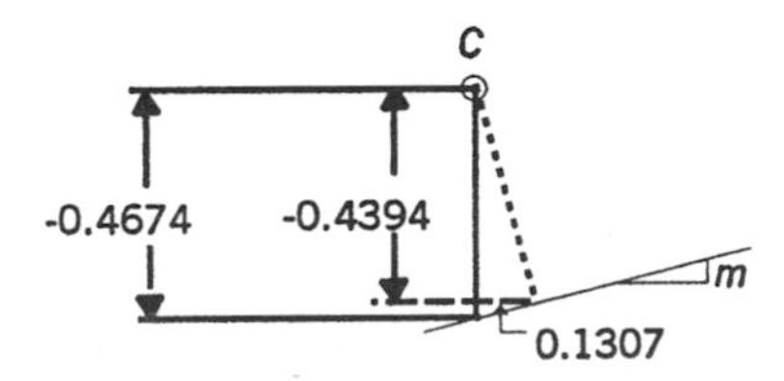

Figure 21.2 Residuals for point C.

$$S_0 = \sqrt{\frac{475.2}{4 - 2}} = \pm 15.4$$

21.4 TWO-DIMENSIONAL COORDINATE TRANSFORMATION BY GENERAL LEAST SQUARES

As was presented in Chapter 17, two-dimensional coordinate transformations are commonly used to convert points from one two-dimensional coordinate system to another. Again, the general least-squares method is a more appropriate method for these transformations since the coordinates in both systems are measurements that contain errors.

21.4.1 Two-Dimensional Conformal Coordinate Transformation

The two-dimensional conformal coordinate transformation, presented in Chapter 17, has four unknowns, consisting of a scale factor, rotation angle, and two translations. Equations (17.5) express this transformation, and they are repeated here for convenience. The transformation equations after dropping the residuals, are

$$\begin{aligned} X &= ax - by + c \\ Y &= bx + ay + d \end{aligned} \tag{21.15}$$

Equations (21.15) can be rearranged as

$$\begin{aligned} F &= ax - by + c - X = 0 \\ G &= bx + ay + d - Y = 0 \end{aligned} \tag{21.16}$$

Since the coordinates from both the xy and XY systems contain errors, Equations (21.16) are rewritten as

$$\begin{aligned} F(x,y,X,Y) &= a(x + v_x) - b(y + v_y) + c - (X + v_x) = 0 \\ G(x,y,X,Y) &= b(x + v_x) + a(y + v_y) + d - (Y + v_Y) = 0 \end{aligned} \tag{21.17}$$

These equations are nonlinear in terms of their measurements and residuals. They are solved by linearizing the equations and iterating to a solution. The partial derivatives with respect to each unknown are

$$
\begin{array}{lll}
\dfrac{\partial F}{\partial x} = a & \dfrac{\partial F}{\partial y} = -b & \dfrac{\partial F}{\partial X} = -1 \\
\dfrac{\partial F}{\partial a} = x & \dfrac{\partial F}{\partial b} = -y & \dfrac{\partial F}{\partial c} = 1 \\
\dfrac{\partial G}{\partial x} = b & \dfrac{\partial G}{\partial y} = a & \dfrac{\partial G}{\partial Y} = -1 \\
\dfrac{\partial G}{\partial a} = y & \dfrac{\partial G}{\partial b} = x & \dfrac{\partial G}{\partial d} = 1
\end{array}
\tag{21.18}
$$

Using the partial derivatives in Equation (21.18), a matrix representing Equating (2.17) for each point can be built as

$$
\begin{bmatrix} a_0 & -b_0 & -1 & 0 \\ b_0 & a_0 & 0 & -1 \end{bmatrix}
\begin{bmatrix} v_x \\ v_y \\ v_X \\ v_Y \end{bmatrix}
+
\begin{bmatrix} x & -y & 1 & 0 \\ y & x & 0 & 1 \end{bmatrix}
\begin{bmatrix} da \\ db \\ dc \\ dd \end{bmatrix}
=
\begin{bmatrix} X - (a_0x - b_0y + T_x) \\ Y - (b_0x + a_0y + T_y) \end{bmatrix}
\tag{21.19}
$$

For a redundant system, the matrices are solved following the matrix procedures outlined in Section 21.3.

Example 21.2 Four fiducial points are digitized from an aerial photo, and their measured (*x*,*y*) and control (*X*,*Y*) values are listed in Table 21.1. The standard deviations of these measurements are also listed. What are the most probable values for the transformation parameters, and the resulting residuals?

SOLUTION: In this problem, initial approximations for *a*, *b*, *c*, and *d* must be computed. These values can be found using a standard least-squares adjustment or by solving the system with only two points. The former procedure was demonstrated in Example 17.1. Using standard least squares, initial approximations for the parameters are determined to be

Table 21.1 Data for two-dimensional conformal coordinate transformation

Point	X	S_X	Y	S_Y	x	S_x	y	S_y
1	−113.000	± 0.002	0.003	± 0.002	0.7637	± 0.026	5.9603	± 0.028
3	0.001	± 0.002	112.993	± 0.002	5.0620	± 0.024	10.5407	± 0.030
5	112.998	± 0.002	0.003	± 0.002	9.6627	± 0.028	6.2430	± 0.022
7	0.001	± 0.002	−112.999	± 0.002	5.3500	± 0.024	1.6540	± 0.026

$$a_0 = 25.386458$$

$$b_0 = -0.8158708$$

$$c_0 = -137.216$$

$$d_0 = -150.600$$

Now the B, Q, J, and K matrices described in Section 21.3 can be formed. They are listed below (note that the numbers are rounded to three decimal places for publication purposes only):

$$B =$$

$$\begin{bmatrix} 25.386 & 0.816 & -1 & 0 & 0.000 & 0.000 & 0 & 0 & 0.000 & 0.000 & 0 & 0 & 0.000 & 0.000 & 0 & 0 \\ -0.816 & 25.386 & 0 & -1 & 0.000 & 0.000 & 0 & 0 & 0.000 & 0.000 & 0 & 0 & 0.000 & 0.000 & 0 & 0 \\ 0.000 & 0.000 & 0 & 0 & 25.386 & 0.816 & -1 & 0 & 0.000 & 0.000 & 0 & 0 & 0.000 & 0.000 & 0 & 0 \\ 0.000 & 0.000 & 0 & 0 & -0.816 & 25.386 & 0 & -1 & 0.000 & 0.000 & 0 & 0 & 0.000 & 0.000 & 0 & 0 \\ 0.000 & 0.000 & 0 & 0 & 0.000 & 0.000 & 0 & 0 & 25.386 & 0.816 & -1 & 0 & 0.000 & 0.000 & 0 & 0 \\ 0.000 & 0.000 & 0 & 0 & 0.000 & 0.000 & 0 & 0 & -0.816 & 25.386 & 0 & -1 & 0.000 & 0.000 & 0 & 0 \\ 0.000 & 0.000 & 0 & 0 & 0.000 & 0.000 & 0 & 0 & 0.000 & 0.000 & 0 & 0 & 25.386 & 0.816 & -1 & 0 \\ 0.000 & 0.000 & 0 & 0 & 0.000 & 0.000 & 0 & 0 & 0.000 & 0.000 & 0 & 0 & -0.816 & 25.386 & 0 & -1 \end{bmatrix}$$

$$J = \begin{bmatrix} 0.764 & -5.960 & 1 & 0 \\ 5.960 & 0.764 & 0 & 1 \\ 5.062 & -10.541 & 1 & 0 \\ 10.541 & 5.062 & 0 & 1 \\ 9.663 & -6.243 & 1 & 0 \\ 6.243 & 9.663 & 0 & 1 \\ 5.350 & -1.654 & 1 & 0 \\ 1.654 & 5.350 & 0 & 1 \end{bmatrix} \qquad K = \begin{bmatrix} -0.03447270 \\ -0.08482509 \\ 0.11090026 \\ 0.13190015 \\ -0.18120912 \\ -0.00114251 \\ 0.04999940 \\ -0.02329275 \end{bmatrix}$$

Also, the Q matrix is

$$Q = \begin{bmatrix} 0.026^2 & & & & & & & & & & \text{(Zeros)} \\ & 0.028^2 \\ & & 0.002^2 \\ & & & 0.002^2 \\ & & & & 0.024^2 \\ & & & & & 0.030^2 \\ & & & & & & 0.002^2 \\ & & & & & & & 0.002^2 \\ & & & & & & & & 0.028^2 \\ & & & & & & & & & 0.022^2 \\ & & & & & & & & & & 0.002^2 \\ & & & & & & & & & & & 0.002^2 \\ & & & \text{(Zeros)} & & & & & & & & & 0.024^2 \\ & & & & & & & & & & & & & 0.026^2 \\ & & & & & & & & & & & & & & 0.002^2 \\ & & & & & & & & & & & & & & & 0. \end{bmatrix}$$

The solution for the first iteration is

$$X = \begin{bmatrix} -0.000124503 \\ -0.000026212 \\ -0.000325016 \\ -0.000029546 \end{bmatrix}$$

Adding these corrections to the initial approximations gives

$$a = 25.38633347$$
$$b = -0.815897012$$
$$c = -137.2163$$
$$d = -150.6000$$

In the next iteration, only minor corrections occur, and thus the system has converged to a solution. The residuals and reference variance are computed as before. Although the solution has changed only slightly from the standard least-squares method, it properly considers the fact that each observation equation contains four measurements. Of course, once the transformation parameters have been determined, any points that exist only in the xy system can be transformed into the XY system by substitution into Equations (21.15). This part of the problem is not demonstrated in this example. In fact, for the remainder of the chapter only the adjustment model is developed.

21.4.2 Two-Dimensional Affine Coordinate Transformation

As discussed in Section 17.5, the main difference between conformal and affine transformations is that the latter allows for differing scales along the x and y axes, and also accounts for nonorthogonality in the axes. This results in six parameters. Equations (17.9) express the affine transformation, and they are repeated here for convenience.

$$\begin{aligned} X &= ax + by + c \\ Y &= dx + ey + f \end{aligned} \tag{21.20}$$

Equations (21.20) can be rewritten as

$$\begin{aligned} F(x,y,X,Y) &= ax + by + c - X = 0 \\ G(x,y,X,Y) &= dx + ey + f - Y = 0 \end{aligned} \tag{21.21}$$

Again, Equation (21.21) consists of observations in both the x and y, and

x and y coordinates, and thus it is more appropriate to use the general least-squares method. Therefore, these equations can be rewritten as

$$\begin{aligned} F(x,y,X,Y) &= a(x + v_x) + b(y + v_y) + c - (X + v_X) = 0 \\ G(x,y,X,Y) &= d(x + v_x) + e(y + v_y) + f - (Y + v_Y) = 0 \end{aligned} \tag{21.22}$$

For each point, the linearized equations in matrix form are

$$\begin{bmatrix} a_0 & b_0 & -1 & 0 \\ d_0 & e_0 & 0 & -1 \end{bmatrix} \begin{bmatrix} v_x \\ v_y \\ v_X \\ v_Y \end{bmatrix} + \begin{bmatrix} x & y & 1 & 0 & 0 & 0 \\ 0 & 0 & 0 & x & y & 1 \end{bmatrix} \begin{bmatrix} Da \\ Db \\ Dc \\ Dd \\ De \\ Df \end{bmatrix}$$

$$= \begin{bmatrix} X - (a_0 x + b_0 y + c_0) \\ Y - (d_0 x + e_0 y + f_0) \end{bmatrix} = \begin{bmatrix} 0 \\ 0 \end{bmatrix} \tag{21.23}$$

Two observations equations, like those of Equation (21.23), result for each control point. Since there are six unknown parameters, three control points are needed for a unique solution. With more than three, a redundant system exists, and the solution is obtained following the least-squares procedures outlined in Section 21.3.

21.4.3 Two-Dimensional Projective Transformation

As discussed in Section 17.6, the two-dimensional projective coordinate transformation converts a projection of one plane coordinate system into another nonparallel plane system. Equations (17.12) for this transformation were developed in Section 17.6 and are repeated here for convenience.

$$\begin{aligned} X &= \frac{a_1 x + b_1 y + c_1}{a_3 x + b_3 y + 1} \\ Y &= \frac{a_2 x + b_2 y + c_2}{a_3 x + b_3 y + 1} \end{aligned} \tag{21.24}$$

Unlike the conformal and affine types, this transformation is nonlinear in its standard form. In their general form, the projective equations become

$$\begin{aligned} F(x,y,X,Y) &= \frac{a_1(x + v_x) + b_1(y + v_y) + c_1}{a_3(x + v_x) + b_3(y + v_y) + 1} - (X - v_{X'}) = 0 \\ G(x,y,X,Y) &= \frac{a_2(x + v_x) + b_2(y + v_y) + c_2}{a_3(x + v_x) + b_3(y + v_y) + 1} - (Y - v_Y) = 0 \end{aligned} \tag{21.25}$$

Again a linearized form is needed for Equations (21.25). The partial derivatives for the a's, b's and c's were given in Section 17.6, and the remaining partial derivatives are given below:

$$\frac{\partial F}{\partial x} = \frac{a_1(b_3y + 1) - a_3(b_1y + c_1)}{(a_3x + b_3y + 1)^2}$$

$$\frac{\partial F}{\partial y} = \frac{b_1(a_3x + 1) - b_3(a_1x + c_1)}{(a_3x + b_3y + 1)^2}$$

$$\frac{\partial G}{\partial x} = \frac{a_2(b_3y + 1) - a_3(b_2y + c_2)}{(a_3x + b_3y + 1)^2}$$

$$\frac{\partial G}{\partial y} = \frac{b_2(a_3x + 1) - b_3(a_2x + c_2)}{(a_3x + b_3y + 1)^2}$$

In matrix form, the linearized equations for each point are

$$\begin{bmatrix} \dfrac{\partial F}{\partial x} & \dfrac{\partial F}{\partial y} & -1 & 0 \\ \dfrac{\partial G}{\partial x} & \dfrac{\partial G}{\partial y} & 0 & -1 \end{bmatrix} \begin{bmatrix} v_x \\ v_y \\ v_X \\ v_Y \end{bmatrix} + \begin{bmatrix} \dfrac{\partial F}{\partial a_1} & \dfrac{\partial F}{\partial b_1} & \dfrac{\partial F}{\partial c_1} & 0 & 0 & 0 & \dfrac{\partial F}{\partial a_3} & \dfrac{\partial F}{\partial b_3} \\ 0 & 0 & 0 & \dfrac{\partial G}{\partial a_2} & \dfrac{\partial G}{\partial b_2} & \dfrac{\partial G}{\partial c_2} & \dfrac{\partial G}{\partial a_3} & \dfrac{\partial G}{\partial b_3} \end{bmatrix} \begin{bmatrix} da_1 \\ db_1 \\ dc_1 \\ da_2 \\ db_2 \\ dc_2 \\ da_3 \\ db_3 \end{bmatrix} = \begin{bmatrix} 0 \\ 0 \end{bmatrix} \tag{21.26}$$

Equation (21.26) gives the observation equations for the two-dimensional projective transformation for one control point. Since there are eight unknown parameters, four control points are needed for a unique solution. More than four control points yields a redundant system that can be solved following the steps outlined in Section 21.3.

21.5 THREE-DIMENSIONAL CONFORMAL COORDINATE TRANSFORMATION BY GENERAL LEAST SQUARES

As explained in Section 17.7, this coordinate transformation converts points from one three-dimensional coordinate system to another. Equations (17.14) express this transformation, and the matrix form of those equations is

$$X = SMx + T \tag{21.27}$$

where the individual matrices are as defined in Section 17.7.

Equation (17.14) gives detailed expressions for the three-dimensional coordinate transformation. Note that these equations involve six observations, *xyz* and *XYZ*. In general least squares, these equations can be rewritten as

$$
\begin{aligned}
f(x,y,z,X,Y,Z) &= S\,[m_{11}(x+v_x) + m_{12}(y+v_y) + m_{13}(z+v_z)] \\
&\quad - (X+v_X) = 0 \\
g(x,y,z,X,Y,Z) &= S\,[m_{21}(x+v_x) + m_{22}(y+v_y) + m_{23}(z+v_z)] \\
&\quad - (Y+v_Y) = 0 \\
h(x,y,z,X,Y,Z) &= S\,[m_{31}(x+v_x) + m_{32}(y+v_y) + m_{33}(z+v_z)] \\
&\quad - (Z+v_Z) = 0
\end{aligned}
\tag{21.28}
$$

Equations (21.28) are for a single point, and again, they are nonlinear. They can be expressed in linearized matrix form as

$$
\begin{bmatrix}
\frac{\partial f}{\partial x} & \frac{\partial f}{\partial y} & \frac{\partial f}{\partial z} & -1 & 0 & 0 \\
\frac{\partial g}{\partial x} & \frac{\partial g}{\partial y} & \frac{\partial g}{\partial z} & 0 & -1 & 0 \\
\frac{\partial h}{\partial x} & \frac{\partial h}{y} & \frac{\partial h}{\partial z} & 0 & 0 & -1
\end{bmatrix}
\begin{bmatrix} v_x \\ v_y \\ v_z \\ v_X \\ v_Y \\ v_Z \end{bmatrix}
+
\begin{bmatrix}
\frac{\partial f}{\partial S} & 0 & \frac{\partial f}{\partial \phi} & \frac{\partial f}{\partial \kappa} & 1 & 0 & 0 \\
\frac{\partial g}{\partial S} & \frac{\partial g}{\partial \omega} & \frac{\partial g}{\partial \phi} & \frac{\partial g}{\partial \kappa} & 0 & 1 & 0 \\
\frac{\partial h}{\partial S} & \frac{\partial h}{\partial \omega} & \frac{\partial h}{\partial \phi} & \frac{\partial h}{\partial \kappa} & 0 & 0 & 1
\end{bmatrix}
\begin{bmatrix} dS \\ d\omega \\ d\phi \\ d\kappa \\ dT_X \\ dT_Y \\ dT_Z \end{bmatrix}
=
\begin{bmatrix} 0 \\ 0 \\ 0 \end{bmatrix}
\tag{21.29}
$$

where

$$\frac{\partial f}{\partial x} = Sm_{11} \qquad \frac{\partial f}{\partial y} = Sm_{12} \qquad \frac{\partial f}{\partial z} = Sm_{13}$$

$$\frac{\partial g}{\partial x} = Sm_{21} \qquad \frac{\partial g}{\partial y} = Sm_{22} \qquad \frac{\partial g}{\partial z} = Sm_{23}$$

$$\frac{\partial h}{\partial x} = Sm_{31} \qquad \frac{\partial h}{\partial y} = Sm_{32} \qquad \frac{\partial h}{\partial z} = Sm_{33}$$

The remaining partial derivatives were given in Section 17.7. This system is solved using the methods discussed in Section 21.3.

Example 21.3 Estimated errors were added to the control coordinates in Example 17.4. The control data are repeated in Table 21.2 and standard deviations of the control coordinates, needed to form the Q matrix, are also listed. Output from solving this problem using the program ADJUST is listed. Note that the solution differs somewhat from the one obtained by standard least squares in Example 17.4.

SOLUTION

```
3D Coordinate Transformation
Using Generalized Least Squares Method
-----------------------------------------------------------------
Measured Points
-----------------------------------------------------------------
Point         x          y         z        Vx        Vy        Vz
-----------------------------------------------------------------
    1  1094.883    820.085   109.821   -0.001   -0.001   -0.000
    2   503.891   1598.698   117.685   -0.002    0.000    0.000
    3  2349.343    207.658   151.387   -0.001    0.000    0.000
    4  1395.320   1348.853   215.261    0.001    0.000   -0.001
```

Table 21.2 Control data for three-dimensional conformal coordinate transformation

Point	X	Y	Z	S_X	S_Y	S_Z
1	10037.81	5262.09	772.04	0.05	0.06	0.05
2	10956.68	5128.17	783.00	0.04	0.06	0.09
3	8780.08	4840.29	782.62	0.02	0.04	0.02
4	10185.80	4700.21	851.32	0.03	0.05	0.03

Control Points

Point	X	VX	Y	VY	Z	VZ
1	10037.810	−0.063	5262.090	−0.026	772.040	−0.001
2	10956.680	−0.019	5128.170	0.063	783.000	−0.027
3	8780.080	0.000	4840.290	0.038	782.620	−0.001
4	10185.800	0.032	4700.210	−0.085	851.320	0.007

Transformation Coefficients

```
Scale = 0.94996 ±0.00002
Omega = 2°17′00.0″ ±0°00′26.7″
  Phi = −0°33′05.6″ ±0°00′06.1″
Kappa = 224°32′11.5″ ±0°00′07.7″
   Tx = 10233.855 ±0.066
   Ty = 6549.964 ±0.055
   Tz = 720.867 ±0.219

Reference Standard Deviation: 1.293
Degrees of Freedom: 5
Iterations: 3
```

Transformed Coordinates

Point	X	SX	Y	SY	Z	SZ
1	10037.874	0.082	5262.116	0.063	772.041	0.275
2	10956.701	0.087	5128.106	0.070	783.027	0.286
3	8780.080	0.098	4840.251	0.087	782.622	0.314
4	10185.767	0.096	4700.296	0.072	851.313	0.324
5	10722.016	0.075	5691.210	0.062	766.067	0.246
6	10043.245	0.074	5675.887	0.060	816.857	0.246

PROBLEMS

Note: For problems below requiring least-squares adjustment, if a computer program is not distinctly specified for use in the problem, it is expected that the least-squares algorithm will be solved using the program MATRIX which is included within the diskette supplied with the book.

21.1 Solve Problem 10.8 using the general least-squares method. Assign all coordinates a standard deviation of 1.

21.2 Do Problem 10.10 using the general least-squares method. Assign the computed coordinates the following standard deviations.

Station	S_X	S_Y
A	0.001	0.001
B	0.020	0.012
C	0.020	0.013
D	0.020	0.021
E	0.020	0.027
F	0.020	0.028
G	0.020	0.035
H	0.020	0.040
I	0.020	0.046

21.3 Solve Problem 10.12 using the general least-squares method. Assign all coordinates a standard deviation of 0.1.

21.4 Do Example 17.1 using the general least-squares method. Assume estimated standard deviations for the coordinates of $S_{Easting} = S_{Northing} = \pm 0.05$, and $S_x = S_y = \pm 0.005$.

21.5 Do Problem 17.1 using the general least-squares method. Assume estimated standard deviations for all the coordinates of ± 0.05. Compare this solution with that of Problem 21.4.

21.6 Solve Problem 17.3 using the general least-squares method. Assume estimated standard deviations for all the measured coordinates of ± 0.02 mm, and assume ± 0.005 mm for all the control coordinates.

21.7 Given the following measurements in mm:

	Measured Coordinates				Control Coordinates			
Point	x	y	S_x	S_y	X	Y	S_X	S_Y
6	103.555	−103.670	0.01	0.01	103.951	−103.969	0.002	0.002
7	0.390	−112.660	0.01	0.01	0.001	−112.999	0.002	0.002
3	0.275	111.780	0.01	0.01	0.001	112.993	0.002	0.002
4	103.450	102.815	0.01	0.01	103.956	103.960	0.002	0.002
5	112.490	−0.395	0.01	0.01	112.998	0.003	0.002	0.002
32	18.565	−87.580						
22	−5.790	2.035						
12	6.840	95.540						
13	86.840	102.195						
23	93.770	2.360						
33	92.655	−90.765						

Using the general least-squares method, transform the points from the measured system to the control system using:

(a) a two-dimensional conformal transformation.

(b) a two-dimensional affine transformation.

(c) a two-dimensional projective transformation.

21.8 Given the following data in mm:

	Measured Coordinates			Control Coordinates		
Point	x	y	z	X	Y	Z
1	607.54	501.63	469.09	390.13	499.74	469.32
2	589.98	632.36	82.81	371.46	630.95	81.14
3	643.65	421.28	83.50	425.43	419.18	82.38
4	628.58	440.51	82.27			
5	666.27	298.16	98.29			
6	632.59	710.62	103.01			

Using the general least-squares method, transform the points from the measured system to the control system using a three-dimensional conformal coordinate transformation. Assume that all coordinates have estimated standard deviations of ± 0.05 mm.

21.9 Solve Problem 17.8 using general least squares. Assume that all the control coordinates have estimated standard deviations of 0.05 ft.

21.10 Do Problem 21.1 using the program ADJUST.

Solve each problem with the program ADJUST using the standard and general least-squares adjustment options. Compare and explain any differences in the solutions.

21.11 Problem 21.6

21.12 Problem 21.7

21.13 Problem 21.8

Programming Problems

Develop a spreadsheet or program that constructs the B, Q, J, and K matrices for each transformation.

21.14 A two-dimensional conformal coordinate transformation

21.15 A two-dimensional affine coordinate transformation

21.16 A two-dimensional projective coordinate transformation

21.17 A three-dimensional conformal coordinate transformation

22

COMPUTER OPTIMIZATION

22.1 INTRODUCTION

Large amounts of computer time and memory are used when performing least-squares adjustments. This is due to the fact that as the number of unknowns increase in an adjustment, the storage requirements and time consumed in doing the numerical operations both increase rapidly. For example, a 25-station horizontal least-squares adjustment that has 50 distance and 50 angle observations, the coefficient matrix would have dimensions of 100 rows and 50 columns. If this adjustment were done in double precision,[1] it would require 40,000 bytes of storage for the coefficient matrix alone. The weight matrix would require an additional 80,000 bytes of storage. Also, at least two additional intermediate matrices[2] must be computed during the solution. From this example it is easy to see that large quantities of computer time and memory are used in the least-squares adjustment. Thus when writing least-squares software, it is desirable to take advantage of storage and computing optimization techniques. In this chapter some of these techniques are described.

22.2 STORAGE OPTIMIZATION

Many matrices used in a surveying adjustment are large but sparse. Using the example above, a single row of the coefficient matrix for a distance obser-

[1] The storage requirements of a double-precision number is eight bytes.

[2] The intermediate matrices developed are A^T and A^TW or J^T and J^TW, depending on whether the adjustment is linear or nonlinear.

vation would require 50-elements row for its four nonzero elements. In fact, the entire coefficient matrix is very sparsely populated by nonzero values. Similarly, the normal matrix is always symmetric, and thus nearly half its storage is used by duplicate entries. It is relatively easy to take advantage of these conditions to reduce the storage requirements.

In the normal matrix, only its upper or lower triangular portion need be saved. In this storage scheme, the two-dimensional matrix is saved as a vector. An example of a 4 × 4 normal matrix is shown in Figure 22.1. Its upper and lower triangular portions are shown separated and their elements numbered for reference. The vector on the right of Figure 22.1 shows the storage scheme. This scheme eliminates the need to save the duplicate entries of the normal matrix but requires some form of *mathematical mapping* between the original matrix indices of row i and column j and the vector's index of row i. For the upper triangular portion of the matrix, it can be shown that the vector index for any (i,j) element is computed as

$$\text{Index}(i,j) = 1/2\ [j \times (j - 1)] + i \tag{22.1}$$

Using the function given in Equation (22.1), the vector index for element (2,3) of the normal matrix would be computed as:

$$\text{Index}(2,3) = 1/2 \times [3 \times (3 - 1)] + 2 = 5$$

An equivalent mapping formula for the lower triangular portion of the matrix is given as

$$\text{Index}(i,j) = 1/2\ [i \times (i - 1)] + j \tag{22.2}$$

Equation (22.1) or (22.2) can be used to compute the storage location for each element in the upper or lower portion of the normal matrix, respectively.

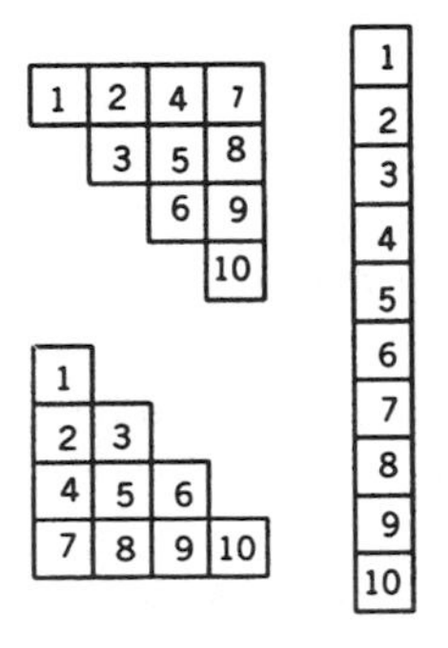

Figure 22.1 Normal matrix.

Of course, it requires additional computing time to map each element's location every time it is used. One method of minimizing the additional computation time is to utilize a *mapping table*. The mapping table is an integer vector that identifies the storage location for the initial element of each row or column. To reduce total computation time, the indices are stored as one number less than their actual value. For the example matrix stored in Figure 22.1, it would have a mapping table of

VI
0
1
3
6
10

Although the storage location of the first element is 1, it can be seen above that it is identified as 0. Similarly, every other mapping index is one number less than its actual location in the vector, as shown in Figure 22.1. The mapping table is found using the pseudocode shown in Table 22.1.

By using this type of mapping table, the storage location for the first element of any upper triangular column is computed as

$$\text{Index}(i,j) = \text{VI}(j) + i \tag{22.3}$$

Using the mapping table, the storage location for element (2,3) is

$$\text{Index}(2,3) = \text{VI}(3) + 2 = 3 + 2 = 5$$

The equivalent formula for the lower triangular portion is

$$\text{Index}(i,j) = \text{VI}(i) + j \tag{22.4}$$

Obviously this method of indexing an element's position requires additional storage in a mapping table, but the additional storage is offset by the decreased computation time originally created by the operations $1/2[j(j - 1)]$ or, equivalently, $1/2[i(i - 1)]$. Since the mapping table replaces

Table 22.1 Creation of a mapping table

Vi(1) = 0
For i going from 2 to the number of unknowns: $\text{VI}(i) = \text{VI}(i - 1) + i - 1$

the more time-consuming multiplication and division operations of Equations (22.1) and (22.2), it is prudent to use this index method whenever memory is available. Although every compiler and machine are different, a set of times to compare indexing methods are shown in Table 22.2. The best times were achieved by minimizing both the number of operations and calls to the mapping function. Thus in method C the mapping table was created to be a global variable in the program and the table was accessed directly as needed. For comparison, the time to access the matrix elements directly using the loop shown in method D was 1.204 seconds. Thus method C uses only slightly more time than method D but helps avoid the additional memory that is required by the full matrix required for method D. Meanwhile, method A is slower than method C by a factor of 3. Since the matrix elements must be accessed repeatedly during a least-squares solution, this difference can significantly reduce the total solution time.

The additional overhead cost of having the mapping table is minimized by declaring it as an integer array so that each element requires only two bytes of storage. In the earlier example of 25 stations, this results in 100 bytes of additional storage for the mapping table. Speed versus storage is a decision that every programmer will face continually.

22.3 DIRECT FORMATION OF THE NORMAL EQUATIONS

The basic weighted observation equation form is $WAX = WL + WV$. The least-squares solution for this equation is $X = (A^TWA)^{-1}(A^TWL) = N^{-1}C$,

Table 22.2 Comparison of indexing methods

Method	Time (sec)	Extra Storage
A	4.932	0
B	3.937	2 bytes per element
C	1.638	2 bytes per element
D	1.204	Full matrix required

Full A: Index = j*(j − 1)/2 + i
Function B: Index = VI(j) + i;

Method A: Using function A

```
i from 1 to 1000
  j from i to 1000
    k = Index(i,j)
```

Method B: Using function B

```
i from 1 to 1000
  j from i to 1000
    k = Index(i,j)
```

Method C: Direct access to mapping table

```
i from 1 to 1000
  j from i to 1000
    k = VI(j) + I
```

Method D: Direct matrix element access.

```
i from 1 to 1000
  j from i to 1000
    k = A[i,j]
```

where A^TWA is the normal equations matrix N, and A^TWL is the constants matrix C. If the weight matrix is diagonal, formation of the A, W, L, and A^TW matrices are unnecessary when building the normal equations, and thus the storage requirements for them can be eliminated. This is accomplished by forming the normal matrix directly from the observations. The tabular method in Section 10.8 showed the feasibility of this method. Notice in Table 10.2 that the contribution of each observation to the normal matrix is computed individually and subsequently, added. This shows that there is no need to form the coefficient, constants, weight, or any intermediate matrices when deriving the normal matrix. Conceptually, this is developed as follows:

Step 1: Zero the normal and constants matrices.

Step 2: Zero a single row of the coefficient matrix.

Step 3: Based on the values of a single row of the coefficient matrix, add the proper value to the appropriate normal and constant elements.

Step 4: Repeat steps 3 and 4 for all observations.

This procedure works in all situations that involve a diagonal weight matrix. Procedural modifications can be developed for weight matrices with limited correlation between the measurements. Computer algorithms in BASIC, FORTRAN, C, and Pascal for this method are shown in Table 22.3.

22.4 CHOLESKY DECOMPOSITION

Having formed only a triangular portion of the normal matrix, a Cholesky decomposition can be used to greatly reduce the time needed to find their solution. This procedure takes advantage of the fact that the normal matrix is always a positive definite[3] matrix. Due to this property, it can be expressed as the product of a lower triangular matrix and its transpose; that is,

$$N = LU = LL^T = U^TU$$

where L is a lower triangular matrix of the form:

$$L = \begin{bmatrix} l_{11} & 0 & 0 & \cdots & 0 \\ l_{21} & l_{22} & 0 & \cdots & 0 \\ l_{31} & l_{32} & l_{33} & & 0 \\ \vdots & \vdots & \vdots & \ddots & 0 \\ l_{n1} & l_{n2} & l_{n3} & \cdots & l_{nn} \end{bmatrix}$$

[3] A positive definite square matrix, A, has the property that $X^TAX > 0$ for all nonzero vectors, X.

Table 22.3 Algorithms for building the normal equations directly from their observations

BASIC Language

```
FOR i = 1 to unknown
 ix = Vi(i): ixi = ix + i
 N(ixi) = N(ixi) + A(i)^2 * W(i)
 C(i) = C(i) + A(i)*W(i)*L(i)
 For j = i + 1 to unknown
  N(ix + j) = N(ix + j) + A(i)*W(i)*A(j)
 Next j
Next i
```

FORTRAN Language

```
 Do 100 i = 1, unknown
  ix = Vi(i)
  ixi = ix + i
  N(ixi) = N(ixi) + A(i)**2 * W(i)
  C(i) = C(i) + A(i)*W(i)*L(i)
  Do 100 j = i + 1, unknown
   N(ix + j) = N(ix + j) + A(i)*W(i)*A(j)
100 Continue
```

C Language

```
for (i = 1; i <= unknown; i++) {
 ix = vi[i]; ixi = ix + i;
 n[ixi] = n[ixi] + a[i]*a[i] * w[i];
 c[i] = c[i] + a[i]*w[i]*l[i];
 for (j = i + 1; j <= unknown; j++)
  n[ix + j] = n[ix + j] + a[i]*w[i]*a[j];
}//for i
```

Pascal Language

```
for i := 1 to unknown do begin
 ix := Vi[i]; ixi := ix + i;
 N[ixi] := N[ixi] + Sqr(A[i]) * W[i];
 C[i] := C[i] + A[i]*W[i]*L[i];
 for j := i + 1 to unknown do
  N[ix + j] := N[ix + j] + A[i]*W[i]*A[j]
end; {for i}
```

When the normal matrix is stored as a lower triangular matrix, the matrix can be factored using the following procedure:

For $i = 1, 2, \ldots$, number of unknowns, compute

$$l_{ii} = \left(l_{ii} - \sum_{k=1}^{i-1} l_{ik}^2 \right)^{1/2} \tag{22.5}$$

For $j = i + 1, i + 2, \ldots$, number of unknowns, compute

$$l_{ji} = \frac{l_{ji} - \sum_{k=1}^{i-1} l_{ik} l_{jk}}{l_{ii}}$$

The procedures above are shown in code form in Table 22.4.

22.5 FORWARD AND BACK SOLUTIONS

Being able to factor the normal matrix into triangular matrices has the advantage that the matrix solution can be obtained without the use of an inverse. The equivalent triangular matrices representing the normal equations are

$$\begin{aligned}
(A^TWA)X &= NX = LUX = LL^TX \\
&= \begin{bmatrix} l_{11} & 0 & 0 & \cdots & 0 \\ l_{21} & l_{22} & 0 & \cdots & 0 \\ l_{31} & l_{32} & l_{33} & & 0 \\ \vdots & \vdots & \vdots & \ddots & 0 \\ l_{n1} & l_{n2} & l_{n3} & \cdots & l_{nn} \end{bmatrix} \begin{bmatrix} l_{11} & l_{21} & l_{31} \cdots & l_{n1} \\ 0 & l_{22} & l_{32} \cdots & l_{n2} \\ 0 & 0 & l_{33} \cdots & l_{n3} \\ \vdots & \vdots & \ddots & \vdots \\ 0 & 0 & 0 \cdots & l_{nn} \end{bmatrix} \begin{bmatrix} x_1 \\ x_2 \\ x_3 \\ \vdots \\ x_n \end{bmatrix} \\
&= \begin{bmatrix} c_1 \\ c_2 \\ c_3 \\ \vdots \\ c_n \end{bmatrix} = C
\end{aligned} \tag{22.6}$$

Equation (22.6) can be rewritten as $LY = C$, where $L^TX = Y$. From this, the solution for Y can be found by taking advantage of the triangular form of L. This is known as a *forward substitution*. Steps involved in forward substitution follow.

Step 1: Solve for y_1 as:

$$y_1 = \frac{c_1}{l_{11}}$$

Table 22.4 Computer algorithms for computing Cholesky factors of a normal matrix

BASIC Language

```
FOR i = 1 TO unknown
 ix = vi(i): ixi = ix + i: s = 0#
  FOR k = 1 TO i - 1
    s = s + N(ix + k)^2: NEXT k
  N(ixi) = SQR(N(ixi) - s)
  FOR j = i + 1 TO unknown
    s = 0#: jx = vi(j)
    FOR k = 1 TO i - 1
     s = s + N(jx + k) * N(ix + k): NEXT k
   N(jx + i) = (N(jx + i) - s)/N(ixi)
 NEXT j
NEXT I
```

FORTRAN Language

```
   Do 30 i = 1, unknown
    ix = Vi(i)
    i1 = i - 1
    S = 0.0
    Do 10 k = 1, i1
10   s = s + N(ix + k)**2
    N(ix + i) = Sqrt(N(ix + i) - s)
    Do 30 j = i + 1, Unknown
     s = 0.0
     Do 20 k = 1, i1
20     s = s + N(Vi[j] + k) * N(ix + k)
     N(Vi[j] + i) = (N(Vi[j] + i) - s)/N(ix + i)
30 Continue
```

C Language

```
for (i = 1; i <= unknown; i++){
 ix = vi[i]; ixi = ix + i; s = 0.0;
 for (k = 1; k < i; k++)
  s = s + n[ix + k]*n[ix + k];
 n[ixi] = sqrt(n[ixi] - s);
 for (j = i + 1; j <= unknown; j++){
  s = 0; jx = vi[j];
  for (k = 1; k < i; k++)
   s = s + n[jx + k] * n[ix + k];
  n[jx + i] = (n[jx + i] - s)/n[ixi];
 }//for j
}//for i
```

Pascal Language

```
For i := 1 to unknown do Begin
 ix := Vi[i]; ixi := ix + i; S := 0;
 For k := 1 to Pred(i) do
  S := S + Sqr(N[ix + k]);
 N[ixi] := Sqrt(N[ixi] - S);
 For j := Succ(i) to unknown do Begin
  S := 0; jx := Vi[j];
  For k := 1 to Pred(i) do
   S := S + N[jx + k]*N[ix + k];
  N[jx + i] := (N[jx + i] - S)/N[ixi]
 End; {for j}
End; {for i}
```

Step 2: Substitute this value into row 2 and compute y_2 as

$$y_2 = \frac{c_2 - l_{21}y_1}{l_{22}}$$

Step 3: Repeat this procedure until all values for y are found using the algorithm

$$y_i = \frac{c_i - \sum_{k=1}^{i-1} l_{ik}y_k}{l_{ii}}$$

Having determined Y, the solution for the matrix system $L^TX = Y$ is computed in a manner similar to that above. However, this time the solution starts at the lower right corner and proceeds up the matrix L^T. This is called a *back substitution*, which is done with the following steps:

Step 1: Compute x_n as

$$x_n = \frac{y_n}{l_{nn}}$$

Step 2: Solve for x_{n-1} as

$$x_{n-1} = \frac{y_{n-1} - l_{nn}x_n}{l_{(n-1)(n-1)}}$$

Step 3: Repeat this procedure until all unknowns are computed using the algorithm

$$x_k = \frac{y_k - \sum_{j=k+1}^{n} l_{kj}y_j}{l_{kk}}$$

In this process, once the original values in the constant matrix are accessed and changed, they are not needed again, and thus the original C matrix can be overwritten with Y and X matrices so that the entire process requires no additional storage. In fact, this method of solution also requires fewer operations than solving $X = (A^TWA)^{-1}A^TWL$ directly. Table 22.5 lists the computer codes for these algorithms.

Table 22.5 Computer algorithms for forward and backward substitutions

BASIC Language

```
REM Forward Substitution
For i = 1 to Unknown
 ix = Vi(i)
 C(i) = C(i)/N(ix + i)
 For j = i + 1 to Unknown
  C(j) = C(j) − N(ix + j) * C(i)
 Next j
Next i
Rem Backward Substitution
For i = Unknown to 1 Step −1
 ix = Vi(i)
 For j = Unknown To i + 1 Step −1
  C(i) = N(ix + j) * C(j)
 Next j
 C(i) = C(i)/N(ix + i)
Next i
```

FORTRAN Language

```
C Forward Substitution
 Do 100 i = 1, Unknown
  ix = Vi(i)
  C(i) = C(i)/N(ix + i)
  Do 100 j = i + 1, Unknown
   C(j) = C(j) − N(ix + j) * C(i)
100 Continue
C Backward Substitution
 Do 110 i = Unknown, 1, −1
  ix = Vi(i)
  Do 120 j = Unknown, i + 1, −1
120  C(i) = C(i) − N(Ix + j) * C(j)
    C(i) = C(i)/N(Ix + i)
110 Continue
```

C Language

```
//Forward Substitution
for (i = 1; i <= unknown; i++){
 ix = vi[i];
 c[i] = c[i]/n[ix + i];
 for (j = i + 1; j <= unknown; j++){
  c[j] = c[j] − n[ix + j]*c[i];
 } //for j
} //for j
//Backward Substitution
for (i = unknown; i >= 1; i−−){
 ix = vi[i];
 for (j = unknown; j >= i + 1; j−−){
  c[i] = c[i] − n[ix + j] * c[j];
 } //for j
 c[i] = c[i]/n[ix + i];
} //for i
```

Pascal Language

```
{Forward Substitution}
For i := 1 to Unknown Do Begin
 ix := Vi[i];
 C[i] := C[i]/N[ix + i];
 For j := i + 1 to Unknown Do
  C[j] := C[j] − N[ix + j] * C[i];
End; {for i}
 {Backward Substitution}
 For i := Unknown DownTo 1 do Begin
  ix := Vi[i];
  For j := Unknown DownTo i + 1 Do
   C[i] := C[i] − N[Ix + j] * C[j];
  C[i] := C[i]/N[Ix + i];
End; {for i}
```

22.6 USING THE CHOLESKY FACTOR TO FIND THE INVERSE OF THE NORMAL MATRIX

If necessary, the original normal matrix inverse can be found with the Cholesky factor. To derive this matrix, the inverse of the Cholesky factor is computed. The normal matrix inverse is the product of this inverse times its transpose. The Cholesky factor inverse is determined using the algorithm in Table 22.6.

Table 22.6 Pseudocode for computing the inverse of a Cholesky decomposed matrix

```
for i going from the number of unknowns down to 1
 for k going from number of unknowns down to i + 1
  for j going from i + 1 to k, sum the product N(i, j) * N(j,k)
  N(i,k) = -S/N(i,i)
 N(i,i) = 1/N(i,i)
```

Code for this algorithm is shown in Table 22.7.

22.7 SPARSENESS AND OPTIMIZATION OF THE NORMAL MATRIX

In the least-squares adjustment of most surveying and photogrammetry problems, it is known that certain locations in the normal matrix will contain zeros. The network shown in Figure 22.2 can be used to demonstrate this fact. In that figure, assume that distances and angles were measured for every line and arc, respectively. The following observations are made about the network's connectivity. Station 1 is connected to stations 2 and 8 by distance observations and is also connected to stations 2, 3, and 8 by angles. Notice that its connection to station 3 is due to angles turned at both 2 and 8. Since these angles directly connect stations 1 and 2, it follows that if the coordinates of 3 change, so will the coordinates of 1, and thus with respect to station 1, the normal matrix can be expected to have nonzero elements corresponding to the stations 1, 2, 3, and 8. Conversely, the positions corresponding to stations 4, 5, 6, and 7 will have zeros.

Using this analysis with station 3, because it is connected to stations 1, 2, 4, 7, and 8 by angles, zero elements can be expected in the normal matrix corresponding to stations 5 and 6. This analysis can be made for each station. The resulting symbolic normal matrix representation is shown in Figure 22.3.

A process known as *reordering the unknowns* can minimize both storage and computational time. Examine the matrix shown in Figure 22.4 that results from placing the unknowns of Figure 22.3 in the order of 6, 5, 4, 7, 3, 1, 2, and 8. This new matrix has its nonzero elements immediately adjacent to the diagonal elements, and thus the known zero elements are grouped together and appear in the leftmost columns of their respective matrix rows. By modifying the mapping table, storing the known zero elements of the original matrix in Figure 22.3 can be avoided, as is shown in the column matrix of Figure 22.4.

This storage scheme requires a mapping table that provides the storage location for the first nonzero element of each row. Since the diagonal elements are always nonzero, they can be stored in a separate column matrix without

Table 22.7 Computer algorithms to find the inverse of a Cholesky factored matrix

BASIC Language

```
{Inverse}
For i = Unknown To 1 Step -1
 ix = Vi(i): ixi = ix + i
 For k = Unknown To i + 1 Step -1
  S = 0!
  For j = i + 1 To k
   S = S + N(ix + j) * N(Vi(j) + k)
  Next j
  N(ix + k) = -S/N(ixi)
 Next k
 N(ixi) = 1.0/N(ixi)
Next i

{Inverse * Transpose of Inverse}
For j = 1 to Unknown
  ixj = Vi(j)
 For j = 1 to Unknown
   S = 0!
  For i = k to Unknown
   S = S + N(Vi(k) + i) * N(ixj + i)
  Next i
  N(ixj + k) = S
 Next k
Next j
```

FORTRAN Language

```
C Inverse
  Do 10 i = Unknown, 1, -1
     ix = Vi(i)
     ixi = ix + i
     Do 11 k = Unknown, i + 1, -1
        S = 0.0
        Do 12 j = i + 1, k
12        S = S + N(ix + j) * N(Vi(j) +
          N(ix + k) = -S/N(ixi)
11     Continue
       N(ixi) = 1.0/N(ixi)
10 Continue

C Inverse * Transpose of Inverse}
  Do 20 j = 1, Unknown
     ixj = Vi(j)
     Do 21 k = j, Unknown
        S = 0.0
        Do 22 i = k, Unknown
22            S = S + N(Vi(k) + i) * N(i
+ i)
        N(ixj + k) = S
21     Continue
20 Continue
```

C Language

```
{Inverse}
for (i = unknown; i >= 1; i--){
 ix = vi[i]; ixi = ix + i;
 for (k = unknown; k >= i + 1; k--){
  s = 0.0;
  for (j = i + 1; j <= k; j ++)
   s = s + n[ix + j] * n[vi[j] + k];
  n[ix + k] = -s/n[ixi];
 } //for k
 n[ixi] = 1.0/n[ixi];
}//for i

{inverse * transpose of inverse}
for (j = 1; j <= unknown; j++){
 ixj = vi[j];
 for (k = j; k <= unknown; k++){
  s = 0.0;
  For (i = k; i <= unknown; i++)
   s = s + n[vi[k] + i] * n[ixj + i];
  n[ixj + k] = s
 }//for k
}//for j
```

Pascal Language

```
{Inverse}
For i := Unknown Down To 1 do Begin
 ix := Vi[i]; ixi := ix + i;
 For k := Unknown Down To i + 1 Do Beg
  S := 0.0;
  For j := i + 1 To k Do
   S := S + N[ix + j] * N[Vi[j] + k];
  N[ix + k] := -S/N[ixi];
 End; {For k}
 N[ixi] := 1.0/N[ixi];
End; {For i}

{Inverse * Transpose of Inverse}
For j := 1 to Unknown Do Begin
  ixj := Vi[j];
 For k := j to Unknown Do Begin
   S := 0.0;
  For i := k to Unknown Do
   S := S + N[Vi[k] + i] * N[ixj + i];
  N[ixj + k] := S
 End; {For k}
End; {For j}
```

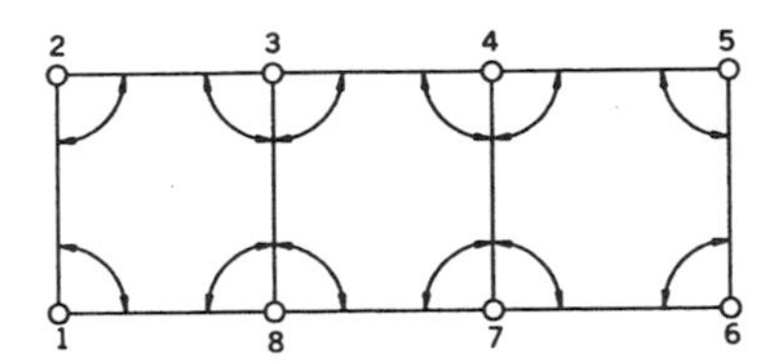

Figure 22.2 Series of connected traverses.

loss of efficiency. Ignoring the fact that each station has two unknowns, the mapping of the first element for each row is 1, 1, 2, 4, 7, 9, 10, 12, and 17. This mapping scheme enables finding the starting position for each row and allows for the determination of the off-diagonal length of the row. That is, row 1 starts at position 1 of the column matrix, but has a length of zero $(1 - 1)$, indicating that there are no off-diagonal elements in this row. Row 2 starts at position 1 of the column matrix and has a length of $(2 - 1) = 1$. Row 5 starts at 7 and has a length of $(9 - 7) = 2$. By optimizing for matrix sparseness and symmetry, only 24 $(16 + 8)$ elements of the original 64-element matrix need be stored.

This storage savings, when exploited, can also result in a savings of computational time. To understand this, first examine how the Cholesky factorization procedure processes the normal matrix. For a lower triangular matrix factorization process, Figure 22.5 shows the manner in which elements are accessed. Notice when a particular column is modified, the elements to the left of this column are used. Also notice that no rows above the corresponding diagonal element are used.

To see how the reorganized sparse matrix can be exploited in the factorization process, the processing steps must be understood. In Figure 22.4 the

	1	2	3	4	5	6	7	8
1	X	X	X	O	O	O	O	X
2	X	X	X	O	O	O	O	X
3	X	X	X	X	O	O	X	X
4	O	O	X	X	X	X	X	X
5	O	O	O	X	X	X	X	O
6	O	O	O	X	X	X	X	O
7	O	O	X	X	X	X	X	X
8	X	X	X	X	O	O	X	X

Figure 22.3 Normal matrix.

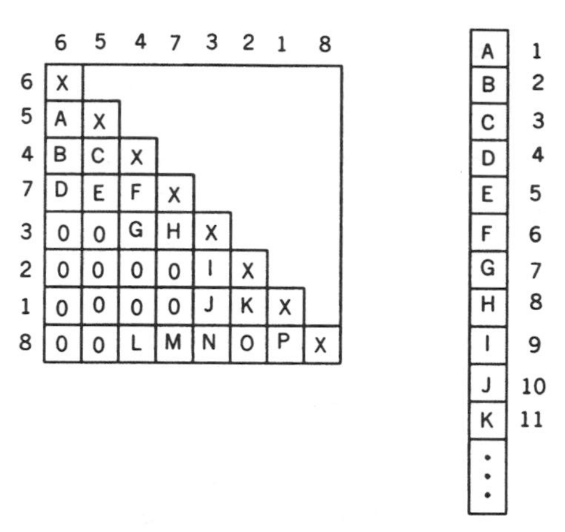

Figure 22.4 Reordered matrix.

first two elements of row 5 are known zeros, and thus each summation loop in the factorization process should start with the third column. If the zeros in the off-diagonals are rearranged such that they occur in the leftmost columns of the rows, it is possible to avoid that portion of the operations for the rows. Notice how the rows near the lower portion of the matrix are optimized to minimize the computational effort. A savings of eight multiplications is obtained for computations of column 5. In a large system, the cost savings of this optimization technique can be enormous. A complete discussion of these techniques is presented by George and Lui (1981) (see the Bibliography). A comparison of the operations performed when computing column 5 of Figure 22.4 in a Cholesky decomposition of both a full and optimized solution routine are shown in Table 22.8. Note that computations for rows 6 and 7 do not exist since these rows had known zeros in the columns before column 5 after the reordering process. If only the more time-consuming multiplication and division operations are counted, the optimized solution requires five operations as compared to 19 in the nonoptimized (full) solution.

Several methods have been developed to optimize the ordering of the unknowns. Two of the more well-known reordering schemes are the *reverse*

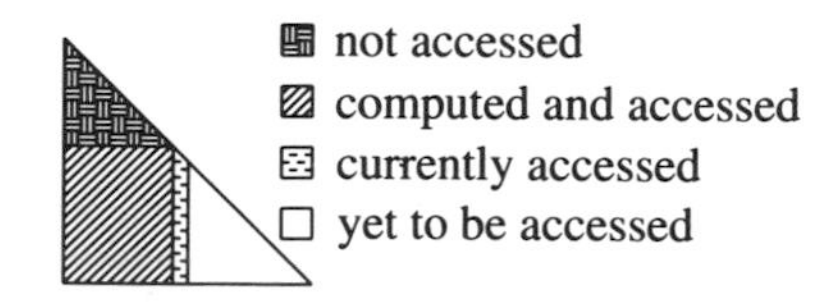

Figure 22.5 Computation of Cholesky factor.

Table 22.8 Comparison of number of operations in computing one column

Full Operations	Optimized Operations
1. $s = n_{51}^2 + n_{52}^2 + \cdots + n_{54}^2$	$s = n_{53}^2 + n_{54}^2$
2. n_{55} = square root of $(n_{55} - s)$	n_{55} = square root on $(n_{55} - s)$
3. $j = 6$	$j = 6$
(a) $s = n_{61}*n_{51} + n_{62}*n_{52} + \cdots + n_{64}*n_{54}$	No operations necessary
(b) $n_{65} = (n_{65} - s)/n_{55}$	
$j = 7$	$j = 7$
(a) $s = n_{71}*n_{51} + n_{72}*n_{52} + \cdots + n_{74}*n_{54}$	No operations necessary
(b) $n_{75} = (n_{75} - s)/n_{55}$	
$j = 8$	$j = 8$
(a) $s = n_{81}*n_{51} + n_{82}*n_{52} + \cdots + n_{84}*n_{54}$	$s = n_{83}*n_{53} + n_{84}*n_{54}$
(b) $n_{85} = (n_{85} - s)/n_{55}$	$n_{85} = (n_{85} - s)/n_{55}$

Cuthill–Mckee and the *banker's algorithm.* Both of these algorithms reorder stations based on their connectivity. The example of Figure 22.2 will be used to demonstrate the banker's algorithm. In this example, a connectivity matrix is first developed. This is simply a list of stations that are connected to a given station by either an angle or distance. For instance, station 1 is connected to stations 2 and 8 by distances, and additionally to station 3 by angles. The complete connectivity of the example above is listed in the first two columns of Table 22.9. In addition to the connectivity, the algorithms also need to know the number of connected points for each station. This is known as the *station's degree.*

To start the reordering, select the station with the lowest degree. In this example, that could be either station 1, 2, 5, or 6. For demonstration purposes, begin with station 6. This station is now removed from each station's connectivity matrix, and thus stations 4, 5, and 7 have their degrees reduced by one. This is shown in column 6. Now select the next station with the smallest

Table 22.9 Connectivity matrix

Station	Connectivity	Degree	6	5	4	7	3	1	2
1	2,3,8	3	3	3	3	3	2	—	—
2	1,3,8	3	3	3	3	3	2	1	—
3	1,2,4,7,8	5	5	5	4	3	—	—	—
4	3,5,6,7,8	5	4	3	—	—	—	—	—
5	4,6,7	3	2	—	—	—	—	—	—
6	4,5,7	3	—	—	—	—	—	—	—
7	3,4,5,6,8	5	4	3	2	—	—	—	—
8	1,2,3,7	5	5	5	4	3	2	1	0

degree. Since station 5 had its degree reduced to 2 when station 6 was selected, it now has the smallest degree. This process is continued; that is, the next station with the smallest degree is selected and is removed from the connectivity of the remaining stations. If two or more stations have the smallest degree, the most recently changed station with the smallest degree is chosen. In this example, this happened after choosing station 5. Station 4 was selected over stations 1 and 2 simply because it just had its degree reduced. Although station 7 qualified equally to station 4, station 4 was higher on the list and was selected for that reason. This happened again after selecting station 7. Station 3 was the next choice since it was the first station reduced to a degree of 3. Note that the final selection of station 8 was omitted from the list.

A similar optimized reordering can be found by starting with station 1, 2, or 5. This is left as an exercise. For a computer algorithm of this reordering, the reader should refer to the NOAA Technical Memorandum NGS 4 by Richard Snay; the computer algorithm for the *reverse Cuthill–McKee* is given in the book by George and Lui (see the Bibliography).

PROBLEMS

22.1 Explain why computer optimization techniques are necessary for doing large least-squares adjustments.

22.2 When using double precision, how much computer memory is required in forming the J, W, K, J^T, J^TW, J^TWK, and J^TWJ matrices for a 20-station adjustment having 50 total observations?

22.3 If the normal matrix of Problem 22.2 were stored in lower triangular form and computed directly from the observations, how much storage would be required?

22.4 Discuss how computational savings are created when solving the normal equations by Cholesky factorization and substitution.

22.5 Write a comparison of an operations table similar to that in Table 22.8 for column 3 of the matrix shown in Figure 22.4.

22.6 Same question as Problem 22.5 except with column 6.

22.7 Using the station reordering algorithm presented in Section 22.7, develop a connectivity matrix and reorder the stations when starting with station 1. Draw the normal matrix of the newly reordered station's, which is similar to the sketch shown in Figure 22.4.

22.8 Same as Problem 22.7 except start the reordering with station 2.

22.9 Same as Problem 22.7 except start the reordering with station 5.

Programming Problem

22.10 Develop a least-squares program similar to those developed in Chapter 15 that builds the normal equations directly from the observations and uses a Cholesky factorization procedure to find the solution.

APPENDIX A

INTRODUCTION TO MATRICES

A.1 INTRODUCTION

Matrix algebra provides at least two important advantages: (1) it enables reducing complicated systems of equations to simple expressions that can be visualized and manipulated more easily, and (2) it provides a systematic, mathematical method for solving problems that is well adapted to computers. Problems are frequently encountered in surveying, geodesy, and photogrammetry that require the solution of large systems of equations. This book deals specifically with the analysis and adjustment of redundant measurements that must satisfy certain geometric conditions. This frequently results in large equation systems which when solved according to the least-squares method yield most probable estimates for adjusted observations and unknown parameters. As will be demonstrated, matrix methods are particularly well suited for least-squares computations, and in this book they are used for analyzing and solving these equation systems.

A.2 DEFINITION OF A MATRIX

A matrix is a set of numbers or symbols arranged in a square or rectangular array of m rows and n columns. The arrangement is such that certain defined mathematical operations can be performed in a systematic and efficient manner. As an example of a matrix representation, consider the following system of three linear equations involving three unknowns:

$$
\begin{aligned}
a_{11}x_1 + a_{12}x_2 + a_{13}x_3 &= c_1 \\
a_{21}x_1 + a_{22}x_2 + a_{23}x_3 &= c_2 \\
a_{31}x_1 + a_{32}x_2 + a_{33}x_3 &= c_3
\end{aligned}
\tag{A.1}
$$

In Equation (A.1), the a's are coefficients of the unknowns x's, and the c's are the constant terms. The system above can be represented in summation notation as

$$\sum_{i=1}^{3} a_{1i}x_i = c_1$$

$$\sum_{i=1}^{3} a_{2i}x_i = c_2$$

$$\sum_{i=1}^{3} a_{3i}x_i = c_3$$

It can also be represented in matrix form as

$$
\begin{bmatrix} a_{11} & a_{12} & a_{13} \\ a_{21} & a_{22} & a_{23} \\ a_{31} & a_{32} & a_{33} \end{bmatrix}
\begin{bmatrix} x_1 \\ x_2 \\ x_3 \end{bmatrix}
=
\begin{bmatrix} c_1 \\ c_2 \\ c_3 \end{bmatrix}
\tag{A.2}
$$

In turn, Equation (A.2) can be represented in compact matrix notation as

$$AX = C \tag{A.3}$$

In Equation (A.3), capital letters (A, X, and C) are used to denote a matrix or an array of numbers or symbols. From simplified equation (A.3), it is immediately obvious that matrix methods provide a compact shorthand notation convenient for handling large systems of equations.

A.3 SIZE OR DIMENSIONS OF A MATRIX

The size or dimension of a matrix is specified by its number of rows m and its number of columns n. Thus the following matrix ${}_2D^3$, is a 2×3 matrix, or there are two rows and three columns (i.e., $m = 2$, $n = 3$). Notice that the subscript indicates the number of rows and the superscript indicates the number of columns in the notation ${}_2D^3$:

$$ {}_2D^3 = \begin{bmatrix} d_{11} & d_{12} & d_{13} \\ d_{21} & d_{22} & d_{23} \end{bmatrix} = \begin{bmatrix} 3 & 2 & 6 \\ 5 & 4 & 1 \end{bmatrix} $$

Also, the matrix E below is a 3×2 matrix:

$$ {}_3E^2 = \begin{bmatrix} e_{11} & e_{12} \\ e_{21} & e_{22} \\ e_{31} & e_{32} \end{bmatrix} = \begin{bmatrix} 7 & 1 \\ 4 & 3 \\ 2 & 8 \end{bmatrix} $$

Note that the position of an element in a matrix is defined by a double subscript and that a lowercase letter is used to designate any particular element within a matrix. Thus $d_{23} = 1$ is in row 2 and column 3 of the above matrix D. In general, the subscript ij indicates an element's position in a matrix, where i represents the row and j the column.

A.4 TYPES OF MATRICES

Several different types of matrices exist, as described below. Various symbols can be used to designate them as illustrated.

1. *Column matrix.* The number of rows can be any positive integer, but the number of columns is one.

$$ A = \begin{bmatrix} 3 \\ -2 \\ 5 \end{bmatrix} $$

2. *Row matrix.* The number of columns can be any positive integer but the number of rows is one.

$$ A = [6 \quad -4 \quad 2] $$

3. *Rectangular matrix.* The number of rows and columns are m and n, respectively, where m and n are any positive integers.

$$ A = \begin{bmatrix} 3 & 2 & 6 \\ 5 & 4 & 1 \end{bmatrix} $$

4. *Square matrix.* The number of rows equals the number of columns.

$$ A = \begin{bmatrix} 4 & 2 & -5 \\ -7 & 3 & 4 \\ 6 & -1 & 9 \end{bmatrix} $$

A square matrix, for which the determinant is zero, is termed *singular*. If the determinant is nonzero, it is termed *nonsingular*. (The determinant of a matrix is described in Section B.3).

5. *Symmetric matrix.* The matrix is mirrored about the main diagonal going from top left to bottom right (i.e., element a_{ij} = element a_{ji}). A symmetric matrix is always a square matrix.

$$A = \begin{bmatrix} 2 & -4 & 6 \\ -4 & 7 & 3 \\ 6 & 3 & 5 \end{bmatrix}$$

6. *Diagonal matrix.* Only the elements on the main diagonal are not zero. The diagonal matrix is always a square matrix.

$$A = \begin{bmatrix} 7 & 0 & 0 \\ 0 & -3 & 0 \\ 0 & 0 & 6 \end{bmatrix}$$

7. *Unit matrix.* A diagonal matrix with ones along the main diagonal. It is also called the identity matrix and is usually identified by the symbol I.

$$I = \begin{bmatrix} 1 & 0 & 0 \\ 0 & 1 & 0 \\ 0 & 0 & 1 \end{bmatrix}$$

8. *Transpose of a matrix.* It is obtained by interchanging rows and columns (i.e., element a_{ij}^T = element a_{ji}). Thus the dimensions of A^T are the reverse of the dimensions of A. If $A = \begin{bmatrix} 2 & 4 & 7 \\ 5 & 3 & 1 \end{bmatrix}$, then the transpose of A, denoted A^T, is

$$A^T = \begin{bmatrix} 2 & 5 \\ 4 & 3 \\ 7 & 1 \end{bmatrix}$$

A.5 MATRIX EQUALITY

Two matrices are said to be equal only when they are equal element by element. Thus the two matrices must be the same size or have the same dimensions.

$$A = \begin{bmatrix} 1 & 7 & 6 \\ 4 & 3 & 2 \end{bmatrix} = B = \begin{bmatrix} 1 & 7 & 6 \\ 4 & 3 & 2 \end{bmatrix}$$

A.6 ADDITION OR SUBTRACTION OF MATRICES

Matrices can be added or subtracted, but to do so, they must have the same dimensions. If two matrices have equal dimensions, they are said to be *conformable* for addition or subtraction. In adding or subtracting matrices, elements from each unique row/column position of the two matrices are added or subtracted systematically and the sum or difference is placed in the same unique row/column location of the resulting matrix. The following example illustrates this procedure.

$$\begin{aligned} {}_2A^3 + {}_2B^3 &= \begin{bmatrix} 7 & 3 & -1 \\ 2 & -5 & 6 \end{bmatrix} + \begin{bmatrix} 1 & 5 & 6 \\ -4 & -2 & 3 \end{bmatrix} = \begin{bmatrix} 8 & 8 & 5 \\ -2 & -7 & 9 \end{bmatrix} \\ &= {}_2C^3 \end{aligned}$$

Assuming that two matrices are conformable for addition or subtraction, the following are true:

(a) $A + B = B + A$ (commutative law)
(b) $A + (B + C) = (A + B) + C$ (associative law)

A.7 SCALAR MULTIPLICATION OF A MATRIX

Matrices can be multiplied by a scalar (i.e., a constant). Let k be any scalar quantity; then

$$kA = Ak$$

The following are examples:

$$4 \times \begin{bmatrix} 3 & -1 \\ 2 & 6 \\ 4 & 7 \\ 5 & 3 \end{bmatrix} = \begin{bmatrix} 3 & -1 \\ 2 & 6 \\ 4 & 7 \\ 5 & 3 \end{bmatrix} \times 4 = \begin{bmatrix} 12 & -4 \\ 8 & 24 \\ 16 & 28 \\ 20 & 12 \end{bmatrix}$$

As illustrated in the example above, each element of the matrix A is multiplied by the scalar k to obtain the elements of C. Note that $4 \times A = A + A + A + A = A \times 4$.

A.8 MATRIX MULTIPLICATION

If matrix A is to be *postmultiplied* by matrix B (i.e., the product of $A \times B$ determined), the number of columns in matrix A must equal the number of rows in matrix B. This is a basic requirement for matrix multiplication. When this condition is satisfied, A and B are said to be *conformable* for multiplication. The product C will have the same number of rows as A and the same number of columns as B. Thus the following multiplications are possible:

$$ {}_4A^2 \times {}_2B^3 = {}_4C^3 $$

$$ {}_1A^3 \times {}_3B^1 = {}_1C^1 $$

$$ {}_3A^3 \times {}_3B^1 = {}_3C^1 $$

and these multiplications are not possible:

$$ {}_2B^3 \times {}_4A^2 $$

$$ {}_6A^2 \times {}_6B^3 $$

To demonstrate the process of matrix multiplication, consider the following example:

$$ {}_2A^3 \times {}_3B^2 = \begin{bmatrix} a_{11} & a_{12} & a_{13} \\ a_{21} & a_{22} & a_{23} \end{bmatrix} \times \begin{bmatrix} b_{11} & b_{12} \\ b_{21} & b_{22} \\ b_{31} & b_{32} \end{bmatrix} = \begin{bmatrix} c_{11} & c_{12} \\ c_{21} & c_{22} \end{bmatrix} = {}_2C^2 $$

The elements c_{ij} of matrix C are the total sums obtained by successively multiplying each element in row i of matrix A by the elements in column j of matrix B and then summing these products. Thus for the example above,

$$ c_{11} = a_{11} \times b_{11} + a_{12} \times b_{21} + a_{13} \times b_{31} = \sum_{i=1}^{3} a_{1i}b_{i1} $$

$$ c_{12} = a_{11} \times b_{12} + a_{12} \times b_{22} + a_{13} \times b_{32} = \sum_{i=1}^{3} a_{1i}b_{i2} $$

$$ c_{21} = a_{21} \times b_{11} + a_{22} \times b_{21} + a_{23} \times b_{31} = \sum_{i=1}^{3} a_{2i}b_{i1} $$

$$c_{22} = a_{21} \times b_{12} + a_{22} \times b_{22} + a_{23} \times b_{32} = \sum_{i=1}^{3} a_{2i} b_{i2}$$

The process above is seen more easily with a numerical example.

$$\begin{vmatrix} 1 & 2 & 3 \\ 4 & 2 & 7 \end{vmatrix} \times \begin{vmatrix} 4 & 8 \\ 6 & 2 \\ 5 & 3 \end{vmatrix} = \begin{vmatrix} c_{11} & c_{12} \\ c_{21} & c_{22} \end{vmatrix} = \begin{vmatrix} 31 & 21 \\ 63 & 57 \end{vmatrix}$$

$$c_{11} = 1 \times 4 + 2 \times 6 + 3 \times 5 = 31$$

$$c_{12} = 1 \times 8 + 2 \times 2 + 3 \times 3 = 21$$

$$c_{21} = 4 \times 4 + 2 \times 6 + 7 \times 5 = 63$$

$$c_{22} = 4 \times 8 + 2 \times 2 + 7 \times 3 = 57$$

The student should now verify the matrix representation of Equations (A.1) and (A.2). Notice that the product of a unit matrix, I, and a conformable matrix, A (one with the same number of rows as I), equals the original matrix A. Thus

$${}_2I^2 \times {}_2A^2 = \begin{bmatrix} 1 & 0 \\ 0 & 1 \end{bmatrix} \times \begin{bmatrix} 5 & 6 \\ 7 & 8 \end{bmatrix} = \begin{bmatrix} 5 & 6 \\ 7 & 8 \end{bmatrix} = {}_2A^2$$

Assuming that matrices A, B, and C are conformable for the operations indicated, the following are true:

(c) $A(B + C) = AB + AC$ (first distributive law)
(d) $(A + B)C = AC + BC$ (second distributive law)
(e) $A(BC) = (AB)\,C$ (associative law)
(f) $(AB)^T = B^T A^T$

The following cautions are also stated:

(g) AB is not generally equal to BA, and BA may not even be conformable.
(h) If $AB = 0$, neither A nor B necessarily $= 0$.
(i) If $AB = AC$, B does not necessarily $= C$.

Let

$$A = \begin{bmatrix} 2 & 1 \\ 1 & 2 \end{bmatrix} \quad B = \begin{bmatrix} -1 & -1 \\ 2 & 2 \end{bmatrix} \quad C = \begin{bmatrix} 1 & 1 \\ 1 & 1 \end{bmatrix}.$$

Example of (c):

$$\begin{aligned} A(B + C) &= \begin{bmatrix} 2 & 1 \\ 1 & 2 \end{bmatrix} \times \begin{bmatrix} 0 & 0 \\ 3 & 3 \end{bmatrix} = \begin{bmatrix} 3 & 3 \\ 6 & 6 \end{bmatrix} = AB + AC \\ &= \begin{bmatrix} 0 & 0 \\ 3 & 3 \end{bmatrix} + \begin{bmatrix} 3 & 3 \\ 3 & 3 \end{bmatrix} = \begin{bmatrix} 3 & 3 \\ 6 & 6 \end{bmatrix}^1 \end{aligned}$$

Example of (d):

$$\begin{aligned} (A + B)C &= \begin{bmatrix} 1 & 0 \\ 3 & 4 \end{bmatrix} \times \begin{bmatrix} 1 & 1 \\ 1 & 1 \end{bmatrix} = \begin{bmatrix} 1 & 1 \\ 7 & 7 \end{bmatrix} = AC + BC \\ &= \begin{bmatrix} 3 & 3 \\ 3 & 3 \end{bmatrix} + \begin{bmatrix} -2 & -2 \\ 4 & 4 \end{bmatrix} = \begin{bmatrix} 1 & 1 \\ 7 & 7 \end{bmatrix} \end{aligned}$$

Example of (e):

$$\begin{aligned} A(BC) &= \begin{bmatrix} 2 & 1 \\ 1 & 2 \end{bmatrix} \times \begin{bmatrix} -2 & -2 \\ 4 & 4 \end{bmatrix} = \begin{bmatrix} 0 & 0 \\ 6 & 6 \end{bmatrix} = (AB)C \\ &= \begin{bmatrix} 0 & 0 \\ 3 & 3 \end{bmatrix} + \begin{bmatrix} 1 & 1 \\ 1 & 1 \end{bmatrix} = \begin{bmatrix} 0 & 0 \\ 6 & 6 \end{bmatrix} \end{aligned}$$

Example of (f): Let

$$A = \begin{bmatrix} 2 & 6 & 4 \\ 1 & 2 & 7 \end{bmatrix} \quad \text{and} \quad B = \begin{bmatrix} 3 & 2 \\ 9 & 0 \\ 1 & 3 \end{bmatrix}$$

Then

$$AB = \begin{bmatrix} 64 & 16 \\ 28 & 23 \end{bmatrix} \quad \text{and} \quad (AB)^T = \begin{bmatrix} 64 & 28 \\ 16 & 23 \end{bmatrix}$$

Also,

[1] The multiplication symbol, ×, is generally not written in matrix equations. This convention of not using × will be followed in this book.

$$B^TA^T = \begin{bmatrix} 3 & 9 & 1 \\ 2 & 0 & 3 \end{bmatrix} \begin{bmatrix} 2 & 1 \\ 6 & 2 \\ 4 & 7 \end{bmatrix} = \begin{bmatrix} 64 & 28 \\ 16 & 23 \end{bmatrix}$$

Example of (g):

$$AB = \begin{bmatrix} 0 & 0 \\ 3 & 3 \end{bmatrix} \neq \begin{bmatrix} -3 & -3 \\ 6 & 6 \end{bmatrix} = BA$$

Example of (h): Let

$$A = \begin{bmatrix} 2 & 2 \\ 1 & 1 \end{bmatrix} \quad \text{and} \quad B = \begin{bmatrix} -1 & -2 \\ 1 & 2 \end{bmatrix}$$

Then $AB = 0$, but neither A nor $B = 0$.

Example of (i): Let

$$A = \begin{bmatrix} 2 & 2 \\ 1 & 1 \end{bmatrix}, \quad B = \begin{bmatrix} 1 & 2 \\ 1 & 3 \end{bmatrix}, \quad \text{and} \quad C = \begin{bmatrix} 4 & 10 \\ -2 & -5 \end{bmatrix}$$

Then $AB = AC$, but $B \neq C$ where

$$AB = \begin{bmatrix} 4 & 10 \\ 2 & 5 \end{bmatrix} = AC$$

A.9 COMPUTER ALGORITHMS FOR MATRIX OPERATIONS

It should be apparent that addition, subtraction, and multiplication of large matrices involves many arithmetic operations. These are very tedious when done by hand but can be done quickly with a computer. In this section, general mathematical expressions are developed for performing these operations using a computer. These general mathematical expressions, when programmed for computer solution, are called *algorithms*.

A.9.1 Addition or Subtraction of Two Matrices

Consider the two matrices A and B that are shown in Figure A.1. Note that matrices A and B are *conformable* for addition. Find the sum of the two matrices and place the results in C.

Step 1: Add the first element (a_{11}) of the A matrix to the first element (b_{11}) of B, placing the result in the first element (c_{11}) of C. Repeat this process

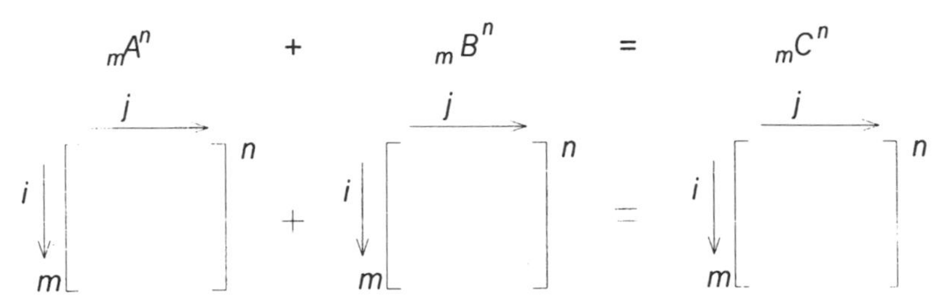

Figure A.1 Addition of matrices.

for each of the successive columns along the first row of A and B, or from $j = 1$ to $j = n$.

Step 2: Iterate step 1 for each successive row of the matrices, or for rows increasing from $i = 1$ to $i = m$. Table A.1 shows this entire operation of adding two matrices in the four computer languages of BASIC, C, FORTRAN, and Pascal.

A.9.2 Matrix Multiplication

Consider the two matrices A and B that are shown in Figure A.2. Again these matrices, A and B, are conformable for multiplication. Find the product AB and place the results in C.

Step 1: Sum the products (A row $i = 1$) $\times$ (B column $k = 1$), with the elements of A and B being increased successively from $j = 1$ to $j = p$. Place the result in c_{ik}, or c_{11}, for this first step. Mathematically, this step is represented as

Table A.1 Addition algorithm in BASIC, C, FORTRAN, and Pascal

BASIC Language

```
1000 For i = 1 to M
1010  For j = 1 to N
1020  C(i,j) = A(i,j) + B(i,j)
1030  Next j
1040 Next i
```

FORTRAN Language

```
     Do 100 I = 1,M
      Do 100 J = 1,N
       C(i,j) = A(i,j) + B(i,j)
100  Continue
```

C Language

```
for (i = 0; i < m; i++)
 for (j = 0; j < n; j++)
  C[i][j] = A[i][j] + B[i][j];
```

Pascal Language

```
For i := 1 to M do
 For j := 1 to N do
  C[i,j] := A[i,j] + B[i,j];
```

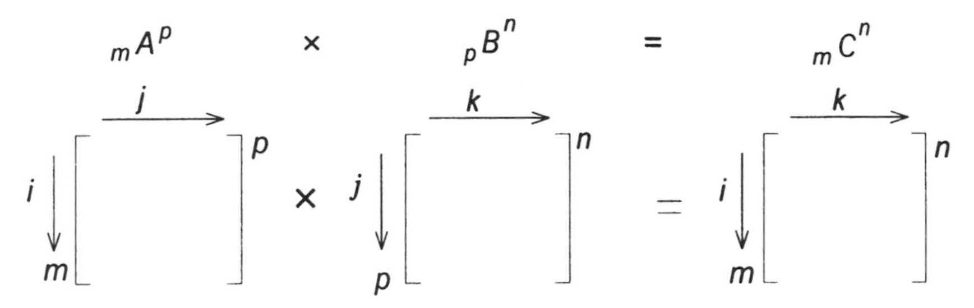

Figure A.2 Multiplication of matrices.

$$c_{ik} = \sum_{j=1}^{p} a_{ij} \times b_{jk} \qquad \text{with } i = 1 \text{ and } k = 1$$

Step 2: Increase k successively by 1, repeat step 1, and place the results in c_{ik}. Continue increasing k and repeating step 1 until $k = p$.

Step 3: Increase i from one to two, and repeat steps 1 and 2 in entirety. Upon completion with $i = 2$, increment i by one and repeat steps 1 and 2 again in entirety. Continue this process through $i = m$. This completes the matrix multiplication.

This operation is shown Table A.2 in the four languages of BASIC, C, FORTRAN, and Pascal. As a final note on multiplication, it is essential that the order in which the matrices are multiplied be specified.

Table A.2 Multiplication algorithm in BASIC, C, FORTRAN, and Pascal

BASIC Language

```
For i = 1 to M
 For k = 1 to N
  C(i,k) = 0.0
  For j = 1 to P
   C(i,k) = C(i,k) + A(i,j)*B(j,k)
  Next j: Next k: Next i
```

C Language

```
for (i = 0; i < m; i++)
 for (k = 0; k < n; k++){
  C[i][k] = 0;
  for (j = 0; j < p; j++)
   C[i][k] = C[i][k] + A[i][j]*B[j][k]
 } //for k
```

FORTRAN Language

```
    Do 100 I = 1,M
     Do 100 K = 1,N
      C(i,k) = 0.0
      Do 100 J = 1,P
       C(i,k) = C(i,k) + A(i,j)*B(j,k)
100 Continue
```

Pascal Language

```
For i := 1 to M do
 For k := 1 to N do Begin
  C[i,k] := 0.0;
  For j := 1 to P do
   C[i,k] := C[i,k] + A[i,j]*B[j,k]
 End; {for k}
```

A.10 USE OF THE MATRIX PROGRAM

A software package called MATRIX is included on the diskette that is provided with this book. It includes all the matrix operations that will be necessary to study the subject of adjustment computations and solve the after-chapter homework problems herein that require matrix manipulation. Instructions for use of the software are given in Appendix F.

PROBLEMS

A.1 Suppose that the following system of linear equations is to be represented in matrix form as $AX = B$. Formulate the three matrices.

$$
\begin{aligned}
2x_1 + x_2 + 5x_3 + x_4 &= 5 \\
x_1 + x_2 - 3x_3 + 4x_4 &= -1 \\
3x_1 + 6x_2 - 2x_3 + x_4 &= 8 \\
2x_1 + 2x_2 + 2x_3 - 3x_4 &= 2
\end{aligned}
$$

A.2 For Problem A.1, find the product of A^TA.

A.3 Do the following matrix operations:

(a) $$\begin{bmatrix} 1 & 2 & -1 & 0 \\ 4 & 0 & 2 & 1 \\ 2 & -5 & 1 & 2 \end{bmatrix} + \begin{bmatrix} 3 & -4 & 1 & 2 \\ 1 & 5 & 0 & -3 \\ 2 & -2 & 3 & -1 \end{bmatrix}$$

(b) $$\begin{bmatrix} 1 & 2 & -1 & 0 \\ 4 & 0 & 2 & 1 \\ 2 & -5 & 1 & 2 \end{bmatrix} - \begin{bmatrix} 3 & -4 & 1 & 2 \\ 1 & 5 & 0 & 3 \\ 2 & -2 & 3 & -1 \end{bmatrix}$$

(c) $$3\begin{bmatrix} 1 & 2 & -1 & 0 \\ 4 & 0 & 2 & 1 \\ 2 & -5 & 1 & 2 \end{bmatrix}$$

(d) $$\begin{bmatrix} 1 & 2 & 1 \\ 4 & 0 & 2 \end{bmatrix} \times \begin{bmatrix} 3 & -4 \\ 1 & 5 \\ -2 & 2 \end{bmatrix}$$

A.4 Solve Problem A.2 using the program MATRIX.

A.5 Solve Problem A.3 using the program MATRIX.

A.6 Let

$$A = \begin{bmatrix} 2 & 0 \\ 3 & 1 \end{bmatrix} \quad B = \begin{bmatrix} 4 & -1 \\ 0 & 2 \end{bmatrix} \quad C = \begin{bmatrix} 1 & 0 \\ 0 & 1 \end{bmatrix} \quad D = \begin{bmatrix} 0 & 0 \\ 0 & 0 \end{bmatrix}$$

(a) Find AB. **(c)** Find CB. **(e)** Find B^3. **(g)** Find C^3.
(b) Find B^2. **(d)** Find BA. **(f)** Find DB.

A.7 Do Problem A.6 using the program MATRIX.

A.8 Multiply the following, if possible. If not possible, give reasons why the multiplication cannot be done.

(a) $\begin{bmatrix} 2 & 1 \\ 4 & 0 \end{bmatrix} [3 \quad 2]$

(b) $[3 \quad 1 \quad 3] \begin{bmatrix} 4 \\ 0 \\ 9 \end{bmatrix}$

(c) $\begin{bmatrix} 2 & 3 & 4 & 4 \\ 1 & 0 & -1 & 6 \\ 0 & 1 & 2 & 9 \end{bmatrix} \begin{bmatrix} 0 & 2 \\ 3 & 1 \\ 1 & 0 \\ 0 & -1 \end{bmatrix}$

(d) $\begin{bmatrix} 2 \\ 0 \end{bmatrix} [3 \quad -1]$

A.9 Expand the following summations into their equivalent algebraic expressions.

(a) $\sum_{k=1}^{5} k$

(b) $\sum_{k=3}^{7} (k - 2)$

(c) $\sum_{k=1}^{4} a_k$

(d) $\sum_{k=1}^{3} a_{2k} a_{k3}$

(e) $\sum_{i=1}^{3} \sum_{j=1}^{4} (a_{ij} + b_{ij})$

(f) $\sum_{i=1}^{2} \sum_{j=1}^{2} \sum_{k=1}^{3} a_{ik} b_{kj}$

A.10 Write the algebraic expressions for the system of equations represented by the following matrix notation.

$$\begin{bmatrix} 2 & 0 \\ 1 & 3 \\ 4 & 2 \end{bmatrix} \begin{bmatrix} x_1 \\ x_2 \end{bmatrix} = \begin{bmatrix} 2 \\ 1 \\ 3 \end{bmatrix}$$

A.11 Show that in general, $(AB)^T = B^T A^T$.

Programming Problems

A.12 Write a program that will read and write the elements of a matrix, compute the transpose of the matrix, and write the solution. (*Hint*: Place the reading, writing, and transposition codes in separate subroutines/procedures/functions to be called from the main program.)

A.13 Select one of the matrix addition codes from Table A.1, and enter it and any necessary supporting code to solve Problem A.3(a). (*Hint*: Place the code in Table A.1 in a separate subroutine/procedure/function to be called from the main program.)

A.14 Select one of the matrix multiplication codes from Table A.2, and enter it and any necessary supporting code to solve Problem A.3(d). (*Hint*: Place the code in Table A.2 in a separate subroutine/procedure/function to be called from the main program.)

A.15 Write a program that will read and write a file of matrices, and do the matrix operations of transposition, addition, subtraction, and multiplication. Demonstrate the program by doing Problems A.2 and A.3(a), (b), and (d). (*Hint:* Place the reading, writing, transposition, addition, subtraction, and multiplication, codes in separate subroutines/procedures/functions to be called from the main program.)

APPENDIX B

SOLUTION OF EQUATIONS BY MATRIX METHODS

B.1 INTRODUCTION

As stated in Appendix A, an advantage offered by matrix algebra is its adaptability to computer usage. Using matrix algebra, large systems of simultaneous linear equations can be programmed for computer solution using only a few systematic steps. The simplicities of programming matrix additions and multiplications, for example, were presented in Section A.9. To solve a system of equations using matrix methods, it is first necessary to define and compute the inverse matrix.

B.2 INVERSE MATRIX

If a square matrix is nonsingular (its determinant is not zero), it possesses an inverse matrix. The inverse of matrix A, symbolized as A^{-1}, is defined as

$$A^{-1}A = I \tag{B.1}$$

where I is the identity matrix.

When a system of simultaneous linear equations consisting of n equations and involving n unknowns is expressed as $AX = B$, the coefficient matrix (A) is a square matrix of dimensions $n \times n$. Consider this system of linear equations:

$$AX = B \tag{B.2}$$

Premultiplying both sides of matrix equation (B.2) by A^{-1} gives

$$A^{-1}AX = A^{-1}B$$

Reducing yields

$$IX = A^{-1}B$$

$$X = A^{-1}B \tag{B.3}$$

Thus the inverse is used to find the matrix of unknowns, X. The following points should be emphasized regarding matrix inversions:

1. Square matrices have inverses, with the exception noted below.
2. When the determinant of a matrix is zero, the matrix is said to be singular and its inverse cannot be found.
3. The inversion of even a small matrix is a tedious and time-consuming operation when done by hand. However, when done by a computer, the inverse can be found quickly and easily.

B.3 INVERSE OF A 2 × 2 MATRIX

Several general methods are available for finding a matrix inverse. Two shall be considered herein. Before proceeding with general cases, however, consider the specific case of finding the inverse for a 2 × 2 matrix using simple elementary matrix operations. Let any 2 × 2 matrix be symbolized as A. Also, let

$$A = \begin{bmatrix} a & b \\ c & d \end{bmatrix} \quad \text{and} \quad A^{-1} = \begin{bmatrix} w & x \\ y & z \end{bmatrix}$$

By applying Equation (B.1), and recalling the definition of an identity matrix I as given in Section A.4, it is possible to calculate w, x, y, and z in terms of a, b, c, and d of A^{-1}. Substituting in the appropriate values gives

$$\begin{bmatrix} a & b \\ c & d \end{bmatrix} \begin{bmatrix} w & x \\ y & z \end{bmatrix} = \begin{bmatrix} 1 & 0 \\ 0 & 1 \end{bmatrix}$$

By matrix multiplication

$$aw + by = 1 \qquad (a)$$

$$ax + bz = 0 \qquad (b)$$

$$cw + dy = 0 \qquad (c)$$

$$cx + dz = 1 \qquad (d)$$

The determinant of A is symbolized as $\|A\|$ and is equal to $ad - bc$.

$$\left\{\begin{array}{c} \text{From } (a)\ y = \dfrac{1 - aw}{b};\ \text{From } (c)\ y = -\dfrac{cw}{d} \\ \text{Then } \dfrac{1 - aw}{b} = -\dfrac{cw}{d};\ \text{Reducing: } w = \dfrac{d}{da - bc} = \dfrac{d}{\|A\|} \end{array}\right\}$$

$$\left\{\begin{array}{c} \text{From } (b)\ z = -\dfrac{ax}{b};\ \text{From } (d)\ z = \dfrac{1 - cx}{d} \\ \text{Then } -\dfrac{ax}{b} = \dfrac{1 - cx}{d};\ \text{Reducing: } x = \dfrac{b}{-da + bc} = -\dfrac{b}{\|A\|} \end{array}\right\}$$

$$\left\{\begin{array}{c} \text{From } (a)\ w = \dfrac{1 - by}{a};\ \text{From } (c)\ w = -\dfrac{dy}{c} \\ \text{Then } \dfrac{1 - by}{a} = -\dfrac{dy}{c};\ \text{Reducing: } y = \dfrac{c}{-ad + cb} = -\dfrac{c}{\|A\|} \end{array}\right\}$$

$$\left\{\begin{array}{c} \text{From } (b)\ x = -\dfrac{bz}{a};\ \text{From } (d)\ x = \dfrac{1 - dz}{c} \\ \text{Then } -\dfrac{bz}{a} = \dfrac{1 - dz}{c};\ \text{Reducing: } z = \dfrac{a}{ad - bc} = \dfrac{a}{\|A\|} \end{array}\right\}$$

Thus for any 2×2 matrix composed of the elements $\begin{bmatrix} a & b \\ c & d \end{bmatrix}$, its inverse is simply

$$\begin{bmatrix} \dfrac{d}{\|A\|} & -\dfrac{b}{\|A\|} \\ -\dfrac{c}{\|A\|} & \dfrac{a}{\|A\|} \end{bmatrix} = \frac{1}{\|A\|}\begin{bmatrix} d & -b \\ -c & a \end{bmatrix}$$

Example B.1 If $A = \begin{bmatrix} 2 & 3 \\ 4 & 1 \end{bmatrix}$, find A^{-1}.

SOLUTION

$$A^{-1} = -\frac{1}{10}\begin{bmatrix} 1 & -3 \\ -4 & 2 \end{bmatrix}$$

A check on the inverse can be obtained by testing it against its definition, or $A^{-1}A = I$. Thus:

$$-\frac{1}{10}\begin{bmatrix} 1 & -3 \\ -4 & 2 \end{bmatrix}\begin{bmatrix} 2 & 3 \\ 4 & 1 \end{bmatrix} = \begin{bmatrix} 1 & 0 \\ 0 & 1 \end{bmatrix} = I$$

B.4 INVERSES BY ADJOINTS

The inverse of matrix A can be found using the method of adjoints with the following equation:

$$A^{-1} = \frac{\text{adjoint of } A}{\text{determinant of } A} = \frac{\text{adjoint of } A}{\|A\|} \tag{B.4}$$

The *adjoint of A* is obtained by first replacing each matrix element by its signed minor or *cofactor*, and then transposing the resulting matrix. The cofactor of element a_{ij} equals $(-1)^{i+j}$ times the numerical value of the determinant for the remaining elements after row i and column j have been removed from the matrix. This procedure is illustrated in Figure B.1, where the cofactor of a_{12} is

$$(-1)^{1+2}(a_{21}a_{33} - a_{31}a_{23}) = a_{31}a_{23} - a_{21}a_{33}$$

Using this procedure, the inverse of the following A matrix is found:

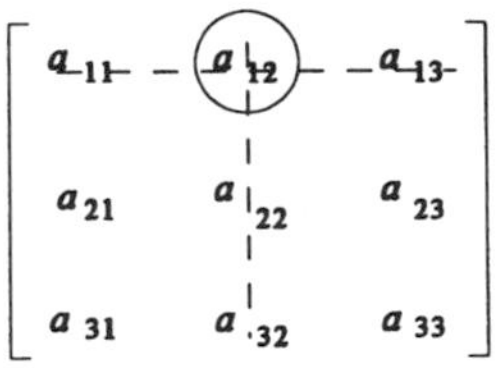

$$\begin{bmatrix} a_{11} & a_{12} & a_{13} \\ a_{21} & a_{22} & a_{23} \\ a_{31} & a_{32} & a_{33} \end{bmatrix}$$

Figure B.1 Cofactor of the a_{12} element.

$$A = \begin{bmatrix} a_{11} & a_{12} & a_{13} \\ a_{21} & a_{22} & a_{23} \\ a_{31} & a_{32} & a_{33} \end{bmatrix} = \begin{bmatrix} 4 & 3 & 2 \\ 3 & 4 & 1 \\ 2 & 3 & 4 \end{bmatrix}$$

For this A matrix, the cofactors are calculated as follows:

$$\text{cofactor of } a_{11} = (-1)^2 (4 \times 4 - 1 \times 3) = 13$$

$$\text{cofactor of } a_{21} = (-1)^3 (3 \times 4 - 2 \times 3) = -6$$

$$\text{cofactor of } a_{31} = (-1)^4 (3 \times 1 - 2 \times 4) = -5$$

$$\text{cofactor of } a_{12} = (-1)^3 (3 \times 4 - 1 \times 2) = -10$$

$$\vdots$$

Following the procedure above, the matrix of cofactors is

$$\text{matrix of cofactors} = \begin{bmatrix} 13 & -10 & 1 \\ -6 & 12 & -6 \\ -5 & 2 & 7 \end{bmatrix}$$

Transposing this cofactor matrix produces the following adjoint of A:

$$\text{adjoint of } A = \begin{bmatrix} 13 & -6 & -5 \\ -10 & 12 & 2 \\ 1 & -6 & 7 \end{bmatrix}$$

The determinant of A is the sum of the products of the elements in the first row of the original matrix times their respective cofactors. Since the cofactors were already obtained in the previous step, this simplifies to

$$\|A\| = 4(13) + 3(-10) + 2(1) = 24$$

The inverse of A is now calculated as

$$A^{-1} = \frac{1}{24} \begin{bmatrix} 13 & -6 & -5 \\ -10 & 12 & 2 \\ 1 & -6 & 7 \end{bmatrix} = \begin{bmatrix} 13/24 & -1/4 & -5/24 \\ -5/12 & 1/2 & 1/12 \\ 1/24 & -1/4 & 7/24 \end{bmatrix}$$

Again, a check on the arithmetical work is obtained by using the definition of an inverse:

$$A^{-1}A = \begin{bmatrix} 13/24 & -1/4 & -5/24 \\ -5/12 & 1/2 & 1/12 \\ 1/24 & -1/4 & 7/24 \end{bmatrix} \begin{bmatrix} 4 & 3 & 2 \\ 3 & 4 & 1 \\ 2 & 3 & 4 \end{bmatrix} = \begin{bmatrix} 1 & 0 & 0 \\ 0 & 1 & 0 \\ 0 & 0 & 1 \end{bmatrix} = I$$

B.5 INVERSES BY ROW TRANSFORMATIONS

1. The multiplication of every element in any row by a nonzero scalar
2. The addition (or subtraction) of the elements in any row to the elements of any other row
3. Combinations of 1 and 2.

If elementary row transformations are successively performed on A such that A is transformed into I, and if throughout the procedure the same row transformations are also done to the same rows of the identity matrix I, the I matrix will be transformed into A^{-1}. This procedure is illustrated with the same matrix used to demonstrate the method of adjoints.

Initially, the original matrix and the identity matrix are listed side by side:

$$\begin{array}{c} \qquad A \qquad\qquad\qquad I \\ \left[\begin{array}{ccc|ccc} 4 & 3 & 2 & 1 & 0 & 0 \\ 3 & 4 & 1 & 0 & 1 & 0 \\ 2 & 3 & 4 & 0 & 0 & 1 \end{array}\right] \end{array}$$

With the following three row transformations performed on A and I, they are transformed into matrices A_1 and I_1, respectively:

1. Multiply row 1 of matrices A and I by $1/a_{11}$ or $1/4$. Place the results in row 1 of A_1 and I_1, respectively. This converts a_{11} of matrix A_1 to 1, as shown below.

$$\left[\begin{array}{ccc|ccc} 1 & 3/4 & 1/2 & 1/4 & 0 & 0 \\ 3 & 4 & 1 & 0 & 1 & 0 \\ 2 & 3 & 4 & 0 & 0 & 1 \end{array}\right]$$

2. Multiply row 1 of matrices A_1 and I_1 by a_{21} or 3. Subtract the resulting row from row 2 of matrices A and I and place the difference in row 2 of A_1 and I_1, respectively. This converts a_{21} of A_1 to zero.
3. Multiply row 1 of matrices A_1 and I_1 by a_{31} or 2. Subtract the resulting row from row 3 of matrices A and I and place the difference in row 3 of A_1 and I_1, respectively. This changes a_{31} of A_1 to zero.

After doing these operations, the transformed matrices A_1 and I_1 are

$$\begin{array}{c} \qquad\qquad A_1 \qquad\qquad\qquad\qquad I_1 \\ \left[\begin{array}{ccc|ccc} 1 & 3/4 & 1/2 & 1/4 & 0 & 0 \\ 0 & 7/4 & -1/2 & -3/4 & 1 & 0 \\ 0 & 3/2 & 3 & -1/2 & 0 & 1 \end{array}\right] \end{array}$$

Notice that the first column of A_1 has been made equivalent to the first column of a 3×3 identity matrix as a result of these three row transformations. For matrices having more than three rows, this same general procedure would be followed for each row to convert the first element in each row of A_1 to zero with exception to first row of A.

Next the following three elementary row transformations are done on matrices A_1 and I_1 to transform them into matrices A_2 and I_2:

1. Multiply row 2 of A_1 and I_1 by $1/a_{22}$ or 4/7, and place the results in row 2 of A_2 and I_2. This converts a_{22} to 1, as shown below.

$$\left[\begin{array}{ccc|ccc} 1 & 3/4 & 1/2 & 1/4 & 0 & 0 \\ 0 & 1 & -2/7 & -3/7 & 4/7 & 0 \\ 0 & 3/2 & 3 & -1/2 & 0 & 1 \end{array}\right]$$

2. Multiply row 2 of A_2 and I_2 by a_{12} or 3/4. Subtract the resulting row from row 1 of A_1 and I_1 and place the difference in row 1 of A_2 and I_2, respectively.
3. Multiply row 2 of A_2 and I_2 by a_{32}, or 3/2. Subtract the resulting row from row 3 of A_1 and I_1 and place the difference in row 3 of A_2 and I_2, respectively.

After doing these operations, the transformed matrices A_2 and I_2 are:

$$\begin{array}{c} \qquad A_2 \qquad\qquad\qquad\qquad I_2 \\ \left[\begin{array}{ccc|ccc} 1 & 0 & 5/7 & 4/7 & -3/7 & 0 \\ 0 & 1 & -2/7 & -3/7 & 4/7 & 0 \\ 0 & 0 & 24/7 & 1/7 & -6/7 & 1 \end{array}\right] \end{array}$$

Notice that after this second series of steps is completed, the second column of A_2 conforms to column two of a 3×3 identity matrix. Again, for matrices having more than three rows, this same general procedure would be followed for each row, to convert the second element in each row (except the second row) of A_2 to zero.

Finally, the following three row transformations are applied to matrices A_2 and I_2 to transform them into matrices A_3 and I_3. These three steps are:

1. Multiply row 3 of A_2 and I_2 by $1/a_{33}$ or 7/24 and place the results in row 3 of A_3 and I_3, respectively. This converts a_{33} to 1, as shown below.

$$\left[\begin{array}{ccc|ccc} 1 & 0 & 5/7 & 4/7 & -3/7 & 0 \\ 0 & 1 & -2/7 & -3/7 & 4/7 & 0 \\ 0 & 0 & 1 & 1/24 & -1/4 & 7/24 \end{array}\right]$$

2. Multiply row 3 of A_2 and I_2 by a_{13} or 5/7. Subtract the results from row 1 of A_2 and I_2 and place the difference in row 1 of A_3 and I_3, respectively.
3. Multiply row 3 of A_2 and I_2 by a_{23} or $-2/7$. Subtract the results from row 2 of A_2 and I_2 and place the difference in row 2 of A_3 and I_3.

Following these operations, the transformed matrices A_3 and I_3 are

$$\begin{array}{c} \qquad A_3 \qquad\qquad\qquad I_3 = A^{-1} \\ \left[\begin{array}{ccc|ccc} 1 & 0 & 0 & 13/24 & -1/4 & -5/24 \\ 0 & 1 & 0 & -5/12 & 1/2 & 1/12 \\ 0 & 0 & 1 & 1/24 & -1/4 & 7/24 \end{array}\right] \end{array}$$

Notice that through these nine elementary row transformations, the original A matrix is transformed into the identity matrix and the original identity matrix is transformed into A^{-1}. Also note that A^{-1} obtained by this method agrees exactly with the inverse obtained by the method of adjoints. This is because any nonsingular matrix has a unique inverse.

It should be obvious that the quantity of work involved in inverting matrices greatly increases with the matrix size, since the number of necessary row transformations is equal to the square of the number of rows or columns. Because of this, it is not considered practical to invert large matrices by hand. This work is more conveniently done with a computer. Since the procedure of elementary row transformations is systematic, it is easily programmed. Table B.1 shows algorithms written in the BASIC, C, FORTRAN, and Pascal programming languages for calculating the inverse of any $n \times n$ nonsingular matrix A. Students should review the code in their preferred language to gain familiarity with the computer procedures.

B.6 NUMERICAL EXAMPLE

Example B.2 Suppose that an EDM instrument is placed at point A in Figure B.2 and a reflector is successively placed at B, C, and D. The observed values AB, AC, and AD are shown in the figure. Calculate the unknowns X_1, X_2, and X_3 by matrix methods. The measured values are

Table B.1 Inverse algorithm in BASIC, C, FORTRAN, and Pascal

BASIC Language

```
REM INVERT A MATRIX
FOR k = 1 TO n
 FOR j = 1 TO n
  IF j<>k THEN A(k,j) = A(k,j)/A(k,k)
 NEXT j
 A(k,k) = 1/A(k,k)
 FOR i = 1 TO n
  IF i<>k THEN
   FOR j = 1 TO n
    IF j<>k THEN A(i,j) = A(i,j) - A(i,k)*A(k,j)
   NEXT j
   A(i,k) = -A(i,k)*A(k,k)
  END IF
 NEXT i: NEXT k
```

C Language

```
for (k = 0; k < n; k++){
 for (j = 0; j < n; j++)
  if (j! = k) A[k][j] = A[k][j]/A[k][k];
 A[k][k] = 1.0/A[k][k];
 for (i = 0; i < n; i++)
  if (i! = k){
   for (j = 0; j < n; j++)
    if (j! = k) A[i][j] = A[i][j] - A[i][k]*A[k][j];
   A[i][k] = -A[i][k]*A[k][k];
  } //if i<>k
} //for k
```

FORTRAN Language

```
      Do 560 k = 1,N
       Do 520 j = 1,N
        If (j.NE.k) Then
         A(k,j) = A(k,j)/A(k,k)
520    Continue
       A(K,K) = 1.0/A(K,K)
       Do 560 i = 1,N
        If (i.EQ.k) Then GOTO 560
         Do 550 j = 1,N
          If (j.NE.k) Then
          A(i,j) = A(i,j) - A(i,k)*A(k,j)
550      Continue
        A(i,k) = -A(i,k) * A(k,k)
560   Continue
```

Pascal Language

```
For k := 1 to N do Begin
 For j := 1 to N do
  If (j<>k) then A[k,j] := A[k,j]/A[k,k];
 A[k,k] := 1.0/A[k,k];
 For i := 1 to N do
  If (i<>k) then Begin
   For j := 1 to N do
    If (j<>k) then A[i,j] := A[i,j] - A[i,k]*A[k,j];
   A[i,k] := -A[i,k]*A[k,k];
  End; {If i<>k}
End; {for k}
```

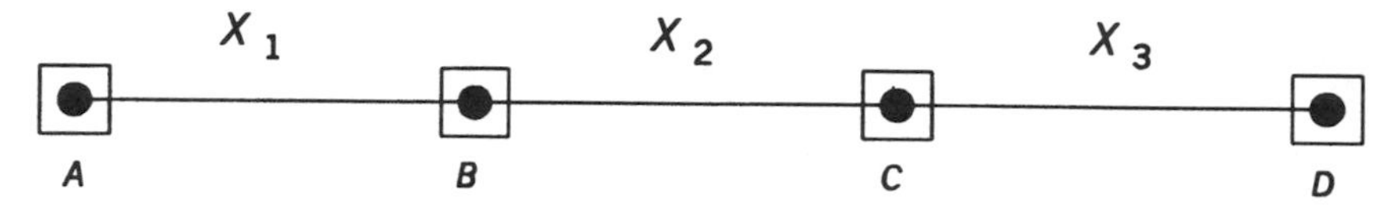

Figure B.2 Measured distances.

$$AB = 125.27$$

$$AC = 259.60$$

$$AD = 395.85$$

SOLUTION: Formulate the basic equations:

$$1X_1 + 0X_2 + 0X_3 = 125.27$$

$$1X_1 + 1X_2 + 0X_3 = 259.60$$

$$1X_1 + 1X_2 + 1X_3 = 395.85$$

Represented in matrix notation, these equations are $AX = L$. In this matrix equation, the individual matrices are:

$$A = \begin{bmatrix} 1 & 0 & 0 \\ 1 & 1 & 0 \\ 1 & 1 & 1 \end{bmatrix} \qquad X = \begin{bmatrix} x_1 \\ x_2 \\ x_3 \end{bmatrix} \quad L = \begin{bmatrix} 125.27 \\ 259.60 \\ 395.85 \end{bmatrix}$$

The solution in matrix notation is $X = A^{-1} L$.

Using elementary row transformation, the inverse of A is

$$\left[\begin{array}{ccc|ccc} 1 & 0 & 0 & 1 & 0 & 0 \\ 1 & 1 & 0 & 0 & 1 & 0 \\ 1 & 1 & 1 & 0 & 0 & 1 \end{array}\right] \rightarrow \left[\begin{array}{ccc|ccc} 1 & 0 & 0 & 1 & 0 & 0 \\ 0 & 1 & 0 & -1 & 1 & 0 \\ 0 & 0 & 1 & 0 & -1 & 1 \end{array}\right]$$

Solving $X = A^{-1}L$, the unknowns are

$$X = A^{-1}L = \begin{bmatrix} 1 & 0 & 0 \\ -1 & 1 & 0 \\ 0 & -1 & 1 \end{bmatrix} \begin{bmatrix} 125.27 \\ 259.60 \\ 395.85 \end{bmatrix} = \begin{bmatrix} 125.27 \\ 134.33 \\ 136.25 \end{bmatrix}$$

$$X_1 = 125.27$$

$$X_2 = 134.33$$

$$X_3 = 136.25$$

PROBLEMS

B.1 Explain when a 2×2 matrix has no inverse.

B.2 Find the inverse of A using the method of adjoints:

$$A = \begin{bmatrix} 3 & -1 & -1 \\ -1 & 3 & -1 \\ -1 & -1 & 3 \end{bmatrix}$$

B.3 Find the inverse of A in Problem B.2 using elementary row transformations.

B.4 Solve the following system of linear equations using matrix methods:

$$x + 5y = -8$$

$$-x - 2y = -1$$

B.5 Solve the following system of linear equations using matrix methods:

$$x + y - z = -8$$

$$3x - y + z = -4$$

$$-x + 2y + 2z = 21$$

B.6 Compute the inverses of the following matrices:

$$A = \begin{bmatrix} 8 & 5 \\ 3 & 12 \end{bmatrix} \qquad B = \begin{bmatrix} 16 & 2 \\ -8 & 3 \end{bmatrix}$$

B.7 Compute the inverses of the following matrices:

$$A = \begin{vmatrix} 3 & -1 & 0 \\ -1 & 3 & -1 \\ 0 & -1 & 3 \end{vmatrix} \qquad B = \begin{vmatrix} 4 & 3 & 7 \\ -1 & 0 & 4 \\ 2 & 8 & 10 \end{vmatrix}$$

B.8 Compute the inverses of the following matrices:

$$A = \begin{bmatrix} 13 & -6 & 0 \\ -6 & 18 & -6 \\ 0 & -6 & 16 \end{bmatrix} \qquad B = \begin{bmatrix} 1 & 2 & 6 \\ 2 & -3 & 4 \\ 0 & 6 & -12 \end{bmatrix}$$

B.9 Solve the following matrix system.

$$\begin{bmatrix} 13 & -6 & 0 \\ -6 & 18 & -6 \\ 0 & -6 & 16 \end{bmatrix} \begin{bmatrix} A \\ B \\ C \end{bmatrix} = \begin{bmatrix} 740.02 \\ 612.72 \\ 1072.22 \end{bmatrix}$$

Use the program MATRIX to do each problem.

B.10 Problem B.4

B.11 Problem B.5

B.12 Problem B.6

B.13 Problem B.7

B.14 Problem B.8

B.15 Problem B.9

Programming Problem

B.16 Select one of the coded matrix inverse routines from Table B.1, enter the code into a computer and use it to solve Problem B.7. (*Hint:* Place the code in Table B.1 in a separate subroutine/function/procedure to be called from the main program.)

B.17 Add a block of code to the inverse routine in the language of your choice that will inform the user when a matrix is singular.

B.18 Write a program that reads and writes a file with a nonsingular matrix, finds its inverse; and writes the results. Use this program to solve Problem B.7. (*Hint:* Place the reading, writing, and inversing code in separate subroutines/functions/procedures to be called from the main program. Provide a way to identify each matrix in the output file.)

B.19 Write a program that reads and writes a file containing a system of equations written in matrix form, solves the system using matrix operations, and writes the solution. Use this program to solve Problem B.9. (*Hint:* Place the reading, writing, and inversing code in separate subroutines/functions/procedures to be called from the main program. Provide a way to identify each matrix in the output file.)

APPENDIX C

NONLINEAR EQUATIONS AND TAYLOR'S THEOREM

C.1 INTRODUCTION

In adjustment computations it is frequently necessary to deal with nonlinear equations. For example, some observation equations relate measured quantities to unknown parameters through the transcendental functions of sine, cosine, or tangent, while others relate them through terms raised to second- and higher-order powers. The task of solving a system of nonlinear equations is formidable. To facilitate the solution, a first-order Taylor series approximation may be used to create a set of linear equations. The equations can then be solved by matrix methods discussed in Appendix B.

C.2 TAYLOR SERIES LINEARIZATION OF NONLINEAR EQUATIONS

Suppose that the following equation relates a measured value L to its unknown parameters x and y through nonlinear coefficients, as

$$L = f(x,y) \tag{C.1}$$

By Taylor's theorem, the equation is represented as

$$\begin{aligned} L = f(x,y) = f(x_0,y_0) &+ \frac{(\partial L/\partial x)_0\, dx}{1!} + \frac{(\partial^2 L/\partial x^2)_0\, dx^2}{2!} + \cdots \\ &+ \frac{(\partial^n L/\partial x^n)_0}{n!} + \frac{(\partial L/\partial y)_0\, dy}{1!} + \frac{(\partial^2 L/\partial y^2)_0\, dy^2}{2!} + \cdots \\ &+ \frac{(\partial^n L/\partial y^n)_0\, dy^n)_0\, dy^n}{n!} + R \end{aligned} \tag{C.2}$$

In Equation (C.2), x_0 and y_0 are approximations of x and y, $f(x_0,y_0)$ is the nonlinear function evaluated at these approximations, R is the remainder; and dx and dy are corrections to the initial approximations, such that

$$\begin{aligned} x &= x_0 + dx \\ y &= y_0 + dy \end{aligned} \tag{C.3}$$

A more exact Taylor series approximation is obtained by increasing the value of n in Equation (C.2). However, as the order of each successive term increases, its significance in the overall expression decreases. If all terms containing derivatives higher than the first are dropped, the following linear expression is obtained:

$$L = f(x,y) = f(x_0,y_0) + \left(\frac{\partial L}{\partial x}\right)_0 dx + \left(\frac{\partial L}{\partial y}\right)_0 dy \tag{C.4}$$

Once the initial approximations are selected, the only unknowns in Equation (C.4) are the corrections dx and dy. Of course, by dropping the higher-order terms from the Taylor series, Equation (C.4) becomes only a good approximation of the original equation. However, in the solution, an iterative procedure can be followed that yields accurate answers. This iterative procedure uses the following steps:

Step 1: Determine initial approximations for the unknowns. They may be obtained by guessing or from actual measurements, but the closer the initial approximations are to the final solution, the faster it will be obtained. For some problems, initial approximations can be obtained from graphical solutions or computed from available data or measurements. For others, the determination of initial approximations can involve considerable computational effort.

Step 2: Substitute the initial approximations into Equation (C.4) and solve for the corrections dx and dy.

Step 3: Calculate revised values of x and y using Equations (C.3).

Step 4: Using these newly revised values for x and y, repeat steps 2 and 3.

Step 5: Continue the procedure above until the corrections dx and dy are small enough to bring x and y within tolerable accuracy. When this occurs, the solution is said to have *converged*.

C.3 NUMERICAL EXAMPLE

To clarify this procedure further, a numerical example will be solved.

Example C.1 Linearize the following pair of nonlinear equations, containing the two unknowns x and y.

$$F(x,y) = x + y - 2y^2 = -4$$

$$G(x,y) = x^2 + y^2 = 8$$

SOLUTION: Determine the partial derivative for each equation with respect to each unknown.

$$\frac{\partial F}{\partial x} = 1 \qquad \frac{\partial F}{\partial y} = 1 - 4y$$

$$\frac{\partial G}{\partial x} = 2x \qquad \frac{\partial G}{\partial y} = 2y$$

Compute initial approximations for each unknown. An estimate of $x = 1$ and $y = 1$ is used initially for the approximations.

First Iteration
Write the linearized equations in the form of Equation (C.4).

$$F(x,y) = 1 + 1 - 2(1)^2 + dx + [1 - 4(1)]\ dy = -4$$

$$G(x,y) = (1)^2 + (1)^2 + 2(1)\ dx + 2(1)\ dy = 8$$

From the two equations above, solve for the unknowns dx and dy according to Equation (C.3):

$$dx = 1.25 \quad \text{and} \quad dy = 1.75$$

Using this solution, determine updated values for x and y:

$$x = x_0 + dx = 1.00 + 1.25 = 2.25$$

$$y = y_0 + dy = 1.00 + 1.75 = 2.75$$

Second Iteration
Continue the procedure demonstrated for the first iteration.

$$F = 2.25 + 2.75 - 2(2.75)^2 + dx + [1 - 4(2.75)]\ dy = -4$$

$$G = (2.25)^2 + (2.75)^2 + 2(2.25)\ dx + 2(2.75)\ dy = 8$$

From the two equations above, $dx = -0.25$ and $dy = -0.64$, from which

$$x = x_0 + dx = 2.25 - 0.25 = 2.00$$

$$y = y_0 + dy = 2.75 - 0.64 = 2.11$$

Third Iteration

$$F = 2 + 2.11 - 2(2.11)^2 + dx + [1 - 4(2.11)]\ dy = -4$$

$$G = (2)^2 + (2.11)^2 + 2(2)\ dx + 2(2.11)\ dy = 8$$

From the two equations above, $dx = 0.00$ and $dy = -0.11$, from which

$$x = x_0 + dx = 2.00 + 0.00 = 2.00$$

$$y = y_0 + dy = 2.11 - 0.11 = 2.00$$

Fourth Iteration

$$F = 2 + 2 - 2(2)^2 + dx + [1 - 4(2)]\ dy = -4$$

$$G = (2)^2 + (2)^2 + 2(2)\ dx + 2(2)\ dy = 8$$

Using the two equations above, the corrections to x and y are zero to the nearest hundredth. Thus the solution has converged and the values of $x = 2.00$ and $y = 2.00$ are the desired unknowns. Note that the initial values for the initial approximations were relatively poor, and thus four iterations were required. Had better estimates been made (say, $x_0 = 2.1$ and $y_0 = 1.9$), the solution would have converged in one or two iterations and saved computing time.

C.4 USING MATRICES TO SOLVE NONLINEAR EQUATIONS

The example of Section C.3 could be solved using matrix methods. Again, however, as in the algebraic approach, the equations must be linearized using Taylor's series. To facilitate linearization using Taylor's theorem, the *Jacobian* matrix (a matrix consisting of the partial derivatives taken with respect to the unknown variables) is formed. This is the coefficient matrix of the linearized equations. The Jacobian matrix for the example of Section C.3 is

$$J = \begin{bmatrix} \dfrac{\partial F}{\partial x} & \dfrac{\partial F}{\partial y} \\ \dfrac{\partial G}{\partial x} & \dfrac{\partial G}{\partial y} \end{bmatrix}$$

In the Jacobian matrix above, the first column contains the partial derivative for each equation with respect to x, and the second column contains each the partial derivative of each equation with respect to y. The linearized form of the equations may then be expressed in matrix notation as

$$JX = K \tag{C.5}$$

In Equation (C.5), J is the Jacobian matrix, X the matrix of unknown corrections dx and dy, and K the matrix of constants. Specifically for the example of Section C.3, these matrices are

$$J = \begin{bmatrix} 1 & 1 - 4y \\ 2x & 2y \end{bmatrix} \qquad X = \begin{bmatrix} dx \\ dy \end{bmatrix} \qquad K = \begin{bmatrix} -4 - F(x_0,y_0) \\ 8 - G(x_0,y_0) \end{bmatrix}$$

where $F(x_0,y_0)$ and $G(x_0,y_0)$ are the equations F and G solved at the initial approximations of x_0 and y_0.

Beginning with a set of initial approximations x_0 and y_0, the J and K matrices of Equation (C.5) are formed, and then X is computed using the matrix methods presented in Appendix B. Having updated the unknowns according to Equations (C.3), the J and K matrices are formed again and the solution for X computed again. This procedure is iterated until convergence is achieved.

C.5 A SIMPLE MATRIX EXAMPLE

Example C.2 Find the solution of the nonlinear system of equations shown below using matrix methods.

$$F(x,y) = x^2 + 3xy - 4y^2 = 6$$

$$G(x,y) = x + xy - y^2 = 3$$

SOLUTION: The partial derivatives of functions F and G with respect to the unknown's, x and y, are:

$$\frac{\partial F}{\partial x} = 2x + 3y \qquad \frac{\partial F}{\partial y} = 3x - 8y$$

$$\frac{\partial G}{\partial x} = 1 + y \qquad \frac{\partial G}{\partial y} = x - 2y$$

Thus, the Jacobian matrix is

$$J = \begin{bmatrix} \frac{\partial F}{\partial x} & \frac{\partial F}{\partial y} \\ \frac{\partial G}{\partial x} & \frac{\partial G}{\partial y} \end{bmatrix} = \begin{bmatrix} 2x + 3y & 3x - 8y \\ 1 + y & x - 2y \end{bmatrix}$$

and the system of equations to solve is:

$$\begin{bmatrix} 2x + 3y & 3x - 8y \\ 1 + y & x - 2y \end{bmatrix} \begin{bmatrix} dx \\ dy \end{bmatrix} = \begin{bmatrix} 6 - F(x_0,y_0) \\ 3 - G(x_0,y_0) \end{bmatrix}$$

First Iteration
Using $x_0 = 3$ and $y_0 = 0$ gives

$$\begin{bmatrix} 6 & 9 \\ 1 & 3 \end{bmatrix} \begin{bmatrix} dx \\ dy \end{bmatrix} = \begin{bmatrix} 6 - 9 \\ 3 - 3 \end{bmatrix} = \begin{bmatrix} -3 \\ 0 \end{bmatrix}$$

The determinant for the Jacobian matrix above is $[3(6)] - [9(1)] = 9$ and thus the matrix solution is

$$\begin{bmatrix} dx \\ dy \end{bmatrix} = \frac{1}{9} \begin{bmatrix} 3 & -9 \\ -1 & 6 \end{bmatrix} \begin{bmatrix} -3 \\ 0 \end{bmatrix} = \begin{bmatrix} -1.0 \\ 0.3 \end{bmatrix}$$

Applying Equation (C.3), initial approximations for a second iteration are

$$\begin{bmatrix} x \\ y \end{bmatrix} = \begin{bmatrix} x_0 \\ y_0 \end{bmatrix} + \begin{bmatrix} dx \\ dy \end{bmatrix} = \begin{bmatrix} 3 \\ 0 \end{bmatrix} + \begin{bmatrix} -1.0 \\ 0.3 \end{bmatrix} = \begin{bmatrix} 2.0 \\ 0.3 \end{bmatrix}$$

Second Iteration

$$\begin{bmatrix} 4.9 & 3.6 \\ 1.3 & 1.4 \end{bmatrix} \begin{bmatrix} dx \\ dy \end{bmatrix} = \begin{bmatrix} 6 - 5.44 \\ 3 - 2.51 \end{bmatrix} = \begin{bmatrix} 0.56 \\ 0.49 \end{bmatrix}$$

From this dx and dy are found to be 0.45 and 0.77, respectively. This makes the approximations of x and y for the third iteration 1.55 and 1.07, respectively. The procedures are followed until the final solution for x and y are found to be 2.00 and 1.00, respectively. Again, fewer iterations would have been required if the initial approximations had been closer to the final values.

C.6 A PRACTICAL EXAMPLE

Example C.3 Assume that the x and y coordinates of three points on a circle have been measured. Their coordinates are (9.4, 5.6), (7.6, 7.2), and (3.8, 4.8), respectively. The equation for a circle with center (h,k) and radius r is

$(x - h)^2 + (y - k)^2 = r^2$. Determine the coordinates of the center of the circle and its radius.

SOLUTION: The equation of a circle is rewritten as $C(h,k,r) = (x - h)^2 + (y - k)^2 - r^2 = 0$. The partial derivatives with respect to the unknowns h, k, and r are

$$\frac{\partial C}{\partial h} = -2(x - h) \qquad \frac{\partial C}{\partial k} = -2(y - k) \qquad \frac{\partial C}{\partial r} = -2r$$

For each point measured, one equation is written, resulting in a system of three equations and three unknowns. The general linearized form of these equations expressed using matrices is

$$\begin{bmatrix} \frac{\partial C_1}{\partial h} & \frac{\partial C_1}{\partial k} & \frac{\partial C_1}{\partial r} \\ \frac{\partial C_2}{\partial h} & \frac{\partial C_2}{\partial k} & \frac{\partial C_2}{\partial r} \\ \frac{\partial C_3}{\partial h} & \frac{\partial C_3}{\partial k} & \frac{\partial C_3}{\partial r} \end{bmatrix} \begin{bmatrix} dh \\ dk \\ dr \end{bmatrix} = \begin{bmatrix} 0 - [(x_1 - h_0)^2 + (y_1 - k_0)^2 - r_0^2] \\ 0 - [(x_2 - h_0)^2 + (y_2 - k_0)^2 - r_0^2] \\ 0 - [(x_3 - h_0)^2 + (y_3 - k_0)^2 - r_0^2] \end{bmatrix} \tag{C.6}$$

After taking partial derivatives, Equation (C.6) becomes

$$\begin{bmatrix} -2(x_1 - h_0) & -2(y_1 - k_0) & -2r_0 \\ -2(x_2 - h_0) & -2(y_2 - k_0) & -2r_0 \\ -2(x_3 - h_0) & -2(y_3 - k_0) & -2r_0 \end{bmatrix} \begin{bmatrix} dh \\ dk \\ dr \end{bmatrix} = \begin{bmatrix} -[(x_1 - h_0)^2 + (y_1 - k_0)^2 - r_0^2] \\ -[(x_2 - h_0)^2 + (y_2 - k_0)^2 - r_0^2] \\ -[(x_3 - h_0)^2 + (y_3 - k_0)^2 - r_0^2] \end{bmatrix} \tag{C.7}$$

Equations (C.7) can be simplified by multiplying each side by $-1/2$. The resulting equations are

$$\begin{bmatrix} (x_1 - h_0) & (y_1 - k_0) & r_0 \\ (x_2 - h_0) & (y_2 - k_0) & r_0 \\ (x_3 - h_0) & (y_3 - k_0) & r_0 \end{bmatrix} \begin{bmatrix} dh \\ dk \\ dr \end{bmatrix} = \begin{bmatrix} 0.5[(x_1 - h_0)^2 + (y_1 - k_0)^2 - r_0^2] \\ 0.5[(x_2 - h_0)^2 + (y_2 - k_0)^2 - r_0^2] \\ 0.5[(x_3 - h_0)^2 + (y_3 - k_0)^2 - r_0^2] \end{bmatrix} \tag{C.8}$$

Assuming approximate initial values for h, k, and r as 7, 4.5, and 3, respectively, Equations (C.8) are

$$\begin{bmatrix} (9.4 - 7) & (5.6 - 4.5) & 3 \\ (7.6 - 7) & (7.2 - 4.5) & 3 \\ (3.8 - 7) & (4.8 - 4.5) & 3 \end{bmatrix} \begin{bmatrix} dh \\ dk \\ dr \end{bmatrix} = \begin{bmatrix} 0.5[(9.4 - 7)^2 + (5.6 - 4.5)^2 - 3^2] \\ 0.5[(7.6 - 7)^2 + (7.2 - 4.5)^2 - 3^2] \\ 0.5[(3.8 - 7)^2 + (4.8 - 4.5)^2 - 3^2] \end{bmatrix} \tag{C.9}$$

Simplifying Equations (C.9) yields

$$\begin{bmatrix} 2.4 & 1.1 & 3 \\ 0.6 & 2.7 & 3 \\ -3.2 & 0.3 & 3 \end{bmatrix} \begin{bmatrix} dh \\ dk \\ dr \end{bmatrix} = \begin{bmatrix} -1.015 \\ -0.675 \\ 0.665 \end{bmatrix}$$

Solving this system gives

$$\begin{bmatrix} dh \\ dk \\ dr \end{bmatrix} = \begin{bmatrix} -0.28462 \\ -0.10769 \\ -0.07115 \end{bmatrix}$$

After applying these changes to the initial approximations, updated values for h, k, and r of 6.7154, 4.3923, and 2.9288, respectively, are obtained. The second iteration results in corrections of 0, 0, and 0.014945. Since the correction for r is still comparatively large, the iteration process must be continued. After the third iteration, suitable convergence was achieved. The final values for h, k, and r are 6.72, 4.39, and 2.94, respectively, which are within 0.00001 of a perfect solution.

Sometimes, more than one method is available for solving a problem. For example, in the preceding problem, an alternative linear form of the equation of a circle could have been used. That equation is $x^2 + y^2 + 2dx + 2ey + f = 0$, where the center of the circle is at $(-d,-e)$ and the circle's radius is $\sqrt{d^2 + e^2 - f}$. Note that the equation is linear in terms of its unknowns (d,e,f), and thus iterations are not necessary in solving for the unknowns. Writing a rearranged form of this linear equation for each of three measured sets of (x,y) coordinates yields

$$\begin{aligned} 2dx_1 + 2ey_1 + f &= -(x_1^2 + y_1^2) \\ 2dx_2 + 2ey_2 + f &= -(x_2^2 + y_2^2) \\ 2dx_3 + 2ey_3 + f &= -(x_3^2 + y_3^2) \end{aligned} \tag{C.10}$$

Equations (C.10) can in turn be represented in matrix notation as

$$\begin{bmatrix} 2x_1 & 2y_1 & 1 \\ 2x_2 & 2y_2 & 1 \\ 2x_3 & 2y_3 & 1 \end{bmatrix} \begin{bmatrix} d \\ e \\ f \end{bmatrix} = \begin{bmatrix} -(x_1^2 + y_1^2) \\ -(x_2^2 + y_2^2) \\ -(x_3^2 + y_3^2) \end{bmatrix}$$

Solving this matrix system, the center of the circle is again found to be (6.72, 4.39) and its radius is determined to be 2.94.

In this appendix, the Taylor's series has been applied to solve for the unknowns in nonlinear equations. Many equations in surveying, geodesy, and photogrammetry are nonlinear. In surveying, examples include the distance and angle formulas that are nonlinear in terms of station coordinates. Taylor's series is used to linearize these equations, and find least-squares solutions. Thus when performing least-squares adjustments of horizontal measurements, the techniques presented in this appendix must be used in the solutions.

PROBLEMS

C.1 Solve for unknowns x and y in the following nonlinear equations using Taylor's theorem. (Use $x_0 = 5$ and $y_0 = 5$ for initial approximations.)

$$x^2y - 3x^2 = 75$$

$$x^2 - y = 19$$

C.2 Solve for the unknown values of x, y, and z in the following three nonlinear equations using the Taylor series. (Use $x_0 = y_0 = z_0 = 2$ for initial approximations.)

$$x^2 - y^2 + 2xy + z = 4$$

$$-x + y + z = 4$$

$$-2x^2 - y + z^3 = 23$$

C.3 Use the program MATRIX to solve Problem C.1.

C.4 Use the program MATRIX to solve Problem C.2.

C.5 Find center and radius of a circle using the equation $(x - h)^2 + (y - k)^2 = r^2$ given the coordinates of points A, B, and C, on the circle. Follow the procedures discussed in Section C.5. Use initial approximations of $h_0 = 5$, $k_0 = 4$, and $r_0 = 2$ for the first iteration.

A: (7.2,5.2) B: (4.0,6.4) C: (4.0,2.4)

C.6 Repeat Problem C.5 using the linear equation: $x^2 + y^2 + 2dx + 2ey + f = 0$.

C.7 Repeat Problem C.5 using the points A: (0.50,−0.70), B: (1.00,0.00), C: (0.70,0.70). Use initial approximations of $h_0 = 0$, $k_0 = 0$, and $r_0 = 1$.

C.8 Repeat Problem C.7 using the linear equation $x^2 + y^2 + 2dx + 2ey + f = 0$.

C.9 Use the program ADJUST to solve Problem C.5.

C.10 Use the program ADJUST to solve Problem C.7.

C.11 The distance formula between two stations i and j is

$$D_{ij} = \sqrt{(x_j - x_i)^2 + (y_j - y_i)^2}$$

Write the linearized form of this equation in terms of the variables x_i, y_i, x_j, and y_j.

C.12 The azimuth formula between two stations i and j is

$$\alpha_{ij} = \tan^{-1}\left(\frac{x_j - x_i}{y_j - y_i}\right)$$

Write the linearized form of this equation in terms of the variables x_i, y_i, x_j, and y_j.

C.13 The formula for an angle $\angle jik$ is $\alpha_{ik} - \alpha_{ij}$, where α is defined in Problem C.12. Write the linearized form of this equation in terms of the variables x_i, y_i, x_j, y_j, x_k, and y_k.

APPENDIX D

NORMAL ERROR DISTRIBUTION CURVE AND OTHER STATISTICAL TABLES

D.1 DEVELOPMENT FOR NORMAL DISTRIBUTION CURVE EQUATION

In Section 2.4 the histogram and frequency polygon were presented as methods for graphically portraying random error distributions. If a large number of these distributions were examined for sets, measurements in surveying, geodesy, and photogrammetry, it would be found that they conform to normal (or Gaussian) distributions. The general laws governing normal distributions are stated as follows:

1. Positive and negative errors occur with equal probability and equal frequency.
2. Small errors are more common than large errors.
3. Large errors seldom occur, and there is a limit to the size of the greatest random error that will occur in any set of measurements.

A curve that conforms to the laws above, plotted with the size of the error on the abscissa and probability of occurrence on the ordinate, appears as Figure 3.3. This curve is repeated on Figure D.1 and is called the *normal distribution curve*, the *normal curve of error*, or simply the *probability curve*. A smooth curve of this same shape would be obtained if for a very large group of measurements, a histogram were plotted with an infinitesimally small class interval. In this section, the equation for this curve is developed.

Assume that the normal distribution curve is continuous and that the probability of an error occurring between x and $x + dx$ is given by the function $y = f(x)$. Further assume that this is the equation for the probability curve.

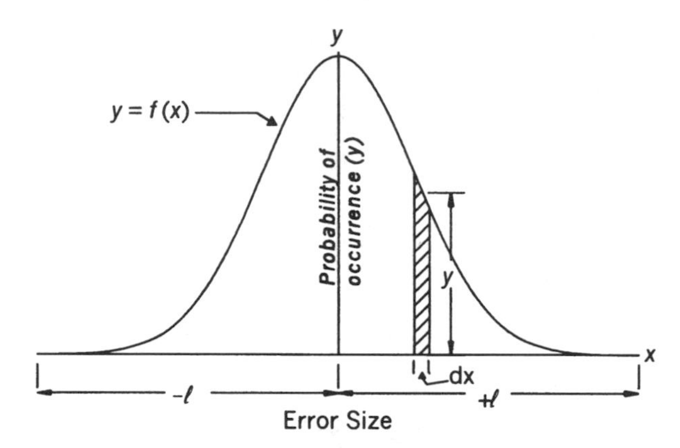

Figure D.1 Normal distribution curve.

The form of $f(x)$ will now be determined. Since as explained in Chapter 3, probabilities are equivalent to areas under the probability curve, the probabilities of errors occurring within the ranges of (x_1 and $x_1 + dx_1$), (x_2 and $x_2 + dx_2$), etc., are $f(x_1)dx_1, f(x_2)dx_2, ..., f(x_n)dx_n$. The total area under the probability curve represents the total probability or simply the integer one. Then for a finite number of possible errors:

$$f(x_1)dx_1 + f(x_2)dx_2 + \cdots + f(x_n)dx_n = 1 \qquad \text{(D.1)}$$

If the total range of errors $x_1, x_2, ..., x_n$ is between $\pm l$, then considering an infinite number of errors that makes the curve continuous, the area under the curve can be set equal to:

$$1 = \int_{-l}^{+l} f(x)\, dx$$

But because the area under the curve from $+l$ to $+\infty$ and from $-l$ to $-\infty$ is essentially zero, the integration limits are extended to $\pm\infty$, as:

$$1 = \int_{-\infty}^{+\infty} f(x)\, dx = \int_{-\infty}^{+\infty} y\, dx \qquad \text{(D.2)}$$

Now suppose that quantity M has been measured, and that it is equal to some function of n unknown parameters $z_1, z_2, ..., z_n$ such that $M = f(z_1, z_2, ..., z_n)$. Also let $x_1, x_2, ..., x_m$ be the errors of m observations $M_1, M_2, ..., M_m$, and let $f(x_1)\, dx_1, f(x_2)\, dx_2, ..., f(x_m)\, dx_m$ be the probabilities of errors falling within the ranges of (x_1 and dx_1), (x_2 and dx_2), etc. By Equation (3.1) the probability P of the simultaneous occurrence of all of these errors is equal to the product of the individual probabilities, thus:

$$P = [f(x_1)\,dx_1]\,[f(x_2)\,dx_2] \cdots [f(x_m)\,dx_m]$$

Then by logs:

$$\begin{aligned}\text{Log } P = {} & \log f(x_1) + \log f(x_2) + \cdots + \log f(x_m) + \log dx_1 \\ & + \log dx_2 + \cdots + \log dx_m \end{aligned} \tag{D.3}$$

The most probable values of the errors will occur when P is maximized or when the log of P is maximized. To maximize a function, it is differentiated with respect to each unknown parameter z, and the results set equal to zero. After logarithmic differentiation of Equation (D.3), the following n equations result (note that the dx's are constants independent of the z's and therefore their differentials with respect to the z's are zero):

$$\begin{aligned}
\frac{1}{P}\frac{\partial P}{\partial z_1} &= \frac{1}{f(x_1)}\frac{df(x_1)}{dx_1}\frac{dx_1}{dz_1} + \frac{1}{f(x_2)}\frac{df(x_2)}{dx_2}\frac{dx_2}{dz_1} + \cdots \\
&\quad + \frac{1}{f(x_m)}\frac{df(x_m)}{dx_m}\frac{dx_m}{dz_1} = 0 \\
\frac{1}{P}\frac{\partial P}{\partial z_2} &= \frac{1}{f(x_1)}\frac{df(x_1)}{dx_1}\frac{dx_1}{dz_2} + \frac{1}{f(x_2)}\frac{df(x_2)}{dx_2}\frac{dx_2}{dz_1} + \cdots \\
&\quad + \frac{1}{f(x_m)}\frac{df(x_m)}{dx_m}\frac{dx_m}{dz_2} = 0 \\
&\ \ \vdots \\
\frac{1}{P}\frac{\partial P}{\partial z_n} &= \frac{1}{f(x_1)}\frac{df(x_1)}{dx_1}\frac{dx_1}{dz_n} + \frac{1}{f(x_2)}\frac{df(x_2)}{dx_2}\frac{dx_2}{dz_n} + \cdots \\
&\quad + \frac{1}{f(x_m)}\frac{df(x_m)}{dx_m}\frac{dx_m}{dz_n} = 0
\end{aligned} \tag{D.4}$$

Now let

$$f'(x) = \frac{df(x)}{dx} \tag{D.5}$$

Substituting Equation (D.5) into Equations (D.4) gives

$$\frac{f'(x_1)\,dx_1}{f(x_1)\,dz_1} + \frac{f'(x_2)\,dx_2}{f(x_2)\,dz_1} + \cdots + \frac{f'(x_m)\,dx_m}{f(x_m)\,dz_1} = 0$$

$$\frac{f'(x_1)\,dx_1}{f(x_1)\,dz_2} + \frac{f'(x_2)\,dx_2}{f(x_2)\,dz_2} + \cdots + \frac{f'(x_m)\,dx_m}{f(x_m)\,dz_2} = 0 \tag{D.6}$$

$$\vdots$$

$$\frac{f'(x_1)\,dx_1}{f(x_1)\,dz_n} + \frac{f'(x_2)\,dx_2}{f(x_2)\,dz_n} + \cdots + \frac{f'(x_m)\,dx_m}{f(x_m)\,dz_n} = 0$$

Thus far $f(x)$ and $f'(x)$ are general, regardless of the number of unknown parameters. Now consider the special case where there is only one unknown z and M_1, M_2, ..., M_m are m observed values of z. If z^* is the true value of the quantity, the errors associated with the observations are

$$x_1 = z^* - M_1, \quad x_2 = z^* - M_2, \ldots, x_m = z^* - M_m \tag{D.7}$$

Differentiating Equations (D.7) gives

$$\frac{dx_1}{dz} = 1 = \frac{dx_2}{dz} = \cdots = \frac{dx_m}{dz} \tag{D.8}$$

Then for this special case, substituting Equations (D.7) and (D.8) into Equations (D.6), they reduce to a single equation:

$$\frac{f'(z^* - M_1)}{f(z^* - M_1)} + \frac{f'(z^* - M_2)}{f(z^* - M_2)} + \cdots + \frac{f'(z^* - M_m)}{f(z^* - M_m)} = 0 \tag{D.9}$$

Equation (D.9) for this special case in consideration is also general for any value of m and for any observed values M_1, M_2, ..., M_m. Thus let the values of M be

$$M_2 = M_3 = \cdots = M_m = M_1 - mN$$

where N is chosen for convenience as $N = (M_1 - M_2)/m$.

The arithmetic mean is the most probable value for this case of a single quantity having been observed several times; therefore, z^* the most probable value in this case is

$$z^* = \frac{M_1 + M_2 + \cdots + M_m}{m}$$

$$z^* = \frac{M_1 + (m - 1)(M_1 - mN)}{m}$$

$$z^* = M_1 - mN + N$$

$$z^* = M_1 - N(m - 1)$$

$$z^* - M_1 = -N(m - 1) = N(1 - m) \tag{D.10}$$

Recall that $N = (M_1 - M_2)/m$, from which $M_1 = mN + M_2$. Substituting into Equation (D.10) gives

$$z^* - (mN + M_2) = N(1 - m)$$

$$z^* - M_2 = N$$

Similarly, since $N = (M_1 - M_3)/m = (M_1 - M_4)/m$, and so on,

$$z^* - M_3 = N$$

$$z^* - M_4 = N$$

$$\vdots$$

Substituting these expressions into Equation (D.9) gives

$$\frac{f'[n(1 - m)]}{f[N(1 - M)]} + \frac{(m - 1)f'(N)}{f(N)} = 0 \tag{D.11}$$

Rearranging yields

$$\frac{f'[N(1 - m)]}{Nf[N(1 - m)](1 - m)} = \frac{f'(N)}{f(N)N} = \text{a constant}$$

because N in this case is a constant. Thus

$$\frac{f'(x)}{xf(x)} = \text{a constant} = K \tag{D.12}$$

Substituting Equation (D.5) into Equation (D.12) yields

$$f'(x) = xf(x)K = \frac{df(x)}{dx}$$

from which $df(x)/f(x) = Kxdx$. Integrating gives

$$\log_e f(x) = \frac{1}{2} Kx^2 + C_1$$

$$f(x) = e^{C_1} e^{Kx^2/2}$$

But letting

$$e^{C_1} = C$$

then

$$f(x) = Ce^{Kx^2/2} \tag{D.13}$$

In Equation (D.13), since $f(x)$ decreases as x increases, and thus the exponent must be negative. Arbitrarily letting

$$h = \sqrt{\frac{K}{2}} \tag{D.14}$$

and incorporating the negative into Equation (D.13), there results

$$f(x) = Ce^{-h^2x^2} \tag{D.15}$$

To find the value of the constant C, substitute Equation (D.15) into Equation (D.2):

$$\int_{-\infty}^{+\infty} Ce^{-h^2x^2}\, dx = 1$$

Also, arbitrarily set $t = hx$, then $dt = hdx$ and $dx = dt/h$, from which, after changing variables, we obtain

$$\frac{C}{h} \int_{-\infty}^{+\infty} e^{-t^2\, dt} = 1$$

The value of the definite integral is $\sqrt{\pi}$, from which[1]

[1] The technique of integrating this nonelementary function is beyond the scope of this book but can be found in advanced references.

$$\frac{C}{h\sqrt{\pi}} = 1$$

$$C = \frac{h}{\sqrt{\pi}} \tag{D.16}$$

Substituting Equation (D.16) into Equation (D.15) given

$$f(x) = \frac{h}{\sqrt{\pi}} e^{-h^2x^2} \tag{D.17}$$

Note that from Equation (D.14) that $h = \sqrt{K/2}$. For the normal distribution $K = 1/\sigma^2$. Substituting this into Equation (D.17) gives

$$f(x) = \frac{1}{\sqrt{2\sigma^2\pi}} e^{-(1/2\sigma^2)x^2} = \frac{1}{\sigma\sqrt{2\pi}} e^{-x^2/2\sigma^2} \tag{D.18}$$

where the terms are as defined for Equation (3.2).

This is the general equation for the probability curve, having been derived in this instance from the consideration of a special case. In Table D.1, which follows, values for areas under the standard normal distribution function from negative infinity to t are tabulated.

Table D.1 Percentage Points for the Standard Normal Distribution Function

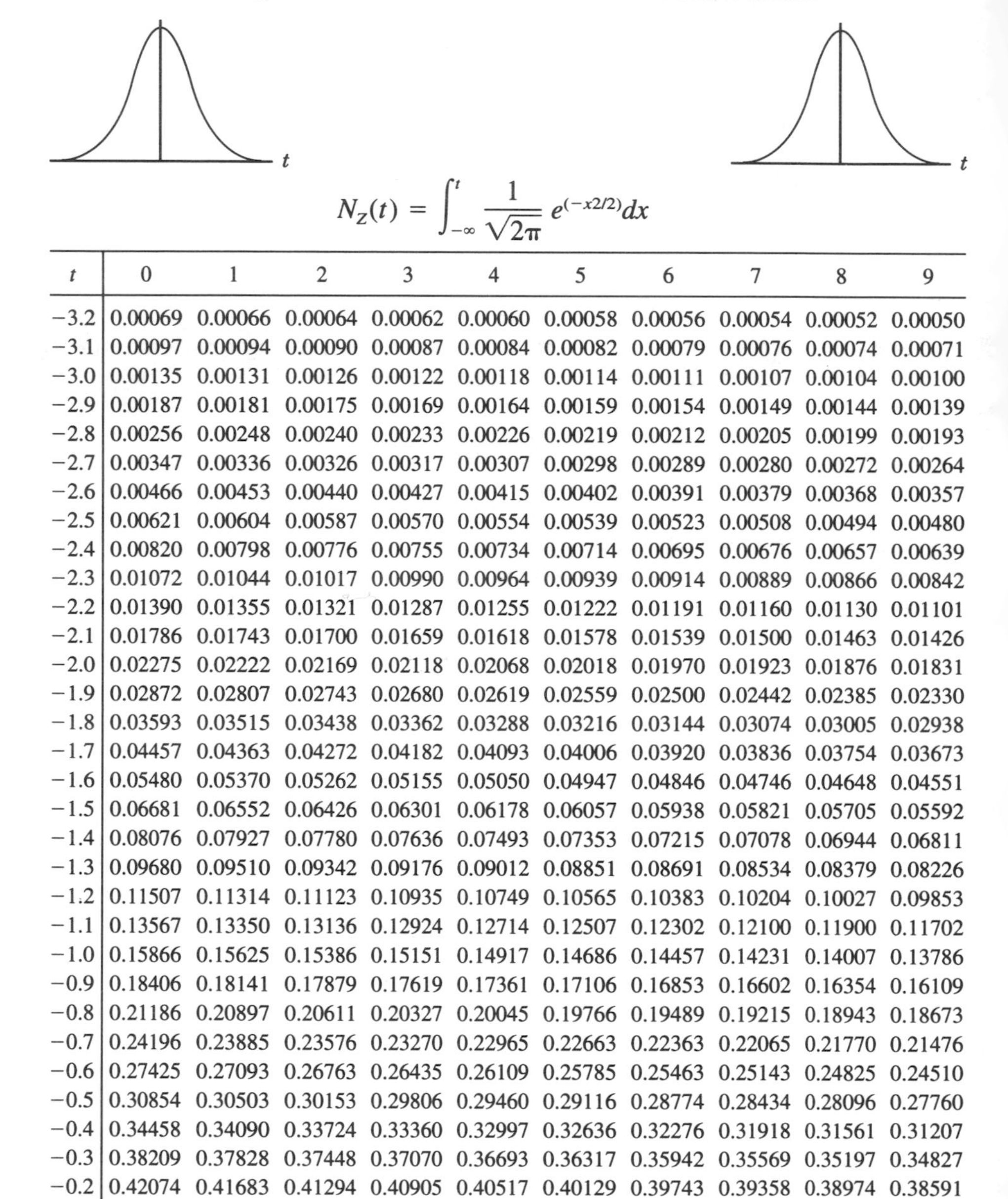

$$N_Z(t) = \int_{-\infty}^{t} \frac{1}{\sqrt{2\pi}} e^{(-x2/2)} dx$$

t	0	1	2	3	4	5	6	7	8	9
−3.2	0.00069	0.00066	0.00064	0.00062	0.00060	0.00058	0.00056	0.00054	0.00052	0.00050
−3.1	0.00097	0.00094	0.00090	0.00087	0.00084	0.00082	0.00079	0.00076	0.00074	0.00071
−3.0	0.00135	0.00131	0.00126	0.00122	0.00118	0.00114	0.00111	0.00107	0.00104	0.00100
−2.9	0.00187	0.00181	0.00175	0.00169	0.00164	0.00159	0.00154	0.00149	0.00144	0.00139
−2.8	0.00256	0.00248	0.00240	0.00233	0.00226	0.00219	0.00212	0.00205	0.00199	0.00193
−2.7	0.00347	0.00336	0.00326	0.00317	0.00307	0.00298	0.00289	0.00280	0.00272	0.00264
−2.6	0.00466	0.00453	0.00440	0.00427	0.00415	0.00402	0.00391	0.00379	0.00368	0.00357
−2.5	0.00621	0.00604	0.00587	0.00570	0.00554	0.00539	0.00523	0.00508	0.00494	0.00480
−2.4	0.00820	0.00798	0.00776	0.00755	0.00734	0.00714	0.00695	0.00676	0.00657	0.00639
−2.3	0.01072	0.01044	0.01017	0.00990	0.00964	0.00939	0.00914	0.00889	0.00866	0.00842
−2.2	0.01390	0.01355	0.01321	0.01287	0.01255	0.01222	0.01191	0.01160	0.01130	0.01101
−2.1	0.01786	0.01743	0.01700	0.01659	0.01618	0.01578	0.01539	0.01500	0.01463	0.01426
−2.0	0.02275	0.02222	0.02169	0.02118	0.02068	0.02018	0.01970	0.01923	0.01876	0.01831
−1.9	0.02872	0.02807	0.02743	0.02680	0.02619	0.02559	0.02500	0.02442	0.02385	0.02330
−1.8	0.03593	0.03515	0.03438	0.03362	0.03288	0.03216	0.03144	0.03074	0.03005	0.02938
−1.7	0.04457	0.04363	0.04272	0.04182	0.04093	0.04006	0.03920	0.03836	0.03754	0.03673
−1.6	0.05480	0.05370	0.05262	0.05155	0.05050	0.04947	0.04846	0.04746	0.04648	0.04551
−1.5	0.06681	0.06552	0.06426	0.06301	0.06178	0.06057	0.05938	0.05821	0.05705	0.05592
−1.4	0.08076	0.07927	0.07780	0.07636	0.07493	0.07353	0.07215	0.07078	0.06944	0.06811
−1.3	0.09680	0.09510	0.09342	0.09176	0.09012	0.08851	0.08691	0.08534	0.08379	0.08226
−1.2	0.11507	0.11314	0.11123	0.10935	0.10749	0.10565	0.10383	0.10204	0.10027	0.09853
−1.1	0.13567	0.13350	0.13136	0.12924	0.12714	0.12507	0.12302	0.12100	0.11900	0.11702
−1.0	0.15866	0.15625	0.15386	0.15151	0.14917	0.14686	0.14457	0.14231	0.14007	0.13786
−0.9	0.18406	0.18141	0.17879	0.17619	0.17361	0.17106	0.16853	0.16602	0.16354	0.16109
−0.8	0.21186	0.20897	0.20611	0.20327	0.20045	0.19766	0.19489	0.19215	0.18943	0.18673
−0.7	0.24196	0.23885	0.23576	0.23270	0.22965	0.22663	0.22363	0.22065	0.21770	0.21476
−0.6	0.27425	0.27093	0.26763	0.26435	0.26109	0.25785	0.25463	0.25143	0.24825	0.24510
−0.5	0.30854	0.30503	0.30153	0.29806	0.29460	0.29116	0.28774	0.28434	0.28096	0.27760
−0.4	0.34458	0.34090	0.33724	0.33360	0.32997	0.32636	0.32276	0.31918	0.31561	0.31207
−0.3	0.38209	0.37828	0.37448	0.37070	0.36693	0.36317	0.35942	0.35569	0.35197	0.34827
−0.2	0.42074	0.41683	0.41294	0.40905	0.40517	0.40129	0.39743	0.39358	0.38974	0.38591
−0.1	0.46017	0.45620	0.45224	0.44828	0.44433	0.44038	0.43644	0.43251	0.42858	0.42465
−0.0	0.50000	0.49601	0.49202	0.48803	0.48405	0.48006	0.47608	0.47210	0.46812	0.46414

Table D.1 Percentage Points for the Standard Normal Distribution Function (continued)

t	0	1	2	3	4	5	6	7	8	9
0.0	0.50000	0.50399	0.50798	0.51197	0.51595	0.51994	0.52392	0.52790	0.53188	0.53586
0.1	0.53983	0.54380	0.54776	0.55172	0.55567	0.55962	0.56356	0.56749	0.57142	0.57535
0.2	0.57926	0.58317	0.58706	0.59095	0.59483	0.59871	0.60257	0.60642	0.61026	0.61409
0.3	0.61791	0.62172	0.62552	0.62930	0.63307	0.63683	0.64058	0.64431	0.64803	0.65173
0.4	0.65542	0.65910	0.66276	0.66640	0.67003	0.67364	0.67724	0.68082	0.68439	0.68793
0.5	0.69146	0.69497	0.69847	0.70194	0.70540	0.70884	0.71226	0.71566	0.71904	0.72240
0.6	0.72575	0.72907	0.73237	0.73565	0.73891	0.74215	0.74537	0.74857	0.75175	0.75490
0.7	0.75804	0.76115	0.76424	0.76730	0.77035	0.77337	0.77637	0.77935	0.78230	0.78524
0.8	0.78814	0.79103	0.79389	0.79673	0.79955	0.80234	0.80511	0.80785	0.81057	0.81327
0.9	0.81594	0.81859	0.82121	0.82381	0.82639	0.82894	0.83147	0.83398	0.83646	0.83891
1.0	0.84134	0.84375	0.84614	0.84849	0.85083	0.85314	0.85543	0.85769	0.85993	0.86214
1.1	0.86433	0.86650	0.86864	0.87076	0.87286	0.87493	0.87698	0.87900	0.88100	0.88298
1.2	0.88493	0.88686	0.88877	0.89065	0.89251	0.89435	0.89617	0.89796	0.89973	0.90147
1.3	0.90320	0.90490	0.90658	0.90824	0.90988	0.91149	0.91309	0.91466	0.91621	0.91774
1.4	0.91924	0.92073	0.92220	0.92364	0.92507	0.92647	0.92785	0.92922	0.93056	0.93189
1.5	0.93319	0.93448	0.93574	0.93699	0.93822	0.93943	0.94062	0.94179	0.94295	0.94408
1.6	0.94520	0.94630	0.94738	0.94845	0.94950	0.95053	0.95154	0.95254	0.95352	0.95449
1.7	0.95543	0.95637	0.95728	0.95818	0.95907	0.95994	0.96080	0.96164	0.96246	0.96327
1.8	0.96407	0.96485	0.96562	0.96638	0.96712	0.96784	0.96856	0.96926	0.96995	0.97062
1.9	0.97128	0.97193	0.97257	0.97320	0.97381	0.97441	0.97500	0.97558	0.97615	0.97670
2.0	0.97725	0.97778	0.97831	0.97882	0.97932	0.97982	0.98030	0.98077	0.98124	0.98169
2.1	0.98214	0.98257	0.98300	0.98341	0.98382	0.98422	0.98461	0.98500	0.98537	0.98574
2.2	0.98610	0.98645	0.98679	0.98713	0.98745	0.98778	0.98809	0.98840	0.98870	0.98899
2.3	0.98928	0.98956	0.98983	0.99010	0.99036	0.99061	0.99086	0.99111	0.99134	0.99158
2.4	0.99180	0.99202	0.99224	0.99245	0.99266	0.99286	0.99305	0.99324	0.99343	0.99361
2.5	0.99379	0.99396	0.99413	0.99430	0.99446	0.99461	0.99477	0.99492	0.99506	0.99520
2.6	0.99534	0.99547	0.99560	0.99573	0.99585	0.99598	0.99609	0.99621	0.99632	0.99643
2.7	0.99653	0.99664	0.99674	0.99683	0.99693	0.99702	0.99711	0.99720	0.99728	0.99736
2.8	0.99744	0.99752	0.99760	0.99767	0.99774	0.99781	0.99788	0.99795	0.99801	0.99807
2.9	0.99813	0.99819	0.99825	0.99831	0.99836	0.99841	0.99846	0.99851	0.99856	0.99861
3.0	0.99865	0.99869	0.99874	0.99878	0.99882	0.99886	0.99889	0.99893	0.99896	0.99900
3.1	0.99903	0.99906	0.99910	0.99913	0.99916	0.99918	0.99921	0.99924	0.99926	0.99929
3.2	0.99931	0.99934	0.99936	0.99938	0.99940	0.99942	0.99944	0.99946	0.99948	0.99950

D.2 OTHER STATISTICAL TABLES

On the remaining pages of this chapter are three often-used statistical tables. Application of these tables and their interpretation are discussed in Chapter 4. The equations used to generate each of these tables are also presented.

D.2.1 χ^2 Distribution

Chi square is a density function for the distribution of sample variances computed from sets with selected degrees of freedom for a population. The use

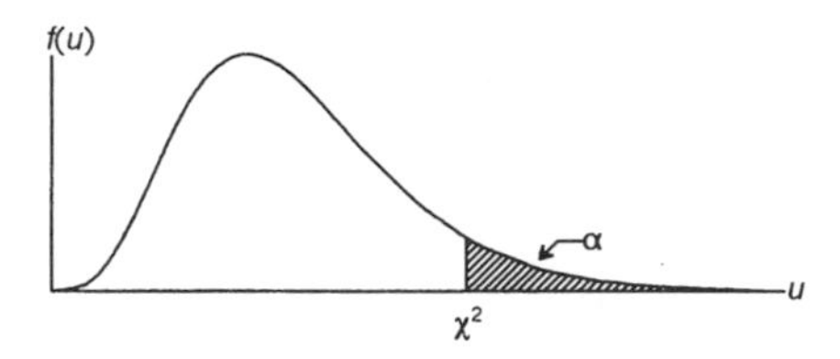

Figure D.2 χ^2 distribution.

of this distribution to construct confidence intervals for the population variance, and to perform hypothesis testing involving the population variance are discussed in Chapter 4. The χ^2 distribution is illustrated Figure D.2.

The χ^2 distribution critical values given in Table D.2 were generated using the following function. Both tails of the distribution were derived with a program using numerical integration routines similar to those available in STATS.

$$\alpha = \int_{\chi^2}^{\infty} \frac{1}{2^{v/2}\Gamma(v/2)}\, u^{(v-2)/2} e^{-u/2}\, du$$

where v is the degrees of freedom and Γ is known as the gamma function, which is defined as

$$\Gamma(v) = \int_0^{\infty} u^{v-1} e^{-u}\, du$$

It is computed as $\Gamma(v) = (v - 1)! = (v - 1)(v - 2)(v - 3) \cdots (3)(2)(1)$.

Table D.2 Critical values for χ^2 distribution

α / v	0.999	0.995	0.990	0.975	0.950	0.900	0.500	0.100	0.050	0.025	0.010	0.005	0.001
1	0.000002	0.000039	0.000157	0.000982	0.004	0.016	0.455	2.705	3.841	5.023	6.634	7.877	10.81
2	0.002	0.01	0.02	0.05	0.10	0.21	1.39	4.61	5.99	7.38	9.21	10.60	13.81
3	0.02	0.07	0.12	0.22	0.35	0.58	2.37	6.25	7.82	9.35	11.34	12.84	16.26
4	0.09	0.21	0.30	0.48	0.71	1.06	3.36	7.78	9.49	11.14	13.28	14.86	18.47
5	0.21	0.41	0.55	0.83	1.15	1.61	4.35	9.24	11.07	12.83	15.09	16.75	20.51
6	0.38	0.68	0.87	1.24	1.64	2.20	5.35	10.64	12.59	14.45	16.81	18.55	22.46
7	0.60	0.99	1.24	1.69	2.17	2.83	6.35	12.02	14.07	16.01	18.48	20.28	24.32
8	0.86	1.34	1.65	2.18	2.73	3.49	7.34	13.36	15.51	17.53	20.09	21.96	26.12
9	1.15	1.74	2.09	2.70	3.33	4.17	8.34	14.68	16.92	19.02	21.67	23.59	27.88
10	1.48	2.16	2.56	3.25	3.94	4.87	9.34	15.99	18.31	20.48	23.21	25.19	29.59
11	1.83	2.60	3.05	3.82	4.58	5.58	10.34	17.28	19.68	21.92	24.72	26.76	31.26
12	2.21	3.07	3.57	4.40	5.23	6.30	11.34	18.55	21.03	23.34	26.22	28.30	32.91
13	2.62	3.57	4.11	5.01	5.89	7.04	12.34	19.81	22.36	24.74	27.69	29.82	34.53
14	3.04	4.08	4.66	5.63	6.57	7.79	13.34	21.06	23.68	26.12	29.14	31.32	36.12
15	3.48	4.60	5.23	6.26	7.26	8.55	14.34	22.31	25.00	27.49	30.58	32.80	37.70
16	3.94	5.14	5.81	6.91	7.96	9.31	15.34	23.54	26.30	28.85	32.00	34.27	39.25
17	4.42	5.70	6.41	7.56	8.67	10.09	16.34	24.77	27.59	30.19	33.41	35.72	40.79
18	4.91	6.27	7.02	8.23	9.39	10.86	17.34	25.99	28.87	31.53	34.81	37.16	42.31
19	5.41	6.84	7.63	8.91	10.12	11.65	18.34	27.20	30.14	32.85	36.19	38.58	43.82
20	5.92	7.43	8.26	9.59	10.85	12.44	19.34	28.41	31.41	34.17	37.57	40.00	45.31
21	6.45	8.03	8.90	10.28	11.59	13.24	20.34	29.62	32.67	35.48	38.93	41.40	46.80
22	6.98	8.64	9.54	10.98	12.34	14.04	21.34	30.81	33.92	36.78	40.29	42.80	48.27
23	7.53	9.26	10.20	11.69	13.09	14.85	22.34	32.01	35.17	38.08	41.64	44.18	49.73
24	8.09	9.89	10.86	12.40	13.85	15.66	23.34	33.20	36.42	39.36	42.98	45.56	51.18
25	8.65	10.52	11.52	13.12	14.61	16.47	24.34	34.38	37.65	40.65	44.31	46.93	52.62
26	9.22	11.16	12.20	13.84	15.38	17.29	25.34	35.56	38.89	41.92	45.64	48.29	54.05
27	9.80	11.81	12.88	14.57	16.15	18.11	26.34	36.74	40.11	43.19	46.96	49.64	55.48
28	10.39	12.46	13.56	15.31	16.93	18.94	27.34	37.92	41.34	44.46	48.28	50.99	56.89
29	10.99	13.12	14.26	16.05	17.71	19.77	28.34	39.09	42.56	45.72	49.59	52.34	58.30
30	11.59	13.79	14.95	16.79	18.49	20.60	29.34	40.26	43.77	46.98	50.89	53.67	59.70
35	14.69	17.19	18.51	20.57	22.47	24.80	34.34	46.06	49.80	53.20	57.34	60.27	66.62
40	17.92	20.71	22.16	24.43	26.51	29.05	39.34	51.81	55.76	59.34	63.69	66.77	73.40
50	24.67	27.99	29.71	32.36	34.76	37.69	49.33	63.17	67.50	71.42	76.15	79.49	86.66
60	31.74	35.53	37.48	40.48	43.19	46.46	59.33	74.40	79.08	83.30	88.38	91.95	99.61
120	77.76	83.85	86.92	91.57	95.70	100.62	119.33	140.23	146.57	152.21	158.95	163.65	173.6

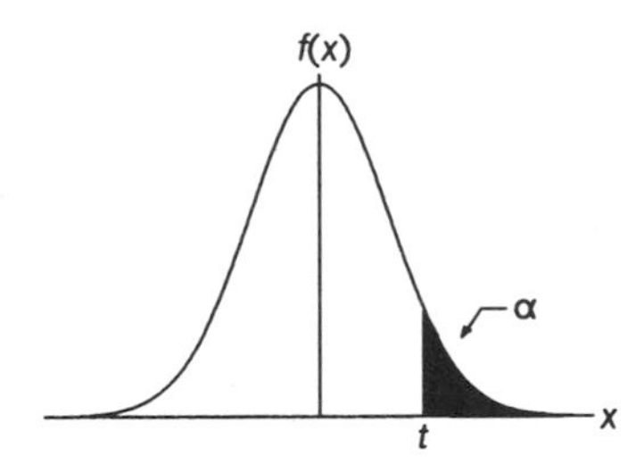

Figure D.3 *t* distribution.

D.2.2 *t* Distribution (Also known as *Student's distribution)*

The t distribution function, shown in Figure D.3, is used to derive confidence intervals for the population mean when the sample set is small. It is also used in hypothesis testing to check the validity of a sample mean against a population mean. The uses for this distribution are discussed in greater detail in Chapter 4.

The t distribution tables were generated using the following function. (Critical t values for the upper tail of the distribution were derived with a program using numerical integration routines similar to those available in STATS.)

$$\alpha = \int_{-\infty}^{t} \frac{\Gamma((v+1)/2)}{\sqrt{v\pi}\,\Gamma(v/2)} \left(1 + \frac{x^2}{v}\right)^{-(v+1)/2} dx$$

where Γ is the gamma function as defined in Section D.2.1, and v is the degrees of freedom in the function. In Table D.3 critical values of t are listed that are required to achieve the percentage points listed in the top row. The distribution is symmetrical, and thus

$$F(-t) = 1 - F(t)$$

Table D.3 Critical values for t distribution

α / v	0.400	0.350	0.300	0.250	0.200	0.150	0.100	0.050	0.025	0.010	0.005	0.001	0.0005
1	0.325	0.510	0.727	1.000	1.376	1.963	3.078	6.314	12.70	31.82	63.64	318.2	636.2
2	0.289	0.445	0.617	0.816	1.061	1.386	1.886	2.920	4.303	6.964	9.925	22.40	31.58
3	0.277	0.424	0.584	0.765	0.978	1.250	1.638	2.353	3.183	4.541	5.842	10.22	12.95
4	0.271	0.414	0.569	0.741	0.941	1.190	1.533	2.132	2.776	3.748	4.604	6.897	8.610
5	0.267	0.408	0.559	0.727	0.920	1.156	1.476	2.015	2.571	3.365	4.032	5.895	6.880
6	0.265	0.404	0.553	0.718	0.906	1.134	1.440	1.943	2.447	3.143	3.708	5.208	5.961
7	0.263	0.402	0.549	0.711	0.896	1.119	1.415	1.895	2.365	2.998	3.500	4.785	5.408
8	0.262	0.399	0.546	0.706	0.889	1.108	1.397	1.860	2.306	2.896	3.356	4.510	5.041
9	0.261	0.398	0.543	0.703	0.883	1.100	1.383	1.833	2.262	2.821	3.250	4.304	4.781
10	0.260	0.397	0.542	0.700	0.879	1.093	1.372	1.812	2.228	2.764	3.169	4.149	4.605
11	0.260	0.396	0.540	0.697	0.876	1.088	1.363	1.796	2.201	2.718	3.106	4.029	4.452
12	0.259	0.395	0.539	0.695	0.873	1.083	1.356	1.782	2.179	2.681	3.055	3.933	4.329
13	0.259	0.394	0.538	0.694	0.870	1.079	1.350	1.771	2.160	2.650	3.012	3.854	4.230
14	0.258	0.393	0.537	0.692	0.868	1.076	1.345	1.761	2.145	2.624	2.977	3.789	4.148
15	0.258	0.393	0.536	0.691	0.866	1.074	1.341	1.753	2.131	2.602	2.947	3.734	4.079
16	0.258	0.392	0.535	0.690	0.865	1.071	1.337	1.746	2.120	2.583	2.921	3.688	4.021
17	0.257	0.392	0.534	0.689	0.863	1.069	1.333	1.740	2.110	2.567	2.898	3.647	3.970
18	0.257	0.392	0.534	0.688	0.862	1.067	1.330	1.734	2.101	2.552	2.878	3.611	3.926
19	0.257	0.391	0.533	0.688	0.861	1.066	1.328	1.729	2.093	2.539	2.861	3.580	3.887
20	0.257	0.391	0.533	0.687	0.860	1.064	1.325	1.725	2.086	2.528	2.845	3.553	3.853
21	0.257	0.391	0.532	0.686	0.859	1.063	1.323	1.721	2.080	2.518	2.831	3.528	3.822
22	0.256	0.390	0.532	0.686	0.858	1.061	1.321	1.717	2.074	2.508	2.819	3.506	3.795
23	0.256	0.390	0.532	0.685	0.858	1.060	1.319	1.714	2.069	2.500	2.807	3.486	3.770
24	0.256	0.390	0.531	0.685	0.857	1.059	1.318	1.711	2.064	2.492	2.797	3.467	3.748
25	0.256	0.390	0.531	0.684	0.856	1.058	1.316	1.708	2.060	2.485	2.787	3.451	3.727
26	0.256	0.390	0.531	0.684	0.856	1.058	1.315	1.706	2.056	2.479	2.779	3.435	3.708
27	0.256	0.389	0.531	0.684	0.855	1.057	1.314	1.703	2.052	2.473	2.771	3.421	3.691
28	0.256	0.389	0.530	0.683	0.855	1.056	1.313	1.701	2.048	2.467	2.763	3.409	3.675
29	0.256	0.389	0.530	0.683	0.854	1.055	1.311	1.699	2.045	2.462	2.756	3.397	3.661
30	0.256	0.389	0.530	0.683	0.854	1.055	1.310	1.697	2.042	2.457	2.750	3.385	3.647
35	0.255	0.388	0.529	0.682	0.852	1.052	1.306	1.690	2.030	2.438	2.724	3.340	3.592
40	0.255	0.388	0.529	0.681	0.851	1.050	1.303	1.684	2.021	2.423	2.704	3.307	3.552
60	0.254	0.387	0.527	0.679	0.848	1.045	1.296	1.671	2.000	2.390	2.660	3.232	3.461
120	0.254	0.386	0.526	0.677	0.845	1.041	1.289	1.658	1.980	2.358	2.617	3.160	3.374
∞	0.253	0.385	0.525	0.675	0.842	1.037	1.282	1.645	1.960	2.326	2.576	3.291	3.300

D.2.3 *F* Distribution (Also known as *Fisher distribution)*

This F distribution function, shown in Figure D.4, is used to derive confidence intervals for the ratio of two population variances. It is also used in hypothesis testing for this same ratio. The uses for this distribution are discussed in Chapter 4.

Critical F values for the upper tail of the distribution were derived with a program using numerical integration routines similar to those available in program STATS. The tables were generated using the following function.

$$\alpha = \int_F^{\infty} \frac{\Gamma((v_1 + v_2)/2)}{\Gamma(v_1/2)\,\Gamma(v_2/2)} \left(\frac{v_1}{v_2}\right)^{v_1/2} \frac{x^{(v_1-2)/2}}{1 + (v_1/v_2)x^{(v_1+v_2)/2}}\, dx$$

where Γ is the gamma function as defined in Section D.2.1, v_1 the numerator degrees of freedom, and v_2 the denominator degrees of freedom. For critical values in the lower tail of the distribution, the following relationship can be used in conjunction with the tabular values given in Table D.4:

$$F_{\alpha,v_1,v_2} = \frac{1}{F_{1-\alpha,v_2,v_1}}$$

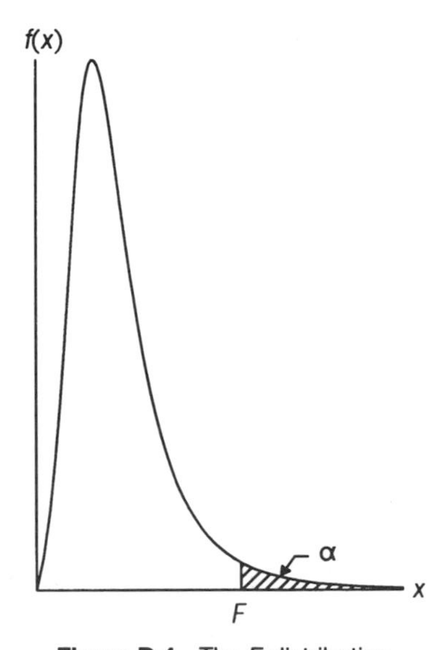

Figure D.4 The *F* distribution.

Table D.4 Critical values for F distribution

$\alpha = 0.20$

v_2 \ v_1	1	2	3	4	5	6	7	8	9	10	12	15	20	24	30	40	60	120
1	9.47	12.00	13.06	13.64	14.01	14.26	14.44	14.58	14.68	14.77	14.90	15.04	15.17	15.24	15.31	15.37	15.44	15.51
2	3.56	4.00	4.16	4.24	4.28	4.32	4.34	4.36	4.37	4.38	4.40	4.42	4.43	4.44	4.45	4.46	4.46	4.47
3	2.68	2.89	2.94	2.96	2.97	2.97	2.97	2.98	2.98	2.98	2.98	2.98	2.98	2.98	2.98	2.98	2.98	2.98
4	2.35	2.47	2.48	2.48	2.48	2.47	2.47	2.47	2.46	2.46	2.46	2.45	2.44	2.44	2.44	2.44	2.43	2.43
5	2.18	2.26	2.25	2.24	2.23	2.22	2.21	2.20	2.20	2.19	2.18	2.18	2.17	2.16	2.16	2.15	2.15	2.14
6	2.07	2.13	2.11	2.09	2.08	2.06	2.05	2.04	2.03	2.03	2.02	2.01	2.00	1.99	1.98	1.98	1.97	1.96
7	2.00	2.04	2.02	1.99	1.97	1.96	1.94	1.93	1.93	1.92	1.91	1.89	1.88	1.87	1.86	1.86	1.85	1.84
8	1.95	1.98	1.95	1.92	1.90	1.88	1.87	1.86	1.85	1.84	1.83	1.81	1.80	1.79	1.78	1.77	1.76	1.75
9	1.91	1.93	1.90	1.87	1.85	1.83	1.81	1.80	1.79	1.78	1.76	1.75	1.73	1.72	1.71	1.70	1.69	1.68
10	1.88	1.90	1.86	1.83	1.80	1.78	1.77	1.75	1.74	1.73	1.72	1.70	1.68	1.67	1.66	1.65	1.64	1.63
11	1.86	1.87	1.83	1.80	1.77	1.75	1.73	1.72	1.70	1.69	1.68	1.66	1.64	1.63	1.62	1.61	1.60	1.59
12	1.84	1.85	1.80	1.77	1.74	1.72	1.70	1.69	1.67	1.66	1.65	1.63	1.61	1.60	1.59	1.58	1.56	1.55
13	1.82	1.83	1.78	1.75	1.72	1.69	1.68	1.66	1.65	1.64	1.62	1.60	1.58	1.57	1.56	1.55	1.53	1.52
14	1.81	1.81	1.76	1.73	1.70	1.67	1.65	1.64	1.63	1.62	1.60	1.58	1.56	1.55	1.53	1.52	1.51	1.49
15	1.80	1.80	1.75	1.71	1.68	1.66	1.64	1.62	1.61	1.60	1.58	1.56	1.54	1.53	1.51	1.50	1.49	1.47
16	1.79	1.78	1.74	1.70	1.67	1.64	1.62	1.61	1.59	1.58	1.56	1.54	1.52	1.51	1.49	1.48	1.47	1.45
17	1.78	1.77	1.72	1.68	1.65	1.63	1.61	1.59	1.58	1.57	1.55	1.53	1.50	1.49	1.48	1.46	1.45	1.43
18	1.77	1.76	1.71	1.67	1.64	1.62	1.60	1.58	1.56	1.55	1.53	1.51	1.49	1.48	1.46	1.45	1.43	1.42
19	1.76	1.75	1.70	1.66	1.63	1.61	1.58	1.57	1.55	1.54	1.52	1.50	1.48	1.46	1.45	1.44	1.42	1.40
20	1.76	1.75	1.70	1.65	1.62	1.60	1.58	1.56	1.54	1.53	1.51	1.49	1.47	1.45	1.44	1.42	1.41	1.39
21	1.75	1.74	1.69	1.65	1.61	1.59	1.57	1.55	1.53	1.52	1.50	1.48	1.46	1.44	1.43	1.41	1.40	1.38
22	1.75	1.73	1.68	1.64	1.61	1.58	1.56	1.54	1.53	1.51	1.49	1.47	1.45	1.43	1.42	1.40	1.39	1.37
23	1.74	1.73	1.68	1.63	1.60	1.57	1.55	1.53	1.52	1.51	1.49	1.46	1.44	1.42	1.41	1.39	1.38	1.36
24	1.74	1.72	1.67	1.63	1.59	1.57	1.55	1.53	1.51	1.50	1.48	1.46	1.43	1.42	1.40	1.39	1.37	1.35
25	1.73	1.72	1.66	1.62	1.59	1.56	1.54	1.52	1.51	1.49	1.47	1.45	1.42	1.41	1.39	1.38	1.36	1.34
26	1.73	1.71	1.66	1.62	1.58	1.56	1.53	1.52	1.50	1.49	1.47	1.44	1.42	1.40	1.39	1.37	1.35	1.33
27	1.73	1.71	1.66	1.61	1.58	1.55	1.53	1.51	1.49	1.48	1.46	1.44	1.41	1.40	1.38	1.36	1.35	1.33
28	1.72	1.71	1.65	1.61	1.57	1.55	1.52	1.51	1.49	1.48	1.46	1.43	1.41	1.39	1.37	1.36	1.34	1.32
29	1.72	1.70	1.65	1.60	1.57	1.54	1.52	1.50	1.49	1.47	1.45	1.43	1.40	1.39	1.37	1.35	1.33	1.31
30	1.72	1.70	1.64	1.60	1.57	1.54	1.52	1.50	1.48	1.47	1.45	1.42	1.39	1.38	1.36	1.35	1.33	1.31
50	1.69	1.66	1.60	1.56	1.52	1.49	1.47	1.45	1.43	1.42	1.39	1.37	1.34	1.32	1.30	1.28	1.26	1.24
60	1.68	1.65	1.59	1.55	1.51	1.48	1.46	1.44	1.42	1.41	1.38	1.35	1.32	1.31	1.29	1.27	1.24	1.22
80	1.67	1.64	1.58	1.53	1.50	1.47	1.44	1.42	1.41	1.39	1.37	1.34	1.31	1.29	1.27	1.25	1.22	1.19
120	1.66	1.63	1.57	1.52	1.48	1.45	1.43	1.41	1.39	1.37	1.35	1.32	1.29	1.27	1.25	1.23	1.20	1.17

(continued)

Table D.4 (Continued)

$\alpha = 0.10$

v_2 \ v_1	1	2	3	4	5	6	7	8	9	10	12	15	20	24	30	40	60	120
1	39.85	49.49	53.59	55.83	57.23	58.20	58.90	59.43	59.85	60.19	60.70	61.21	61.73	61.99	62.26	62.52	62.79	63.05
2	8.53	9.00	9.16	9.24	9.29	9.33	9.35	9.37	9.38	9.39	9.41	9.42	9.44	9.45	9.46	9.47	9.47	9.48
3	5.54	5.46	5.39	5.34	5.31	5.28	5.27	5.25	5.24	5.23	5.22	5.20	5.18	5.18	5.17	5.16	5.15	5.14
4	4.54	4.32	4.19	4.11	4.05	4.01	3.98	3.95	3.94	3.92	3.90	3.87	3.84	3.83	3.82	3.80	3.79	3.78
5	4.06	3.78	3.62	3.52	3.45	3.40	3.37	3.34	3.32	3.30	3.27	3.24	3.21	3.19	3.17	3.16	3.14	3.12
6	3.78	3.46	3.29	3.18	3.11	3.05	3.01	2.98	2.96	2.94	2.90	2.87	2.84	2.82	2.80	2.78	2.76	2.74
7	3.59	3.26	3.07	2.96	2.88	2.83	2.78	2.75	2.72	2.70	2.67	2.63	2.59	2.58	2.56	2.54	2.51	2.49
8	3.46	3.11	2.92	2.81	2.73	2.67	2.62	2.59	2.56	2.54	2.50	2.46	2.42	2.40	2.38	2.36	2.34	2.32
9	3.36	3.01	2.81	2.69	2.61	2.55	2.51	2.47	2.44	2.42	2.38	2.34	2.30	2.28	2.25	2.23	2.21	2.18
10	3.28	2.92	2.73	2.61	2.52	2.46	2.41	2.38	2.35	2.32	2.28	2.24	2.20	2.18	2.16	2.13	2.11	2.08
11	3.23	2.86	2.66	2.54	2.45	2.39	2.34	2.30	2.27	2.25	2.21	2.17	2.12	2.10	2.08	2.05	2.03	2.00
12	3.18	2.81	2.61	2.48	2.39	2.33	2.28	2.24	2.21	2.19	2.15	2.10	2.06	2.04	2.01	1.99	1.96	1.93
13	3.14	2.76	2.56	2.43	2.35	2.28	2.23	2.20	2.16	2.14	2.10	2.05	2.01	1.98	1.96	1.93	1.90	1.88
14	3.10	2.73	2.52	2.39	2.31	2.24	2.19	2.15	2.12	2.10	2.05	2.01	1.96	1.94	1.91	1.89	1.86	1.83
15	3.07	2.70	2.49	2.36	2.27	2.21	2.16	2.12	2.09	2.06	2.02	1.97	1.92	1.90	1.87	1.85	1.82	1.79
16	3.05	2.67	2.46	2.33	2.24	2.18	2.13	2.09	2.06	2.03	1.99	1.94	1.89	1.87	1.84	1.81	1.78	1.75
17	3.03	2.64	2.44	2.31	2.22	2.15	2.10	2.06	2.03	2.00	1.96	1.91	1.86	1.84	1.81	1.78	1.75	1.72
18	3.01	2.62	2.42	2.29	2.20	2.13	2.08	2.04	2.00	1.98	1.93	1.89	1.84	1.81	1.78	1.75	1.72	1.69
19	2.99	2.61	2.40	2.27	2.18	2.11	2.06	2.02	1.98	1.96	1.91	1.86	1.81	1.79	1.76	1.73	1.70	1.67
20	2.97	2.59	2.38	2.25	2.16	2.09	2.04	2.00	1.96	1.94	1.89	1.84	1.79	1.77	1.74	1.71	1.68	1.64
21	2.96	2.57	2.36	2.23	2.14	2.08	2.02	1.98	1.95	1.92	1.87	1.83	1.78	1.75	1.72	1.69	1.66	1.62
22	2.95	2.56	2.35	2.22	2.13	2.06	2.01	1.97	1.93	1.90	1.86	1.81	1.76	1.73	1.70	1.67	1.64	1.60
23	2.94	2.55	2.34	2.21	2.11	2.05	1.99	1.95	1.92	1.89	1.84	1.80	1.74	1.72	1.69	1.66	1.62	1.59
24	2.93	2.54	2.33	2.19	2.10	2.04	1.98	1.94	1.91	1.88	1.83	1.78	1.73	1.70	1.67	1.64	1.61	1.57
25	2.92	2.53	2.32	2.18	2.09	2.02	1.97	1.93	1.89	1.87	1.82	1.77	1.72	1.69	1.66	1.63	1.59	1.56
26	2.91	2.52	2.31	2.17	2.08	2.01	1.96	1.92	1.88	1.86	1.81	1.76	1.71	1.68	1.65	1.61	1.58	1.54
27	2.90	2.51	2.30	2.17	2.07	2.00	1.95	1.91	1.87	1.85	1.80	1.75	1.70	1.67	1.64	1.60	1.57	1.53
28	2.89	2.50	2.29	2.16	2.06	2.00	1.94	1.90	1.87	1.84	1.79	1.74	1.69	1.66	1.63	1.59	1.56	1.52
29	2.89	2.50	2.28	2.15	2.06	1.99	1.93	1.89	1.86	1.83	1.78	1.73	1.68	1.65	1.62	1.58	1.55	1.51
30	2.88	2.49	2.28	2.14	2.05	1.98	1.93	1.88	1.85	1.82	1.77	1.72	1.67	1.64	1.61	1.57	1.54	1.50
50	2.81	2.41	2.20	2.06	1.97	1.90	1.84	1.80	1.76	1.73	1.68	1.63	1.57	1.54	1.50	1.46	1.42	1.38
60	2.79	2.39	2.18	2.04	1.95	1.87	1.82	1.77	1.74	1.71	1.66	1.60	1.54	1.51	1.48	1.44	1.40	1.35
80	2.77	2.37	2.15	2.02	1.92	1.85	1.79	1.75	1.71	1.66	1.63	1.57	1.51	1.48	1.44	1.40	1.36	1.31
120	2.75	2.35	2.13	1.99	1.90	1.82	1.77	1.72	1.68	1.65	1.60	1.55	1.48	1.45	1.41	1.37	1.32	1.26

Table D.4 (Continued)

$\alpha = 0.05$

v_2 \ v_1	1	2	3	4	5	6	7	8	9	10	12	15	20	24	30	40	60	120
1	161.4	199.4	215.6	224.5	230.1	233.9	236.7	238.8	240.5	241.8	243.8	245.9	247.9	249	250	251.1	252.1	253.2
2	18.51	19.00	19.16	19.25	19.30	19.33	19.35	19.37	19.38	19.39	19.41	19.43	19.44	19.45	19.46	19.47	19.48	19.49
3	10.13	9.55	9.28	9.12	9.01	8.94	8.89	8.85	8.81	8.79	8.74	8.70	8.66	8.64	8.62	8.59	8.57	8.55
4	7.71	6.94	6.59	6.39	6.26	6.16	6.09	6.04	6.00	5.96	5.91	5.86	5.80	5.77	5.75	5.72	5.69	5.66
5	6.61	5.79	5.41	5.19	5.05	4.95	4.88	4.82	4.77	4.74	4.68	4.62	4.56	4.53	4.50	4.46	4.43	4.40
6	5.99	5.14	4.76	4.53	4.39	4.28	4.21	4.15	4.10	4.06	4.00	3.94	3.87	3.84	3.81	3.77	3.74	3.70
7	5.59	4.74	4.35	4.12	3.97	3.87	3.79	3.73	3.68	3.64	3.57	3.51	3.44	3.41	3.38	3.34	3.30	3.27
8	5.32	4.46	4.07	3.84	3.69	3.58	3.50	3.44	3.39	3.35	3.28	3.22	3.15	3.12	3.08	3.04	3.01	2.97
9	5.12	4.26	3.86	3.63	3.48	3.37	3.29	3.23	3.18	3.14	3.07	3.01	2.94	2.90	2.86	2.83	2.79	2.75
10	4.96	4.10	3.71	3.48	3.33	3.22	3.14	3.07	3.02	2.98	2.91	2.85	2.77	2.74	2.70	2.66	2.62	2.58
11	4.84	3.98	3.59	3.36	3.20	3.09	3.01	2.95	2.90	2.85	2.79	2.72	2.65	2.61	2.57	2.53	2.49	2.45
12	4.75	3.89	3.49	3.26	3.11	3.00	2.91	2.85	2.80	2.75	2.69	2.62	2.54	2.51	2.47	2.43	2.38	2.34
13	4.67	3.81	3.41	3.18	3.03	2.92	2.83	2.77	2.71	2.67	2.60	2.53	2.46	2.42	2.38	2.34	2.30	2.25
14	4.60	3.74	3.34	3.11	2.96	2.85	2.76	2.70	2.65	2.60	2.53	2.46	2.39	2.35	2.31	2.27	2.22	2.18
15	4.54	3.68	3.29	3.06	2.90	2.79	2.71	2.64	2.59	2.54	2.48	2.40	2.33	2.29	2.25	2.20	2.16	2.11
16	4.49	3.63	3.24	3.01	2.85	2.74	2.66	2.59	2.54	2.49	2.42	2.35	2.28	2.24	2.19	2.15	2.11	2.06
17	4.45	3.59	3.20	2.96	2.81	2.70	2.61	2.55	2.49	2.45	2.38	2.31	2.23	2.19	2.15	2.10	2.06	2.01
18	4.41	3.55	3.16	2.93	2.77	2.66	2.58	2.51	2.46	2.41	2.34	2.27	2.19	2.15	2.11	2.06	2.02	1.97
19	4.38	3.52	3.13	2.90	2.74	2.63	2.54	2.48	2.42	2.38	2.31	2.23	2.16	2.11	2.07	2.03	1.98	1.93
20	4.35	3.49	3.10	2.87	2.71	2.60	2.51	2.45	2.39	2.35	2.28	2.20	2.12	2.08	2.04	1.99	1.95	1.90
21	4.32	3.47	3.07	2.84	2.68	2.57	2.49	2.42	2.37	2.32	2.25	2.18	2.10	2.05	2.01	1.96	1.92	1.87
22	4.30	3.44	3.05	2.82	2.66	2.55	2.46	2.40	2.34	2.30	2.23	2.15	2.07	2.03	1.98	1.94	1.89	1.84
23	4.28	3.42	3.03	2.80	2.64	2.53	2.44	2.37	2.32	2.27	2.20	2.13	2.05	2.01	1.96	1.91	1.86	1.81
24	4.26	3.40	3.01	2.78	2.62	2.51	2.42	2.36	2.30	2.25	2.18	2.11	2.03	1.98	1.94	1.89	1.84	1.79
25	4.24	3.39	2.99	2.76	2.60	2.49	2.40	2.34	2.28	2.24	2.16	2.09	2.01	1.96	1.92	1.87	1.82	1.77
26	4.22	3.37	2.98	2.74	2.59	2.47	2.39	2.32	2.27	2.22	2.15	2.07	1.99	1.95	1.90	1.85	1.80	1.75
27	4.21	3.35	2.96	2.73	2.57	2.46	2.37	2.31	2.25	2.20	2.13	2.06	1.97	1.93	1.88	1.84	1.79	1.73
28	4.20	3.34	2.95	2.71	2.56	2.45	2.36	2.29	2.24	2.19	2.12	2.04	1.96	1.91	1.87	1.82	1.77	1.71
29	4.18	3.33	2.93	2.70	2.55	2.43	2.35	2.28	2.22	2.18	2.10	2.03	1.94	1.90	1.85	1.81	1.75	1.70
30	4.17	3.32	2.92	2.69	2.53	2.42	2.33	2.27	2.21	2.16	2.09	2.01	1.93	1.89	1.84	1.79	1.74	1.68
50	4.03	3.18	2.79	2.56	2.40	2.29	2.20	2.13	2.07	2.03	1.95	1.87	1.78	1.74	1.69	1.63	1.58	1.51
60	4.00	3.15	2.76	2.53	2.37	2.25	2.17	2.10	2.04	1.99	1.92	1.84	1.75	1.70	1.65	1.59	1.53	1.47
80	3.96	3.11	2.72	2.49	2.33	2.21	2.13	2.06	2.00	1.95	1.88	1.79	1.70	1.65	1.60	1.54	1.48	1.41
120	3.92	3.07	2.68	2.45	2.29	2.18	2.09	2.02	1.96	1.91	1.83	1.75	1.66	1.61	1.55	1.50	1.43	1.35

Table D.4 (Continued)

$\alpha = 0.025$

v_2 \ v_1	1	2	3	4	5	6	7	8	9	10	12	15	20	24	30	40	60	120
1	647.8	799.5	8642	899.6	921.8	937.1	948.2	956.7	963.3	968.6	976.7	984.9	993.1	997.2	1001	1006	1010	1014
2	38.51	39.00	39.17	39.25	39.30	39.33	39.36	39.37	39.39	39.40	39.41	39.43	39.45	39.46	39.46	39.47	39.48	39.48
3	17.44	16.04	15.44	15.10	14.88	14.73	14.62	14.54	14.47	14.42	14.34	14.25	14.17	14.12	14.08	14.04	13.99	13.95
4	12.22	10.65	9.98	9.60	9.36	9.20	9.07	8.98	8.90	8.84	8.75	8.66	8.56	8.51	8.46	8.41	8.36	8.31
5	10.01	8.43	7.76	7.39	7.15	6.98	6.85	6.76	6.68	6.62	6.52	6.43	6.33	6.28	6.23	6.18	6.12	6.07
6	8.81	7.26	6.60	6.23	5.99	5.82	5.70	5.60	5.52	5.46	5.37	5.27	5.17	5.12	5.07	5.01	4.96	4.90
7	8.07	6.54	5.89	5.52	5.29	5.12	4.99	4.90	4.82	4.76	4.67	4.57	4.47	4.41	4.36	4.31	4.25	4.20
8	7.57	6.06	5.42	5.05	4.82	4.65	4.53	4.43	4.36	4.30	4.20	4.10	4.00	3.95	3.89	3.84	3.78	3.73
9	7.21	5.71	5.08	4.72	4.48	4.32	4.20	4.10	4.03	3.96	3.87	3.77	3.67	3.61	3.56	3.51	3.45	3.39
10	6.94	5.46	4.83	4.47	4.24	4.07	3.95	3.85	3.78	3.72	3.62	3.52	3.42	3.37	3.31	3.26	3.20	3.14
11	6.72	5.26	4.63	4.28	4.04	3.88	3.76	3.66	3.59	3.53	3.43	3.33	3.23	3.17	3.12	3.06	3.00	2.94
12	6.55	5.10	4.47	4.12	3.89	3.73	3.61	3.51	3.44	3.37	3.28	3.18	3.07	3.02	2.96	2.91	2.85	2.79
13	6.41	4.97	4.35	4.00	3.77	3.60	3.48	3.39	3.31	3.25	3.15	3.05	2.95	2.89	2.84	2.78	2.72	2.66
14	6.30	4.86	4.24	3.89	3.66	3.50	3.38	3.29	3.21	3.15	3.05	2.95	2.84	2.79	2.73	2.67	2.61	2.55
15	6.20	4.76	4.15	3.80	3.58	3.41	3.29	3.20	3.12	3.06	2.96	2.86	2.76	2.70	2.64	2.59	2.52	2.46
16	6.11	4.69	4.08	3.73	3.50	3.34	3.22	3.12	3.05	2.99	2.89	2.79	2.68	2.63	2.57	2.51	2.45	2.38
17	6.04	4.62	4.01	3.66	3.44	3.28	3.16	3.06	2.98	2.92	2.82	2.72	2.62	2.56	2.50	2.44	2.38	2.32
18	5.98	4.56	3.95	3.61	3.38	3.22	3.10	3.01	2.93	2.87	2.77	2.67	2.56	2.50	2.44	2.38	2.32	2.26
19	5.92	4.51	3.90	3.56	3.33	3.17	3.05	2.96	2.88	2.82	2.72	2.62	2.51	2.45	2.39	2.33	2.27	2.20
20	5.87	4.46	3.86	3.51	3.29	3.13	3.01	2.91	2.84	2.77	2.68	2.57	2.46	2.41	2.35	2.29	2.22	2.16
21	5.83	4.42	3.82	3.48	3.25	3.09	2.97	2.87	2.80	2.73	2.64	2.53	2.42	2.37	2.31	2.25	2.18	2.11
22	5.79	4.38	3.78	3.44	3.22	3.05	2.93	2.84	2.76	2.70	2.60	2.50	2.39	2.33	2.27	2.21	2.14	2.08
23	5.75	4.35	3.75	3.41	3.18	3.02	2.90	2.81	2.73	2.67	2.57	2.47	2.36	2.30	2.24	2.18	2.11	2.04
24	5.72	4.32	3.72	3.38	3.15	2.99	2.87	2.78	2.70	2.64	2.54	2.44	2.33	2.27	2.21	2.15	2.08	2.01
25	5.69	4.29	3.69	3.35	3.13	2.97	2.85	2.75	2.68	2.61	2.51	2.41	2.30	2.24	2.18	2.12	2.05	1.98
26	5.66	4.27	3.67	3.33	3.10	2.94	2.82	2.73	2.65	2.59	2.49	2.39	2.28	2.22	2.16	2.09	2.03	1.95
27	5.63	4.24	3.65	3.31	3.08	2.92	2.80	2.71	2.63	2.57	2.47	2.36	2.25	2.19	2.13	2.07	2.00	1.93
28	5.61	4.22	3.63	3.29	3.06	2.90	2.78	2.69	2.61	2.55	2.45	2.34	2.23	2.17	2.11	2.05	1.98	1.91
29	5.59	4.20	3.61	3.27	3.04	2.88	2.76	2.67	2.59	2.53	2.43	2.32	2.21	2.15	2.09	2.03	1.96	1.89
30	5.57	4.18	3.59	3.25	3.03	2.87	2.75	2.65	2.57	2.51	2.41	2.31	2.20	2.14	2.07	2.01	1.94	1.87
50	5.34	3.97	3.39	3.05	2.83	2.67	2.55	2.46	2.38	2.32	2.22	2.11	1.99	1.93	1.87	1.80	1.72	1.64
60	5.29	3.93	3.34	3.01	2.79	2.63	2.51	2.41	2.33	2.27	2.17	2.06	1.94	1.88	1.82	1.74	1.67	1.58
80	5.22	3.86	3.28	2.95	2.73	2.57	2.45	2.35	2.28	2.21	2.11	2.00	1.88	1.82	1.75	1.68	1.60	1.51
120	5.15	3.80	3.23	2.89	2.67	2.52	2.39	2.30	2.22	2.16	2.05	1.94	1.82	1.76	1.69	1.61	1.53	1.43

(continued)

Table D.4 (Continued)

$\alpha = 0.01$

v_2 \ v_1	1	2	3	4	5	6	7	8	9	10	12	15	20	24	30	40	60	120
1	4052	5000	5403	5625	5764	5859	5928	5982	6022	6056	6106	6157	6209	6235	6261	6287	6313	6339
2	98.5	99.0	99.2	99.2	99.3	99.3	99.4	99.4	99.4	99.4	99.4	99.4	99.4	99.5	99.5	99.5	99.5	99.5
3	34.1	30.8	29.5	28.7	28.2	27.9	27.7	27.5	27.3	27.2	27.1	26.9	26.7	26.6	26.5	26.4	26.3	26.2
4	21.2	18.0	16.7	16.0	15.5	15.2	15.0	14.8	14.7	14.6	14.4	14.2	14.0	13.9	13.8	13.7	13.7	13.6
5	16.25	13.27	12.06	11.39	10.97	10.67	10.46	10.29	10.16	10.05	9.89	9.72	9.55	9.47	9.38	9.29	9.20	9.11
6	13.74	10.92	9.78	9.15	8.75	8.47	8.26	8.10	7.98	7.87	7.72	7.56	7.40	7.31	7.23	7.14	7.06	6.97
7	12.24	9.55	8.45	7.85	7.46	7.19	6.99	6.84	6.72	6.62	6.47	6.31	6.16	6.07	5.99	5.91	5.82	5.74
8	11.26	8.65	7.59	7.01	6.63	6.37	6.18	6.03	5.91	5.81	5.67	5.52	5.36	5.28	5.20	5.12	5.03	4.95
9	10.56	8.02	6.99	6.42	6.06	5.80	5.61	5.47	5.35	5.26	5.11	4.96	4.81	4.73	4.65	4.57	4.48	4.40
10	10.04	7.56	6.55	5.99	5.64	5.39	5.20	5.06	4.94	4.85	4.71	4.56	4.41	4.33	4.25	4.17	4.08	4.00
11	9.64	7.21	6.22	5.67	5.32	5.07	4.89	4.74	4.63	4.54	4.40	4.25	4.10	4.02	3.94	3.86	3.78	3.69
12	9.33	6.93	5.95	5.41	5.06	4.82	4.64	4.50	4.39	4.30	4.16	4.01	3.86	3.78	3.70	3.62	3.54	3.45
13	9.07	6.70	5.74	5.21	4.86	4.62	4.44	4.30	4.19	4.10	3.96	3.82	3.66	3.59	3.51	3.43	3.34	3.25
14	8.86	6.51	5.56	5.04	4.69	4.46	4.28	4.14	4.03	3.94	3.80	3.66	3.51	3.43	3.35	3.27	3.18	3.09
15	8.68	6.36	5.42	4.89	4.56	4.32	4.14	4.00	3.89	3.80	3.67	3.52	3.37	3.29	3.21	3.13	3.05	2.96
16	8.53	6.23	5.29	4.77	4.44	4.20	4.03	3.89	3.78	3.69	3.55	3.41	3.26	3.18	3.10	3.02	2.93	2.84
17	8.40	6.11	5.18	4.67	4.34	4.10	3.93	3.79	3.68	3.59	3.46	3.31	3.16	3.08	3.00	2.92	2.83	2.75
18	8.28	6.01	5.09	4.58	4.25	4.01	3.84	3.71	3.60	3.51	3.37	3.23	3.08	3.00	2.92	2.84	2.75	2.66
19	8.18	5.93	5.01	4.50	4.17	3.94	3.77	3.63	3.52	3.43	3.30	3.15	3.00	2.92	2.84	2.76	2.67	2.58
20	8.09	5.85	4.94	4.43	4.10	3.87	3.70	3.56	3.46	3.37	3.23	3.09	2.94	2.86	2.78	2.69	2.61	2.52
21	8.01	5.78	4.87	4.37	4.04	3.81	3.64	3.51	3.40	3.31	3.17	3.03	2.88	2.80	2.72	2.64	2.55	2.46
22	7.94	5.72	4.82	4.31	3.99	3.76	3.59	3.45	3.35	3.26	3.12	2.98	2.83	2.75	2.67	2.58	2.50	2.40
23	7.88	5.66	4.76	4.26	3.94	3.71	3.54	3.41	3.30	3.21	3.07	2.93	2.78	2.70	2.62	2.54	2.45	2.35
24	7.82	5.61	4.72	4.22	3.90	3.67	3.50	3.36	3.26	3.17	3.03	2.89	2.74	2.66	2.58	2.49	2.40	2.31
25	7.77	5.57	4.68	4.18	3.85	3.63	3.46	3.32	3.22	3.13	2.99	2.85	2.70	2.62	2.54	2.45	2.36	2.27
26	7.72	5.53	4.64	4.14	3.82	3.59	3.42	3.29	3.18	3.09	2.96	2.81	2.66	2.58	2.50	2.42	2.33	2.23
27	7.67	5.49	4.60	4.11	3.78	3.56	3.39	3.26	3.15	3.06	2.93	2.78	2.63	2.55	2.47	2.38	2.29	2.20
28	7.63	5.45	4.57	4.07	3.75	3.53	3.36	3.23	3.12	3.03	2.90	2.75	2.60	2.52	2.44	2.35	2.26	2.17
29	7.60	5.42	4.54	4.04	3.73	3.50	3.33	3.20	3.09	3.00	2.87	2.73	2.57	2.49	2.41	2.33	2.23	2.14
30	7.56	5.39	4.51	4.02	3.70	3.47	3.30	3.17	3.07	2.98	2.84	2.70	2.55	2.47	2.39	2.30	2.21	2.11
50	7.17	5.06	4.20	3.72	3.41	3.19	3.02	2.89	2.78	2.70	2.56	2.42	2.27	2.18	2.10	2.01	1.91	1.80
60	7.08	4.98	4.13	3.65	3.34	3.12	2.95	2.82	2.72	2.63	2.50	2.35	2.20	2.12	2.03	1.94	1.84	1.73
80	6.96	4.88	4.04	3.56	3.26	3.04	2.87	2.74	2.64	2.55	2.42	2.27	2.12	2.03	1.94	1.85	1.75	1.63
120	6.85	4.79	3.95	3.48	3.17	2.96	2.79	2.66	2.56	2.47	2.34	2.19	2.03	1.95	1.86	1.76	1.66	1.53

(continued)

Table D.4 (Continued)

$\alpha = 0.005$

v_2 \ v_1	1	2	3	4	5	6	7	8	9	10	12	15	20	24	30	40	60	120
1	16211	20000	21615	22500	23056	23437	23715	23925	24091	24224	24426	24630	24836	24940	25044	25148	25253	253591
2	198.5	199.0	199.2	199.2	199.3	199.4	199.4	199.4	199.4	199.4	199.4	199.4	199.4	199.5	199.5	199.5	199.5	199.5
3	55.55	49.80	47.47	46.19	45.39	44.84	44.43	44.13	43.88	43.69	43.39	43.06	42.78	42.62	42.47	42.31	42.15	41.99
4	31.33	26.28	24.26	23.15	22.46	21.97	21.62	21.35	21.14	20.97	20.70	20.44	20.17	20.03	19.89	19.75	19.61	19.47
5	22.77	18.31	16.53	15.55	14.94	14.51	14.20	13.96	13.77	13.62	13.38	13.15	12.90	12.78	12.66	12.53	12.40	12.27
6	18.62	14.54	12.91	12.03	11.46	11.07	10.79	10.57	10.39	10.25	10.03	9.81	9.59	9.47	9.36	9.24	9.12	9.00
7	16.23	12.40	10.88	10.05	9.52	9.16	8.89	8.68	8.51	8.38	8.18	7.97	7.75	7.64	7.53	7.42	7.31	7.19
8	14.68	11.04	9.60	8.80	8.30	7.95	7.69	7.50	7.34	7.21	7.01	6.81	6.61	6.50	6.40	6.29	6.18	6.06
9	13.61	10.10	8.72	7.96	7.47	7.13	6.88	6.69	6.54	6.42	6.23	6.03	5.83	5.73	5.62	5.52	5.41	5.30
10	12.82	9.43	8.08	7.34	6.87	6.54	6.30	6.12	5.97	5.85	5.66	5.47	5.27	5.17	5.07	4.97	4.86	4.75
11	12.22	8.91	7.60	6.88	6.42	6.10	5.86	5.68	5.54	5.42	5.24	5.05	4.86	4.76	4.65	4.55	4.44	4.34
12	11.75	8.51	7.23	6.52	6.07	5.76	5.52	5.35	5.20	5.09	4.91	4.72	4.53	4.43	4.33	4.23	4.12	4.01
13	11.37	8.19	6.93	6.23	5.79	5.48	5.25	5.08	4.94	4.82	4.64	4.46	4.27	4.17	4.07	3.97	3.87	3.76
14	11.06	7.92	6.68	6.00	5.56	5.26	5.03	4.86	4.72	4.60	4.43	4.25	4.06	3.96	3.86	3.76	3.66	2.55
15	10.79	7.70	6.48	5.80	5.37	5.07	4.85	4.67	4.54	4.42	4.25	4.07	3.88	3.79	3.69	3.58	3.48	3.37
16	10.57	7.51	6.30	5.64	5.21	4.91	4.69	4.52	4.38	4.27	4.10	3.92	3.73	3.64	3.54	3.44	3.33	3.22
17	10.38	7.35	6.16	5.50	5.07	4.78	4.56	4.39	4.25	4.14	3.97	3.79	3.61	3.51	3.41	3.31	3.21	3.10
18	10.21	7.21	6.03	5.37	4.96	4.66	4.44	4.28	4.14	4.03	3.86	3.68	3.50	3.40	3.30	3.20	3.10	2.99
19	10.07	7.09	5.92	5.27	4.85	4.56	4.34	4.18	4.04	3.93	3.76	3.59	3.40	3.31	3.21	3.11	3.00	2.89
20	9.94	6.99	5.82	5.17	4.76	4.47	4.26	4.09	3.96	3.85	3.68	3.50	3.32	3.22	3.12	3.02	2.92	2.81
21	9.83	6,89	5.73	5.09	4.68	4.39	4.18	4.01	3.88	3.77	3.60	3.43	3.24	3.15	3.05	2.95	2.84	2.73
22	9.72	6.81	5.65	5.02	4.61	4.32	4.11	3.94	3.81	3.70	3.54	3.36	3.18	3.08	2.98	2.88	2.77	2.66
23	9.63	6.73	5.58	4.95	4.54	4.26	4.05	3.88	3.75	3.64	3.47	3.30	3.12	3.02	2.92	2.82	2.71	2.60
24	9.55	6.66	5.52	4.89	4.49	4.20	3.99	3.83	3.69	3.59	3.42	3.25	3.06	2.97	2.87	2.77	2.66	2.55
25	9.47	6.60	5.46	4.83	4.43	4.15	3.94	3.78	3.64	3.54	3.37	3.20	3.01	2.92	2.82	2.72	2.61	2.50
26	9.40	6.54	5.41	4.79	4.38	4.10	3.89	3.73	3.60	3.49	3.33	3.15	2.97	2.87	2.77	2.67	2.56	2.45
27	9.34	6.49	5.36	4.74	4.34	4.06	3.85	3.69	3.56	3.45	3.28	3.11	2.93	2.83	2.73	2.63	2.52	2.41
28	9.28	6.44	5.32	4.70	4.30	4.02	3.81	3.65	3.52	3.41	3.25	3.07	2.89	2.79	2.69	2.59	2.48	2.37
29	9.23	6.39	5.28	4.66	4.26	3.98	3.77	3.61	3.48	3.38	3.21	3.04	2.86	2.76	2.66	2.56	2.45	2.33
30	9.18	6.35	5.24	4.62	4.23	3.95	3.74	3.58	3.45	3.34	3.18	3.01	2.82	2.73	2.63	2.52	2.42	2.30
50	8.62	5.90	4.83	4.23	3.85	3.58	3.38	3.22	3.09	2.99	2.82	2.65	2.47	2.37	2.27	2.16	2.05	1.93
60	8.49	5.79	4.73	4.14	3.76	3.49	3.29	3.13	3.01	2.90	2.74	2.57	2.39	2.29	2.19	2.08	1.96	1.83
80	8.33	5.66	4.61	4.03	3.65	3.39	3.19	3.03	2.91	2.80	2.64	2.47	2.29	2.19	2.08	1.97	1.85	1.72
120	8.18	5.54	4.50	3.92	3.55	3.28	3.09	2.93	2.81	2.71	2.54	2.37	2.19	2.09	1.98	1.87	1.75	1.61

Table D.4 (Continued)

$\alpha = 0.001$

v_2 \ v_1	1	2	3	4	5	6	7	8	9	10	12	15	20	24	30	40	60	120
1	405269	500004	540387	562506	576412	585943	592881	598151	602292	605625	610676	615772	620913	623504	626107	628720	631345	633980
2	998.5	999.0	999.1	999.2	999.3	999.3	999.4	999.4	999.4	999.4	999.4	999.4	999.5	999.5	999.5	999.5	999.5	999.5
3	167.0	148.5	141.1	137.1	134.6	132.9	131.6	130.6	129.9	129.3	128.3	127.4	126.4	125.9	125.5	125.0	124.5	124.0
4	74.14	61.25	56.18	53.44	51.71	50.53	49.66	49.00	48.47	48.05	47.41	46.76	46.10	45.77	45.43	45.09	44.75	44.40
5	47.18	37.12	33.20	31.09	29.75	28.83	28.16	27.65	27.24	26.92	26.42	25.91	25.39	25.13	24.87	24.60	24.33	24.06
6	35.51	27.00	23.70	21.92	20.80	20.03	19.46	19.03	18.69	18.41	17.99	17.56	17.12	16.90	16.67	16.44	16.21	15.98
7	29.25	21.69	18.77	17.20	16.21	15.52	15.02	14.63	14.33	14.08	13.71	13.32	12.93	12.73	12.53	12.33	12.12	11.91
8	25.41	18.49	15.83	14.39	13.48	12.86	12.40	12.05	11.77	11.54	11.91	10.84	10.48	10.30	10.11	9.92	9.73	9.53
9	22.86	16.39	13.90	12.56	11.71	11.13	10.70	10.37	10.11	9.89	9.57	9.24	8.90	8.72	8.55	8.37	8.19	8.00
10	21.04	14.91	12.55	11.28	10.48	9.93	9.52	9.20	8.96	8.75	8.45	8.13	7.80	7.64	7.47	7.30	7.12	6.94
11	19.69	13.81	11.56	10.35	9.58	9.05	8.66	8.35	8.12	7.92	7.63	7.32	7.01	6.85	6.68	6.52	6.35	6.18
12	18.64	12.97	10.80	9.63	8.89	8.38	8.00	7.71	7.48	7.29	7.00	6.71	6.40	6.25	6.09	5.93	5.76	5.59
13	17.82	12.31	10.21	9.07	8.35	7.86	7.49	7.21	6.98	6.80	6.52	6.23	5.93	5.78	5.63	5.47	5.30	5.14
14	17.14	11.78	9.73	8.62	7.92	7.44	7.08	6.80	6.58	6.40	6.13	5.85	5.56	5.41	5.25	5.10	4.94	4.77
15	16.59	11.34	9.34	8.25	7.57	7.09	6.74	6.47	6.26	6.08	5.81	5.54	5.25	5.10	4.95	4.80	4.64	4.47
16	16.12	10.97	9.01	7.94	7.27	6.80	6.46	6.19	5.98	5.81	5.55	5.27	4.99	4.85	4.70	4.54	4.39	4.23
17	15.72	10.66	8.73	7.68	7.02	6.56	6.22	5.96	5.75	5.58	5.32	5.05	4.78	4.63	4.48	4.33	4.18	4.02
18	15.38	10.39	8.49	7.46	6.81	6.35	6.02	5.76	5.56	5.39	5.13	4.87	4.59	4.45	4.30	4.15	4.00	3.84
19	15.08	10.16	8.28	7.27	6.62	6.18	5.85	5.59	5.39	5.22	4.97	4.70	4.43	4.29	4.14	3.99	3.84	3.68
20	14.82	9.95	8.10	7.10	6.46	6.02	5.69	5.44	5.24	5.08	4.82	4.56	4.29	4.15	4.00	3.86	3.70	3.54
21	14.59	9.77	7.94	6.95	6.32	5.88	5.56	5.31	5.11	4.95	4.70	4.44	4.17	4.03	3.88	3.74	3.58	3.42
22	14.38	9.61	7.80	6.81	6.19	5.76	5.44	5.19	4.99	4.83	4.58	4.33	4.06	3.92	3.78	3.63	3.48	3.32
23	14.20	9.47	7.67	6.70	6.08	5.65	5.33	5.09	4.89	4.73	4.48	4.23	3.96	3.82	3.68	3.53	3.38	3.22
24	14.03	9.34	7.55	6.59	5.98	5.55	5.23	4.99	4.80	4.64	4.39	4.14	3.87	3.74	3.59	3.45	3.29	3.14
25	13.88	9.22	7.45	6.49	5.89	5.46	5.15	4.91	4.71	4.56	4.31	4.06	3.79	3.66	3.52	3.37	3.22	3.06
26	13.74	9.12	7.36	6.41	5.80	5.38	5.07	4.83	4.64	4.48	4.24	3.99	3.72	3.59	3.44	3.30	3.15	2.99
27	13.61	9.02	7.27	6.33	5.73	5.31	5.00	4.76	4.57	4.41	4.17	3.92	3.66	3.52	3.38	3.23	3.08	2.92
28	13.50	8.93	7.19	6.25	5.66	5.24	4.93	4.69	4.50	4.35	4.11	3.86	3.60	3.46	3.32	3.18	3.02	2.86
29	13.39	8.85	7.12	6.19	5.59	5.18	4.87	4.64	4.45	4.29	4.05	3.80	3.54	3.41	3.27	3.12	2.97	2.81
30	13.29	8.77	7.05	6.12	5.53	5.12	4.82	4.58	4.39	4.24	4.00	3.75	3.49	3.36	3.22	3.07	2.92	2.76
40	12.61	8.25	6.59	5.70	5.13	4.73	4.44	4.21	4.02	3.87	3.64	3.40	3.14	3.01	2.87	2.73	2.57	2.41
50	12.22	7.96	6.34	5.46	4.90	4.51	4.22	4.00	3.82	3.67	3.44	3.20	2.95	2.82	2.68	2.53	2.38	2.21
60	11.97	7.77	6.17	5.31	4.76	4.37	4.09	3.86	3.69	3.54	3.32	3.08	2.83	2.69	2.55	2.41	2.25	2.08
120	11.38	7.32	5.78	4.95	4.42	4.04	3.77	3.55	3.38	3.24	3.02	2.78	2.53	2.40	2.26	2.11	1.95	1.77

APPENDIX E

CONFIDENCE INTERVALS FOR THE MEAN

Table E.1 represents 1000 95% confidence intervals constructed from sample sets selected from a population with a mean (μ) of 25.4 and a standard error (σ) of 1.3. These data are discussed in Chapter 4. (*Note:* Intervals with asterisks fail to include μ.)

Table E.1 Confidence Intervals for the Mean

(24.21, 25.94)	(24.77, 26.22)	(24.71, 26.01)	(25.43, 27.01)*	(24.22, 26.19)	(25.12, 26.56)	(24.70, 26.54)	(24.50, 25.95)
(23.99, 25.39)*	(25.33, 26.45)	(24.62, 25.89)	(24.79, 26.16)	(24.85, 26.11)	(23.92, 25.67)	(25.30, 26.84)	(24.71, 26.39)
(24.58, 26.10)	(24.47, 25.50)	(25.14, 26.56)	(24.79, 25.72)	(24.22, 25.79)	(25.03, 26.27)	(24.63, 25.83)	(25.09, 26.84)
(24.80, 26.18)	(24.95, 26.07)	(24.61, 26.27)	(25.06, 26.29)	(24.69, 26.25)	(24.64, 25.88)	(24.96, 26.15)	(25.04, 25.92)
(24.59, 26.09)	(24.99, 26.15)	(24.65, 26.29)	(25.08, 26.28)	(24.74, 26.02)	(25.19, 26.43)	(24.54, 25.73)	(24.17, 25.94)
(24.86, 25.81)	(24.40, 25.92)	(24.86, 25.80)	(24.58, 25.91)	(23.69, 25.25)*	(24.64, 25.77)	(24.88, 26.12)	(24.63, 25.92)
(24.84, 25.92)	(24.97, 26.47)	(25.13, 26.38)	(24.39, 26.32)	(24.15, 26.16)	(24.80, 26.01)	(24.87, 26.54)	(24.66, 25.95)
(24.64, 25.93)	(24.79, 26.18)	(24.68, 25.93)	(24.17, 25.83)	(24.29, 26.14)	(24.43, 25.62)	(24.58, 25.99)	(25.11, 26.33)
(24.85, 26.26)	(25.11, 26.44)	(24.58, 25.99)	(24.89, 26.25)	(24.79, 26.36)	(25.23, 26.42)	(25.30, 26.45)	(25.09, 26.53)
(25.23, 26.34)	(24.58, 26.35)	(24.81, 25.88)	(24.37, 25.62)	(24.79, 26.07)	(24.82, 26.19)	(24.66, 26.56)	(24.56, 26.12)
(25.11, 26.18)	(24.73, 25.83)	(24.46, 25.51)	(24.92, 25.87)	(24.46, 26.45)	(24.40, 25.61)	(24.49, 26.45)	(24.61, 26.02)
(25.20, 26.34)	(24.94, 25.99)	(24.53, 25.90)	(24.73, 26.07)	(24.56, 25.75)	(24.49, 25.85)	(24.49, 26.24)	(25.28, 26.26)
(24.26, 25.48)	(24.50, 26.26)	(25.11, 26.56)	(23.96, 25.61)	(24.94, 26.46)	(24.63, 26.31)	(24.37, 25.95)	(24.67, 26.18)
(24.56, 26.17)	(24.48, 25.76)	(24.72, 26.02)	(25.63, 26.40)*	(24.82, 26.09)	(24.97, 26.02)	(24.51, 25.78)	(24.88, 26.18)
(24.72, 25.91)	(24.94, 26.24)	(24.91, 26.39)	(25.46, 26.82)*	(24.84, 26.37)	(25.05, 26.07)	(24.72, 25.78)	(25.20, 26.60)
(24.83, 26.09)	(24.62, 25.96)	(25.10, 26.29)	(24.78, 26.24)	(24.50, 26.30)	(24.60, 26.20)	(25.38, 26.62)	(24.86, 26.20)
(24.31, 25.89)	(24.35, 25.73)	(24.67, 26.07)	(24.86, 26.58)	(24.52, 25.55)	(24.81, 26.47)	(24.45, 26.32)	(23.94, 25.77)
(24.76, 25.56)	(25.04, 26.20)	(24.25, 26.06)	(24.33, 25.69)	(24.99, 26.16)	(25.13, 26.43)	(24.68, 26.26)	(24.69, 25.68)
(25.11, 26.45)	(24.69, 26.00)	(24.45, 25.42)	(24.72, 26.50)	(24.54, 25.80)	(24.57, 25.76)	(25.10, 26.16)	(24.60, 25.90)
(24.21, 25.86)	(25.00, 26.52)	(25.05, 26.55)	(24.47, 25.76)	(24.76, 26.07)	(24.80, 25.95)	(24.45, 25.89)	(24.82, 26.86)
(25.16, 26.21)	(24.62, 26.01)	(24.45, 25.88)	(25.60, 26.70)*	(24.32, 25.83)	(25.11, 26.12)	(24.59, 25.81)	(24.11, 25.86)
(24.53, 25.86)	(24.89, 26.21)	(24.52, 25.83)	(25.08, 26.22)	(25.22, 26.71)	(24.25, 25.39)*	(24.73, 26.34)	(24.67, 26.24)
(24.84, 26.31)	(24.68, 25.55)	(24.69, 26.28)	(24.82, 26.02)	(25.29, 26.44)	(24.84, 26.21)	(24.83, 26.03)	(24.90, 26.26)
(24.84, 26.14)	(25.13, 26.52)	(25.07, 26.44)	(24.52, 25.88)	(24.62, 26.04)	(24.84, 25.88)	(24.34, 25.71)	(24.25, 25.51)
(24.50, 25.88)	(24.49, 25.73)	(25.20, 26.23)	(24.64, 26.30)	(24.73, 26.16)	(25.16, 26.30)	(24.66, 25.87)	(25.20, 26.59)
(24.44, 25.85)	(24.29, 25.29)*	(24.46, 25.94)	(24.46, 25.45)	(24.66, 25.78)	(24.07, 25.62)	(24.48, 25.88)	(24.99, 26.61)
(24.41, 25.97)	(25.04, 26.23)	(24.32, 25.57)	(24.76, 25.85)	(24.60, 26.03)	(25.04, 26.50)	(24.39, 26.27)	(24.97, 26.13)
(25.41, 26.81)*	(24.17, 25.61)	(25.10, 26.48)	(24.78, 26.18)	(24.85, 26.01)	(24.87, 25.88)	(24.62, 26.00)	(25.21, 26.39)
(24.98, 26.33)	(24.11, 26.09)	(24.74, 26.05)	(24.11, 25.65)	(24.47, 25.83)	(24.85, 26.32)	(24.37, 25.63)	(24.66, 25.95)

(24.83, 26.27)	(25.07, 26.29)	(25.24, 26.56)	(24.13, 25.65)	(25.41, 26.47)*	(24.64, 26.04)	(24.42, 25.94)	(24.61, 25.61)
(25.07, 26.38)	(24.90, 26.44)	(24.44, 25.72)	(24.63, 25.95)	(25.28, 26.37)	(24.63, 25.96)	(24.63, 25.71)	(24.98, 26.38)
(24.61, 26.21)	(24.93, 26.12)	(24.41, 25.79)	(24.95, 26.06)	(24.92, 25.88)	(24.69, 26.10)	(24.84, 26.22)	(25.40, 26.67)
(24.41, 25.51)	(24.93, 26.11)	(24.90, 25.91)	(25.18, 26.44)	(24.90, 26.48)	(25.06, 26.12)	(25.23, 26.18)	(24.87, 26.06)
(24.36, 25.85)	(24.45, 25.68)	(24.84, 25.91)	(24.83, 26.40)	(24.69, 26.04)	(24.68, 26.11)	(24.95, 26.17)	(24.94, 26.00)
(24.56, 25.98)	(25.19, 26.24)	(24.27, 25.82)	(24.98, 26.00)	(24.85, 25.98)	(25.00, 25.82)	(24.72, 25.99)	(24.98, 26.48)
(25.19, 26.85)	(24.38, 25.92)	(24.59, 26.19)	(24.76, 26.42)	(24.56, 26.00)	(24.57, 26.29)	(24.60, 26.12)	(24.99, 26.18)
(24.23, 26.01)	(24.28, 25.56)	(24.14, 25.37)*	(24.64, 26.29)	(24.96, 26.40)	(24.89, 26.33)	(24.43, 25.11)*	(24.88, 26.53)
(25.17, 26.78)	(24.52, 26.20)	(24.29, 25.86)	(24.26, 25.97)	(24.74, 26.17)	(24.81, 26.20)	(24.97, 26.15)	(24.85, 26.43)
(24.42, 25.86)	(24.61, 26.04)	(25.13, 26.18)	(24.09, 25.73)	(24.83, 26.19)	(25.18, 26.53)	(25.01, 26.66)	(24.52, 25.97)
(25.43, 26.53)*	(25.00, 26.32)	(24.86, 25.70)	(24.52, 25.65)	(24.67, 26.00)	(25.17, 26.69)	(24.73, 25.98)	(24.58, 25.93)
(24.85, 26.42)	(25.49, 26.78)*	(25.52, 26.38)*	(24.14, 25.61)	(25.03, 26.43)	(24.38, 26.29)	(25.16, 26.73)	(24.92, 26.35)
(24.24, 26.00)	(24.00, 25.11)*	(24.85, 26.03)	(24.73, 26.02)	(25.19, 26.92)	(24.70, 26.18)	(24.85, 26.40)	(24.84, 26.31)
(24.75, 26.11)	(24.82, 26.07)	(24.62, 25.84)	(25.42, 26.48)*	(24.75, 26.03)	(24.32, 25.70)	(25.19, 26.45)	(24.71, 26.31)
(24.42, 25.94)	(24.71, 26.39)	(24.15, 25.84)	(24.72, 25.98)	(25.22, 26.51)	(24.16, 25.76)	(25.18, 25.94)	(25.04, 26.87)
(24.50, 25.73)	(24.96, 26.19)	(25.35, 26.41)	(24.58, 25.92)	(24.23, 25.64)	(25.18, 26.28)	(24.36, 26.29)	(24.18, 25.74)
(25.11, 26.24)	(24.14, 25.70)	(24.94, 25.85)	(24.59, 25.96)	(25.02, 26.60)	(24.67, 26.26)	(24.83, 26.38)	(24.43, 25.38)*
(25.06, 26.28)	(24.26, 25.69)	(24.30, 25.42)	(25.03, 26.56)	(24.47, 26.10)	(24.52, 25.77)	(25.12, 26.44)	(25.55, 26.84)*
(24.53, 26.19)	(24.77, 26.25)	(25.15, 26.49)	(24.96, 26.12)	(25.13, 26.32)	(24.69, 25.99)	(24.63, 26.15)	(25.37, 26.48)
(24.56, 26.21)	(24.44, 25.76)	(24.27, 25.50)	(25.09, 26.03)	(24.76, 26.01)	(24.63, 26.14)	(24.36, 25.59)	(24.36, 26.04)
(24.82, 25.75)	(24.68, 25.98)	(25.01, 26.78)	(24.83, 26.36)	(25.39, 26.46)	(24.39, 25.52)	(25.07, 26.11)	(24.97, 26.26)
(25.05, 26.31)	(24.89, 26.20)	(24.12, 25.66)	(24.79, 26.43)	(24.98, 26.15)	(25.05, 26.57)	(24.81, 26.08)	(24.32, 25.69)
(24.76, 26.37)	(25.30, 26.54)	(25.05, 26.20)	(24.38, 25.72)	(24.89, 26.37)	(24.34, 25.68)	(24.64, 25.75)	(24.54, 26.08)
(24.52, 26.21)	(24.85, 25.86)	(25.16, 25.95)	(24.71, 25.92)	(24.65, 25.36)*	(25.00, 26.48)	(24.58, 25.61)	(24.66, 26.26)
(24.74, 25.92)	(24.45, 25.68)	(24.89, 26.23)	(24.80, 26.04)	(24.84, 26.24)	(24.15, 25.62)	(25.48, 27.00)*	(25.00, 26.34)
(24.71, 26.55)	(25.01, 26.44)	(24.67, 26.20)	(24.94, 26.35)	(24.79, 26.36)	(24.98, 26.31)	(24.61, 26.21)	(23.94, 25.88)
(24.55, 25.94)	(24.79, 26.08)	(24.94, 25.90)	(24.73, 25.87)	(24.40, 25.68)	(25.24, 26.39)	(25.18, 26.61)	(24.42, 25.61)
(24.36, 25.77)	(24.35, 25.70)	(24.85, 26.06)	(24.56, 25.51)	(24.23, 25.69)	(24.41, 25.71)	(24.52, 25.89)	(24.53, 25.55)
(25.57, 26.72)*	(24.85, 26.11)	(24.98, 26.41)	(24.71, 26.00)	(25.26, 26.38)	(24.71, 26.36)	(24.66, 25.99)	(24.89, 25.86)

(continued)

Table E.1 ***(continued)***

(25.46, 26.24)*	(25.25, 26.39)	(24.70, 26.22)	(24.59, 25.88)	(24.43, 25.54)	(25.09, 26.50)	(24.83, 26.22)	(24.87, 26.11)
(24.33, 26.14)	(25.00, 26.19)	(23.88, 25.22)*	(24.42, 26.05)	(24.68, 26.02)	(24.93, 26.18)	(24.85, 26.20)	(25.52, 26.77)*
(25.00, 25.95)	(24.51, 26.03)	(24.24, 25.97)	(24.92, 26.35)	(24.45, 26.33)	(24.94, 26.08)	(24.07, 26.38)	(24.84, 26.27)
(24.95, 26.22)	(24.94, 26.28)	(24.75, 26.00)	(24.78, 26.37)	(25.45, 27.04)*	(24.20, 25.91)	(24.71, 26.08)	(24.14, 25.83)
(25.14, 26.48)	(24.63, 26.28)	(24.91, 25.91)	(25.08, 26.23)	(24.47, 25.85)	(24.54, 26.04)	(24.95, 26.74)	(25.00, 26.40)
(24.60, 25.65)	(24.75, 26.31)	(24.23, 26.01)	(24.94, 26.36)	(24.66, 26.01)	(23.84, 25.91)	(24.78, 26.42)	(25.13, 26.34)
(24.97, 26.37)	(24.42, 26.73)	(24.63, 25.53)	(24.48, 25.73)	(24.44, 25.66)	(24.14, 25.11)*	(24.84, 26.33)	(25.05, 26.14)
(24.18, 25.96)	(24.59, 25.61)	(24.73, 25.86)	(24.63, 25.97)	(24.09, 25.90)	(25.47, 26.45)*	(24.69, 25.92)	(24.58, 25.77)
(24.67, 25.80)	(24.59, 26.10)	(23.87, 25.38)*	(24.93, 26.46)	(24.69, 26.16)	(24.66, 25.94)	(24.48, 25.95)	(24.33, 26.04)
(24.30, 25.75)	(25.33, 26.18)	(24.18, 26.00)	(24.58, 26.18)	(25.04, 26.25)	(24.33, 26.15)	(24.58, 26.13)	(24.32, 25.70)
(24.39, 25.48)	(24.89, 26.44)	(24.79, 26.14)	(24.62, 26.04)	(25.29, 26.58)	(24.81, 26.29)	(24.35, 25.28)*	(25.32, 26.60)
(24.04, 25.44)	(24.83, 26.22)	(24.43, 25.72)	(24.38, 26.20)	(24.63, 25.63)	(24.40, 26.17)	(25.09, 26.50)	(24.93, 26.45)
(24.89, 26.28)	(24.05, 25.43)	(24.78, 26.06)	(24.50, 25.94)	(24.54, 26.01)	(24.97, 25.89)	(24.58, 25.87)	(24.86, 26.11)
(24.76, 26.25)	(24.37, 25.53)	(24.85, 25.86)	(24.26, 25.48)	(25.14, 26.41)	(24.17, 25.84)	(24.77, 26.03)	(25.26, 26.53)
(25.08, 26.29)	(24.60, 26.08)	(25.24, 26.50)	(24.79, 26.10)	(25.17, 26.53)	(24.85, 25.88)	(24.79, 26.18)	(24.64, 26.11)
(24.52, 25.99)	(24.89, 26.14)	(24.29, 25.84)	(24.68, 25.99)	(24.41, 26.17)	(25.18, 26.57)	(24.10, 25.54)	(25.23, 26.47)
(24.97, 26.44)	(24.59, 25.63)	(24.80, 26.00)	(24.91, 26.29)	(24.56, 25.92)	(24.65, 26.02)	(24.34, 25.42)	(24.86, 26.32)
(24.54, 25.79)	(25.13, 26.68)	(24.60, 25.78)	(24.49, 25.87)	(24.94, 26.19)	(25.13, 26.39)	(24.47, 25.60)	(24.61, 25.77)
(24.74, 26.67)	(25.03, 26.54)	(24.69, 26.59)	(24.84, 26.37)	(24.81, 26.35)	(24.64, 25.95)	(24.89, 26.07)	(24.50, 25.91)
(24.34, 25.83)	(25.10, 26.12)	(24.87, 26.48)	(25.00, 26.54)	(24.64, 26.00)	(24.71, 26.09)	(24.43, 26.05)	(24.54, 25.69)
(24.37, 25.61)	(24.49, 25.97)	(24.48, 25.60)	(24.65, 26.73)	(25.18, 26.41)	(25.09, 26.44)	(24.95, 26.40)	(24.64, 26.01)
(24.58, 25.81)	(24.47, 25.66)	(24.39, 25.72)	(24.12, 25.71)	(24.58, 25.85)	(25.77, 27.28)*	(24.67, 25.86)	(24.78, 26.43)
(24.44, 26.16)	(24.70, 26.25)	(24.48, 25.68)	(24.32, 25.64)	(24.65, 25.50)	(24.75, 26.21)	(24.25, 25.69)	(24.10, 25.83)
(24.72, 26.08)	(24.73, 26.17)	(25.44, 26.34)*	(24.43, 25.75)	(25.27, 26.74)	(24.13, 25.49)	(25.04, 26.28)	(24.63, 25.80)
(24.27, 25.94)	(24.93, 26.73)	(24.43, 25.84)	(24.29, 25.52)	(24.55, 26.35)	(25.13, 26.54)	(24.27, 25.63)	(25.17, 26.45)
(25.09, 26.67)	(24.44, 25.88)	(24.59, 26.63)	(24.92, 26.37)	(25.10, 26.29)	(24.57, 26.27)	(24.49, 25.90)	(24.76, 26.08)
(24.64, 26.13)	(24.46, 25.84)	(24.31, 25.89)	(25.34, 26.66)	(24.87, 26.31)	(24.17, 25.47)	(25.28, 26.57)	(24.75, 26.14)
(24.44, 26.00)	(24.46, 25.63)	(24.75, 26.33)	(25.03, 26.03)	(24.87, 26.36)	(24.45, 25.83)	(24.85, 26.15)	(24.12, 25.75)

(25.53, 26.94)*	(23.89, 25.24)*	(24.78, 26.12)	(24.95, 26.19)	(24.81, 26.05)	(24.76, 26.13)	(24.38, 25.77)	(24.45, 26.24)
(25.72, 26.82)*	(24.36, 25.75)	(24.63, 26.25)	(24.65, 26.19)	(25.04, 25.80)	(25.01, 26.66)	(24.48, 25.96)	(24.88, 26.48)
(24.45, 26.06)	(25.19, 26.42)	(24.64, 26.18)	(24.95, 26.25)	(24.43, 26.35)	(24.89, 26.11)	(24.19, 25.61)	(24.52, 26.26)
(24.95, 26.13)	(25.17, 26.46)	(24.70, 26.19)	(24.59, 25.94)	(24.71, 26.06)	(24.84, 26.45)	(24.41, 25.60)	(24.76, 25.96)
(24.83, 26.42)	(24.62, 26.12)	(24.39, 25.41)	(25.18, 26.22)	(24.75, 26.33)	(24.47, 26.37)	(24.45, 25.74)	(25.19, 26.37)
(24.75, 26.15)	(25.33, 26.63)	(24.69, 25.90)	(24.67, 26.29)	(24.89, 26.24)	(24.35, 26.27)	(24.95, 26.01)	(24.77, 25.94)
(24.69, 25.95)	(24.74, 26.22)	(24.61, 26.15)	(25.10, 26.15)	(24.95, 26.40)	(24.46, 26.08)	(25.07, 26.68)	(25.45, 26.28)*
(24.88, 26.52)	(25.04, 26.48)	(24.71, 25.90)	(24.86, 26.18)	(24.45, 25.96)	(24.82, 25.94)	(24.61, 26.09)	(24.87, 26.67)
(24.32, 25.93)	(24.74, 26.00)	(24.42, 25.72)	(25.14, 26.53)	(24.74, 25.97)	(24.35, 26.08)	(24.70, 25.75)	(25.11, 26.29)
(24.55, 25.95)	(24.33, 25.56)	(24.22, 25.61)	(24.59, 26.25)	(25.21, 26.31)	(25.03, 26.40)	(25.09, 26.54)	(24.58, 25.72)
(25.18, 26.27)	(24.10, 25.13)*	(24.64, 25.69)	(25.16, 26.15)	(24.95, 26.03)	(24.91, 26.08)	(24.83, 25.95)	(24.49, 25.69)
(24.60, 26.16)	(24.83, 26.07)	(25.63, 26.64)*	(24.77, 26.01)	(24.41, 25.84)	(25.38, 26.21)	(24.53, 25.86)	(24.17, 25.80)
(24.69, 25.83)	(25.00, 26.41)	(25.06, 26.19)	(24.32, 26.34)	(25.11, 26.42)	(23.99, 25.83)	(24.21, 25.43)	(24.84, 25.99)
(24.89, 26.16)	(24.99, 26.73)	(24.85, 26.24)	(24.91, 26.12)	(24.79, 26.03)	(24.52, 25.83)	(24.74, 26.21)	(24.37, 26.01)
(24.95, 26.36)	(24.37, 25.92)	(24.81, 26.10)	(24.89, 26.40)	(24.80, 25.90)	(24.60, 26.12)	(24.84, 26.32)	(25.02, 26.19)
(24.84, 25.95)	(25.06, 26.51)	(25.09, 25.96)	(25.44, 26.34)*	(25.09, 26.27)	(25.19, 26.86)	(24.97, 26.15)	(25.31, 26.21)
(24.65, 26.09)	(24.96, 26.36)	(24.88, 26.18)	(24.94, 26.05)	(24.70, 25.78)	(25.10, 26.22)	(24.63, 26.06)	(24.77, 26.36)
(24.86, 26.81)	(24.22, 25.71)	(24.98, 26.46)	(24.29, 25.43)	(24.53, 26.00)	(24.58, 26.31)	(24.27, 25.58)	(25.25, 26.68)
(24.65, 26.09)	(25.23, 26.49)	(25.25, 26.83)	(24.97, 25.94)	(24.88, 26.13)	(24.37, 25.41)	(25.37, 26.31)	(24.80, 26.43)
(24.73, 26.32)	(24.42, 25.81)	(24.63, 25.70)	(24.66, 25.43)	(25.58, 26.41)*	(24.22, 26.04)	(24.43, 26.08)	(24.81, 26.14)
(25.14, 26.47)	(24.92, 26.19)	(24.48, 25.91)	(25.13, 26.42)	(24.15, 25.45)	(24.84, 26.14)	(24.77, 26.08)	(25.22, 26.36)
(24.27, 25.36)*	(24.37, 25.56)	(25.35, 26.84)	(24.73, 26.10)	(24.80, 26.20)	(24.35, 25.81)	(24.30, 25.74)	(24.15, 25.64)
(24.70, 26.04)	(24.88, 26.14)	(24.45, 25.86)	(24.30, 26.08)	(24.66, 25.43)	(24.46, 26.12)	(24.38, 25.55)	(25.13, 26.01)
(24.60, 26.25)	(25.52, 26.51)*	(24.68, 26.20)	(24.87, 26.18)	(24.57, 25.65)	(25.60, 26.77)*	(24.91, 26.44)	(24.88, 26.14)
(24.20, 25.79)	(24.52, 25.70)	(25.58, 26.83)*	(25.15, 25.94)	(24.87, 26.34)	(24.70, 26.12)	(24.40, 25.56)	(24.93, 26.21)
(24.62, 25.87)	(24.54, 26.05)	(24.07, 25.80)	(25.01, 26.40)	(25.08, 26.52)	(24.15, 25.80)	(24.59, 26.22)	(24.65, 25.90)
(25.21, 26.21)	(24.85, 26.46)	(25.22, 26.74)	(25.07, 26.45)	(24.05, 25.59)	(24.08, 25.41)	(24.24, 25.36)*	(24.67, 26.08)
(24.40, 25.75)	(25.29, 26.31)	(24.51, 26.07)	(24.80, 26.23)	(24.82, 26.58)	(24.81, 26.06)	(24.74, 26.04)	(24.67, 25.76)
(24.79, 25.71)	(23.90, 25.01)*	(24.61, 26.39)	(24.89, 26.53)	(24.86, 26.49)	(24.54, 25.86)	(24.54, 26.25)	(24.67, 25.82)

(continued)

Table E.1 ***(continued)***

(24.22, 25.61)	(24.39, 26.25)	(24.81, 26.08)	(25.41, 26.35)*	(24.62, 25.97)	(25.02, 26.13)	(24.39, 25.76)	(24.53, 25.72)
(24.44, 25.73)	(24.75, 26.44)	(25.03, 25.84)	(24.53, 26.10)	(25.05, 26.12)	(25.21, 26.39)	(24.37, 25.81)	(25.16, 26.25)
(24.46, 25.86)	(24.92, 26.30)	(24.65, 26.25)	(24.85, 26.22)	(24.22, 25.46)	(25.24, 26.42)	(24.63, 25.97)	(24.43, 25.65)
(24.92, 25.76)	(24.65, 26.04)	(24.96, 26.33)	(24.97, 26.55)	(25.30, 26.55)	(25.15, 26.62)	(24.82, 25.93)	(25.05, 26.51)
(25.02, 26.02)	(25.01, 26.38)	(24.71, 26.19)	(24.45, 25.26)*	(24.55, 26.16)	(24.90, 26.09)	(24.71, 26.29)	(25.14, 26.43)
(24.98, 26.21)	(24.78, 26.11)	(24.82, 26.44)	(24.75, 26.34)	(25.21, 26.27)	(24.55, 25.71)	(24.94, 26.42)	(24.34, 25.64)
(25.13, 26.29)	(24.36, 25.74)	(24.38, 25.42)	(24.83, 26.40)	(24.23, 25.42)	(24.34, 25.89)	(24.29, 26.05)	(24.01, 25.61)
(24.80, 26.15)	(24.77, 26.08)	(24.52, 26.03)	(24.42, 25.97)	(25.47, 26.93)*	(24.56, 25.93)	(24.53, 25.74)	(24.72, 26.14)
(25.34, 26.62)	(24.29, 25.38)*	(24.89, 26.23)	(24.89, 26.19)	(24.85, 26.08)	(24.84, 26.24)	(24.89, 26.15)	(24.60, 25.76)
(25.38, 26.61)	(25.28, 26.67)	(24.76, 26.06)	(25.10, 26.19)	(24.65, 25.90)	(24.55, 26.19)	(24.96, 26.05)	(24.48, 25.98)

APPENDIX F

DOCUMENTATION FOR SOFTWARE

F.1 INTRODUCTION

This section provides documentation for three computer programs: STATS, ADJUST, and MATRIX. Each of these has valuable applications in analyzing errors and performing adjustments with measured data. Many end-of-chapter problems in this book call for the use of these programs. All three software programs are DOS compatible and will run on any 286 or higher machine with a math coprocessor. They each have varying disk requirements but are designed to run from either a disk drive or a hard disk. For those wishing to install these programs on their hard disk, an *INSTALL* program has been provided. Since these programs are not compressed in any manner, they may also be copied to the hard disk or another diskette. Each program has different system memory requirements; although all will run in less than 640 kilobytes. To install files do the following:

DOS: At the DOS prompt, type A:\INSTALL.
Windows 3.1: In Program Manager, choose File, Run.
Type A:\INSTALL and click OK.

Icons have been provided for users who want to create Program Items and Groups in the Windows environment.

To run the programs from your hard drive in DOS, type the program file name (adjust.exe, matrix.exe, or stats.exe) at the C:\ADJCOMPS> prompt. To run the programs in Windows, double click on the program file name (adjust.exe, matrix.exe, or stats.exe) in the ADJCOMPS directory in File Manager.

F.2 SIMILARITIES IN THE PROGRAMS STATS AND ADJUST

Two of the programs have similar user interfaces. Since these program are so similar in their screen methods, this section was developed to familiarize users with the methodologies employed in the software. In addition to this common section, items that are specific to the programs are covered in detail in Sections F.3 and F.4.

F.2.1 Screen Components

The three visible components of the programs are shown in Figure F.1. They are the *menu bar* at the screen's top, the *desktop*, and the *status line* at the screen's bottom. The time of day given by the computer's clock and the available memory are shown on the right top and bottom of the screen, respectively.

F.2.2 Menus

The menu bar is the primary means of access to all commands. A highlighted menu title indicates that the item is active. If the menu command is followed by a (▶), the command leads to another menu. A command followed by a (...) leads to a dialogue box. A command followed by neither a (▶) or (...) leads directly to an action. Some commands will be lighter in color, indicating that they are not currently active. No action will occur by selecting these commands. These commands become active given some conditions. For example, the cut command will not become active until a block of text is selected.

Here's how to choose a menu item:

1. Press *Alt-F10* or *F10*. This makes the menu bar active.

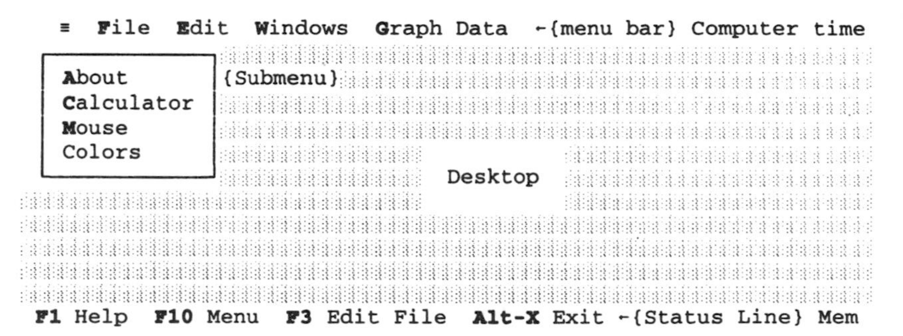

Figure F.1 Stats screen components.

2. Use the arrow keys to select from the menu, and press *Enter*; or as a shortcut to this step, just press the highlighted letter of the menu title. For example, from the menu bar, press *F* or *Alt F* to select the **File** menu.
3. Use the arrow keys (↑↓) to select the desired option and press *Enter*, or alternatively, press the highlighted letter to select the option. For example, press *O* to **O**pen a file or press *F3* from anywhere to open a file. As shown in Figure F.5, the **F3** key is indicated in the **File** menu as a hot key to the option **Open**.

At this point, the program either carries out the command, displays a dialogue box, or displays another menu.

Mouse Support. If a mouse is available, it can be used to choose commands. The process is:

1. Click the desired menu title to display the menu.
2. Click the desired command.
3. Alternatively, click the appropriate menu and drag the highlight bar to the desired menu item.

Hot Keys. The program offers a number of shortcuts to commands. These are highlighted to the right of the menu item and in the status line. Some commands can be accessed anywhere in the programs, and some are active only when in a menu or dialogue box.

F.2.3 Windows Menu

The editing screens are displayed in windows (memory allowing) with only one window *active* at a time. The active window is the top window. Any command selected or text typed applies only to the active window. Windows can be resized, moved, zoomed, tiled, or cascaded using the commands listed in Figure F.2. Some of these commands are also available using *hotkeys*. That is, to zoom a window, click your left mouse button in the window to make it the active window, and then press *F5* on the keyboard. These hot keys are also shown in Figure F.2 There are several types of windows, but most share the items shown in Figure F.3.

Menu Commands. The windows menu provides control of the edit windows. Windows may be zoomed, selected, moved, or resized. Windows will cover the entire screen by selecting *Zoom* from the menu or pressing *F5*. Change windows by pressing the key combination *Alt-#*, where # represents the window number (numbers are displayed only in the first nine windows)

```
Windows

Size/move  Ctrl-F5
Zoom            F5
Tile
Cascade
Next            F6
Previous  Shift-F6
Close       Alt-F3
```

Figure F.2 Windows menu.

or by selecting *Next* and *Previous* from the menu. A window can be sized by dragging the *resize corner* in the lower right corner of the window, or moved by selecting the *size/move* option and using the *Ctrl-arrow keys*. A window also can be moved by clicking and dragging the mouse on the upper left corner of the window. If more than one window is open, they may be tiled or cascaded. Tiled and cascaded windows can be selected by clicking the mouse on any portion of the window. Close windows by clicking the [·] in the upper left corner of the window or by pressing *Alt-F3*. If the file in the window was modified, the user will be prompted to save the file before closing the window.

F.2.4 Dialogue Boxes

If a menu or command requires further instructions from the user, a dialogue box will appear. When working in the dialogue box, most items may be

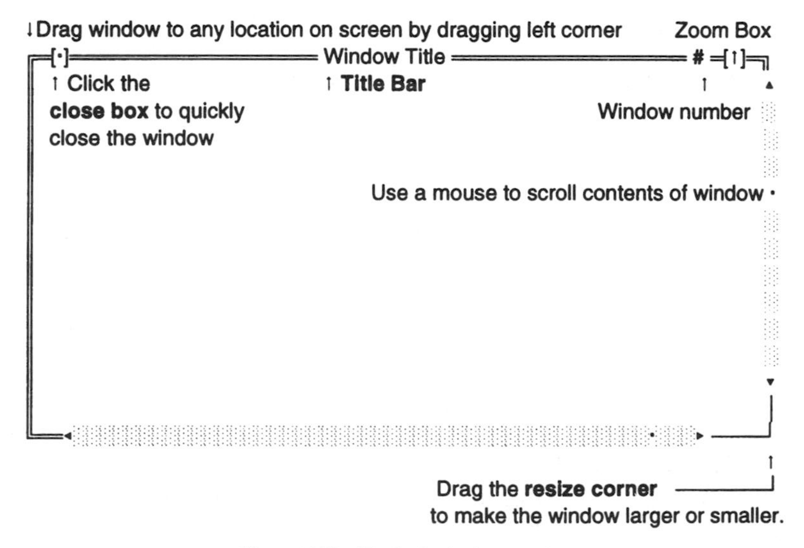

Figure F.3 Typical window parts.

selected using either radio buttons (·), check boxes **[X]**, input boxes, or action buttons. Each item is selected by using the **Tab** →| or shift tab keys, by *hot keys,* or by clicking the mouse on the item.

F.2.5 System

The system menu contains information *about* the package, a *calculator*, and options for *mouse* and *color* screen support. Note that the mouse and color options will return to their original settings between work sessions. Press *Alt-space bar* to activate this menu.

Mouse Options

1. The mouse buttons may be reversed and speed changed during current work session.

Color Options

1. Change colors in the package during current work session.

Note that the text portion of the screen will show the current setting for foreground and background colors. Care should be used when selecting a different color combination.

F.2.6 Help (F1)

Context-sensitive help is available at any time during a work session by pressing **F1**. To select specific help on an option, highlight the menu option and press **F1** or select **Help Index** under the system menu, '≡'. Highlighted items are windows to related items in the context-sensitive help screen. An example help screen is shown in Figure F.4.

F.2.7 File

The file menu (Figure F.5) provides the file commands:

1. **O**pen an existing or **N**ew file. A file may be opened whether or not it exists. Using the new command, an empty, untitled edit window will appear. With the *New* command, a file name is required when the window is closed or saved.

   A file dialogue (Figure F.6) will appear to **O**pen a file. A file can be selected by typing in its name or by selecting the file from the directory list. A history of the most recently selected files can be viewed and selected by depressing the [↓] key. By typing in the full path to the file

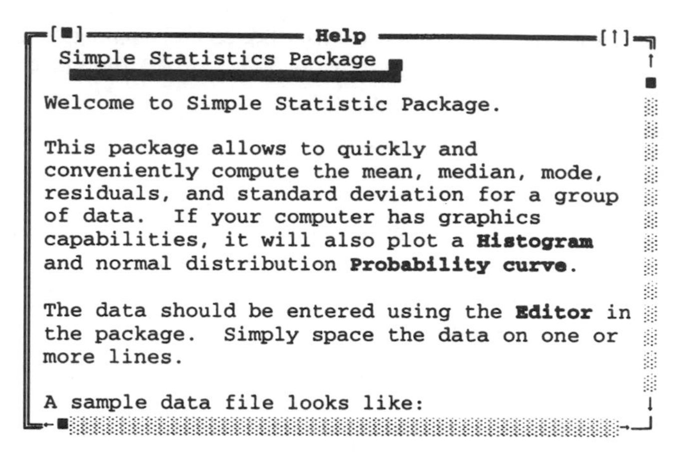

Figure F.4 Typical help screen.

or selecting either the parent (..\) or child directories, it is possible to select a file in a different directory.

2. **S**ave (S**a**ve as...) a file. A file can be saved at any time by pressing the **F2** key. When a file is printed or statistics are computed, the active edit window file will be saved automatically.
3. **E**rase... a file. Enables a file to be erased by selecting it from a file dialogue box. A prompt will ask for confirmation of the file erasure.
4. **W**here is... a file. Search for a file or set of files on current drive using any legal DOS file specification. For instance, to find files with the "DAT" extensions, the file specification of **.DAT* should be provided at the prompt. A window of files will be displayed. Up to 50 files will be displayed in the window.

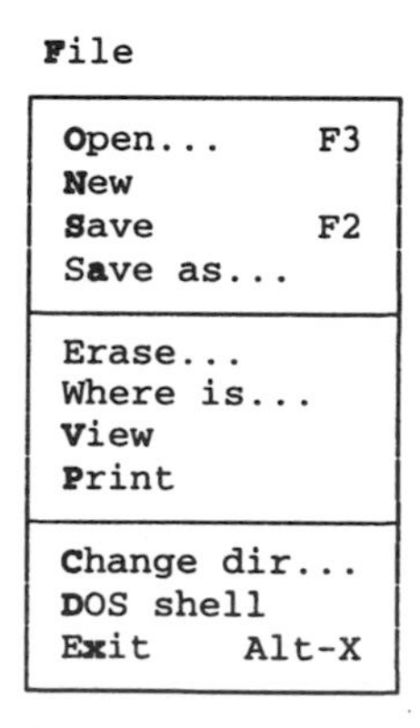

Figure F.5 File menu.

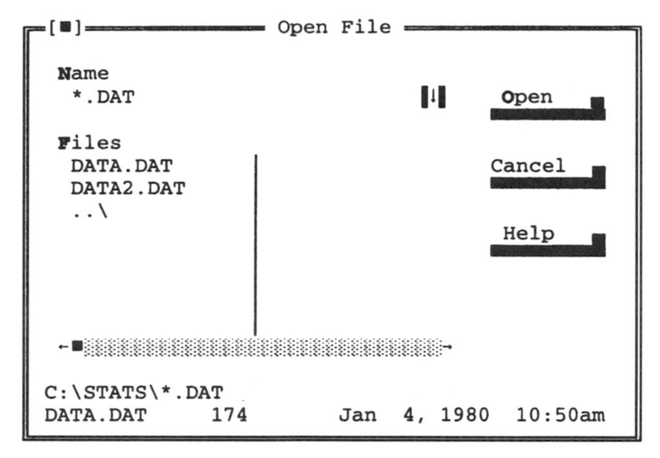

Figure F.6 File open dialogue.

5. **V**iew a DOS file. View any DOS file in a scrolling window by selecting the view option.
6. **P**rint a file. This program offer access to selected printer options that are saved between printings. Note that not all printer options shown in Figure F.7 are supported by every printer, and thus the user must be aware of which options are available for the printer bcing used. To view the list of supported options, users should refer to the printer manual. The HP Laserjet II also supports several font cartridges. A dialogue listing the supported cartridges is presented when this printer is selected.

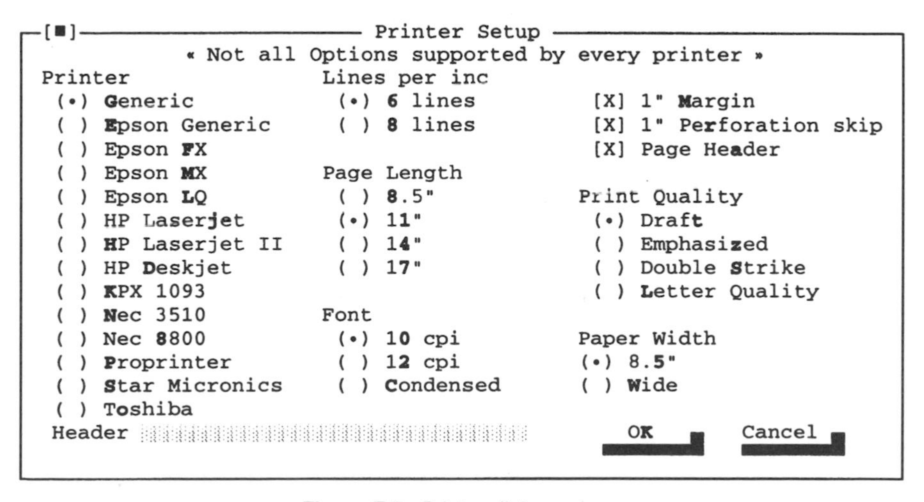

Figure F.7 Printer dialogue box.

For instance, select the **M** cartridge to have listings printed using the prestige elite fonts. Note that since many of these fonts have proportional spacing, formatted listings may lose their appearance. A file printing can be interrupted or discontinued by pressing the escape key, *Esc*.

7. **C**hange dir ... (change default directory) It is possible to change drives or directories on a drive from a tree structure shown in a dialogue box using this command. To change drives or directories, press arrow keys or click the left button of the mouse on the desired directory
8. **D**os shell. Temporarily leave the program to execute DOS commands with the DOS shell command. Due to memory constraints, not all DOS programs are accessible through the shell command. To return to the program, type EXIT at the computer's DOS prompt.
9. E**x**it. Leave the program using the E**x**it command. The hot key to this command is **Alt- X**. When leaving the programs, the software will check for open, modified edit windows. If any are found, the programs will provide a prompt to save these files.

F.2.8 Edit

The edit menu provides access to the editing commands shown in Figure F.8. Text can be blocked by placing the cursor at the beginning or end of the desired text segment, pressing *Ctrl-K-B*, and moving the cursor to the end or beginning of the block. The blocked text will be highlighted. Blocking text will activate the **c**ut and c**o**py menu options.

1. **S**earch menu. The search menu contains the following:
   a. **F**ind. Any text string.
   b. **R**eplace Text. Replace a text string with another string.
   c. **A**gain. Perform previous **F**ind or **R**eplace.
2. **U**ndo. With the **u**ndo command, previous editing changes can be reversed.

```
Edit
 ----------------------
| Search               |
| Undo                 |
|----------------------|
| Cut     Shift-Del    |
| Copy     Ctrl-Ins    |
| Paste   Shift-Ins    |
| Show clipboard       |
|----------------------|
| Clear    Ctrl-Del    |
 ----------------------
```

Figure F.8 Edit menu.

3. Cut/Copy. Text is cut or copied into the clipboard.
4. Show Clipboard. Edit cut or copied text by selecting *show clipboard.*
5. **P**aste. Text from the clipboard can be pasted back into the edit window. The text in the clipboard may be inserted into the edit window by placing the cursor at the desired location and selecting the **P**aste command.
6. Clear Clipboard. Clear the clipboard of any text.

F.3 STATS

STATS is a program that will determine the mean, median, mode, and standard deviation for a set of data, as well as the necessary data for both a histogram and normal distribution plot. The program can also compute the critical values for the *Student's t*, χ^2, and F distributions using any area between 0.0001 and 0.9999 and degrees of freedom between 1 and 200, and the *t value* for any percent probable error between 0.001% and 99.999% from the *standard normal distribution.* Documentation for the individual applications in STATS follows.

F.3.1 Compute Statistics

The data file for this package may be created by the editor in this package or any editor capable of creating ASCII files. Data should be separated by spaces or lines as shown in the sample data file at the end of this section. The applications contained in STATS are shown in Figure F.9 and are discussed below.

1. **S**tatistics (Alt-S). This option will save any modified, open edit window. A file dialogue box will be presented for entry of the data file name. The output file will have the data file name and directory with an '**.OUT**' extension. Following the file entry, the user will be asked for the number of class intervals to divide the data. It will then compute

```
Compute Statistics

Statistics ...          Alt-S
-----------------------------
Normal distribution ...
t-statistic ...
Chi-square statistic ...
F-statistic ...
```

Figure F.9 Statistics menu.

the mean, median, mode, standard deviation, range, residuals, frequency histogram, and normal distribution curve coordinates. The results of the computations will be written in an output file.

2. Normal Distribution. This option computes the *t value* for the any percent probable error between 0.001 and 99.999% for the standard normal error distribution.
3. ***t*** Statistic. This option computes the critical value for the *t* distribution function for a specified area (confidence level) and degrees of freedom (v) less than 200.
4. χ^2—Chi-Square Statistic. This option computes the critical value for the χ^2 distribution function for a specified area (confidence level) and degrees of freedom (v) less than 200.
5. **F** Distribution. This option computes the critical value for the *F* distribution function for a specified area (confidence level) with given numerator (v_1) and denominator (v_2) degrees of freedom less than 200.

Sample Data File

6.5 2.1 4.4 6.5 2.1 4.4 4.7 8.6 4.7 8.6 5.0 4.9 4.0 3.4 5.6 4.7 2.7
2.4 2.7 2.2 5.2 5.3 4.7 6.8 4.1 5.3 7.6 2.4 2.1 4.6 4.3 3.0 4.1 6.1
4.2 6.5 2.1 6.5 2.1 4.4 6.5 2.1 4.4 3.4 4.5 5.6 3.6 2.3 2.4 2.5 2.6

F.4 ADJUST

ADJUST is a program that contains several least-squares adjustment options. It uses the same interface as STATS explained in Section F.2 of this Appendix. The menu shown in Figure F.10 lists several least-squares adjustment packages available in ADJUST. The *coordinate computations* menu contains the forward, inverse, and intersection problems plus NAD 27 and NAD 83 state

```
Programs
+--------------------------------------+
| Coordinate computations            ▸ |
+--------------------------------------+
| Coordinate Transformations...      ▸ |
+--------------------------------------+
| Point fits                         ▸ |
+--------------------------------------+
| Estimated Errors...                ▸ |
+--------------------------------------+
| Check a horizontal data file         |
| Horizontal adjustment...             |
| Vertical adjustment...               |
| GPS data                           ▸ |
+--------------------------------------+
```

Figure F.10 Programs menu.

plane coordinate computations, universal transverse Mercator computations, and geocentric coordinate computations. In the *coordinate transformation* submenu, there are two-dimensional conformal, affine, and projective coordinate transformations plus a three-dimensional conformal coordinate transformation. These transformations can be done using either the standard or general least-squares methods of adjustment. The *point fits* submenu also has programs that will fit a set of coordinates to a line, circle, or parabola using both standard and general least-squares methods. The *estimated errors* sub menu contains programs that will compute estimated standard deviations for surveying observations using the methods discussed in Chapters 6 and 8 of this book. The *horizontal adjustment* option contains programs for least-squares adjustment of horizontal surveying observations. The *vertical adjustment* is a differential leveling least-squares adjustment. Finally, the *GPS* data option leads to the submenu options of *loop misclosure checks* and *baseline adjustment.*

F.4.1 Coordinate Computations

This submenu, shown in Figure F.11, contains several useful coordinate computations options that can be used to compute initial approximations for the horizontal least-squares adjustment. Included in this menu are the options for both *forward* and *inverse* plane coordinate computations, forward and inverse *state plane* coordinate computations for both NAD 27 and NAD 83, forward and inverse computations for the *universal transverse Mercator* projection, forward and inverse computations for geocentric coordinates using Clarke 1866, WGS 72, GRS 80, or WGS 84 ellipsoids, and several intersection computation options under the *coordinate geometry* submenu.

Notice to User of Adjust: ADJUST provides intelligent file reading. That is, the maximum number of decimal places entered by the user into a data file will determine the number of decimal places in the output. For example, in the horizontal least-squares adjustment, if the user enters the coordinates as whole numbers only, then the program will print the adjusted values in

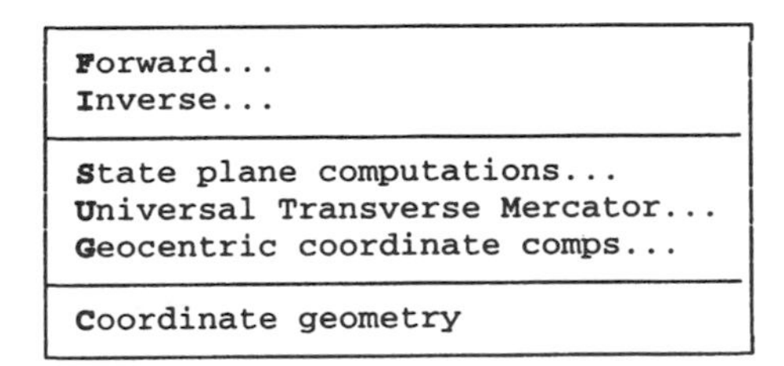

Figure F.11 Coordinate computations submenu.

whole numbers. Because of this feature, the user should be enter the number of decimal places that is desired in the final solution.

Forward. The dialogue box shown in Figure F.12 is presented to the user when the *forward* menu option is selected. This option will compute the coordinates of points for either a radial or a traverse survey. For traverse observations, the *traverse* option should be selected in the dialogue box. A full discussion of the options use for both types of surveys is given below. Select the *Save to file* option to save the computed coordinates in a file.

User note: All angles in this option are entered in DDD.MMSS format. That is, an angular value of 38°14′25.6″ should be entered as 38.14256.

In either radial or traverse surveys, the dialogue box shown in Figure F.12 will return (*cycle*) after each computation. Also in both options, the initial backsight azimuth can be entered by supplying coordinates of the backsight station or the azimuth of the backsight line. In radial surveys, this azimuth is held fixed for the remaining computation cycles. In traverse, the azimuth of the backsight station is advanced to the next station occupied (the station computed from the first cycle). For example, if station *B* is computed from station *A* in the first cycle, the azimuth value for the second cycle will be the direction of *BA*. Similarly in the traverse option, the occupied station will assume the coordinates of the foresight station. That is, if station *B* is the foresight station in the first computation cycle, station *B*'s coordinates will appear in the occupied station's position for the second cycle.

After the information for the dialogue shown in Figure F.12 is entered correctly, the *OK* button should be pressed. Then both distance and angle dialogue boxes will be presented. With some of the traverse examples given in the book, no backsight station coordinates or azimuth are given. In these cases, generally the first line of the traverse has a fixed azimuth. This azimuth can be entered in the program by either entering the backsight azimuth as

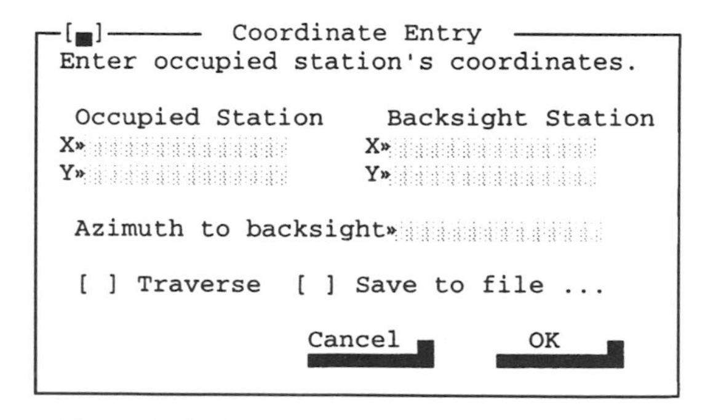

Figure F.12 Forward computations dialogue box.

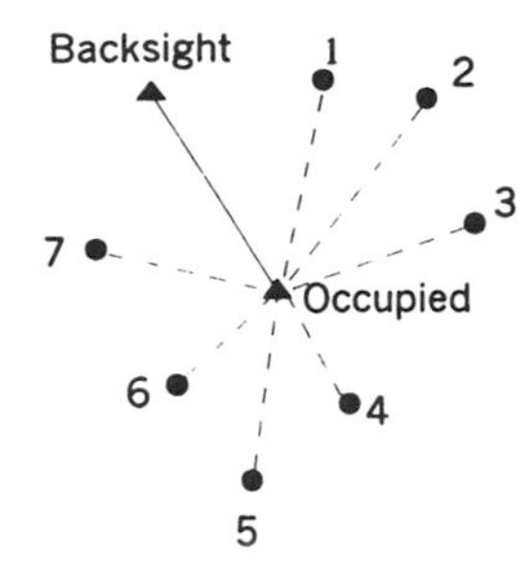

Figure F.13 Radial survey.

0.0000 and entering the first angle as the course azimuth, or by entering the course azimuth as the backsight azimuth and entering the first angle as 0.0000.

Radial Computations. In radial surveys the distances and angles to several stations are measured from a single occupied station. In this option the initial azimuth/backsight station and occupied station coordinate values remain fixed. For example, in Figure F.13, stations 1, 2, 3, 4, 5, 6, and 7 can be computed consecutively without ever needing to repeat entry of the occupied station coordinate values or the backsight coordinates/azimuth. Thus entry of distances and angles to each station measured can be continued.

Traverse Computations. As stated earlier, in traverse computations the occupied station and backsight azimuth are advanced along the traverse. Using Figure F.14 as an example, if station 1 is supplied as the starting station and the direction of course $1A$ as the backsight azimuth, the clockwise angle at station 1 from the backsight line $1A$ to station 2 and the distance of the course $\overline{12}$ would be entered. On the next computation cycle, the occupied station would be station 2, and the backsight azimuth would be that of course $\overline{21}$. At this point, angle 1–2–3 and course distance $\overline{23}$ would be entered to compute the coordinate values for station 3. Station 3 will then become the next

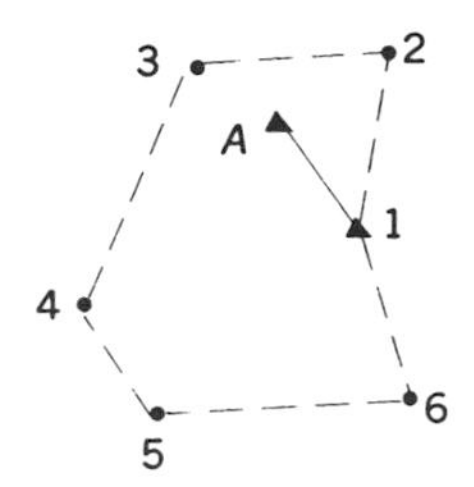

Figure F.14 Traverse survey.

```
┌─[■]── Geocentric Computations ──────────┐
│                                          │
│  ( ) Clarke 1866 - NAD 27                │
│  ( ) WGS 72                              │
│  ( ) GRS 80 - NAD 83                     │
│  (•) WGS 84                              │
│                                          │
│  ( ) File entry   (•) Screen entry       │
└──────────────────────────────────────────┘
```

Figure F.15 Ellipsoid selection in geocentric coordinate computations.

occupied station. This procedure can be repeated until the coordinates for stations 4, 5, and 6 are determined.

Inverse. The inverse option allows computation of the distance and azimuth of a line by supplying the coordinate value for an occupied and sighted station. As with the forward computations, the resulting computational values can be written to a file for printing.

State Plane Coordinates. As discussed earlier in the book, horizontal surveys that extend over a large areas must be transformed onto some map projection. ADJUST includes NAD 27, NAD 83, UTM 27, and UTM 83 map projections for this use. Both forward and inverse computational options are provided. The state plane projections include all 50 states plus some U.S. territories. Latitude and longitude are both entered as positive angles. The computed values include the mapping angle and scale factor for each station.

Geocentric Coordinate Computations. In this option the user can compute geocentric coordinates for a station from latitude, longitude, and ellipsoid height or do the inverse computations to get latitude, longitude, and ellipsoid height from geocentric coordinates. As seen in Figure F.15, the user can select from four different defined ellipsoids. This program can convert either a file of coordinates or a single point in screen entry. The file format for forward computations is shown in Table F.1, where each line has the format

ID, latitude, longitude, ellipsoid height

where **ID** is a maximum of 10 characters containing no commas or spaces;

Table F.1 Sample data file for forward geocentric coordinate computations

A	45	23	15.5678	− 92	15	36.0783	274.915
B	45	21	54.6894	− 92	13	34.5891	290.073
C	45	24	06.9813	− 92	14	32.5467	295.861

Table F.2 Sample data file for inverse geocentric cocordinate computations

A	−176951.58156	−4483726.79225	4517904.79298
B	−174380.10344	−4485616.73492	4516161.58287
C	−175526.89900	−4482666.53823	4519034.34493

latitude is given in degrees, minutes, and seconds format with positive numbers for the northern hemisphere and negative numbers for the southern hemisphere; **longitude** is given in degrees, minutes, and seconds format with positive numbers for the eastern hemisphere and negative numbers for the western hemisphere; and **ellipsoid height** is given as a real number in meters.

Similarly, the file format for the inverse computations is shown in Table F.2, where each line has the format of

ID, X, Y, Z

where **ID** is a maximum of 10 characters containing no commas or spaces, and X, Y, and Z are the geocentric coordinates for the station in meters.

Coordinate Geometry. The coordinate geometry option, shown in Figure F.16, provides the user with several intersection options. These options enable quick computation of initial approximations for horizontal survey least-squares adjustments. The use of dialogue boxes to enter the necessary measurement values is self-explanatory. All angular values should be entered in DDD.MMSS format. That is, an 47°56′12.5″ angle would be entered as 47.56125.

F.4.2 Coordinate Transformations

This menu, shown in Figure F.17, will transform points in one coordinate system to another system using the standard least-squares method as presented in Chapters 12 and 19. All three two-dimensional adjustments use data files having the same format. The conformal adjustment needs a minimum of two

```
Distance intersection...
Direction intersection...
Angle–Angle intersection...
Dir–Distance intersection...
Angle–Distance intersection..
Resection...
```

Figure F.16 Coordinate geometry options.

```
Coordinate Transformations

2D Conformal...
2D Affine...
2D Projective...
3D Conformal...

1 General LSQ 2D Conformal...
2 General LSQ 2D Affine...
3 General LSQ 2D Projective..
4 General LSQ 3D Conformal...
```

Figure F.17 Coordinate transformation submenu.

points with coordinates known in both systems. The affine transformation needs three points with coordinates known in both systems. The projective adjustment needs four points known in both systems. The data file format for the two-dimensional adjustments is

Id {10 char} X, Y [SX, SY], where X,Y are the coordinates of control points in the final system.

Id {10 char} x, y, where x,y are the original system coordinates for the control points.

Id {10 char} x y are the additional points to be transformed after the adjustment.

Sample Data File

```
Data for Two-Dimensional Conformal Transformation {Title}
4                                     # of common points}
1 −113.000 0.003                {Id, X, Y − control points}
3 0.001 112.993                   . [SX, SY necessary for]
5 112.998 0.003                   . [generalized adjustment]
7 0.001 −112.999                  .
1 0.7637 5.9603                   {Id, x, y [optional Sx,Sy]
3 5.0620 10.5407                 .[measure points]
5 9.6627 6.2430                   .
7 5.3500 1.6540                   .
16 1.7530 2.4127  {Id, x, y of points to be transformed}
17 5.3143 2.5193                  .
36 1.7463 9.3537                  .
```

All three two-dimensional coordinate transformations can also be adjusted using the general least-squares method. This procedure is covered in Chapter 21. For weighting, it requires standard deviations in both coordinate systems.

Format File

Title {String of 80 characters}.

Number of common points.

Id {10 char} X Y SX SY x y Sx Sy, where X, Y, SX, SY are the control point coordinates and their standard deviations in the final system and x, y, Sx, Sy are the control point coordinates and their standard deviations in the original system.

Id{10 char} x y are the additional points to be transformed after the adjustment.

Sample Data File

```
Data for Two-Dimensional Conformal Transformation {Title}
4                    {# of common points}
1 −113.000 0.003 0.002 0.002 0.7637 5.9603 0.005 0.005
3 0.001 112.993 0.002 0.002 5.0620 10.5407 0.005 0.005    .
5 112.998 0.003 0.002 0.002 9.6627 6.2430 0.005 0.005    .
7 0.001 -112.999 0.002 0.002 5.3500 1.6540 0.005 0.005    .
16 1.7530 2.4127     {Id, x, y of points to be transformed}
17 5.3143 2.5193           .
36 1.7463 9.3537           .
37 5.3290 9.4630           .
38 8.9250 9.5770           .
```

The three-dimensional coordinate transformation is used to transform points from an *x*, *y*, and *z* three-dimensional coordinate system to another *X*, *Y*, and *Y* system. A common scale factor is used in all directions. The adjustment requires a minimum of

Two control points known in both X and Y, plus

Three control points known in Z;

Options are available to do the standard least-squares adjustment as covered in Chapter 17 or use the general least-squares method as covered in Chapter 21. Both options use the same file format, with the exception of the standard deviations necessary for the control coordinates in the generalized method.

File Format. Note that Spaces and commas are treated as format delimiters in the file. Station Id's must not contain spaces or commas and are limited to 10 characters.

Title Line: 80 characters
Number of horizontal control, number of vertical control
Horizontal control points: Id, X,Y {generalized LSQ:SX, SY}
Vertical control points: Id, Z {generalized LSQ: SZ}
Points to transform: Id, x, y, z {optional: Sx, Sy, Sz}

Sample File

```
Three Dimensional Coordinate Transformation for use with ei
ther standard or generalized LSQ}
4 4
1 8941.52 6671.68 0.142 0.057 {Id, X, Y, SX, SY}
2 8815.15 5749.51 0.082 0.181
3 8510.00 7924.94 0.043 0.161
4 8383.76 6516.54 0.059 0.100
1 761.20 0.111 {Id, Z, SZ}
2 846.30 0.182
3 818.91 0.120
4 853.90 0.054
1 1094.89 820.09 809.72 0.1 0.1 0.1 {Id, x, y, z, Sx, Sy, Sz}
2 503.26 1598.69 917.68 0.1 0.1 0.1
3 2349.35 207.67 851.38 0.1 0.1 0.1
4 1395.32 1348.86 915.27 0.1 0.1 0.1
```

F.4.3 Point Fitting

This menu, shown in Figure F.18, will fit a set of points with two-dimensional coordinates x, y to the best-fit line, circle, or parabola using either the standard or general least-squares methods. These adjustments all use a common data file.

```
Line...
Circle...
Parabola...
--------------------------
1 General LSQ line...
2 General LSQ circle...
3 General LSQ parabola...
```

Figure F.18 Point fit submenu.

File Format

Id {10 char}, X, Y, (Sx, Sy) {No spaces or commas in the point Id} One point per line.

Sample File

```
A 3.00  4.50  0.020  0.015  {standard  deviations  are  op-
tional}
B 4.25 4.25 0.023 0.036
C 5.50 5.50 0.033 0.028
D 8.00 5.50 0.016 0.019
```

F.4.4 Estimated Errors

This program computes the estimated errors for either horizontal or vertical data files using the equations discussed in Chapters 6 and 8. Upon selection of the file, the user will be presented with a dialogue box to enter the appropriate estimated errors for the file type. The program will create a new observation file that is appropriate for a least-squares adjustment. The file format for a horizontal file is

Title line {80 characters}
#-distances #-angles #-azimuths #-control stations #-stations
Control stations
  ID {10 characters with no spaces} X Y {optional: Sx Sy}
Unknown stations
  ID {10 characters with no spaces} X Y {optional: Sx Sy}
Distance observations
  From To Distance
Angle observations
  backsight occupied foresight Angle (° ′ ″)
Azimuth observations
  From To Azimuth (° ′ ″)σ″

F.4.5 Check a Horizontal Data File

This program compares the difference between the measured values for the observations and their computed values against a user-supplied mutliplier times the standard deviation of the observation. This procedure is used to find large errors in horizontal data files prior to a least-squares adjusment and is discussed in detail in Chapter 20. To use the option, the user must supply the name of the horizontal data file, which has the format discussed in Section F.4.6., and the multiplication factor to use in this comparison. Any computed difference that is greater than this factor is flagged with NO in the PASSED column of the output file.

F.4.6 Horizontal Adjustment

This program performs a least-squares adjustment on a file of horizontal measurements, with the adjustment results being placed in a file having an ".OUT" extension. (See Chapters 13 through 21 for discussions on this topic.) The program allows the user to set several adjustment options, shown in Figure F.19. These options perform the following functions: (1) set the maximum number of iterations; (2) adjust the control stations; (3) compute adjusted observation standard deviations; (4) compute either standard error ellipses or ellipses at any alpha confidence level; (5) do a chi-squared test of the reference variance; (6) statistically check for blunders in the data; (7) print the J, K, X (for each iteration), and Q_{xx} matrices; (8) create an AutoCad script file, including the stations, angles, distances, azimuths, and error ellipses of the adjusted file; and (9) provide a simulated adjustment as discussed in Chapter 20. The software also allows the user to set the confidence level (alpha) for the statistical test or error ellipses, and enter the rejection criteria number for statistical blunder detection. The print matrices option is useful to check hand computations.

File Format. Note that entries in the file may be separated by commas or spaces. Enter all angles in D M S format. For example, enter 238°34′23″ as 238 34 23. Standard deviations for observations can be entered following each observation. If not listed, the program will, by default:

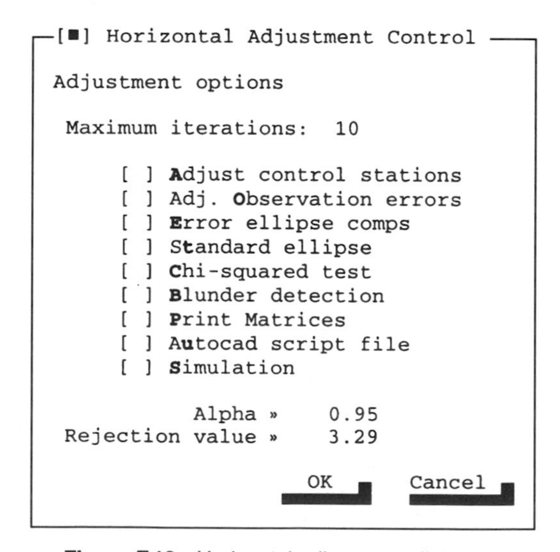

Figure F.19 Horizontal adjustment dialogue.

1. Assign all distances a standard deviation of 1.
2. Assign all angles a standard deviation of 1/(average sight distance).
3. Assign all azimuths a standard deviation of 0.001.

Note: For best results always assign observations standard deviations that are based on field conditions, instruments, and operator abilities, as presented in Chapter 6.

Title Line: e.g. Job number 123

#-Distances #-Angles #-Azimuths #-Control #-Stations [including control]

List of control stations: (Station ID's can not have spaces or commas)
 Station ID (10 char), X, Y (reals), [optional: Sx, Sy (reals)}

List of approximate coordinates for unknown stations:
 Station ID (10 char), X, Y (reals)

List of distance observations:
 Occupied Station   Sighted Station   Distance [S]

List of angle observations:
 Back. Station   Occ. Station   Fore. Station   Deg Min Sec [S″]

List of azimuth observations:
 Occupied station   Sighted station   Deg Min Sec [S″]

Example Horizontal Data File with Standard Deviations

```
Example 13.3 horizontal data
5 5 1 1 5
A 10000.00 10000.00 0.001 0.001
B 10125.66 10255.96
C 10716.31 10102.44
D 10523.62 9408.37
E 10517.55 9611.34
A B 285.10 0.05
B C 610.45 0.05
C D 720.48 0.05
D E 203.00 0.05
E A 647.02 0.05
B A E 100 44 30 15
C B A 101 35 00 15
D C B 89 05 30 15
E D C 17 12 00 15
A E D 231 24 30 15
A B 26 10 00 1
```

F.4.7 Vertical Adjustment

This menu, shown in Figure F.20, performs a least-squares adjustment on a file of differential leveling measurements with the adjustment results being placed in a file having an ".OUT" extension (see Chapter 11 for a discussion of this topic). The program will adjust the control stations, do a χ^2 hypothesis test of the reference variance, and statistically check for blunders in the data. The software also allows setting the confidence level (α) for the statistical test, and entering the rejection criteria number for the statistical blunder detection. If the *print matrices* option is selected, the software will list the A, L, and Q_{xx} matrices used in the adjustment. This option is useful for those doing these computations by hand.

File Formats

Title Line: e.g. Job number 123

#-Benchmarks #-Differences in elevation

List of benchmark stations:

  Station (10 characters with no spaces or commas), Elevation [S_{elev}]

List of observed elevation differences:

  From To (10 characters with no spaces or commas) Delta Elev [S]

The standard deviation for an observation is optional. If no user-specified standard deviation is given, the program will assign all elevation differences a value of 1.

Note: For best results always assign measurements standard deviations that are based on field conditions, instruments, and operator abilities, as discussed in Chapter 8.

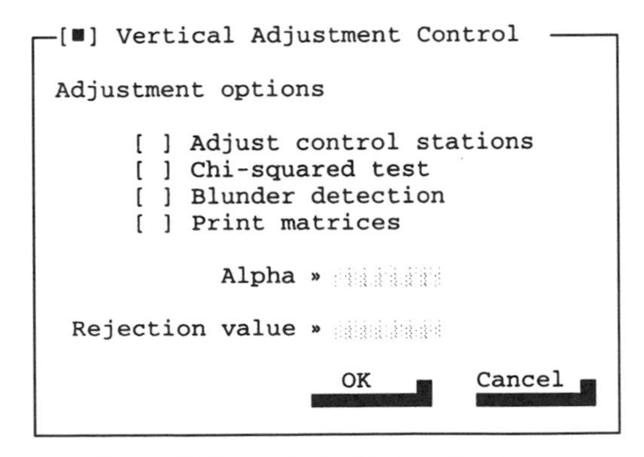

Figure F.20 Vertical adjustment options.

Example Differential Level Data File with Standard Deviations

```
Weighted Example Problem
2 7
BmX 100.00 0.001
BmY 107.50 0.001
Bmx A 5.10 2
A BMY 2.34 1.732
BMY C -1.25 1.414
C BMX -6.13 1.732
A B -0.68 1.414
BMY B -3.00 1.414
B C 1.70 1.414
```

F.4.8 GPS Baseline Adjustment

The GPS data option leads to the *baseline adjustment* program. Its option menu, shown in Figure F.21, sets the adjustment options when performing a least-squares adjustment on a file of GPS baseline vectors. The results of this adjustment are placed in a file having an ".OUT" extension (see Chapter 16 for a discussion on this topic). By selecting the option shown in Figure F.21, the program can create a file of geocentric coordinates, compute estimated errors of the adjusted baseline vectors, use techniques discussed in Chapter 20 to do blunder detection, and do a *chi-square* test of the reference variance at any level of significance (alpha). This geocentric coordinate file can be used with the *geocentric coordinate computations* option discussed previously to compute equivalent geodetic coordinates. If the *print matrices* option is selected, the program will write the selected matrices to the file developed

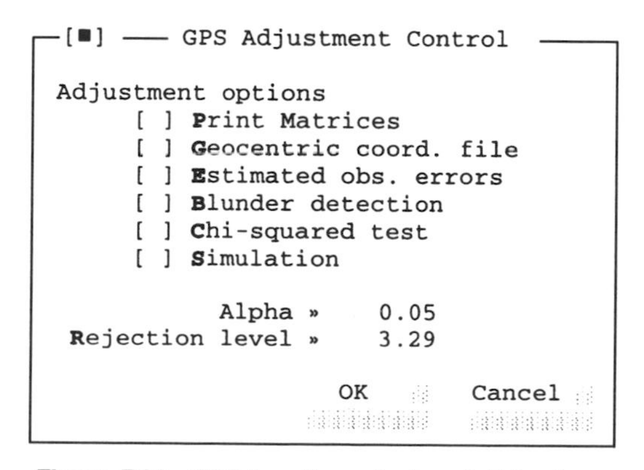

Figure F.21 GPS baseline adjustment dialog box.

for the adjustment. This option is useful for those doing those computations by hand.

File Format

Title Line: e.g. Job number 123
#-Control Stations #-Baselines
List of control stations:
    Station ID X Y Z
List of observed baselines:
    From To dx dy dz σ_{xx} σ_{xy} σ_{xz} σ_{yy} σ_{yz} σ_{zz}

Note that station identifiers can have up to 10 characters without any spaces or commas.

Example File

```
GPS Baseline Vectors
1 6
1 593898.88770 -4856214.54563 4078710.70586
1 2 678.034 1206.714 1325.735 0.50989E-5 -0.14005E-4 0.69288E-5 0.74402E-4 -0.34455E-4 0.20187E-4
2 3 -579.895 145.342 254.820 0.34047E-5 0.20572E-5 -0.30370E-6 0.20152E-4 -0.11478E-4 0.18736E-4
1 3 98.138 1352.039 1580.564 0.65187E-5 -0.11632E-6 -0.38114E-6 0.3845E-4 -0.12971E-4 0.29259E-4
4 3 677.758 553.527 552.978 0.93477E-5 -0.14279E-4 0.87764E-5 0.2954E-4 -0.18535E-4 0.14705E-4
4 3 677.756 553.533 552.975 0.917085E-5 -0.14158E-4 0.85708E-5 0.30109E-4 -0.18629E-4 0.14600E-4
4 1 579.615 -798.509 -1027.609 0.12174E-4 -0.18250E-4 0.11419E-4 0.36754E-4 -0.23237E-4 0.18910E-4
```

A GPS baseline adjustment can be simulated by selecting the simulation option in Figure F.21. This option is useful for planning and designing GPS surveys. The format for a simulation file is as follows:

File Format

Title Line: e.g. Job number 123
#-Control Stations #-Unknown Stations #-Baselines
List of control stations:
    Station ID X Y Z
List of unknown stations:
    Station ID X Y Z
List of observed baselines including the endpoint stations and the estimated covariance matrix:
    From To σ_{xx} σ_{xy} σ_{xz} σ_{yy} σ_{yz} σ_{zz}

Note that station identifiers can have up to 10 characters without any spaces or commas.

Example File

```
GPS Baseline Adjustment Simulation
1 3 5
1 593898.88770 -4856214.54563 4078710.70586
2 594576.92228 -4855007.83637 4080036.44646
3 593997.02665 -4854862.49906 4080291.27904
4 593319.27040 -4855416.03108 4079738.30582
1 2 0.50989E-5 -0.14005E-4 0.69288E-5 0.7440E-4 -0.34455E-4 0.20187E-4
2 3 0.34047E-5 0.20572E-5 -0.30370E-6 0.2015E-4 -0.11478E-4 0.18736E-4
1 3 0.65187E-5 -0.11632E-6 -0.38114E-6 0.3845E-4 -0.12971E-4 0.29259E-4
4 3 0.93477E-5 -0.14279E-4 0.87764E-5 0.2954E-4 -0.18535E-4 0.14705E-4
2 4 0.93477E-5 -0.14279E-4 0.87764E-5 0.2954E-4 -0.18535E-4 0.14705E-4
1 4 0.12174E-4 -0.18250E-4 0.11419E-4 0.3675E-4 -0.23237E-4 0.18910E-4
```

F.4.9 GPS Loop Misclosure Checks

The GPS data menu also leads to the program option for checking GPS loop misclosures. The use of this option is discussed in Chapter 16. The format of the data file for this option is

Number of GPS baselines
Baseline measurements:
  Line # occupied station sighted station dx dy dz
Loop closures to be computed:
  Loops are identified by baseline numbers.

Sample File. See Figure F.22.

```
14 {Number of baselines]
1 A C 11644.2232 3601.2165 3399.2550 {Line #, occupied, sighted, dx, dy, dz}
2 A E -5321.7164 3634.0754 3173.6652
```

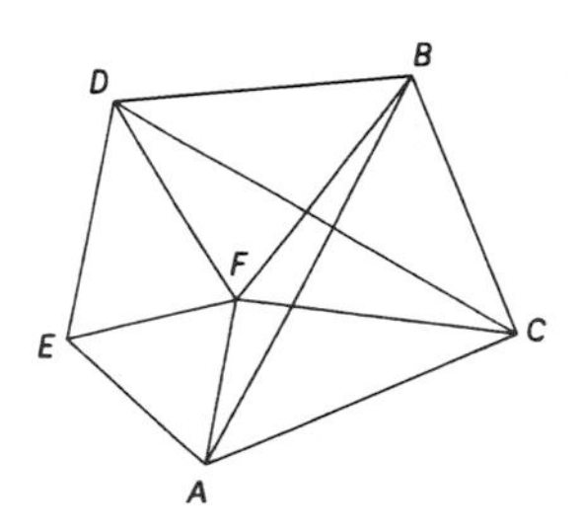

Figure F.22 Figure for sample file.

```
3 B C 3960.5442 -6681.2467 -7279.0148
4 B D -11167.6076 -394.5204 -907.9593
5 D C 15128.1647 -6286.7054 -6371.0583
6 D E -1837.7459 -6253.8534 -6596.6697
7 F A -1116.4523 -4596.1610 -4355.8962
8 F C 10527.7852 -994.9377 -956.6246
9 F E -6438.1364 -962.0694 -1182.2305
10 F D -4600.3787 5291.7785 5414.4311
11 F B 6567.2311 5686.2926 6322.3917
12 B F -6567.2310 -5686.3033 -6322.3807
13 A F 1116.4577 4596.1553 4355.9141
14 A B 7683.6620 10282.4234 10678.2972
 1  8 13              Loop using baselines 1, 8 and 13
 7  9  2
 9 10  6
10  4 12
11  8  3
 1  3 14
 5 10  8
```

F.5 MATRIX

Matrix is a program containing a collection of matrix operations. It is capable of processing up to twenty 50 × 50 matrices when enough memory is available on the user's computer. It will reduce this number when insufficient memory is available.

F.5.1 Overview

MATRIX allows the user to manipulate mathematically matrices that appear on the screen's right side. The number, name, and dimensions of the matrices are displayed on the right side of the screen. Initially, matrices are read from a data file. To select a matrix for processing, either enter the matrix number or click the left mouse button on the appropriate matrix line. To stop a process where multiple matrices are being selected, such as "Write Matrix," press either *Esc, F10*, or click the *exit* prompt at the top of the matrix window. The typical screen in the matrix program is shown in Figure F.23.

F.5.2 Data File Entry

Installing Your Editor. The MATRIX program reads matrices from a single DOS (ASCII) file in which all initial matrices can be entered. The matrix sizes are limited to 50 rows and 50 columns. The file can be created using any DOS editor, including the one provided in *Stats* or *Adjust*. Many word

```
    Matrix Program                                              Ver 2.1

 O  --  Set Options                          #  |      Matrix Name  |  R  |  C  |
 1  --  Write Matrix                        ----+-------------------+-----+-----+
 2  --  Copy Matrix                             |                   |     |     |
 3  --  Rename Matrix                           |                   |     |     |
 4  --  Remove Matrix from List                 |                   |     |     |
                                                |                   |     |     |
 S  --  Scale Matrix                            |                   |     |     |
 5  --  Transpose Matrix                        |                   |     |     |
 6  --  Add/Subtract Matrices                   |                   |     |     |
 7  --  Multiply Matrices                       |                   |     |     |
 8  --  Invert Matrix                           |                   |     |     |
 9  --  Cholesky Decomposition of Matrix        |                   |     |     |
 0  --  Forward/Backward Substitution           |                   |     |     |
                                                |                   |     |     |
 V  --  View a Matrix                           |                   |     |     |
 E  --  Edit File                               |                   |     |     |
 R  --  Read File                               |                   |     |     |
 W  --  Write File                              |                   |     |     |
 L  --  List Matrix File To Printer             |                   |     |     |
 Q  --  Quit                                    |                   |     |     |

Choice »              Press F1 for Help         |                   |     |     |
```

Figure F.23 Matrix menu.

processors are also capable of producing a DOS (ASCII) file. Consult your word processor manual for help in creating DOS files. DOS 5.0 (and higher) provides an excellent editor called EDIT. The editor can be installed in the main menu of matrix by selecting the *Edit* option from the MATRIX menu and providing the complete executable name of the editor. For example, the *EDIT* program in DOS 5.0 is normally installed as *C:\DOS\EDIT.COM.* If the INSTALL program provided on the diskette is used, the STATS program will be loaded automatically as the MATRIX editor.

Entering the File. When entering the software, a prompt will be made for the location of the data file. Supply the appropriate DOS path for the file and press the "*Enter←*" key. After the directory is displayed, enter the name of the data file. The output file will have the data file's name with an extension of "*.OUT*". For example, if DATA.DAT is the data file, the output file will be named DATA.OUT.

F.5.3 Setup Options

Matrices can be written to the screen or a file and in either floating-point or exponential format. *Setup options* controls not only the format style of the numbers but also their field width and number of decimal places. Furthermore, the default output file name can be changed to any legal DOS name.

F.5.4 Formatted Listing of Matrices

It is possible to send a formatted listing to the *Output* file at any time during the work session. To send matrices to the output file, select the *Write Matrix* option from the menu and then either enter the matrix number or click the left mouse button on a matrix. This option allows the user to select more than one matrix. To leave the option, either press *Esc, F10,* or select the *Exit* command at the top of the matrix list with the mouse. To display a matrix on the screen, use the *View a Matrix* option and select the appropriate matrix. To save an entire work session, use the *Write file option* from the main menu.

F.5.5 Supported Matrix Operations

1. **Transpose Matrix**. Simply specify the matrix to transpose, and supply its new name. *Suggestion: Use names that will make sense during later use. For example, use AT to represent the transpose of A.*
2. **Copy Matrix**. Copy any matrix in the matrix screen by selecting the matrix to copy and supplying name of the copied matrix.
3. **Rename Matrix**. Rename any matrix from the matrix screen by selecting the matrix and supplying its new name. *Suggestion: When performing a Cholesky decomposition, either copy the initial matrix and give the matrix an appropriate name [e.g., Chol(A)] or rename the matrix after using this option.*
4. **Remove Matrix**. Delete any matrix when the maximum limit of matrices is reached or when the matrix is no longer needed.
5. **Add/Subtract Matrices**. Add or subtract any two conformable matrices. If the matrices are not conformable for the operation, the program will display an error statement. *Suggestion: Use* $X + Y$ *as the name for the summation X and Y.*
6. **Multiply Matrices**. Multiply two conformable matrices. The program displays an error message when the two matrices cannot be multiplied. *Suggestion: Use* $A \times B$ *for the product matrix of A and B.*
7. **Invert a Matrix**. This option computes the inverse of a matrix. When a matrix does not have an inverse, the program will display an error message. *Suggestion: Use INV(A) for the inverse matrix of A.*
8. **Cholesky Decomposition of a Matrix**. This routine performs a Cholesky decomposition of a matrix, leaving the results in the upper triangular portion of the matrix. When matrices are not positive definite, the program will display an error message. **Warning:** *This routine overwrites the existing matrix whether or not it is successful! Suggestion: Make a copy of the matrix before the decomposition. Use the name CHOL(A) for the Cholesky Decomposed matrix of A.*

9. **Forward/Backward Solution of a Vector**. This routine performs the forward and backward solution using a Cholesky decomposed matrix and the appropriate vector constant. The results are entered into a separate solution vector. The proper procedure is as follows:

   Given: *Ax = L; where the Size(A) = n × n; Size(L) = n × 1; Size(x) = n × 1;*

   a. Perform Cholesky Decomposition of A and rename it CHOL(A).
   b. Perform Forward/Backward by supplying the resultant matrix from step a and the vector L; the results are placed in vector *x*.
10. **View a Matrix**. This routine displays a matrix using the format settings listed in the *Setup Option*.

F.5.6 Format for Matrix Files

Files read by MATRIX must be typed into a DOS file using the following format:

Line 1: name of first matrix

Line 2: number of rows and number of columns

Line 3 to numbers of rows in first matrix:

Enter matrix elements in first matrix one row per line with space delimiters. Continue entering additional matrices in file using the same procedures.

F.5.7 Sample File

```
A
7  3
1  0  0
1  0  0
0  0  1
0  0  1
1 -1  0
0  1  0
0  1 -1
L
7 1
105.10
105.16
106.25
106.13
0.68
104.5
-1.70
```

User Assistance

If you have questions or problems with the installation of your disk, please call our technical support number at (212) 850-6194 between 9 AM and 4 PM Eastern Standard Time, Monday through Friday.

To place additional orders or to request information about other Wiley products, please call (800) 879-4539.

CUSTOMER NOTE: IF THIS BOOK IS ACCOMPANIED BY SOFTWARE, PLEASE READ THE FOLLOWING BEFORE OPENING THE PACKAGE.

This software contains files to help you utilize the models described in the accompanying book. By opening the package, you are agreeing to be bound by the following agreement:

This software product is protected by copyright and all rights are reserved by the author, John Wiley & Sons, Inc., or their licensors. You are licensed to use this software on a single computer. Copying the software to another medium or format for use on a single computer does not violate the U.S. Copyright Law. Copying the software for any other purpose is a violation of the U.S. Copyright Law.

This software product is sold as is without warranty of any kind, either express or implied, including but not limited to the implied warranty of merchantability and fitness for a particular purpose. Neither Wiley nor its dealers or distributors assumes any liability for any alleged or actual damages arising from the use of or the inability to use this software. (Some states do not allow the exclusion of implied warranties, so the exclusion may not apply to you.)

BIBLIOGRAPHY

Aguilar, A. M., Principles of Survey Error Analysis and Adjustment, *Journal of the Surveying and Mapping Division*, ASCE, 99:SU1 (1973).

Amer, F., Theoretical Reliability Studies for Some Elementary Photogrammetric Procedures, *Proceedings of the Aerial Triangulation Symposium*, Department of Surveying, University of Queensland, Brisbane, Australia (1979).

Baarda, W., *Statistical Concepts in Geodesy*, Netherlands Geodetic Commission, Delft, the Netherlands (1967).

Baarda, W., *A Testing Procedure for Use in Geodetic Networks*, Netherlands Geodetic Commission, Delft, the Netherlands (1968).

Barry, B. A., *Engineering Measurements*, John Wiley & Sons, New York (1964).

Benjamin, J. R., and C. A. Cornell, *Probability, Statistics and Decision for Engineers*, McGraw-Hill, New York (1970).

Bjerhammar, A., *Theory of Errors and Generalized Matrix Inverses*, Elsevier Scientific Publishing Co., New York (1973).

Box, G. E. P., W. G. Hunter, and J. S. Hunter, *Statistics for Experimenters*, John Wiley & Sons, New York (1978).

Buckner, R. B., *Surveying Measurements and Their Analysis*, Landmark Enterprises, Rancho Cordova, CA (1983).

Burse, M. L., Profile of a Least Squares Convert, *P.O.B.*, 20:2 (1995).

Conte, S. D., and C. de Boor, *Elementary Numerical Analysis*, 3rd ed., McGraw-Hill, New York (1980).

Cooper, M. A. R., *Fundamentals of Survey Measurement and Analysis*, Crosby, Lockwood Staples, London (1973).

Crandall, K. C., and R. W. Seabloom, *Engineering Fundamentals in Measurements, Probability, Statistics and Dimensions*, McGraw-Hill, New York (1970).

Deming, W. E., *Statistical Adjustment of Data*, Dover Publications, New York (1943).

Dewitt, B. A., An Efficient Memory Paging Scheme for Least Squares Adjustment of Horizontal Surveys, *Surveying and Land Information Systems*, American Congress on Surveying and Mapping 54:3 (1994).

Dracup, J. F., Least Squares Adjustment by the Method of Observation Equations with Accuracy Estimates, *Surveying and Land Information Systems*, American Congress on Surveying and Mapping, 55:2 (1995).

El-Hakim, S. F., The Detection of Gross and Systematic Errors in Combined Adjustment of Terrestrial and Photogrametric Data, *Photogrammetric Engineering and Remote Sensing*, American Society for Photogrammetry and Remote Sensing, 52: 1 (1986).

El-Hakim, S. F., On the Detection of Gross and Systematic Errors in Combined Adjustment of Terrestrial and Photogrametric Data, Commission III, International Archives of Photogrammetry and Remote Sensing, Rio de Janerio, Brazil (1984).

El-Hakim, S. F., A Practical Study of Gross-Error Detection in a Bundle Adjustment, *Canadian Surveyor*, 35:4 (1981).

Franklin, J. N., *Matrix Theory*, Prentice Hall, Upper Saddle River, NJ (1968).

Fubara, D. M. J., Three-Dimensional Adjustment of Terrestrial Geodetic Networks, *Canadian Surveyor*, 26:4 (1972).

Gale, L. A., Theory of Adjustments by Least Squares, *Canadian Surveyor*, 9:1 (1965).

George, A., and J. W.-H. Liu, *Computer Solution of Large Sparse Positive Definite Systems*, Prentice Hall, Upper Saddle River, NJ (1981).

Ghilani, C. D., Predictions in Photo-geodetic Control Extension, *Technical Papers for the 1991 ACSM–ASPRS Annual Convention* (March 1991).

Ghilani, C. D., Some Thoughts on Boundary Survey Measurement Standards, *Surveying and Land Information Systems*, American Congress on Surveying and Mapping, 54:3 (1994).

Ghilani, C. D., A Surveyor's Guide to Practical Least Squares Adjustments, *Surveying and Land Information Systems*, American Congress on Surveying and Mapping, 50:4 (1990).

Hamilton, W. C., *Statistics in Physical Science*, Ronald Press, New York (1964).

Harvey, B. R., *Practical Least Squares and Statistics for Surveyors*, The School of Surveying, University of New South Wales, Sydney, Australia (1994).

Hirvonen, R. A., *Adjustment by Least Squares in Geodesy and Photogrammetry*, Frederick Ungar Publishing Co., New York (1965).

Hoffman-Wellenhof, B., et al., *GPS Theory and Practice*, 2nd ed., Springer-Verlag, New York (1993).

Hogg, R. V., and J. Ledolter, *Applied Statistics for Engineers and Physical Scientists*, Macmillan Publishing Company, New York (1992).

Konecny, G., Classical Concepts of Least Squares Adjustment, *Canadian Surveyor*, 9: 1 (1965).

Kouba, J., Principal Property and Least Squares Adjustment Design, *Canadian Surveyor*, 26:2 (1972).

Kuang, S., A Strategy for GPS Survey Planning: Choice of Optimum Baselines, *Surveying and Land Information Systems*, American Congress on Surveying and Mapping, 54:4 (1994).

Larson, H. J., *Introduction to Probability Theory and Statistical Inference*, John Wiley & Sons, New York (1969).

Leick, A., *GPS Satellite Surveying*, 2nd ed., Wiley-Interscience, an imprint of John Wiley & Sons, New York (1995).

Leininger, J. L., *The Positional Tolerance Question*, *Professional Surveyor*, 16:2 (1996).

Leland, O. M., *Practical Least Squares*, McGraw-Hill, New York (1921).

Maksoudian, Y. L., *Probability and Statistics with Applications*, International Textbook Company, Scranton, PA (1969).

McMillan, K. N., Least Squares—Older and Better Than Barbed Wire, *P.O.B.*, 20:2 (1995).

McMillan, K. N., Least Squares Under the Hood, *P.O.B.*, 20:2 (1995).

Mendenhall, W., and T. Sincich, *Statistics for Engineering and the Sciences*, Dellen Publishing Company, San Francisco (1992).

Mikhail, E. M., *Observations and Least Squares*, University Press of America, Washington, DC (1976).

Mikhail, E. M., and G. Gracie, *Analysis and Adjustment of Survey Measurements*, Van Nostrand Reinhold, New York (1981).

Millbert, K. O., and D. G. Milbert, State Readjustments at the National Geodetic Survey, *Surveying and Land Information Systems*, American Congress on Surveying and Mapping, 54:4 (1994).

Neville, A. M., and J. B. Kennedy, *Basic Statistical Methods*, International Textbook Company, Scranton, PA (1964).

Rainsford, H. F., *Survey Adjustments and Least Squares*, Constable and Co., London (1957).

Schmid, H. H., and E. Schmid, A Generalized Least Squares Solution for Hybrid Measuring Systems, *Canadian Surveyor*, 9:1 (1965).

Schwarz, C. R., The Trouble with Constrained Adjustments, *Surveying and Land Information Systems*, American Congress on Surveying and Mapping, 54:4 (1994).

Schwarz, K. P., E. H. Knickmeyer, and H. Martell, Assessment of Observations Using Minimum Norm Quadratic Unbiased Estimation, *CISM Journal*, Association Canadienne Des Sciences Geomatique 44:1 (1990).

Seber, G. A. F., *Linear Regression Analysis*, John Wiley & Sons, New York (1977).

Sideris, M. G. The Role of the Geoid in One-, Two-, and Three-Dimensional Network Adjustments, *CISM Journal*, Association Canadienne Des Sciences Geomatique, 44:1 (1990).

Snay, R. A., Reducing the Profile of Large Sparse Matrices, *NOAA Technical Memorandum NOS NGS 4* (May 1976); republished in *Bulletin Geodésiqué*, 50:4 (1976).

Sunter, A. B., Statistical Properties of Least Squares Estimates, *Canadian Surveyor*, 15:1 (1966).

Teskey, W. F., and J. W. MacLeod, Application of Statistical Testing to Cadastral Survey Traverses, *CISM Journal*, Association Canadienne Des Sciences Geomatique, 42:1 (1988).

Torge, W., *Geodesy*, Walter de Gruyter, Hawthorne, NY (1992).

Uotila, U. A., Statistical Tests as Guidelines in Analysis of Adjustment of Control Nets, *Surveying and Mapping*, 35:1 (1975).

Uotila, U. A., Useful Statistics for Land Surveyors, *Surveying and Mapping*, 33:1 (1973).

Veress, S. A., *Adjustment by Least Squares*, American Congress on Surveying and Mapping, Washington, DC (1974).

Veress, S. A., Measures of Accuracy for Analysis and Design of Survey, *Surveying and Mapping*, 33:4 (1973).

White, L. A., *Calculus of Observations 380/580*, Western Australian Institute of Technology, Perth (1987).

Wolf, P. R., *Elements of Photogrammetry*, 2nd ed., McGraw-Hill, New York (1983).

Wolf, P. R., Horizontal Position Adjustment, *Surveying and Mapping*, 29:4 (1969).

Wolf, P. R., Matrix Algebra—A Tool for Engineers and Surveyors, *Surveying and Mapping*, 30:3 (1970).

Wolf, P. R. and R. C. Brinker, *Elementary Surveying*, 9th ed., Harper & Row, Cambridge (1994).

INDEX